Theory of Spinors and Its Application in Physics and Mechanics

Vladimir A. Zhelnorovich

Theory of Spinors and Its Application in Physics and Mechanics

Springer

Vladimir A. Zhelnorovich
Institute of Mechanics
Moscow State University
Moscow, Russia

ISBN 978-3-030-27835-9 ISBN 978-3-030-27836-6 (eBook)
https://doi.org/10.1007/978-3-030-27836-6

This book is a revised and updated version of the original Russian edition "Theory of Spinors and its Application in Physics and Mechanics" published by Izdatel'stvo Nauka, Moscow, Russia, 1982.

This Springer imprint is published by the registered company Springer Nature Switzerland AG.
The registered company address is: Gewerbestrasse 11, 6330 Cham, Switzerland

Preface

This book is a reworked and enlarged version of the author's book [82].

The first chapter of the book presents the theory of spinors in n-dimensional (in general case complex) Euclidean spaces. The second chapter contains an exposition of the theory of spinors in Riemannian spaces. The third and fourth chapters are devoted to the theory of spinors and to the methods of their tensor representation in four- and three-dimensional spaces. Along with the material in the book [82], these chapters contain the results of recent papers, related in particular to the use of proper orthonormal tetrads defined by spinors. Some very useful relations are obtained that express the derivatives of the spinor fields in terms of the derivatives of various tensor fields.

The main content of the fifth chapter is the tensor representation of a wide class of relativistically invariant spinor differential equations that contain, as a particular case, the known spinor equations of field theory. As an example of the application of the theory in Chap. 6, we give a series of exact solutions of nonlinear spinor equations used in the theory of elementary particles. In particular, we give a general exact solution of the Einstein–Dirac equations in homogeneous Riemannian spaces and also a series of exact solutions of nonlinear Heisenberg equations. To integrate spinor equations in Riemannian space, a new invariant tetrad gauge is used, which makes it possible to reduce the number of unknown functions in the equations by six units. In the same chapter, with the aid of spinor theory methods, some integrals are given for partial differential equations describing spin fluids in an electromagnetic field; their exact wave solutions are considered.

In Appendix A, on the basis of the variational equation, a closed system of differential equations describing spin liquids interacting with an electromagnetic field is obtained; here, we also study the problem of the invariant definition of the internal energy of the electromagnetic field. In Appendix B, the problem of invariant determination of the energy of a free electromagnetic field in the form of a four-dimensional scalar (an analog of internal energy in the mechanics of a continuous medium) is considered.

The book is basically aimed at physical applications and is destined, above all, for physicists. Therefore, the presentation of the material is based on the use of well-established classical terminology and methods of differential geometry, which requires a minimum of mathematical preparation of the reader, restricted to the first years of university programme. In particular, the knowledge of the basics of tensor algebra and analysis is assumed.

Moscow, Russia
February 2018

Vladimir A. Zhelnorovich

From the Preface to the Book [82] Theory of Spinors and Its Application in Physics and Mechanics (in Russian)

The existence of spinors and spinor representations of orthogonal groups was discovered by Cartan in 1913 [12]. Studies of the theory of spinors were stimulated both by purely mathematical requirements (in the theory of group representations) and by physical applications in quantum mechanics and field theory. At the present time, the journal literature concerned with the algebraic theory of spinors and its applications amounts to thousands of publications. Meanwhile, the problems associated with an invariant description of spinors as objects that do not depend on the choice of a coordinate system are either bypassed or presented in essence incorrectly. The present book, to some extent, fills this gap.

As is known, the finite-dimensional linear representations of orthogonal groups are exhausted by tensor and spinor representations. Therefore, geometric objects associated with representations of orthogonal groups are exhausted by tensors and spinors of various ranks. The theory of tensors and tensor calculus are widely used in modern physics; they represent the basic mathematical formalism of modern physical theories. The theory of spinors at present is mainly used in field theory and quantum mechanics (if we talk about physical applications). In this book, the notion of a spinor and spinor calculus is also used in the mechanics of continuous media.The theory of spinors is expounded in the first three chapters. The main attention is paid here to the concepts of invariant algebraic and geometric relations between spinors and tensors. The classical relations of the theory of spinors, which are presented in the prevalent books, are here only mentioned in passing.

The material of the book has been mainly published in various special journals; in part, it is published for the first time. The book does not aim to give a review of the literature on the issues under consideration; however, such reviews are contained in the cited literature.

Abstract

The book contains a systematic exposition of the theory of spinors in finite-dimensional Euclidean and Riemannian spaces; the application of spinors in field theory and relativistic mechanics of continuous media is considered.

The main mathematical part is connected with the study of invariant algebraic and geometric relations between spinors and tensors. The theory of spinors and the methods of the tensor representation of spinors and spinor equations are specially and thoroughly expounded in four-dimensional and three-dimensional spaces. As an application, we consider an invariant tensor formulation of certain classes of differential spinor equations containing, in particular, the most important spinor equations of field theory and quantum mechanics; exact solutions of the Einstein–Dirac equations, nonlinear Heisberg's spinor equations, and equations for relativistic spin fluids are given. The book contains a large factual material and can be used as a handbook.

The book is intended for specialists in theoretical physics, as well as for students and postgraduate students of physical and mathematical specialties.

Contents

Chapter 1
Spinors in Finite-Dimensional Euclidean Spaces

1.1 Algebra of γ-Matrices

Consider the matrix equation

$$\overset{\circ}{\gamma}_i \overset{\circ}{\gamma}_j + \overset{\circ}{\gamma}_j \overset{\circ}{\gamma}_i = 2\delta_{ij} I, \tag{1.1}$$

in which $\overset{\circ}{\gamma}_i$ are square, generally complex matrices of order 2^ν, ν is a positive integer; the indices i, j, determining the numbers of matrices $\overset{\circ}{\boldsymbol{\gamma}}$, take all integer values from 1 to 2ν; I is the unit matrix of order 2^ν; δ_{ij} are the Kronecker delta:

$$\delta_{ij} = 1, \quad \text{if} \quad i = j,$$
$$\delta_{ij} = 0, \quad \text{if} \quad i \neq j.$$

Let us introduce the matrices $\overset{\circ}{\gamma}_{i_1 i_2 \ldots i_k}$, defined with the aid of the matrices $\overset{\circ}{\gamma}_i$ satisfying Eq. (1.1):

$$\overset{\circ}{\gamma}_{i_1 i_2 \ldots i_k} = \overset{\circ}{\gamma}_{[i_1} \overset{\circ}{\gamma}_{i_2} \cdots \overset{\circ}{\gamma}_{i_k]}. \tag{1.2}$$

Alternation is meant over the indices $i_1 i_2 \ldots i_k$ placed in square brackets in Eq. (1.2) (with division by $k\,!$).

Assuming that the matrices $\overset{\circ}{\gamma}_i$, satisfying Eq. (1.1), do exist, let us establish some general properties of the matrices (1.2) which do not depend on the specific form of $\overset{\circ}{\gamma}_i$.[1]

---

[1]The existence of matrices $\overset{\circ}{\gamma}_i$ satisfying Eq. (1.1) will be shown at the end of the present section.

V. A. Zhelnorovich, *Theory of Spinors and Its Application in Physics and Mechanics*, https://doi.org/10.1007/978-3-030-27836-6_1

1. For any matrix $\overset{\circ}{\gamma}_{i_1 i_2 \ldots i_k}$, defined by equality (1.3), there exists such a matrix $\overset{\circ}{\gamma}_{j_1 j_2 \cdots j_m}$ that the following equality holds:

$$\overset{\circ}{\gamma}_{i_1 i_2 \ldots i_k} \overset{\circ}{\gamma}_{j_1 j_2 \ldots j_m} = -\overset{\circ}{\gamma}_{j_1 j_2 \ldots j_m} \overset{\circ}{\gamma}_{i_1 i_2 \ldots i_k}. \tag{1.3}$$

This property may be proved by directly pointing out, for each matrix $\overset{\circ}{\gamma}_{i_1 i_2 \ldots i_k}$, the corresponding matrices that anti-commute with it. If the number of indices $i_1 i_2 \ldots i_k$ is even, $k = 2m$, then, for example, any matrix $\overset{\circ}{\gamma}_j$ with j coinciding with one of the indices $i_1 i_2 \ldots i_k$, anticommutes with $\overset{\circ}{\gamma}_{i_1 i_2 \ldots i_k}$. If the number of indices $i_1 i_2 \ldots i_k$ is odd, $k = 2m + 1$, then any matrix $\overset{\circ}{\gamma}_j$ with j that does not coincide with any of the indices $i_1 i_2 \ldots i_k$, anticommutes with $\overset{\circ}{\gamma}_{i_1 i_2 \ldots i_k}$.

2. The square of a matrix $\overset{\circ}{\gamma}_{i_1 i_2 \ldots i_k}$ is proportional to the unit matrix,

$$\left(\overset{\circ}{\gamma}_{i_1 i_2 \ldots i_k}\right)^2 = (-1)^{\frac{1}{2}k(k-1)} I. \tag{1.4}$$

Equality (1.4) directly follows from Eq. (1.1) and definition (1.2).

3. All matrices $\overset{\circ}{\gamma}_{i_1 i_2 \ldots i_k}$ for $k = 1, 2, \ldots, 2\nu$ have zero traces. Indeed, taking into account that any square matrices A, B, C satisfy the identity $\operatorname{tr}(ABC) = \operatorname{tr}(CAB)$, we find with (1.4):

$$\begin{aligned} \operatorname{tr} \overset{\circ}{\gamma}_{i_1 i_2 \ldots i_k} &= (-1)^{\frac{1}{2}m(m-1)} \operatorname{tr} \left(\overset{\circ}{\gamma}_{i_1 i_2 \ldots i_k} \overset{\circ}{\gamma}_{j_1 j_2 \ldots j_m} \overset{\circ}{\gamma}_{j_1 j_2 \ldots j_m}\right) \\ &= (-1)^{\frac{1}{2}m(m-1)} \operatorname{tr} \left(\overset{\circ}{\gamma}_{j_1 j_2 \ldots j_m} \overset{\circ}{\gamma}_{i_1 i_2 \ldots i_k} \overset{\circ}{\gamma}_{j_1 j_2 \ldots j_m}\right). \end{aligned} \tag{1.5}$$

Choosing in (1.5), as the matrix $\overset{\circ}{\gamma}_{j_1 j_2 \ldots j_m}$, a matrix satisfying the relation (1.3), we continue the equality (1.5):

$$\operatorname{tr} \overset{\circ}{\gamma}_{i_1 i_2 \ldots i_k} = -(-1)^{\frac{1}{2}m(m-1)} \operatorname{tr} \left(\overset{\circ}{\gamma}_{j_1 j_2 \ldots j_m} \overset{\circ}{\gamma}_{j_1 j_2 \ldots j_m} \overset{\circ}{\gamma}_{i_1 i_2 \ldots i_k}\right) = -\operatorname{tr} \overset{\circ}{\gamma}_{i_1 i_2 \ldots i_k}.$$

The latter equality implies

$$\operatorname{tr} \overset{\circ}{\gamma}_{i_1 i_2 \ldots i_k} = 0. \tag{1.6}$$

4. The trace of a product of two matrices $\overset{\circ}{\gamma}_{j_1 j_2 \ldots j_m}$ and $\overset{\circ}{\gamma}_{i_1 i_2 \ldots i_k}$ is nonzero only in the case when these matrices are the same (or differ in sign due to permutation of indices when $k = m$).

Indeed, if the matrices $\overset{\circ}{\gamma}_{j_1 j_2 \ldots j_m}$ and $\overset{\circ}{\gamma}_{i_1 i_2 \ldots i_k}$ are different, then their product, due to Eq. (1.1), reduces to some matrix $\overset{\circ}{\gamma}_{q_1 q_2 \ldots q_n}$ whose trace is, according to (1.6), equal to zero. If the matrices $\overset{\circ}{\gamma}_{j_1 j_2 \ldots j_m}$ and $\overset{\circ}{\gamma}_{i_1 i_2 \ldots i_k}$ are the same, then, according to

Eq. (1.4), their product is proportional to the unit matrix, and its trace is nonzero. The aforesaid may be written in the form of the equalities

$$\begin{aligned} &\mathrm{tr}\,\big(\overset{\circ}{\gamma}_{i_1 i_2 \ldots i_k} \overset{\circ}{\gamma}_{j_1 j_2 \ldots j_m}\big) = 0, \quad \text{if} \quad k \neq m, \\ &\mathrm{tr}\,\big(\overset{\circ}{\gamma}_{i_1 i_2 \ldots i_k} \overset{\circ}{\gamma}_{j_1 j_2 \ldots j_k}\big) = (-1)^{\frac{1}{2}k(k-1)} k!\, 2^{\nu} \delta^{j_1}_{[i_1} \delta^{j_2}_{i_2} \cdots \delta^{j_k}_{i_k]}. \end{aligned} \tag{1.7}$$

The factor $k!$ in the right-hand side of the second equation (1.7) is connected with the alternation performed over the indices $i_1 i_2 \ldots i_k$ in this formula.

5. The system of matrices

$$I, \quad \overset{\circ}{\gamma}_i, \quad \overset{\circ}{\gamma}_{i_1 i_2}, \quad \ldots, \quad \overset{\circ}{\gamma}_{i_1 i_2 \ldots i_{2\nu}} \qquad (i_1 < i_2 < \cdots < i_{2\nu}) \tag{1.8}$$

is linearly independent.

Indeed, consider the equation

$$\alpha I + \sum_{k=1}^{2\nu} \alpha^{i_1 i_2 \ldots i_k} \overset{\circ}{\gamma}_{i_1 i_2 \ldots i_k} = 0, \tag{1.9}$$

in which all coefficients $\alpha^{i_1 i_2 \ldots i_k}$ are antisymmetric in all indices, $\alpha^{i_1 i_2 \ldots i_k} = \alpha^{[i_1 i_2 \ldots i_k]}$. In Eq. (1.9) and in what follows, we everywhere (unless otherwise indicated) use the summing rule, according to which summing is performed by two repeated indices over all values they take.

Taking the trace of Eq. (1.9), we find, taking into account (1.6) that the coefficient α equals to zero, $\alpha = 0$. Taking the trace of Eq. (1.9), multiplied beforehand by the matrix $\overset{\circ}{\gamma}_{j_1 j_2 \ldots j_m}$, taking into account the equalities (1.6) and (1.7), we obtain $\alpha^{j_1 j_2 \dot{j}_m} = 0$. Thus if Eq. (1.9) holds, then all coefficients α, $\alpha^{i_1 i_2 \ldots i_k}$ in this equation are zero, which proves linear independence of the system of matrices (1.8).

Evidently, the number of all matrices in (1.8) is equal to

$$1 + C^1_{2\nu} + C^2_{2\nu} + \cdots + C^{2\nu}_{2\nu} = 2^{2\nu}$$

(C^{λ}_{α} is the number of combinations of λ elements from α).

Since the system of $2^{2\nu}$ matrices (1.8) is linearly independent, it is obvious that the minimum order of the matrices $\overset{\circ}{\gamma}_i$ satisfying Eq. (1.1) is equal to 2^{ν} (in this case, the number of matrix elements in $\overset{\circ}{\gamma}_i$ is equal to $2^{2\nu}$). It also evident that there exist solutions of Eq. (1.1) in the form of matrices $\overset{\circ}{\gamma}_i$ of order $s2^{\nu}$, where $s \geqslant 1$ is a positive integer. Such solutions may be taken, e.g., as quasi-diagonal matrices of the form

$$\check{\gamma}_i = \begin{Vmatrix} \overset{\circ}{\gamma}_i & 0 & \dots & 0 \\ 0 & \overset{\circ}{\gamma}_i & \dots & 0 \\ \vdots & \vdots & \ddots & \vdots \\ 0 & 0 & \dots & \overset{\circ}{\gamma}_i \end{Vmatrix},$$

where $\overset{\circ}{\gamma}_i$ are the matrices of order 2^ν satisfying Eq. (1.1), and 0 is the zero matrix of order 2^ν. One can show that all solutions of Eq. (1.1) are exhausted by matrices similar to $\check{\gamma}_i$, i.e., have the form $T\check{\gamma}_i T^{-1}$, where T is any nondegenerate matrix.

Note that all properties of the matrices $\overset{\circ}{\gamma}_i$ satisfying Eq. (1.1), formulated in items 1–5, are independent of the order of the matrices $\overset{\circ}{\gamma}_i$.

It is clear that if the matrices $\overset{\circ}{\gamma}_i$ ($i = 1, 2, \dots, 2\nu$), satisfying Eq. (1.1), have the order 2^ν, then the system of matrices (1.8) forms a basis in the full matrix algebra over the field of complex numbers, whose dimension is $2^{2\nu}$.

6. Due to completeness of the set of matrices (1.8) and their linear independence, any complex matrix Ψ of order 2^ν may be represented in the form

$$\Psi = \frac{1}{2^\nu}\left(FI + \sum_{k=1}^{2\nu} \frac{1}{k!} F^{i_1 i_2 \dots i_k} \overset{\circ}{\gamma}_{i_1 i_2 \dots i_k}\right). \tag{1.10}$$

To determine the coefficients F and $F^{i_1 i_2 \dots i_k}$ in Eq. (1.10), let us multiply it by $\overset{\circ}{\gamma}^{j_1 j_2 \dots j_m}$ and take the trace of the resulting expression. We then obtain

$$\mathrm{tr}\left(\Psi \overset{\circ}{\gamma}^{j_1 j_2 \dots j_m}\right) = \frac{1}{2^\nu}\left[F\,\mathrm{tr}\,\overset{\circ}{\gamma}^{j_1 j_2 \dots j_m} + \sum_{k=1}^{2\nu} \frac{1}{k!} F^{i_1 i_2 \dots i_k}\,\mathrm{tr}\left(\overset{\circ}{\gamma}_{i_1 i_2 \dots i_k} \overset{\circ}{\gamma}^{j_1 j_2 \dots j_m}\right)\right].$$

Hence, using Eqs. (1.7), we find for the coefficients $F^{i_1 i_2 \dots i_k}$:

$$F^{i_1 i_2 \dots i_k} = (-1)^{\frac{1}{2}k(k-1)}\,\mathrm{tr}\left(\Psi \overset{\circ}{\gamma}^{i_1 i_2 \dots i_k}\right). \tag{1.11}$$

Here[2] $\overset{\circ}{\gamma}^{i_1 i_2 \dots i_k} = \overset{\circ}{\gamma}_{i_1 i_2 \dots i_k}$.

Taking the trace of (1.10), we find the coefficient F:

$$F = \mathrm{tr}\,\Psi. \tag{1.12}$$

[2] In what follows, the indices i_k in $\overset{\circ}{\gamma}$ matrices will be treated as the tensor indices in the Euclidean space with the metric tensor determined in an orthonormal basis by components δ_{ij}. Here, we are so far formally using i_k written at different places in order that the equations obtained be tensor equations in an invariant form.

In particular, one can take as Ψ in equality (1.10) a product of any matrices $\overset{\circ}{\gamma}_{i_1 i_2 \ldots i_k}$. In this case, we will write equality (1.10) in the form

$$\overset{\circ}{\gamma}_{i_1 i_2 \ldots i_k} \overset{\circ}{\gamma}_{j_1 j_2 \ldots j_m} = A_{i_1 i_2 \ldots i_k j_1 j_2 \ldots j_m} I + \sum_{q=1}^{2\nu} A_{i_1 i_2 \ldots i_k j_1 j_2 \ldots j_m}^{s_1 s_2 \ldots s_q} \overset{\circ}{\gamma}_{s_1 s_2 \ldots s_q}. \tag{1.13}$$

According to definitions (1.11) and (1.12), for the coefficients A we have

$$A_{i_1 i_2 \ldots i_k j_1 j_2 \ldots j_m} = \frac{1}{2^\nu} \operatorname{tr}\left(\overset{\circ}{\gamma}_{i_1 i_2 \ldots i_k} \overset{\circ}{\gamma}_{j_1 j_2 \ldots j_m}\right),$$

$$A_{i_1 i_2 \ldots i_k j_1 j_2 \ldots j_m}^{s_1 s_2 \ldots s_q} = (-1)^{\frac{1}{2}q(q-1)} \frac{1}{2^\nu} \operatorname{tr}\left(\overset{\circ}{\gamma}_{i_1 i_2 \ldots i_k} \overset{\circ}{\gamma}_{j_1 j_2 \ldots j_m} \overset{\circ}{\gamma}^{s_1 s_2 \ldots s_q}\right).$$

The coefficients A are sums of different products of the Kronecker deltas. A direct calculation shows that Eq. (1.13) may be written in the form [75]

$$\overset{\circ}{\gamma}_{i_1 i_2 \ldots i_k} \overset{\circ}{\gamma}_{j_1 j_2 \ldots j_m} = \sum_{p=\theta}^{\lambda} (-1)^{\frac{1}{2}p(2k-p-1)} \frac{k!\,m!}{p!\,(k-p)!\,(m-p)!} \delta_{s_1 q_1} \cdots \delta_{s_p q_p}$$
$$\times \delta_{[i_1}^{s_1} \cdots \delta_{i_p}^{s_p} \delta_{i_{p+1}}^{s_{p+1}} \cdots \delta_{i_k]}^{s_k} \delta_{[j_1}^{q_1} \cdots \delta_{j_p}^{q_p} \delta_{j_{p+1}}^{q_{p+1}} \cdots \delta_{j_m]}^{q_m} \overset{\circ}{\gamma}_{s_{p+1} \ldots s_k q_{p+1} \ldots q_m}. \tag{1.14}$$

Here,

$$\lambda = \min(k, m),$$

$$\theta = \begin{cases} 0 & \text{if } \ \frac{1}{2}(k+m-2\nu+1) \leqslant 0, \\ \left[\frac{1}{2}(k+m-2\nu+1)\right] & \text{if } \ \frac{1}{2}(k+m-2\nu+1) > 0. \end{cases}$$

The brackets [ ] in the formula for θ here mean the integer part of the number in the brackets. Summing in p in (1.14) is truncated if $k + m - 2p > 2\nu$. To simplify the expression (1.14), we have omitted the indices with zero number and take $\overset{\circ}{\gamma}_{s_0} = \overset{\circ}{\gamma}_{q_0} = I$.

Let us note that the right-hand side of Eq. (1.14) contains the unit matrix only if $k = m$.

In various calculations, it is also convenient to use another form of Eq. (1.14):

$$\overset{\circ}{\gamma}_{i_1 i_2 \ldots i_k} \overset{\circ}{\gamma}^{j_1 j_2 \ldots j_m} = \sum_{p=\theta}^{\lambda} (-1)^{\frac{1}{2}p(2k-p-1)} \frac{k!\,m!}{p!\,(k-p)!\,(m-p)!}$$
$$\times \delta_{[i_1}^{[j_1} \cdots \delta_{i_p}^{j_p} \overset{\circ}{\gamma}_{i_{p+1} \ldots i_k]}{}^{j_{p+1} \ldots j_m]}. \tag{1.15}$$

It has been denoted here

$$\overset{\circ}{\gamma}_{i_{p+1}\ldots i_k}{}^{j_{p+1}\ldots j_m} = \overset{\circ}{\gamma}_{i_{p+1}\ldots i_k j_{p+1}\ldots j_m}.$$

Let us present Eqs. (1.15) to be used below for different values of the numbers k, m:

For $m = 1$:

$$\begin{aligned}
&\overset{\circ}{\gamma}_i \overset{\circ}{\gamma}_j = \overset{\circ}{\gamma}_{ij} + \delta_{ij} I,\\
&\overset{\circ}{\gamma}_{i_1 i_2 \ldots i_k} \overset{\circ}{\gamma}_j = \overset{\circ}{\gamma}_{i_1 i_2 \ldots i_k j} + k(-1)^{k-1} \delta_{j[i_1} \overset{\circ}{\gamma}_{i_2 \ldots i_k]}, \quad k = 2, 3, , \ldots, 2\nu - 1,\\
&\overset{\circ}{\gamma}_{i_1 i_2 \ldots i_{2\nu}} \overset{\circ}{\gamma}_j = -2\nu \delta_{j[i_1} \overset{\circ}{\gamma}_{i_2 \ldots i_{2\nu}]}.
\end{aligned} \tag{1.16a}$$

For $k = 1$:

$$\begin{aligned}
&\overset{\circ}{\gamma}_j \overset{\circ}{\gamma}_{i_1 i_2 \ldots i_m} = \overset{\circ}{\gamma}_{j i_1 i_2 \ldots i_m} + m \delta_{j[i_1} \overset{\circ}{\gamma}_{i_2 \ldots i_m]}, \quad m = 2, 3, , \ldots, 2\nu - 1,\\
&\overset{\circ}{\gamma}_j \overset{\circ}{\gamma}_{i_1 i_2 \ldots i_{2\nu}} = 2\nu \delta_{j[i_1} \overset{\circ}{\gamma}_{i_2 \ldots i_{2\nu}]}.
\end{aligned} \tag{1.16b}$$

For $m = 2$ ($k = 3, 4, \ldots, 2\nu - 2$):

$$\begin{aligned}
\overset{\circ}{\gamma}_{i_1} \overset{\circ}{\gamma}{}^{j_1 j_2} &= \overset{\circ}{\gamma}_{i_1}{}^{j_1 j_2} + 2\delta_{i_1}^{[j_1} \overset{\circ}{\gamma}{}^{j_2]},\\
\overset{\circ}{\gamma}_{i_1 i_2} \overset{\circ}{\gamma}{}^{j_1 j_2} &= \overset{\circ}{\gamma}_{i_1 i_2}{}^{j_1 j_2} + 4\delta_{[i_1}^{[j_1} \overset{\circ}{\gamma}{}^{j_2]}{}_{i_2]} - 2\delta_{[i_1}^{j_1} \delta_{i_2]}^{j_2} I,\\
\overset{\circ}{\gamma}_{i_1 i_2 \ldots i_k} \overset{\circ}{\gamma}{}^{j_1 j_2} &= \overset{\circ}{\gamma}_{i_1 i_2 \ldots i_k}{}^{j_1 j_2} + 2k \delta_{[i_1}^{[j_1} \overset{\circ}{\gamma}{}^{j_2]}{}_{i_2 \ldots i_k]} - k(k-1) \delta_{[i_1}^{j_1} \delta_{i_2}^{j_2} \overset{\circ}{\gamma}_{i_3 \ldots i_k]},\\
\overset{\circ}{\gamma}_{i_1 i_2 \ldots i_{2\nu-1}} \overset{\circ}{\gamma}{}^{j_1 j_2} &= 2(2\nu - 1) \delta_{[i_1}^{[j_1} \overset{\circ}{\gamma}{}^{j_2]}{}_{i_2 \ldots i_{2\nu-1}]}\\
&\quad - (2\nu - 1)(2\nu - 2) \delta_{[i_1}^{j_1} \delta_{i_2}^{j_2} \overset{\circ}{\gamma}_{i_3 \ldots i_{2\nu-1}]},\\
\overset{\circ}{\gamma}_{i_1 i_2 \ldots i_{2\nu}} \overset{\circ}{\gamma}{}^{j_1 j_2} &= -2\nu(2\nu - 1) \delta_{[i_1}^{j_1} \delta_{i_2}^{j_2} \overset{\circ}{\gamma}_{i_3 \ldots i_{2\nu}]}.
\end{aligned} \tag{1.16c}$$

For $k = 2$ ($m = 3, 4, \ldots, 2\nu - 2$):

$$\begin{aligned}
\overset{\circ}{\gamma}_{i_1 i_2} \overset{\circ}{\gamma}{}^{j_1} &= \overset{\circ}{\gamma}_{i_1 i_2}{}^{j_1} - 2\delta_{[i_1}^{j_1} \overset{\circ}{\gamma}_{i_2]},\\
\overset{\circ}{\gamma}_{i_1 i_2} \overset{\circ}{\gamma}{}^{j_1 j_2 \ldots j_m} &= \overset{\circ}{\gamma}_{i_1 i_2}{}^{j_1 j_2 \ldots j_m} - 2m \delta_{[i_1}^{[j_1} \overset{\circ}{\gamma}_{i_2]}{}^{j_2 \ldots j_m]}\\
&\quad - m(m-1) \delta_{i_1}^{[j_1} \delta_{i_2}^{j_2} \overset{\circ}{\gamma}{}^{j_3 \ldots j_m]},\\
\overset{\circ}{\gamma}_{i_1 i_2} \overset{\circ}{\gamma}{}^{j_1 j_2 \ldots j_{2\nu-1}} &= -2(2\nu - 1) \delta_{[i_1}^{[j_1} \overset{\circ}{\gamma}_{i_2]}{}^{j_2 \ldots i_{2\nu-1}]}\\
&\quad - (2\nu - 1)(2\nu - 2) \delta_{i_1}^{[j_1} \delta_{i_2}^{j_2} \overset{\circ}{\gamma}{}^{j_3 \ldots j_{2\nu-1}]},\\
\overset{\circ}{\gamma}_{i_1 i_2} \overset{\circ}{\gamma}{}^{j_1 j_2 \ldots i_{2\nu}} &= -2\nu(2\nu - 1) \delta_{i_1}^{[j_1} \delta_{i_2}^{j_2} \overset{\circ}{\gamma}{}^{j_3 \ldots j_{2\nu}]}.
\end{aligned} \tag{1.16d}$$

For $m = 2\nu$ ($k = 1, 2, \ldots, 2\nu - 1$):

$$\overset{\circ}{\gamma}_{i_1 i_2 \ldots i_{2\nu}} \overset{\circ}{\gamma}{}^{j_1 j_2 \ldots i_{2\nu}} = (-1)^{\nu}(2\nu)!\, \delta^{[j_1}_{i_1} \delta^{j_2}_{i_2} \cdots \delta^{j_{2\nu}]}_{i_{2\nu}} I, \tag{1.16e}$$

$$\overset{\circ}{\gamma}_{i_1 i_2 \ldots i_k} \overset{\circ}{\gamma}{}^{j_1 j_2 \ldots i_{2\nu}} = (-1)^{\frac{1}{2}k(k-1)} \frac{(2\nu)!}{(2\nu - k)!} \delta^{[j_1}_{i_1} \delta^{j_2}_{i_2} \cdots \delta^{j_k}_{i_k} \overset{\circ}{\gamma}{}^{j_{k+1} \ldots j_{2\nu}]}.$$

For $k = 2\nu$, $m = 1, 2, \ldots, 2\nu - 1$:

$$\overset{\circ}{\gamma}_{i_1 i_2 \ldots i_{2\nu}} \overset{\circ}{\gamma}{}^{j_1 j_2 \ldots i_m} = (-1)^{\frac{1}{2}m(m+1)} \frac{(2\nu)!}{(2\nu - m)!} \delta^{j_1}_{[i_1} \delta^{j_2}_{i_2} \cdots \delta^{j_m}_{i_m} \overset{\circ}{\gamma}_{i_{m+1} \ldots j_{2\nu}]}. \tag{1.16f}$$

7. The set of matrices consisting of the products of any matrix of system (1.8) by each matrix (1.8) contains, up to the sign, all matrices (1.8).

Indeed, the non-degeneracy and linear independence of matrices (1.8) implies that the system under consideration, consisting of products of $\overset{\circ}{\boldsymbol{\gamma}}$ matrices, is also linearly independent and therefore contains $2^{2\nu}$ matrices. From Eq. (1.15) (or, simpler, directly from definition (1.2) and Eq. (1.1)) it follows that each of these products represents (in any case, up to the sign) one of the matrices in (1.8).

8. Pauli's Identity Let us denote elements of the matrix Ψ in Eq. (1.10) by the symbol $\psi^B{}_A$ and elements of the matrices $\overset{\circ}{\gamma}_{i_1 i_2 \ldots i_k}$ and $\overset{\circ}{\gamma}{}^{i_1 i_2 \ldots i_k}$ by the symbols $\overset{\circ}{\gamma}{}^B{}_{A i_1 i_2 \ldots i_k}$ and $\overset{\circ}{\gamma}{}^B{}_A{}^{i_1 i_2 \ldots i_k}$, respectively:

$$\Psi = \|\psi^B{}_A\|, \qquad \overset{\circ}{\gamma}_{i_1 i_2 \ldots i_k} = \|\overset{\circ}{\gamma}{}^B{}_{A i_1 i_2 \ldots i_k}\|, \qquad \overset{\circ}{\gamma}{}^{i_1 i_2 \ldots i_k} = \|\overset{\circ}{\gamma}{}^B{}_A{}^{i_1 i_2 \ldots i_k}\|,$$

where the first index B in the matrix elements in Ψ and $\overset{\circ}{\boldsymbol{\gamma}}$ denotes the row number, while the second index A is the column number.

With these notations introduced, Eq. (1.10) may be written in the form

$$\psi^B{}_A = \frac{1}{2^\nu}\left(F\delta^B_A + \sum_{k=1}^{2\nu} \frac{1}{k!} F^{i_1 i_2 \ldots i_k} \overset{\circ}{\gamma}{}^B{}_{A i_1 i_2 \ldots i_k} \right), \tag{1.17}$$

while definitions (1.11) and (1.12) for the coefficients F and $F^{i_1 i_2 \ldots i_k}$ have the form

$$F = \psi^A{}_A,$$

$$F^{i_1 i_2 \ldots i_k} = (-1)^{\frac{1}{2}k(k-1)} \overset{\circ}{\gamma}{}^B{}_A{}^{i_1 i_2 \ldots i_k} \psi^A{}_B. \tag{1.18}$$

Let us insert into (1.17) the coefficients F and $F^{i_1 i_2 \ldots i_k}$ according to definitions (1.18):

$$\left[-\delta^C_A \delta^B_D + \frac{1}{2^\nu}\left(\delta^C_D \delta^B_A + \sum_{k=1}^{2\nu} \frac{1}{k!} (-1)^{\frac{1}{2}k(k-1)} \overset{\circ}{\gamma}{}^C{}_D{}^{i_1 i_2 \ldots i_k} \overset{\circ}{\gamma}{}^B{}_{A i_1 i_2 \ldots i_k} \right) \right] \psi^D{}_C = 0.$$

From this equality, due to arbitrariness of the quantities $\psi^D{}_C$, we obtain the following important identity connecting products of $\overset{\circ}{\boldsymbol{\gamma}}$ matrices:

$$\delta^C_A\delta^B_D = \frac{1}{2^\nu}\left(\delta^C_D\delta^B_A + \sum_{k=1}^{2\nu}\frac{1}{k!}(-1)^{\frac{1}{2}k(k-1)}\overset{\circ}{\gamma}{}^C{}_D{}^{i_1 i_2\dots i_k}\overset{\circ}{\gamma}{}^B{}_{A i_1 i_2\dots i_k}\right). \tag{1.19}$$

Identity (1.19) for $\nu = 2$ was obtained by Pauli [49] and is usually called Pauli's identity.

Let us multiply identity (1.19) by $\overset{\circ}{\gamma}{}^M{}_{C j_1 j_2\dots j_m}\overset{\circ}{\gamma}{}^E{}_{B s_1 s_2\dots s_q}$ and sum the result by the indices B and C from 1 to 2ν:

$$\overset{\circ}{\gamma}{}^M{}_{A j_1 j_2\dots j_m}\overset{\circ}{\gamma}{}^E{}_{D s_1 s_2\dots s_q} = \frac{1}{2^\nu}\Bigg[\overset{\circ}{\gamma}{}^M{}_{D j_1 j_2\dots j_m}\overset{\circ}{\gamma}{}^E{}_{A s_1 s_2\dots s_q}$$

$$+\sum_{k=1}^{2\nu}\frac{1}{k!}(-1)^{\frac{1}{2}k(k-1)}\left(\overset{\circ}{\gamma}{}^M{}_{C j_1 j_2\dots j_m}\overset{\circ}{\gamma}{}^C{}_D{}^{i_1 i_2\dots i_k}\right)\left(\overset{\circ}{\gamma}{}^E{}_{B s_1 s_2\dots s_q}\overset{\circ}{\gamma}{}^B{}_{A i_1 i_2\dots i_k}\right)\Bigg].$$

Changing here the products of matrices $\overset{\circ}{\boldsymbol{\gamma}}$ according to Eq. (1.13), we obtain relationships expressing the products $\overset{\circ}{\boldsymbol{\gamma}}{}^M{}_A\overset{\circ}{\boldsymbol{\gamma}}{}^E{}_D$ in terms of sums of different products $\overset{\circ}{\boldsymbol{\gamma}}{}^M{}_D\overset{\circ}{\boldsymbol{\gamma}}{}^E{}_A$ with transposed spinor indices D, A[3]:

$$\overset{\circ}{\gamma}{}^M{}_{A j_1 j_2\dots j_m}\overset{\circ}{\gamma}{}^E{}_{D s_1 s_2\dots s_q} = \sum_{k=0}^{2\nu}\sum_{p=0}^{2\nu}\alpha^{i_1 i_2\dots i_k l_1 l_2\dots l_p}_{j_1 j_2\dots j_m s_1 s_2\dots s_q}\overset{\circ}{\gamma}{}^M{}_{D i_1 i_2\dots i_k}\overset{\circ}{\gamma}{}^E{}_{A l_1 l_2\dots l_p}. \tag{1.20}$$

Here, the coefficients α are determined by the equality

$$\alpha^{i_1 i_2\dots i_k l_1 l_2\dots l_p}_{j_1 j_2\dots j_m s_1 s_2\dots s_q} = \sum_{r=0}^{2\nu}\frac{(-1)^{\frac{1}{2}r(r-1)}}{2^\nu r!}\delta^{t_1 n_1}\delta^{t_2 n_2}\dots\delta^{t_r n_r}$$
$$\times A^{i_1 i_2\dots i_k}_{j_1 j_2\dots j_m t_1 t_2\dots t_r}A^{l_1 l_2\dots l_p}_{s_1 s_2\dots s_q n_1 n_2\dots n_r}.$$

9. Let us introduce the matrix $\overset{\circ}{\gamma}_{2\nu+1}$:

$$\overset{\circ}{\gamma}_{2\nu+1} = \mathrm{i}^\nu\overset{\circ}{\gamma}_1\overset{\circ}{\gamma}_2\cdots\overset{\circ}{\gamma}_{2\nu}. \tag{1.21}$$

[3]The full set of identities (1.20) for 4-dimensional pseudo-Euclidean space is presented in Appendix C. These identities for 3-dimensional Euclidean space are presented in Chap. 4, see page 205.

Summing the last equations in (1.16a), (1.16b), we obtain

$$\overset{\circ}{\gamma}_j \overset{\circ}{\gamma}_{i_1 i_2 \dots i_{2\nu}} + \overset{\circ}{\gamma}_{i_1 i_2 \dots i_{2\nu}} \overset{\circ}{\gamma}_j = 0.$$

Hence it follows that the matrix $\overset{\circ}{\gamma}_{2\nu+1}$ anticommutes with all matrices $\overset{\circ}{\gamma}_j$:

$$\overset{\circ}{\gamma}_j \overset{\circ}{\gamma}_{2\nu+1} + \overset{\circ}{\gamma}_{2\nu+1} \overset{\circ}{\gamma}_j = 0, \quad j = 1, 2, \dots, 2\nu.$$

The first equality in (1.16e) also implies that $\overset{\circ}{\gamma}_{2\nu+1} \overset{\circ}{\gamma}_{2\nu+1} = I$. Therefore, Eq. (1.1), defining the 2ν matrices $\overset{\circ}{\gamma}_j$, also holds if we suppose that the indices i, j take values from 1 to $2\nu + 1$,

$$\overset{\circ}{\gamma}_i \overset{\circ}{\gamma}_j + \overset{\circ}{\gamma}_j \overset{\circ}{\gamma}_i = 2\delta_{ij} I, \qquad i, j = 1, 2, \dots, 2\nu + 1.$$

The set of matrices with an even number of indices

$$I, \quad \overset{\circ}{\gamma}_{i_1 i_2}, \quad \dots, \quad \overset{\circ}{\gamma}_{i_1 i_2 \dots i_{2\nu}} \qquad (i_1 < i_2 < \dots < i_{2\nu}) \tag{1.22}$$

and the set of matrices with an odd number of indices

$$\overset{\circ}{\gamma}_i, \quad \overset{\circ}{\gamma}_{i_1 i_2 i_3}, \quad \dots, \quad \overset{\circ}{\gamma}_{i_1 i_2 i_3 \dots i_{2\nu+1}} \qquad (i_1 < i_2 < \dots < i_{2\nu+1}), \tag{1.23}$$

in which $\overset{\circ}{\gamma}_{i_1 i_2 \dots i_k} = \overset{\circ}{\gamma}_{[i_1} \overset{\circ}{\gamma}_{i_2} \cdots \overset{\circ}{\gamma}_{i_k]}$ and the indices i_k take all integer values from 1 tp $2\nu + 1$, are linearly independent.

Linear independence of the set of matrices (1.22) and (1.23) is proved in a way similar to the corresponding proof for the set of matrices (1.8) in item 5.

Evidently, the number of matrices in set (1.22) is equal to $2^{2\nu}$:

$$C^0_{2\nu+1} + C^2_{2\nu+1} + \dots + C^{2\nu}_{2\nu+1} = 2^{2\nu}.$$

The number of matrices in set (1.23) is also equal to $2^{2\nu}$:

$$C^1_{2\nu+1} + C^3_{2\nu+1} + \dots + C^{2\nu+1}_{2\nu+1} = 2^{2\nu}.$$

Therefore the set of matrices (1.22) and the set of matrices (1.23) form bases in the full matrix algebra of the dimension $2^{2\nu}$.

Due to the completeness and linear independence of the set of matrices (1.22), the elements of any square matrix $\|\psi^B{}_A\|$ of order 2^ν may be represented in the form

$$\psi^B{}_A = \frac{1}{2^\nu} \left(F \delta^B_A + \sum_{k=1}^{\nu} \frac{1}{(2k)!} F^{i_1 i_2 \dots i_{2k}} \overset{\circ}{\gamma}{}^B{}_{A i_1 i_2 \dots i_{2k}} \right). \tag{1.24}$$

For the coefficients F, $F^{i_1 i_2 \dots i_{2k}}$, calculations similar to those conducted in item 6, lead to

$$F = \psi^A{}_A,$$
$$F^{i_1 i_2 \dots i_{2k}} = (-1)^k \overset{\circ}{\gamma}{}^B{}_A{}^{i_1 i_2 \dots i_{2k}} \psi^A{}_B.$$

Using the set of matrices (1.23), we can write for $\psi^B{}_A$:

$$\psi^B{}_A = \frac{1}{2^\nu} \sum_{k=0}^{\nu} \frac{1}{(2k+1)!} F^{i_1 i_2 \dots i_{2k+1}} \overset{\circ}{\gamma}{}^B{}_{A i_1 i_2 \dots i_{2k+1}}, \tag{1.25}$$

where

$$F^{i_1 i_2 \dots i_{2k+1}} = (-1)^k \overset{\circ}{\gamma}{}^B{}_A{}^{i_1 i_2 \dots i_{2k+1}} \psi^A{}_B.$$

10. If the matrix A commutes with all matrices $\overset{\circ}{\gamma}_i$,

$$A \overset{\circ}{\gamma}_i = \overset{\circ}{\gamma}_i A,$$

then A is a multiple of the unit matrix $A = \lambda I$, where λ is a certain, generally complex, number.

To prove this, let us represent the matrix A in the form

$$A = aI + \sum_{k=1}^{2\nu} a^{i_1 i_2 \dots i_k} \overset{\circ}{\gamma}_{i_1 i_2 \dots i_k}, \tag{1.26}$$

where the coefficients $a^{i_1 i_2 \dots i_k}$ are antisymmetric in all indices. We have

$$\overset{\circ}{\gamma}_j A - A \overset{\circ}{\gamma}_j = \sum_{k=1}^{2\nu} a^{i_1 i_2 \dots i_k} \big(\overset{\circ}{\gamma}_j \overset{\circ}{\gamma}_{i_1 i_2 \dots i_k} - \overset{\circ}{\gamma}_{i_1 i_2 \dots i_k} \overset{\circ}{\gamma}_j \big).$$

Here, substituting the differences of products of $\overset{\circ}{\gamma}$ matrices by the formulae

$$\overset{\circ}{\gamma}_j \overset{\circ}{\gamma}_{i_1} - \overset{\circ}{\gamma}_{i_1} \overset{\circ}{\gamma}_j = 2 \overset{\circ}{\gamma}_{j i_1},$$
$$\overset{\circ}{\gamma}_j \overset{\circ}{\gamma}_{i_1 i_2 \dots i_k} - \overset{\circ}{\gamma}_{i_1 i_2 \dots i_k} \overset{\circ}{\gamma}_j = [1 - (-1)^k] \overset{\circ}{\gamma}_{j i_1 i_2 \dots i_k} + k[1 + (-1)^k] \delta_{j[i_1} \overset{\circ}{\gamma}_{i_2 \dots i_k]},$$
$$k = 2, 3, \dots, 2\nu - 1,$$
$$\overset{\circ}{\gamma}_j \overset{\circ}{\gamma}_{i_1 i_2 \dots i_{2\nu}} - \overset{\circ}{\gamma}_{i_1 i_2 \dots i_{2\nu}} \overset{\circ}{\gamma}_j = 4\nu \delta_{j[i_1} \overset{\circ}{\gamma}_{i_2 \dots i_{2\nu}]},$$

which follow from relations (1.16a) and (1.16b), we obtain

$$\overset{\circ}{\gamma}_j A - A\overset{\circ}{\gamma}_j = 2\sum_{m=0}^{\nu-1} a^{i_1 i_2 \dots i_{2m+1}} \overset{\circ}{\gamma}_{j i_1 i_2 \dots i_{2m+1}}$$

$$+4\sum_{m=1}^{\nu} a^{i_1 i_2 \dots i_{2m}} m \delta_{j[i_1} \overset{\circ}{\gamma}_{i_2 \dots i_{2m}]} = 0. \qquad (1.27)$$

The first sum in (1.27) contains only $\overset{\circ}{\gamma}$ matrices with an even number of indices, while the second one only those with an odd number of indices.

Due to linear independence of the set of matrices (1.8), it follows from Eq. (1.27) that all coefficients $a^{i_1 i_2 \dots i_k}$ are equal to zero. Thus only the term with the unit matrix in the expansion (1.26) is different from zero.

11. If the matrices $\overset{\circ}{\gamma}_i$ satisfy Eq. (1.1) and T is an arbitrary nondegenerate matrix of order 2^ν, it is evident that the set of matrices

$$\widetilde{\gamma}_i = T^{-1} \overset{\circ}{\gamma}_i T \qquad (1.28)$$

satisfies the equation

$$\widetilde{\gamma}_i \widetilde{\gamma}_j + \widetilde{\gamma}_j \widetilde{\gamma}_i = 2\delta_{ij} I.$$

It also turns out that any two sets of matrices $\overset{\circ}{\gamma}_i$, $\widetilde{\gamma}_i$, satisfying Eq. (1.1), are always connected by relation (1.28), in which the matrix T is determined up to multiplication by an arbitrary nonzero complex number. This property of solutions of Eq. (1.1) has been named Pauli's theorem.

A simple proof of Pauli's theorem consists in explicitly pointing out the matrix T, corresponding to different sets of matrices $\overset{\circ}{\gamma}_i$, $\widetilde{\gamma}_i$. If we denote matrices (1.8) by the symbol $\overset{\circ}{\gamma}_{\mathcal{A}}$ ($\mathcal{A} = 1, 2, \dots, 2^{2\nu}$) and similar matrices formed from $\widetilde{\gamma}_i$ by the symbol $\widetilde{\gamma}_{\mathcal{A}}$, then the matrix T may be written in the form

$$T = \sum_{\mathcal{A}=1}^{2^{2\nu}} \overset{\circ}{\gamma}_{\mathcal{A}} F \, \widetilde{\gamma}_{\mathcal{A}}^{-1}, \qquad (1.29)$$

where F is some nonzero square matrix of order 2^ν.

Indeed, let us calculate the quantity $\overset{\circ}{\gamma}_i T \, \widetilde{\gamma}_i^{-1}$ (without summing over the index i):

$$\overset{\circ}{\gamma}_i T \, \widetilde{\gamma}_i^{-1} = \sum_{\mathcal{A}=1}^{2^{2\nu}} \overset{\circ}{\gamma}_i \overset{\circ}{\gamma}_{\mathcal{A}} F \left(\widetilde{\gamma}_i \widetilde{\gamma}_{\mathcal{A}}\right)^{-1}. \qquad (1.30)$$

Since, according to property 7 of the matrices $\overset{\circ}{\gamma}_{\mathcal{A}}$, the product $\overset{\circ}{\gamma}_i\overset{\circ}{\gamma}_{\mathcal{A}}$ for all $\mathcal{A}$ again gives all matrices $\overset{\circ}{\gamma}_{\mathcal{A}}$ (at least up to the sign), equality (1.30) can be continued:

$$\overset{\circ}{\gamma}_i T\, \widetilde{\gamma}_i^{-1} = \sum_{\mathcal{A}=1}^{2^{2\nu}} \overset{\circ}{\gamma}_{\mathcal{A}} F\, \widetilde{\gamma}_{\mathcal{A}}^{-1} = T.$$

Hence it follows

$$\overset{\circ}{\gamma}_i T = T\, \widetilde{\gamma}_i. \tag{1.31}$$

Obviously, one can always choose the matrix F in definition (1.29) in such a way that $T \neq 0$ (otherwise the set of matrices $\overset{\circ}{\gamma}_{\mathcal{A}}$ would be linearly dependent). Let us show that, with the corresponding choice of F, the matrix T is nondegenerate, $\det T \neq 0$.

Analogously to (1.29)–(1.31), one obtains that the matrix Q defined by the equality

$$Q = \sum_{\mathcal{A}=1}^{2^{2\nu}} \widetilde{\gamma}_{\mathcal{A}} G \overset{\circ}{\gamma}{}_{\mathcal{A}}^{-1},$$

in which G is some square matrix of order 2^ν, satisfies the equation

$$\widetilde{\gamma}_i\, Q = Q \overset{\circ}{\gamma}_i \tag{1.32}$$

and, under the corresponding choice of G, is nonzero, $Q \neq 0$. Multiplying Eq. (1.32) by T from the right and taking into account Eq. (1.31), we find

$$\widetilde{\gamma}_i\, QT = QT\, \widetilde{\gamma}_i.$$

Therefore, the matrix QT is proportional to the unit matrix:

$$QT = \alpha I. \tag{1.33}$$

From linear independence of the set of matrices $\overset{\circ}{\gamma}_{\mathcal{A}}$ it follows that, for $Q \neq 0$, one can always choose F in such a way that the number α in Eq. (1.33) is nonzero, $\alpha \neq 0$. Indeed, if for any matrix F we had $\alpha = 0$, then Eq. (1.33) would imply

$$QT \equiv \sum_{\mathcal{A}=1}^{2^{2\nu}} (Q\overset{\circ}{\gamma}_{\mathcal{A}} F)\widetilde{\gamma}_{\mathcal{A}}^{-1} = 0,$$

and, due to arbitrariness of F,

$$\sum_{\mathcal{A}=1}^{2^{2\nu}} (Q\overset{\circ}{\gamma}_{\mathcal{A}}) \times \widetilde{\gamma}_{\mathcal{A}}^{-1} = 0. \tag{1.34}$$

But since $Q \neq 0$, all matrices $Q\overset{\circ}{\gamma}_{\mathcal{A}}$ cannot be equal to zero. Therefore equality (1.34) contradicts the linear independence of the set of matrices $\overset{\circ}{\gamma}_{\mathcal{A}}$. Thus one can always choose the matrix F is such a way that $\alpha \neq 0$.

From Eq. (1.33) it follows that, for the corresponding choice of F, the matrix T is nondegenerate, and there exists the inverse matrix $T^{-1} = \alpha^{-1} Q$. Therefore, from equality (1.31) it follows (1.28), the equation to be proved.

It remains to show that the matrix T in Eq. (1.28) is determined up to multiplying by an arbitrary nonzero number. Suppose that $\widetilde{T} \neq T$ satisfies the equation

$$\widetilde{\gamma}_i = \widetilde{T}^{-1}\overset{\circ}{\gamma}_i \widetilde{T}. \tag{1.35}$$

Then Eqs. (1.28) and (1.35) imply

$$T^{-1}\overset{\circ}{\gamma}_i T = \widetilde{T}^{-1}\overset{\circ}{\gamma}_i \widetilde{T}, \quad \text{or} \quad \widetilde{T}T^{-1}\overset{\circ}{\gamma}_i = \overset{\circ}{\gamma}_i \widetilde{T}T^{-1},$$

and therefore

$$\widetilde{T}T^{-1} = \lambda I, \quad \lambda \neq 0,$$

so that $\widetilde{T}$ differs from T by only a numerical factor: $\widetilde{T} = \lambda T$, $\lambda \neq 0$. The theorem is proved.

Substituting the matrices $\overset{\circ}{\gamma}_i$ in Eq. (1.28) according to the formula

$$\overset{\circ}{\gamma}_i = -\overset{\circ}{\gamma}{}_{2\nu+1}^{-1}\overset{\circ}{\gamma}_i\overset{\circ}{\gamma}_{2\nu+1},$$

which follows from definition (1.21) of $\overset{\circ}{\gamma}_{2\nu+1}$, we find that the connection between $\overset{\circ}{\gamma}_i$ and $\widetilde{\gamma}_i$ may also be written in the form

$$\widetilde{\gamma}_i = -\overset{*}{T}{}^{-1}\overset{\circ}{\gamma}_i \overset{*}{T},$$

where

$$\overset{*}{T} = \overset{\circ}{\gamma}_{2\nu+1} T.$$

Let us now show that a solution of the matrix equations (1.1) for $\overset{\circ}{\gamma}_i$ does exist for any $\nu \geqslant 1$ and may be realized in the form of the Hermitian matrices

$$(\overset{\circ}{\gamma}_i)^T = (\overset{\circ}{\gamma}_i)^{\cdot}, \tag{1.36}$$

where one can choose the ν matrices $\overset{\circ}{\gamma}_1, \overset{\circ}{\gamma}_2, \ldots, \overset{\circ}{\gamma}_\nu$ to be symmetric and the ν matrices $\overset{\circ}{\gamma}_{\nu+1}, \overset{\circ}{\gamma}_{\nu+2}, \ldots, \overset{\circ}{\gamma}_{2\nu}$ to be antisymmetric:

$$\begin{gathered}(\overset{\circ}{\gamma}_1)^T = \overset{\circ}{\gamma}_1, \quad (\overset{\circ}{\gamma}_2)^T = \overset{\circ}{\gamma}_2, \quad \ldots, \quad (\overset{\circ}{\gamma}_\nu)^T = \overset{\circ}{\gamma}_\nu, \\ (\overset{\circ}{\gamma}_{\nu+1})^T = -\overset{\circ}{\gamma}_{\nu+1}, \quad (\overset{\circ}{\gamma}_{\nu+2})^T = -\overset{\circ}{\gamma}_{\nu+2}, \quad \ldots, \quad (\overset{\circ}{\gamma}_{2\nu})^T = -\overset{\circ}{\gamma}_{2\nu}.\end{gathered} \tag{1.37}$$

In (1.36) and (1.37), the symbol "T" means transposition and an overdot means complex conjugation.

Let us prove the statement formulated.

For $\nu = 1$, it is easy to point out a set of two second-order matrices satisfying Eq. (1.1):

$$\overset{\circ}{\gamma}_1 = \begin{Vmatrix} 1 & 0 \\ 0 & -1 \end{Vmatrix}, \quad \overset{\circ}{\gamma}_2 = \begin{Vmatrix} 0 & -\mathrm{i} \\ \mathrm{i} & 0 \end{Vmatrix}. \tag{1.38}$$

where $\mathrm{i} = \sqrt{-1}$. Evidently, the matrices written are Hermitian, and $\overset{\circ}{\gamma}_1$ is symmetric, while $\overset{\circ}{\gamma}_2$ is antisymmetric.

Assuming that there exists a solution of (1.1) for $\nu = \alpha$, satisfying the conditions (1.36) and (1.37), let us show that there exists a solution of Eq. (1.1) for $\nu = \alpha + 1$, and it satisfies the conditions (1.36) and (1.37).

Let $\overset{\circ}{\gamma}_1, \overset{\circ}{\gamma}_2, \ldots, \overset{\circ}{\gamma}_{2\alpha}$ be a set of 2α Hermitian matrices satisfying Eq. (1.1), in which the indices i and j take values from 1 to 2α. We will assume that the matrices $\overset{\circ}{\gamma}_1, \overset{\circ}{\gamma}_2, \ldots, \overset{\circ}{\gamma}_\alpha$ are symmetric and $\overset{\circ}{\gamma}_{\alpha+1}, \overset{\circ}{\gamma}_{\alpha+2}, \ldots, \overset{\circ}{\gamma}_{2\alpha}$ are antisymmetric. Let us introduce the matrix

$$\overset{\circ}{\gamma}_{2\alpha+1} = \mathrm{i}^\alpha \overset{\circ}{\gamma}_1 \overset{\circ}{\gamma}_2 \cdots \overset{\circ}{\gamma}_{2\alpha}. \tag{1.39}$$

due to Eq. (1.1), the matrix $\overset{\circ}{\gamma}_{2\alpha+1}$ anticommutes with all matrices $\overset{\circ}{\gamma}_i$:

$$\overset{\circ}{\gamma}_i \overset{\circ}{\gamma}_{2\alpha+1} + \overset{\circ}{\gamma}_{2\alpha+1} \overset{\circ}{\gamma}_i = 0, \quad i = 1, 2, \ldots, 2\alpha, \tag{1.40}$$

and the square of the matrix $\overset{\circ}{\gamma}_{2\alpha+1}$ is the unit matrix,

$$\overset{\circ}{\gamma}_{2\alpha+1} \overset{\circ}{\gamma}_{2\alpha+1} = I. \tag{1.41}$$

If the matrices $\overset{\circ}{\gamma}_i$ are Hermitian, then the matrix $\overset{\circ}{\gamma}_{2\alpha+1}$ defined by equality (1.39) is also Hermitian. Indeed, due to hermiticity of the matrices $\overset{\circ}{\gamma}_i$ $(i = 1, 2,\ldots, 2\alpha)$, one can write for the transposed matrix $\overset{\circ}{\gamma}{}^T_{2\alpha+1}$:

$$\overset{\circ}{\gamma}{}^T_{2\alpha+1} = \mathrm{i}^\alpha \overset{\circ}{\gamma}{}^T_{2\alpha} \ldots \overset{\circ}{\gamma}{}^T_2 \overset{\circ}{\gamma}{}^T_1 = \mathrm{i}^\alpha (\overset{\circ}{\gamma}_{2\alpha})^{\cdot} \ldots (\overset{\circ}{\gamma}_2)^{\cdot} (\overset{\circ}{\gamma}_1)^{\cdot} = \mathrm{i}^\alpha (\overset{\circ}{\gamma}_{2\alpha} \cdots \overset{\circ}{\gamma}_2 \overset{\circ}{\gamma}_1)^{\cdot}. \quad (1.42)$$

Since all matrices $\overset{\circ}{\gamma}_i$ in the product (1.39) are different, all of them anticommute with each other according to Eq. (1.1). Permuting the matrices $\overset{\circ}{\gamma}_i$ in equality (1.42), let us continue it:

$$\overset{\circ}{\gamma}{}^T_{2\alpha+1} = \mathrm{i}^\alpha (-1)^{\alpha(2\alpha-1)} (\overset{\circ}{\gamma}_1 \overset{\circ}{\gamma}_2 \cdots \overset{\circ}{\gamma}_{2\alpha})^{\cdot} = (\mathrm{i}^\alpha \overset{\circ}{\gamma}_1 \overset{\circ}{\gamma}_2 \cdots \overset{\circ}{\gamma}_{2\alpha})^{\cdot}.$$

Thus $\overset{\circ}{\gamma}{}^T_{2\alpha+1} = (\overset{\circ}{\gamma}_{2\alpha+1})^{\cdot}$.

Let us now find out the symmetry properties of the matrix $\overset{\circ}{\gamma}_{2\alpha+1}$. We have

$$\overset{\circ}{\gamma}{}^T_{2\alpha+1} = \mathrm{i}^\alpha \overset{\circ}{\gamma}{}^T_{2\alpha} \ldots \overset{\circ}{\gamma}{}^T_2 \overset{\circ}{\gamma}{}^T_1 = \mathrm{i}^\alpha (-1)^\alpha \overset{\circ}{\gamma}_{2\alpha} \ldots \overset{\circ}{\gamma}_2 \overset{\circ}{\gamma}_1.$$

Permuting here the anticommuting matrices $\overset{\circ}{\gamma}_i$, we find:

$$\overset{\circ}{\gamma}{}^T_{2\alpha+1} = \mathrm{i}^\alpha (-1)^\alpha (-1)^{\alpha(2\alpha-1)} \overset{\circ}{\gamma}_1 \overset{\circ}{\gamma}_2 \cdots \overset{\circ}{\gamma}_{2\alpha} = \mathrm{i}^\alpha \overset{\circ}{\gamma}_1 \overset{\circ}{\gamma}_2 \cdots \overset{\circ}{\gamma}_{2\alpha}.$$

Thus the matrix $\overset{\circ}{\gamma}_{2\alpha+1}$ is symmetric $\overset{\circ}{\gamma}{}^T_{2\alpha+1} = \overset{\circ}{\gamma}_{2\alpha+1}$.

Consider the following set of $2(\alpha + 1)$ matrices:

$$\begin{Vmatrix} 0 & I \\ I & 0 \end{Vmatrix}, \quad \begin{Vmatrix} 0 & -\mathrm{i}\overset{\circ}{\gamma}_j \\ \mathrm{i}\overset{\circ}{\gamma}_j & 0 \end{Vmatrix}, \quad j = 1, 2, \ldots, 2\alpha + 1. \quad (1.43)$$

A direct inspection with the aid of Eqs. (1.40) and (1.41) shows that the set of $2(\alpha + 1)$ matrices (1.43) satisfies Eq. (1.1) in which $i, j = 1, 2,\ldots, 2(\alpha + 1)$.

Let us establish the symmetry properties of the set of matrices (1.43). Evidently, the first matrix in (1.43) is Hermitian and symmetric. The other matrices in (1.43) are also Hermitian:

$$\begin{Vmatrix} 0 & -\mathrm{i}\overset{\circ}{\gamma}_j \\ \mathrm{i}\overset{\circ}{\gamma}_j & 0 \end{Vmatrix}^T = \begin{Vmatrix} 0 & \mathrm{i}\overset{\circ}{\gamma}{}^T_j \\ -\mathrm{i}\overset{\circ}{\gamma}{}^T_j & 0 \end{Vmatrix} = \begin{Vmatrix} 0 & \mathrm{i}(\overset{\circ}{\gamma}_j)^{\cdot} \\ -\mathrm{i}(\overset{\circ}{\gamma}_j)^{\cdot} & 0 \end{Vmatrix} = \begin{Vmatrix} 0 & (-\mathrm{i}\overset{\circ}{\gamma}_j)^{\cdot} \\ (\mathrm{i}\overset{\circ}{\gamma}_j)^{\cdot} & 0 \end{Vmatrix}.$$

By virtue of the assumed symmetry properties of the matrices $\overset{\circ}{\gamma}_j$ $(j = 1, 2,\ldots, 2\alpha)$ and due to the symmetric nature of $\overset{\circ}{\gamma}_{2\alpha+1}$, the matrices (1.43) for $j = 1, 2,\ldots,$

α and for $j = 2\alpha + 1$ are antisymmetric:

$$\begin{Vmatrix} 0 & -\mathrm{i}\overset{\circ}{\gamma}_j \\ \mathrm{i}\overset{\circ}{\gamma}_j & 0 \end{Vmatrix}^T = \begin{Vmatrix} 0 & \mathrm{i}\overset{\circ}{\gamma}{}_j^T \\ -\mathrm{i}\overset{\circ}{\gamma}{}_j^T & 0 \end{Vmatrix} = - \begin{Vmatrix} 0 & -\mathrm{i}\overset{\circ}{\gamma}_j \\ \mathrm{i}\overset{\circ}{\gamma}_j & 0 \end{Vmatrix},$$

while for $j = \alpha + 1, \ldots, 2\alpha$ they are symmetric:

$$\begin{Vmatrix} 0 & -\mathrm{i}\overset{\circ}{\gamma}_j \\ \mathrm{i}\overset{\circ}{\gamma}_j & 0 \end{Vmatrix}^T = \begin{Vmatrix} 0 & \mathrm{i}\overset{\circ}{\gamma}{}_j^T \\ -\mathrm{i}\overset{\circ}{\gamma}{}_j^T & 0 \end{Vmatrix} = \begin{Vmatrix} 0 & -\mathrm{i}\overset{\circ}{\gamma}_j \\ \mathrm{i}\overset{\circ}{\gamma}_j & 0 \end{Vmatrix}.$$

We see that half of the matrices in (1.43) are symmetric and another half antisymmetric. Thus matrices (1.43) satisfy Eq. (1.1) and possess the properties (1.36) and (1.37). We have thus proved the previously formulated statement on the existence of solutions of Eq. (1.1).

Since for $\nu = 1$ the order of matrices (1.38) is two, and at transition from $\nu = \alpha$ to $\nu = \alpha + 1$ in the construction of the matrices $\overset{\circ}{\gamma}_j$ considered above their order is doubled, it is evident that the order of the $2(\alpha + 1)$ matrices (1.43) is equal to $2^{\alpha+1}$. It has been proved above that it is the minimal order of the matrices $\overset{\circ}{\gamma}_j$ satisfying Eq. (1.1).

If $\overset{\circ}{\gamma}_i$ is a certain solution of Eq. (1.1), then, evidently, the matrices $-\overset{\circ}{\gamma}{}_i^T$ also form a solution of Eq. (1.1):

$$(-\overset{\circ}{\gamma}{}_i^T)(-\overset{\circ}{\gamma}{}_j^T) + (-\overset{\circ}{\gamma}{}_j^T)(-\overset{\circ}{\gamma}{}_i^T) = 2\delta_{ij} I.$$

It therefore follows from Pauli's theorem that there exists a nondegenerate square matrix $E = \|e_{BA}\|$, defined up to multiplying by an arbitrary nonzero complex number, which satisfies the equation

$$\overset{\circ}{\gamma}{}_i^T = -E\overset{\circ}{\gamma}_i E^{-1}. \tag{1.44}$$

According to Pauli's theorem, any two solutions of Eq. (1.1) are connected by the similari ty transformation (1.28). Let us find out how does the matrix E, defined by Eq. (1.44), change in a transition from the set of matrices $\overset{\circ}{\gamma}_i$ to the set of matrices $\overset{\circ}{\gamma}_i{}'$:

$$\overset{\circ}{\gamma}_i{}' = T^{-1}\overset{\circ}{\gamma}_i T. \tag{1.45}$$

Let E' be defined by the equation

$$(\overset{\circ}{\gamma}_i{}')^T = -E'\overset{\circ}{\gamma}_i{}'(E')^{-1}. \tag{1.46}$$

Substituting, in definition (1.46), the matrices $\overset{\circ}{\gamma}_i{}'$ in terms of $\overset{\circ}{\gamma}_i$ by formula (1.45) and multiplying the resulting equality from the left by $(T^{-1})^T$ and from the right by T^T, we obtain

$$\overset{\circ}{\gamma}{}_i^T = -(T^T)^{-1}E'T^{-1}\overset{\circ}{\gamma}_i T(E')^{-1}T^T. \tag{1.47}$$

Comparing Eqs. (1.44) and (1.47), we find:

$$\overset{\circ}{\gamma}_i E^{-1}(T^T)^{-1}E'T^{-1} = E^{-1}(T^T)^{-1}E'T^{-1}\overset{\circ}{\gamma}_i.$$

Hence, due to the properties of $\overset{\circ}{\gamma}$ matrices formulated in item 10, it follows:

$$E^{-1}(T^T)^{-1}E'T^{-1} = \lambda I,$$

where λ is some nonzero complex number. Evidently, without loss of generality, one can put $\lambda = 1$ (by redefinition $T \to T\lambda^{-1/2}$ leaving Eq. (1.45) unchanged), and then

$$E' = T^T E T. \tag{1.48}$$

Thus if the matrices $\overset{\circ}{\gamma}_i{}'$ and $\overset{\circ}{\gamma}_i$ are connected by a similarity transformation (1.45), then the corresponding matrices E and E' are connected by equality (1.48).

Let us transpose equation (1.44) and multiply the result from the right by $(E^T)^{-1}$ and from the left by E^T. We obtain:

$$\overset{\circ}{\gamma}{}_i^T = -E^T\overset{\circ}{\gamma}_i(E^T)^{-1}. \tag{1.49}$$

Comparing Eqs. (1.49) and (1.44), we find:

$$E\overset{\circ}{\gamma}_i E^{-1} = E^T\overset{\circ}{\gamma}_i(E^T)^{-1}.$$

From the latter equation it follows that the matrix $E^{-1}E^T$ commutes with all $\overset{\circ}{\gamma}_i$:

$$E^{-1}E^T\overset{\circ}{\gamma}_i = \overset{\circ}{\gamma}_i E^{-1}E^T$$

and is therefore proportional to the unit matrix, $E^{-1}E^T = \rho I$, or

$$E^T = \rho E, \tag{1.50}$$

where $\rho \neq 0$. Transposing equation (1.50), we find:

$$E^T = \frac{1}{\rho}E.$$

Thus $\rho = 1/\rho$ and consequently $\rho = \pm 1$. It means that the matrix E defined by Eq. (1.44) is either symmetric or antisymmetric, $E^T = \pm E$.

If one takes the matrices $\overset{\circ}{\gamma}_1, \overset{\circ}{\gamma}_2, \ldots, \overset{\circ}{\gamma}_\nu$ to be symmetric and $\overset{\circ}{\gamma}_{\nu+1}, \overset{\circ}{\gamma}_{\nu+2}, \ldots, \overset{\circ}{\gamma}_{2\nu}$ to be antisymmetric, it is easily seen that one can take for E the matrix

$$E = \lambda \overset{\circ}{\gamma}_1 \overset{\circ}{\gamma}_2 \cdots \overset{\circ}{\gamma}_\nu \tag{1.51}$$

if ν is even and the matrix

$$E = \lambda \overset{\circ}{\gamma}_{\nu+1} \overset{\circ}{\gamma}_{\nu+2} \cdots \overset{\circ}{\gamma}_{2\nu} \tag{1.52}$$

if ν is odd. Here $\lambda \neq 0$ is an arbitrary complex number.

Taking into account the symmetry properties of the matrices $\overset{\circ}{\gamma}_i$ in Eqs. (1.51) and (1.52) for E, it is easy to find that the matrix E is symmetric if the number $\frac{1}{2}\nu(\nu+1)$ is even and is antisymmetric if the number $\frac{1}{2}\nu(\nu+1)$ is odd:

$$E^T = (-1)^{\frac{1}{2}\nu(\nu+1)} E. \tag{1.53}$$

For example, in the case of an even ν, with definition (1.51) we have

$$\begin{aligned} E^T &= \lambda \overset{\circ}{\gamma}{}_\nu^T \cdots \overset{\circ}{\gamma}{}_2^T \overset{\circ}{\gamma}{}_1^T = \lambda \overset{\circ}{\gamma}_\nu \cdots \overset{\circ}{\gamma}_2 \overset{\circ}{\gamma}_1 = \lambda (-1)^{\frac{1}{2}\nu(\nu-1)} \overset{\circ}{\gamma}_1 \overset{\circ}{\gamma}_2 \cdots \overset{\circ}{\gamma}_\nu \\ &= (-1)^{\frac{1}{2}\nu(\nu-1)} E = (-1)^{\frac{1}{2}\nu(\nu+1)} E. \end{aligned}$$

Using Eq. (1.44), let us calculate the result of transposition of a product of arbitrary $\overset{\circ}{\gamma}$ matrices. We have

$$\left(\overset{\circ}{\gamma}_{i_1} \overset{\circ}{\gamma}_{i_2} \cdots \overset{\circ}{\gamma}_{i_k}\right)^T = \overset{\circ}{\gamma}{}_{i_k}^T \cdots \overset{\circ}{\gamma}{}_{i_2}^T \overset{\circ}{\gamma}{}_{i_1}^T = (-1)^k E \overset{\circ}{\gamma}_{i_k} \cdots \overset{\circ}{\gamma}_{i_2} \overset{\circ}{\gamma}_{i_1} E^{-1}.$$

Performing, in this equality, alternation over the indices i and permuting the indices $i_k, \ldots, i_2, i_1$ in its right-hand side to the order of growing numbers, we obtain:

$$\left(\overset{\circ}{\gamma}_{i_1 i_2 \ldots i_k}\right)^T = (-1)^{\frac{1}{2}k(k+1)} E \overset{\circ}{\gamma}_{i_1 i_2 \ldots i_k} E^{-1}. \tag{1.54}$$

Taking into account the symmetry properties (1.53) of the matrix E, we can rewrite equality (1.54) in the form

$$\left(E \overset{\circ}{\gamma}_{i_1 i_2 \ldots i_k}\right)^T = (-1)^{\frac{1}{2}[\nu(\nu+1)+k(k+1)]} E \overset{\circ}{\gamma}_{i_1 i_2 \ldots i_k}. \tag{1.55}$$

Evidently, the symmetry properties (1.53) and (1.55) are of invariant nature and are not related to a specific choice of the matrices $\overset{\circ}{\gamma}_i$.

Denoting the matrix elements of $E\overset{\circ}{\gamma}_{i_1 i_2 \ldots i_k}$ by the symbol $\overset{\circ}{\gamma}_{BAi_1 i_2 \ldots i_k}$, so that

$$\overset{\circ}{\gamma}_{BAi_1 i_2 \ldots i_k} = e_{BC}\overset{\circ}{\gamma}{}^{C}{}_{Ai_1 i_2 \ldots i_k},$$

we can also write the symmetry properties (1.53) and (1.55) in the following way:

$$e_{BA} = (-1)^{\frac{1}{2}\nu(\nu+1)} e_{AB},$$

$$\overset{\circ}{\gamma}_{BAi_1 i_2 \ldots i_k} = (-1)^{\frac{1}{2}[\nu(\nu+1)+k(k+1)]}\overset{\circ}{\gamma}_{ABi_1 i_2 \ldots i_k}. \tag{1.56}$$

1.2 Spinor Representation of the Orthogonal Group of Transformations of Bases of an Even-Dimensional Complex Euclidean Vector Space

1.2.1 Spinor Representation $SO^+_{2\nu} \to \{\pm S\}$ of the Proper Orthogonal Group

Consider an even-dimensional complex Euclidean vector space $E^+_{2\nu}$ of dimension 2ν, referred to an orthonormal basis $\mathfrak{Э}_i$ $(i = 1, 2, \ldots, 2\nu)$. Let $SO^+_{2\nu}$ be the group of proper orthogonal transformations of the bases $\mathfrak{Э}_i$ of the space $E^+_{2\nu}$

$$\mathfrak{Э}'_i = l^j{}_i \mathfrak{Э}_j, \tag{1.57}$$

defined by the equations

$$l^q{}_i l^m{}_j \delta_{qm} = \delta_{ij}, \quad \det \| l^j{}_i \| = 1. \tag{1.58}$$

If the matrices $\overset{\circ}{\gamma}_i$ satisfy Eq. (1.1), then it follows from the orthogonality condition (1.58) that the matrices $\overset{\circ}{\gamma}{}'_i = l^j{}_i \overset{\circ}{\gamma}_j$ also satisfy Eq. (1.1). Indeed,

$$\overset{\circ}{\gamma}{}'_i \overset{\circ}{\gamma}{}'_j + \overset{\circ}{\gamma}{}'_j \overset{\circ}{\gamma}{}'_i = l^q{}_i l^m{}_j (\overset{\circ}{\gamma}_q \overset{\circ}{\gamma}_m + \overset{\circ}{\gamma}_m \overset{\circ}{\gamma}_q) = 2 l^q{}_i l^m{}_j \delta_{qm} I = 2\delta_{ij} I.$$

Therefore Pauli's theorem implies that there is a matrix[4] $S = \| S^B{}_A \|$, determined up to multiplying by an arbitrary nonzero complex number and satisfying the equation

$$l^j{}_i \overset{\circ}{\gamma}_j = S^{-1} \overset{\circ}{\gamma}_i S. \tag{1.59}$$

[4] S is a square matrix of order 2^ν with the elements $S^B{}_A$. To simplify the subsequent expressions, as is conventionally done, we will mostly use matrix notations and omit the indices that determine the matrix elements.

Transposing equation (1.59) and, in the resulting equation, substituting the matrices $\overset{\circ}{\gamma}{}_i^T$ according to Eq. (1.44), we find:

$$l^j{}_i \overset{\circ}{\gamma}_j = E^{-1} S^T E \overset{\circ}{\gamma}_i (E^{-1} S^T E)^{-1}.$$

Comparing this equation with (1.59), we obtain

$$\overset{\circ}{\gamma}_i S E^{-1} S^T E = S E^{-1} S^T E \overset{\circ}{\gamma}_i.$$

Thus the matrix $SE^{-1}S^T E$ commutes with all $\overset{\circ}{\gamma}_i$ and is therefore proportional to of the unit matrix,

$$SE^{-1}S^T E = \lambda I.$$

Here, λ is a nonzero complex number. The latter equation may also be written in the form $S^T E S = \lambda E$.

Since S is determined by Eq. (1.59) up to multiplying by an arbitrary complex number, we can normalize S by the condition

$$S^T E S = E \tag{1.60}$$

and then obtain that Eq. (1.59), under the normalization condition (1.60), puts into correspondence to each proper orthogonal transformation $l^j{}_i$ two matrices: S and $-S$.

Let us identify the transformations S and $-S$ and let us consider the pair of transformations S and $-S$ as a single element $\pm S$. On the set of pairs $\pm S$, let us define a product by the equality

$$(\pm S_1)(\pm S_2) \overset{\text{def}}{=} \pm(S_1 S_2), \tag{1.61}$$

where $S_1 S_2$ is the conventional matrix product of the matrices S_1 and S_2.[5]

In a natural way, one also defines a product of matrices T (not necessarily square ones) and pairs of matrices $\{\pm S\}$:

$$T(\pm S) \overset{\text{def}}{=} \pm(TS), \quad (\pm S)T \overset{\text{def}}{=} \pm(ST).$$

It is easily seen that the set $\{\pm S\}$, corresponding to the proper orthogonal group $SO^+_{2\nu}$, is a group with respect to the multiplication (1.61). Let us show that the group $\{\pm S\}$ realizes a representation of the group of proper orthogonal

[5]Evidently, if the set of certain square matrices $\{S\}$ forms a group, then the corresponding set $\{\pm S\}$ also forms a group with respect to the multiplication (1.61) (which is simply the factor group $S/\pm I$).

transformations $SO^+_{2\nu}$. To do so, it is sufficient to show that a product of proper orthogonal transformations from $SO^+_{2\nu}$ is in correspondence with a product of pairs $\pm S$.

Let the proper orthogonal transformation $\vartheta'_i = l^j_{1i}\vartheta_j$ correspond to the pair $\pm S_1$, the proper orthogonal transformation $\vartheta''_k = l^i_{2k}\vartheta'_i$ to the pair $\pm S_2$, and the product transformation $\vartheta''_k = l^j_{1i}l^i_{2k}\vartheta_j$ to the pair $\pm S$. Thus we have

$$\begin{aligned} l^j_{1i}\overset{\circ}{\gamma}_j &= S_1^{-1}\overset{\circ}{\gamma}_i S_1, & S_1^T E S_1 &= E, \\ l^i_{2k}\overset{\circ}{\gamma}_i &= S_2^{-1}\overset{\circ}{\gamma}_k S_2, & S_2^T E S_2 &= E, \\ l^j_{1i}l^i_{2k}\overset{\circ}{\gamma}_j &= S^{-1}\overset{\circ}{\gamma}_k S, & S^T E S &= E. \end{aligned} \tag{1.62}$$

Let us contract the first equality in (1.62) with l^i_{2k} over the index i and, in the right-hand side of the resulting equality, substitute the contraction $l^i_{2k}\overset{\circ}{\gamma}_i$ according to the second equality (1.62). We obtain:

$$l^j_{1i}l^i_{2k}\overset{\circ}{\gamma}_j = l^i_{2k}S_1^{-1}\overset{\circ}{\gamma}_i S_1 = S_1^{-1}S_2^{-1}\overset{\circ}{\gamma}_k S_2 S_1 = (S_2 S_1)^{-1}\overset{\circ}{\gamma}_k(S_2 S_1). \tag{1.63}$$

Comparing (1.63) with the third equality in (1.62), we find

$$(S_2 S_1)^{-1}\overset{\circ}{\gamma}_k(S_2 S_1) = S^{-1}\overset{\circ}{\gamma}_k S,$$

or

$$S(S_2 S_1)^{-1}\overset{\circ}{\gamma}_k = \overset{\circ}{\gamma}_k S(S_2 S_1)^{-1}.$$

Thus the matrix $S(S_2 S_1)^{-1}$ commutes with all matrices $\overset{\circ}{\gamma}_k$ and is therefore proportional to the unit matrix,

$$S(S_2 S_1)^{-1} = \lambda I, \qquad \text{or} \qquad S = \lambda S_2 S_1. \tag{1.64}$$

Using equality (1.64) and the normalization conditions for S_1 and S_2 in (1.62), we calculate the quantity $S^T E S$:

$$S^T E S = (\lambda S_2 S_1)^T E \lambda S_2 S_1 = \lambda^2 S_1^T (S_2^T E S_2) S_1 = \lambda^2 S_1^T E S_1 = \lambda^2 E. \tag{1.65}$$

Comparing Eq. (1.65) with the normalization condition for S in (1.62), we find $\lambda^2 = 1$. It means that $\lambda = 1$ or $\lambda = -1$ and $S = S_2 S_1$ or $S = -S_2 S_1$. Therefore the product $l^j_{1i}l^i_{2k}$ in the proper orthogonal group $SO^+_{2\nu}$ is in correspondence with the product $\pm S = (\pm S_2)(\pm S_1)$ on the set $\{\pm S\}$.

Thus the normalized set $\{\pm S\}$, consisting of the pairs $\pm S$ corresponding to the proper orthogonal group $SO^+_{2\nu}$, forms a group (1.61) which realizes a representation of the group $SO_{2\nu}^{+}$, called the spinor representation. The group $\{\pm S\}$ will be further called the group of spinor transformations.

Let us calculate the matrix S corresponding to small proper orthogonal transformations $l^{j}{}_{i}$. We write a small proper orthogonal transformation $l^{j}{}_{i}$ in the form

$$l^{j}{}_{i} = \delta_{i}^{j} + \delta\varepsilon_{i}{}^{j}, \tag{1.66}$$

where $\delta\varepsilon_{i}{}^{j}$ are small quantities. The orthogonality condition (1.58) for the quantities $\delta\varepsilon_{ij} = \delta\varepsilon_{i}{}^{j}$ gives $\delta\varepsilon_{ij} = -\delta\varepsilon_{ji}$.

Expanding the matrix S in powers of $\delta\varepsilon_{ij}$ and restricting ourselves to first-order small quantities, we find

$$S = I + \frac{1}{2}A^{ij}\delta\varepsilon_{ij}, \qquad A^{ij} = -A^{ji}. \tag{1.67}$$

The quantities A^{ij} in Eq. (1.67), which are matrices of the order 2^{ν}, are called the infinitesimal operators (or generators) of the spinor representation. Inserting expressions (1.66) and (1.67) for S and $l^{j}{}_{i}$ into Eqs. (1.59) and (1.60), we arrive at equations for determining the infinitesimal operators:

$$\begin{aligned} &\left(A^{ij}\right)^{T} = -EA^{ij}E^{-1}, \qquad A^{ij} = -A^{ji}, \\ &-A^{ij}\overset{\circ}{\gamma}{}^{k} + \overset{\circ}{\gamma}{}^{k}A^{ij} + \delta^{kj}\overset{\circ}{\gamma}{}^{i} - \delta^{ki}\overset{\circ}{\gamma}{}^{j} = 0, \end{aligned}$$

which have the solution

$$A^{ij} = \frac{1}{2}\overset{\circ}{\gamma}{}^{ij}. \tag{1.68}$$

Thus the matrix S, corresponding to the small proper orthogonal transformations (1.66), has the form

$$S = I + \frac{1}{4}\overset{\circ}{\gamma}{}^{ij}\delta\varepsilon_{ij}. \tag{1.69}$$

1.2.2 Spinor Representation of the Full Orthogonal Group

Consider the full orthogonal group $O_{2\nu}^{+}$ of transformations $\ni'_{i} = l^{j}{}_{i}\ni_{j}$ of the bases $\ni_{i}$ of the space $E_{2\nu}^{+}$, defined by the equations

$$l^{q}{}_{i}l^{m}{}_{j}\delta_{qm} = \delta_{ij}.$$

There are several different, mutually non-equivalent representations of the full orthogonal group of transformations $O^+_{2\nu}$ which coincide with the above-defined spinor representation on the subgroup of proper orthogonal transformations $SO^+_{2\nu}$.

1. A spinor representation of the $O^+_{2\nu}$ group may be defined using the same equations as those for the group $SO^+_{2\nu}$:

$$l^j{}_i \overset{\circ}{\gamma}_j = S^{-1} \overset{\circ}{\gamma}_i S, \qquad E = S^T E S. \tag{1.70}$$

It is easy to see that the first equation (1.70) implies

$$l^{j_1}{}_{i_1} l^{j_2}{}_{i_2} \cdots l^{j_m}{}_{i_m} \overset{\circ}{\gamma}_{j_1} \overset{\circ}{\gamma}_{j_2} \cdots \overset{\circ}{\gamma}_{j_m} = S^{-1} \overset{\circ}{\gamma}_{i_1} \overset{\circ}{\gamma}_{i_2} \cdots \overset{\circ}{\gamma}_{i_m} S. \tag{1.71}$$

Hence it follows for $m = 2\nu$:

$$\overset{\circ}{\gamma}_{2\nu+1} = S^{-1} \overset{\circ}{\gamma}_{2\nu+1} S \Delta, \qquad \Delta = \det \| l^j{}_i \|. \tag{1.72}$$

Let us calculate the spinor transformation S that corresponds to the reflection transformation

$$\mathcal{Э}'_1 = -\mathcal{Э}_1, \qquad \mathcal{Э}'_\alpha = \mathcal{Э}_\alpha, \qquad \alpha = 2, 3, \ldots, 2\nu. \tag{1.73}$$

According to Eqs. (1.70), we have

$$S^T E S = E, \qquad S \overset{\circ}{\gamma}_1 = -\overset{\circ}{\gamma}_1 S, \qquad S \overset{\circ}{\gamma}_\alpha = \overset{\circ}{\gamma}_\alpha S.$$

This enables us to find the solution for S

$$S = \mathrm{i}^{\nu+1} \overset{\circ}{\gamma}_1 \overset{\circ}{\gamma}_{2\nu+1}. \tag{1.74}$$

2. A spinor representation of the $O^+_{2\nu}$ group may be defined by the equations

$$l^j{}_i \overset{\circ}{\gamma}_j = S^{-1} \overset{\circ}{\gamma}_i S, \qquad E = S^T E S \Delta. \tag{1.75}$$

In this case, Eq. (1.72) also holds, and using it, we find for the spinor representation considered:

$$S^T E \overset{\circ}{\gamma}_{2\nu+1} S = S^T E S S^{-1} \overset{\circ}{\gamma}_{2\nu+1} S = E \overset{\circ}{\gamma}_{2\nu+1}.$$

Therefore, instead of the second equation (1.75) which normalizes the spinor transformations, we can take the equation

$$E \overset{\circ}{\gamma}_{2\nu+1} = S^T E \overset{\circ}{\gamma}_{2\nu+1} S.$$

The reflection transformation (1.73) correspond to a spinor transformation of the form

$$S = \mathrm{i}^{\nu}\overset{\circ}{\gamma}_1\overset{\circ}{\gamma}_{2\nu+1}.$$

3. A spinor representation of the $O^+_{2\nu}$ group may also be defined by equations of the form

$$l^j{}_i\overset{\circ}{\gamma}_j = S^{-1}\overset{\circ}{\gamma}_i S\Delta, \qquad E = S^T E S. \tag{1.76}$$

A calculation of the spinor transformation matrix S, corresponding to the reflection transformation (1.73), gives $S = \mathrm{i}\overset{\circ}{\gamma}_1$.

The spinor representation under consideration is equivalent to the spinor representation defined by Eqs. (1.70) for odd ν and to that defined by Eqs. (1.75) for even ν. Indeed, a direct inspection shows that the spinor transformations ASA^{-1}, where $A = I + \mathrm{i}\overset{\circ}{\gamma}_{2\nu+1}$ and S is defined by Eqs. (1.76), coincide with the spinor transformations defined by Eqs. (1.70) for odd ν and with the spinor transformations defined by Eqs. (1.75) for even ν.

4. One can also define a spinor representation of the $O^+_{2\nu}$ group by the following equation:

$$l^j{}_i\overset{\circ}{\gamma}_j = S^{-1}\overset{\circ}{\gamma}_i S\Delta, \qquad E = S^T E S\Delta. \tag{1.77}$$

The second equation (1.77) is equivalent to the equation

$$E\overset{\circ}{\gamma}_{2\nu+1} = S^T E\overset{\circ}{\gamma}_{2\nu+1} S.$$

Under the reflection transformation (1.73), for the corresponding spinor transformation we have $S = \overset{\circ}{\gamma}_1$.

The spinor representation $O^+_{2\nu} \to \{\pm S\}$, defined according to Eqs. (1.77), is equivalent to a spinor representation defined by Eqs. (1.70) for even ν and to that defined by a spinor representation defined by Eqs. (1.75) for odd ν, since the spinor transformations ASA^{-1}, where $A = I + \mathrm{i}\overset{\circ}{\gamma}_{2\nu+1}$, coincide with the spinor transformations defined by Eqs. (1.70) or (1.75) depending on parity of the number ν.

Each pair of Eqs. (1.70), (1.75), (1.76) and (1.77) completely defines a certain group $\{\pm S\}$ which realizes a representation of the full group of orthogonal transformations of the bases $\mathfrak{I}_i$ of the complex Euclidean space $E^+_{2\nu}$. Evidently, all definitions (1.70), (1.75), (1.76) and (1.77) are identical for the group of proper orthogonal transformations $SO^+_{2\nu}$ and coincide with definition (1.59), (1.60). For the full orthogonal group $O^+_{2\nu}$, the spinor representations defined by Eqs. (1.76) and (1.77), are equivalent to the representations defined by Eqs. (1.70) and (1.75).

Using the property of the $\overset{\circ}{\gamma}_i$ matrices, formulated in item 10 of Sect. 1.1, it is not difficult to show that all spinor representations of the orthogonal group $O_{2\nu}^{+}$ defined above are exact, i.e., there is a one-to-one correspondence between the $O_{2\nu}^{+}$ group and the spinor groups $\{\pm S\}$.

1.2.3 Connection Between Spinor Representations Determined by Different Sets of Matrices E and γ_i

Equations (1.70) or (1.75), (1.76) and (1.77) completely determine the spinor group $\{\pm S\}$ if the matrices E and $\overset{\circ}{\gamma}_i$, entering into these equations, are specified. As has been already noted in Sect. 1.1, arbitrary sets of matrices $E, \overset{\circ}{\gamma}_i$ and $E', \overset{\circ}{\gamma}{}'_i$, satisfying Eqs. (1.1) and (1.44), are connected by the relations

$$\overset{\circ}{\gamma}{}'_i = T^{-1}\overset{\circ}{\gamma}_i T, \quad E' = T^T E T, \tag{1.78}$$

where T is some nondegenerate matrix. The spinor group $\{\pm S'\}$, corresponding to the matrices E' and $\overset{\circ}{\gamma}_i{}'$, is defined by the equations (for definiteness, we are using Eqs. (1.70); for the spinor representations defined by Eqs. (1.75)–(1.77), the subsequent transformations are entirely similar)

$$l^j{}_i \overset{\circ}{\gamma}{}'_j = (S')^{-1}\overset{\circ}{\gamma}{}'_i S', \quad (S')^T E' S' = E'. \tag{1.79}$$

Let us find out a connection between the groups $\{\pm S\}$ and $\{\pm S'\}$. To do so, we replace, in Eq. (1.79), the matrices E' and $\overset{\circ}{\gamma}{}'_i$ with E and $\overset{\circ}{\gamma}_i$ using (1.78). We obtain:

$$l^j{}_i \overset{\circ}{\gamma}_j = T(S')^{-1}T^{-1}\overset{\circ}{\gamma}_i T S' T^{-1}. \tag{1.80}$$

Comparing Eq. (1.80) with the first equation (1.70), we find

$$T(S')^{-1}T^{-1}\overset{\circ}{\gamma}_i T S' T^{-1} = S^{-1}\overset{\circ}{\gamma}_i S,$$

or

$$S T(S')^{-1}T^{-1}\overset{\circ}{\gamma}_i = \overset{\circ}{\gamma}_i S T(S')^{-1}T^{-1}.$$

Hence it follows

$$S T(S')^{-1}T^{-1} = \mu I,$$

where $\mu \neq 0$ is some, generally complex, number. The latter equation leads to a connection between the spinor transformations S and S':

$$S' = \mu^{-1} T^{-1} S T. \tag{1.81}$$

To determine the number μ, we insert E' and S', defined according to (1.78) and (1.81), into the second equation (1.79). After an identical transformation, we find

$$\mu^{-2} S^T E S = E. \tag{1.82}$$

Comparing Eqs. (1.82) and (1.70), we find $\mu^2 = 1$, and consequently $\mu = 1$ or $\mu = -1$. So the spinor groups $\{\pm S\}$ and $\{\pm S'\}$, corresponding to different sets of matrices E and $\overset{\circ}{\gamma}_i$, are connected by the similarity transformation

$$\pm S' = T^{-1}(\pm S)T, \tag{1.83}$$

and thus realize equivalent representations of the full orthogonal group $O^+_{2\nu}$.

1.3 Spinors in Even-Dimensional Complex Euclidean Spaces

Let S_N be a complex linear (vector) space of dimension N, and let $\{\boldsymbol{\varepsilon}_{\mathcal{A}}\}$ be a vector basis of the space S_N. Consider in S_N an arbitrary vector $\psi^{\mathcal{A}}\boldsymbol{\varepsilon}_{\mathcal{A}}$ and the vector $-\psi^{\mathcal{A}}\boldsymbol{\varepsilon}_{\mathcal{A}}$, determined, in the basis $\{\boldsymbol{\varepsilon}_{\mathcal{A}}\}$, by components $\psi^{\mathcal{A}}$ and $-\psi^{\mathcal{A}}$, respectively. Let us identify the vectors $\psi^{\mathcal{A}}\boldsymbol{\varepsilon}_{\mathcal{A}}$ and $-\psi^{\mathcal{A}}\boldsymbol{\varepsilon}_{\mathcal{A}}$, and let us consider the pair of vectors $\psi^{\mathcal{A}}\boldsymbol{\varepsilon}_{\mathcal{A}}$ and $-\psi^{\mathcal{A}}\boldsymbol{\varepsilon}_{\mathcal{A}}$ as a single object $\boldsymbol{\psi} = \pm\psi^{\mathcal{A}}\boldsymbol{\varepsilon}_{\mathcal{A}}$ in the space S_N.[6] Let us also identify the bases $\{\boldsymbol{\varepsilon}_{\mathcal{A}}\}$ and $-\{\boldsymbol{\varepsilon}_{\mathcal{A}}\}$ of the space S_N and the systems of components $\psi^{\mathcal{A}}$ and $-\psi^{\mathcal{A}}$, and let us consider the pairs of bases $\{\boldsymbol{\varepsilon}_{\mathcal{A}}\}$ and $-\{\boldsymbol{\varepsilon}_{\mathcal{A}}\}$ as a single element $\pm\{\boldsymbol{\varepsilon}_{\mathcal{A}}\}$; the pair of systems of components $\psi^{\mathcal{A}}$ and $-\psi^{\mathcal{A}}$ will be considered as a single element $\pm\psi^{\mathcal{A}}$. To each pair of bases $\pm\{\boldsymbol{\varepsilon}_{\mathcal{A}}\}$ and to each pair of systems of components $\pm\psi^{\mathcal{A}}$, we put into one-to-one correspondence the pair of vectors $\boldsymbol{\psi} = \pm\psi^{\mathcal{A}}\boldsymbol{\varepsilon}_{\mathcal{A}}$ of the space S_N.

Consider now a certain group of linear transformations $S = \|S^{\mathcal{B}}{}_{\mathcal{A}}\|$ of bases $\{\boldsymbol{\varepsilon}_{\mathcal{A}}\}$ of the vector space S_N, and let $\{\pm S\}$ be a group which consists of pairs of transformations S and $-S$. On the set of pairs of bases $\pm\{\boldsymbol{\varepsilon}_{\mathcal{A}}\}$ in S_N and on the set of pairs of systems of components $\pm\psi^{\mathcal{A}}$, we define, with the aid of the group $\{\pm S\}$,

[6] Geometrically, such an object determines a non-oriented segment in S_N.

the transformations $\pm\{\boldsymbol{\varepsilon}_{\mathcal{A}}\} \rightarrow \pm\{\boldsymbol{\varepsilon}'_{\mathcal{A}}\}$ and $\pm\psi^{\mathcal{A}} \rightarrow \pm\psi'^{\mathcal{A}}$ by the equalities

$$\pm\,\psi'^{\mathcal{A}} = (\pm\, S^{\mathcal{A}}{}_{\mathcal{B}})(\pm\,\psi^{\mathcal{B}}) \overset{\text{def}}{=} \pm(S^{\mathcal{A}}{}_{\mathcal{B}}\psi^{\mathcal{B}}),$$

$$\pm\,\{\boldsymbol{\varepsilon}'_{\mathcal{A}}\} = (\pm\, Z^{\mathcal{B}}{}_{\mathcal{A}})(\pm\,\boldsymbol{\varepsilon}_{\mathcal{B}}) \overset{\text{def}}{=} \pm\{Z^{\mathcal{B}}{}_{\mathcal{A}}\boldsymbol{\varepsilon}_{\mathcal{B}}\},$$

where $\|Z^{\mathcal{B}}{}_{\mathcal{A}}\| = S^{-1}$.

Evidently, under such a transformation, the pairs of vectors $\pm\psi^{\mathcal{A}}\boldsymbol{\varepsilon}_{\mathcal{A}}$, corresponding to the pairs $\pm\psi^{\mathcal{A}}$ and $\pm\{\boldsymbol{\varepsilon}_{\mathcal{A}}\}$, are invariant:

$$\boldsymbol{\psi} = \pm\psi^{\mathcal{A}}\boldsymbol{\varepsilon}_{\mathcal{A}} = \pm\psi'^{\mathcal{A}}\boldsymbol{\varepsilon}'_{\mathcal{A}}.$$

Consider the case that the dimension N of the space S_N is equal to 2^ν and the group $\{\pm S\}$ is a spinor group that realizes representations of the orthogonal group of transformations of bases of the space $E^+_{2\nu}$. Let us put into correspondence to a certain orthonormal basis $\boldsymbol{\mathcal{9}}_i$ of the Euclidean space $E^+_{2\nu}$, a pair of bases $\pm\{\boldsymbol{\varepsilon}_A\}$ of the space S_{2^ν}, and to each orthonormal basis $\boldsymbol{\mathcal{9}}'_i$ of the space $E^+_{2\nu}$, obtained from $\boldsymbol{\mathcal{9}}_i$ by the orthogonal transformation $l^j{}_i$ according to (1.57), let us put into correspondence a pair of bases $\pm\{\boldsymbol{\varepsilon}'_A\}$ of the space S_{2^ν}, obtained from $\pm\{\boldsymbol{\varepsilon}_A\}$ by the transformation

$$\pm\,\{\boldsymbol{\varepsilon}'_A\} = \pm\{Z^B{}_A\boldsymbol{\varepsilon}_B\}, \tag{1.84}$$

where the matrix $\|Z^B{}_A\| = S^{-1}$ is defined by equalities (1.70) or (1.75), (1.76), and (1.77).

Since different transformations $l^j{}_i$ correspond to different transformations $\pm S$, the established correspondence is a one-to-one correspondence between all orthonormal bases $\boldsymbol{\mathcal{9}}_i$ of the space $E^+_{2\nu}$ and a certain set of pairs of bases $\pm\{\boldsymbol{\varepsilon}'_A\}$ of the space S_{2^ν}. By the correspondence established between the spaces $E^+_{2\nu}$ and S_{2^ν}, under the orthogonal transformation (1.57) of the bases $\boldsymbol{\mathcal{9}}_i$ in $E^+_{2\nu}$, the pairs of bases $\pm\{\boldsymbol{\varepsilon}_A\}$ in S_{2^ν} are subject to the transformation S^{-1} according to Eq. (1.84). Having established the above correspondence between the spaces $E^+_{2\nu}$ and S_{2^ν}, we can consider the object $\boldsymbol{\psi} = \pm\psi^A\boldsymbol{\varepsilon}_A$ as an invariant geometric object in the Euclidean space $E^+_{2\nu}$, and then the transformation $\boldsymbol{\mathcal{9}}'_i = l^j{}_i\boldsymbol{\mathcal{9}}_j$ of a basis in $E^+_{2\nu}$ corresponds to the following transformation of components $\pm\psi^A$:

$$\pm\,\psi'^A = \pm S^A{}_B\psi^B, \tag{1.85}$$

where the matrix $S = \|S^B{}_A\|$ is defined by one of equalities (1.70), (1.75), (1.76) and (1.77).

The invariant geometric object $\boldsymbol{\psi} = \pm\psi^A\boldsymbol{\varepsilon}_A$, in which the pairs of contravariant components $\pm\psi^A$ and the pairs of bases $\pm\{\boldsymbol{\varepsilon}_A\}$ $(A = 1, 2, \ldots, 2^\nu)$ are referred to a certain orthonormal basis $\boldsymbol{\mathcal{9}}_i$ in the Euclidean space $E^+_{2\nu}$ and are transformed under the orthogonal transformation (1.57) of the bases $\mathbf{Э}_i$ according to formulae (1.84), (1.85), is called a first-rank spinor in the complex Euclidean space $E^+_{2\nu}$. The bases $\pm\{\boldsymbol{\varepsilon}_A\}$ are usually called spinbases.

We will say that components ψ^A determine (or represent) the spinor in the spinbasis $\pm\{\boldsymbol{\varepsilon}_A\}$. According to the definitions, components ψ^A and components $-\psi^A$, related to the same spinbasis $\pm\{\boldsymbol{\varepsilon}_A\}$, determine the same spinor $\boldsymbol{\psi}$.

By definition, the covariant components of the spinor ψ_B are given by the equality

$$\psi_B = e_{BA}\psi^A, \tag{1.86}$$

in which e_{BA} are the components of the matrix E defined by Eq. (1.44). Let us denote the components of the inverse matrix E^{-1} by e^{BC}. Then,

$$e_{BA}e^{AC} = \delta^C_B. \tag{1.87}$$

From Eqs. (1.86) and (1.87) it follows

$$\psi^A = e^{AB}\psi_B. \tag{1.88}$$

Let us denote the column of components ψ^A $(A = 1, 2,\ldots, 2^\nu)$ by the symbol ψ and the row of covariant components ψ_A by the symbol $\widetilde{\psi}$. Then definition (1.86) may be written in a matrix form:

$$\widetilde{\psi} = (E\psi)^T = \psi^T E^T = (-1)^{\frac{1}{2}\nu(\nu+1)}\psi^T E.$$

By definition, the covariant components of the spinor $\pm\psi_B$ are transformed, under the transformation (1.57) of the basis $\mathbf{Э}_i$, with the aid of the inverse matrix $S^{-1} = \|Z^B{}_A\|$:

$$\pm\psi'_B = \pm Z^A{}_B\psi_A.$$

In a matrix form, the transformation laws for the contravariant and covariant components of a spinor are written as follows:

$$\pm\psi' = \pm S\psi, \qquad \pm\widetilde{\psi}' = \pm\widetilde{\psi}S^{-1}.$$

The components of a spinor of rank $m + n$ with m contravariant indices and n covariant indices $\psi^{A_1A_2\ldots A_m}_{B_1B_2\ldots B_n}$ $(A_i, B_i = 1, 2,\ldots, 2^\nu)$, under transformation (1.57) of the basis $\mathbf{Э}_i$ in the space $E^+_{2\nu}$, are transformed as a product of the components of a first-rank spinor:

$$\left(\psi^{A_1A_2\ldots A_m}_{B_1B_2\ldots B_n}\right)' = S^{A_1}{}_{D_1}S^{A_2}{}_{D_2}\cdots S^{A_m}{}_{D_m}Z^{C_1}{}_{B_1}Z^{C_2}{}_{B_2}\cdots Z^{C_n}{}_{B_n}\psi^{D_1D_2\ldots D_m}_{C_1C_2\ldots C_n}.$$

According to the definition, the components of odd-rank spinors are defined in an orthonormal basis $\mathbf{\mathit{Э}}_i$ in $E^+_{2\nu}$ up to a common sign, while the components of even-rank spinors are defined uniquely. A transformation of the components of even-rank spinors may be defined uniquely if the identical transformation of the basis $\mathbf{\mathit{Э}}'_i = \mathbf{\mathit{Э}}_i$ in the space $E^+_{2\nu}$ is put into correspondence to an identical transformation of the spinor components, and for nearby transformations of bases in $E^+_{2\nu}$ the sign of even-rank spinor components is defined by continuity.

Let us define the operations of multiplication and contraction of spinors in the spinor space.

We will call a product of two spinors with components $\psi^{A_1\ldots A_m}_{B_1\ldots B_n}$ and $\xi^{C_1\ldots C_k}_{D_1\ldots D_r}$, referred to the same basis $\mathbf{\mathit{Э}}_i$, a spinor determined, in the same basis, by the components

$$\eta^{A_1\ldots A_m C_1\ldots C_k}_{B_1\ldots B_n D_1\ldots D_r} = \psi^{A_1\ldots A_m}_{B_1\ldots B_n}\xi^{C_1\ldots C_k}_{D_1\ldots D_r}.$$

Thus the rank of a product of two spinor is equal to a sum of ranks of the factor spinors. It is easy to show that this definition of a product of spinors is independent of the chosen basis $\mathbf{\mathit{Э}}_i$.

This definition of the product of spinors is unambiguous if ψ and ξ are even-rank spinors or if the ranks of the spinors ψ and ξ have opposite parity.

If ψ and ξ are odd-rank spinors, the above definition puts into correspondence to the two spinors ψ and ξ two even-rank spinors which differ by their signs. The choice of one of these even-rank spinors as a product of odd-rank spinors is equivalent to a choice of the relative sign of the odd-rank spinor components.

A spinor with components

$$\psi^{A_3\ldots A_m} = e_{A_2A_1}\psi^{A_1A_2A_3\ldots A_m} = \psi_A{}^{AA_3\ldots A_m},$$

or

$$\psi^{A_3\ldots A_m} = e_{A_1A_2}\psi^{A_1A_2A_3\ldots A_m} = \psi^A{}_A{}^{A_3\ldots A_M},$$

referred to the same basis as $\psi^{A_1A_2A_3\ldots A_m}$, is called a contraction of the spinor with components $\psi^{A_1A_2A_3\ldots A_m}$ with respect to the indices A_1 and A_2.

Evidently, if m is even and the components of the spinor $e_{A_1A_2}$ are antisymmetric, then the components $\psi_A{}^{AA_3\ldots A_m}$ and $\psi^A{}_A{}^{A_3\ldots A_M}$ determine two even-rank spinors which differ by sign. If m is odd, then the components $\psi_A{}^{AA_3\ldots A_m}$ and $\psi^A{}_A{}^{A_3\ldots A_M}$ determine the same spinor, since odd-rank spinor components are defined up to a common sign.

A contraction over covariant indices is performed using the components e^{BA}:

$$\psi_{B_3\ldots B_n} = e^{B_2B_1}\psi_{B_1B_2B_3\ldots B_n} = \psi^B{}_{BB_3\ldots B_n}$$

or

$$\psi_{B_3\ldots B_n} = e^{B_1 B_2}\psi_{B_1 B_2 B_3\ldots B_n} = \psi_B{}^B{}_{B_3\ldots B_n}.$$

Objects having components with both spinor and tensor indices will be called spintensors.

Consider a spintensor the components $\overset{\circ}{\gamma}_i = \|\overset{\circ}{\gamma}{}^B{}_{Ai}\|$ in an orthonormal basis $\mathcal{E}_i$ of the space $E^+_{2\nu}$. In the basis $\mathcal{E}'_i = l^j{}_i\mathcal{E}_j$, its component have the form

$$\overset{\circ}{\gamma}_i{}' = l^j{}_i S\overset{\circ}{\gamma}_j S^{-1}. \tag{1.89}$$

It is easy to see that definitions (1.70) and (1.75) imply

$$l^j{}_i S\overset{\circ}{\gamma}_j S^{-1} = \overset{\circ}{\gamma}_i. \tag{1.90}$$

Comparing equalities (1.89) and (1.90), we find $\overset{\circ}{\gamma}_i{}' = \overset{\circ}{\gamma}_i$. Thus the values of the transformed components of the spintensor $\left(\overset{\circ}{\gamma}{}^B{}_{Ai}\right)'$ in the basis $\mathcal{E}'_i$ coincide with those of the components of the spintensor $\overset{\circ}{\gamma}{}^B{}_{Ai}$ in the basis $\mathcal{E}_i$. Therefore, for spinor representations defined by Eqs. (1.70) and (1.75), the components of $\overset{\circ}{\gamma}{}^B{}_{Ai}$ may be considered as those of a spintensor with one covariant tensor index, one covariant spinor index and one contravariant spinor index, which is invariant under the transformations (1.57). In the same way one obtains that, for spinor representations defined by Eqs. (1.76) and (1.77), the equality $\overset{\circ}{\gamma}_i{}' = \Delta\overset{\circ}{\gamma}_i$ is valid, and it implies that in this case $\overset{\circ}{\gamma}_i$ are invariant under continuous transformations of $\mathcal{E}_i$ and change their sign under reflection transformations.

From Eq. (1.72) (which holds for all spinor representations considered above) it follows that the spintensor defined by the matrix $\overset{\circ}{\gamma}_{2\nu+1} = \|(\overset{\circ}{\gamma}_{2\nu+1})^B{}_A\|$ is invariant under continuous transformations of the basis $\mathcal{E}_i$ and changes its sign under reflection transformations,

$$\overset{\circ}{\gamma}{}'_{2\nu+1} = S\overset{\circ}{\gamma}_{2\nu+1}S^{-1} = \overset{\circ}{\gamma}_{2\nu+1}\Delta. \tag{1.91}$$

In the same way, from the normalization condition in definitions (1.70) and (1.76) it follows that the components of $E = \|e_{BA}\|$ ($E^{-1} = \|e^{BA}\|$) are covariant (contravariant) components of a second-rank spinor, invariant under transformations (1.57):

$$E' = (S^{-1})^T E S^{-1} = E.$$

For spinor representations defined by Eqs. (1.75) or (1.77), the components of E are invariant under only continuous transformations (1.57), while under reflection transformations they change their sign; in this case, the components of the second-rank spinor $E\overset{\circ}{\gamma}_{2\nu+1}$ are invariant under all transformations (1.57):

$$E' = (S^{-1})^T E S^{-1} = E\Delta,$$
$$(E\overset{\circ}{\gamma}_{2\nu+1})' = (S^{-1})^T E\overset{\circ}{\gamma}_{2\nu+1} S^{-1} = E\overset{\circ}{\gamma}_{2\nu+1}.$$

With the aid of the components of the invariant spinor E, the indices of the components of spinors of any rank are raised and lowered (juggled) by the scheme (1.86) and (1.88). Therefore the second-rank spinor E is called the metric spinor.[7] Let us notice that the contraction in Eqs. (1.86) and (1.88) is performed over the second index of e^{BA} and e_{BA} (which is significant for second-rank spinors if $e_{BA} = -e_{AB}$).

The index juggling operation in the components of spinors in the spaces $E^+_{2\nu}$ for odd $\frac{1}{2}\nu(\nu+1)$ is essentially different from tensor index juggling since, for odd $\frac{1}{2}\nu(\nu+1)$ (in particular, in the two- and four-dimensional spaces E^+_2 and E^+_4), the components of the metric spinor e_{BA} are, according to (1.53), antisymmetric. This brings certain peculiarity to the spinor transformation formalism. In particular, for odd $\frac{1}{2}\nu(\nu+1)$, it is necessary to take into account the equality

$$\psi^A \chi_A = -\psi_A \chi^A, \tag{1.92}$$

which follows from the antisymmetric nature of the metric spinor e_{BA}. In the same case, the following equality holds:

$$e^{BA} e^{CD} e_{AD} = -e^{BC}, \tag{1.93}$$

which does not hold for the components of the tensor $g_{ij} = \left(\mathfrak{Э}_i, \mathfrak{Э}_j\right)$ used for index juggling in tensor components.

If, in the spinbasis $\pm\{\boldsymbol{\varepsilon}_A\}$, the spinor components transformation corresponding to an orthogonal transformation of a basis $\mathfrak{Э}_i$ is determined by the matrix S, then, in the spinbasis $\pm\{\widetilde{\boldsymbol{\varepsilon}}_A\} = \pm\{T^B{}_A \boldsymbol{\varepsilon}_B\}$, the spinor components transformation that corresponds to the same orthogonal transformation of the basis $\mathfrak{Э}_i$, will be determined by the matrices $\widetilde{S} = T^{-1} S T$, $T = \|T^B{}_A\|$. But, as has been shown (see Eq. (1.83)), if the group of spinor transformations $\{\pm S\}$ is associated with the matrices of the spintensors E and $\overset{\circ}{\gamma}_i$, then the group $\{\pm T^{-1} S T\}$ is associated with the matrices of the spintensors $\widetilde{E} = T^T E T$, $\widetilde{\gamma}_i = T^{-1}\overset{\circ}{\gamma}_i T$. Thus the choice of

[7] One could evidently take as a metric spinor the second-rank spinor with covariant components, defined by the matrix $E\overset{\circ}{\gamma}_{2\nu+1}$. If one requires that the metric spinor should be invariant under all transformations of the basis in the Euclidean space $E^+_{2\nu}$, then, for spinor representations defined by Eqs. (1.70) and (1.76), one should take the spinor E as a metric spinor, while for spinor representations defined by Eqs. (1.75) and (1.77), the spinor $E\overset{\circ}{\gamma}_{2\nu+1}$ should be taken.

a certain set of spintensors E, $\overset{\circ}{\gamma}_i$ corresponds to the choice of a certain spinbasis $\pm\{\boldsymbol{\varepsilon}_A\}$ and a certain group of spinor transformations $\{\pm S\}$. Therefore one can say that a spinor in the space $E^+_{2\nu}$ is determined by specifying the components $\pm\psi^A$ referred to a certain orthonormal basis $\boldsymbol{\ni}_i$ in $E^+_{2\nu}$ and by specifying the invariant spintensors E and $\overset{\circ}{\gamma}_i$ which determine the spinor transformations $\pm S$ and the spinbases $\pm\{\boldsymbol{\varepsilon}_A\}$.

Let us note that specifying only the invariant spintensors $\overset{\circ}{\gamma}_i$ determines the spinbasis $\pm\{\boldsymbol{\varepsilon}_A\}$ up to multiplying $\boldsymbol{\varepsilon}_A$ by an arbitrary nonzero complex number. Indeed, assuming that, in the two different spinbases $\pm\{\boldsymbol{\varepsilon}_A\}$ and $\pm\{\boldsymbol{\varepsilon}'_A\}$, the spintensors $\overset{\circ}{\gamma}_i$ are the same, we find by writing the transformation of the spintensors $\overset{\circ}{\gamma}_i$ in a transition from the spinbasis $\pm\{\boldsymbol{\varepsilon}_A\}$ to the spinbasis $\pm\{\boldsymbol{\varepsilon}'_A\}$:

$$S\overset{\circ}{\gamma}_i S^{-1} = \overset{\circ}{\gamma}_i .$$

From this it follows, by virtue of the properties of the matrices $\overset{\circ}{\gamma}_i$ considered in item 10 of Sect. 1.1, that the transformation S is proportional to the unit matrix, $S = \lambda I$, where λ is an arbitrary nonzero complex number.

1.4 Connection Between Even-Rank Spinors and Tensors

It is easily seen that, while in Eqs. (1.17) and (1.18) $\psi^B{}_A$ are components of a second-rank spinor, the quantity F is an invariant (at any rate, it is invariant under continuous orthogonal transformations of the basis $\boldsymbol{\ni}_i$), while $F^{i_1 i_2 \cdots i_k}$ are components of a rank-k tensor, antisymmetric with respect to all indices. Equations (1.17) and (1.18) realize a linear nondegenerate connection between the components of an arbitrary second-rank spinor and those of the tensors F, $F^{i_1 i_2 \cdots i_k}$. Thus, in the complex Euclidean space $E^+_{2\nu}$, a second-rank spinor with the components $\psi^B{}_A$ is equivalent to a set of tensors consisting of a scalar, a vector and antisymmetric tensors with ranks up to 2ν inclusive:

$$\|\psi^B{}_A\| \sim \left\{F, F^i, F^{i_1 i_2}, \ldots, F^{i_1 i_2 \ldots i_{2\nu}}\right\}.$$

Raising the index A in Eq. (1.17) using the metric spinor E, one can write Eq. (1.17) in the form

$$\psi^{BA} = \frac{1}{2^\nu}\left[(-1)^{\frac{1}{2}\nu(\nu+1)} F e^{BA} + \sum_{k=1}^{2\nu} \frac{1}{k!} F^{i_1 i_2 \cdots i_k} \overset{\circ}{\gamma}{}^{BA}_{i_1 i_2 \cdots i_k}\right], \tag{1.94}$$

to be used in what follows. For the coefficients F, $F^{i_1 i_2 \cdots i_k}$ in Eq. (1.94), we have

$$F = e_{BA}\psi^{BA} = \psi^{A}{}_{A},$$

$$F^{i_1 i_2 \cdots i_k} = (-1)^k \overset{\circ}{\gamma}{}_{BA}^{i_1 i_2 \cdots i_k} \psi^{BA} = (-1)^{\frac{1}{2}k(k-1)} \overset{\circ}{\gamma}{}^{A}{}_{B}{}^{i_1 i_2 \cdots i_k} \psi^{B}{}_{A}. \tag{1.95}$$

If, in Eqs. (1.95), the components of the second-rank spinor ψ^{BA} are symmetric, $\psi^{BA} = \psi^{AB}$, then, due to the symmetry properties (1.53), (1.55) of the spintensors E and $\overset{\circ}{\gamma}_{i_1 i_2 \cdots i_k}$, some of the components of the tensors F and $F^{i_1 i_2 \cdots i_k}$ turn to zero:

$$F = 0 \quad \text{if} \quad \frac{1}{2}\nu(\nu+1) \quad \text{is odd},$$

$$F^{i_1 i_2 \cdots i_k} = 0 \quad \text{if} \quad \frac{1}{2}[\nu(\nu+1) + k(k+1)] \quad \text{is odd}.$$

If, in Eqs. (1.94) and (1.95), the components of the second-rank spinor ψ^{BA} are antisymmetric, $\psi^{BA} = -\psi^{AB}$, then, due to the symmetry properties (1.53) and (1.55), we find:

$$F = 0 \quad \text{if} \quad \frac{1}{2}\nu(\nu+1) \quad \text{is even},$$

$$F^{i_1 i_2 \cdots i_k} = 0 \quad \text{if} \quad \frac{1}{2}[\nu(\nu+1) + k(k+1)] \quad \text{is even}.$$

It is clear that the components of a spinor of any even rank may be expanded, for each pair of indices, in the set of invariant spintensors E^{-1}, $\overset{\circ}{\gamma}_i E^{-1}, \ldots,$ $\overset{\circ}{\gamma}_{i_1 i_2 \ldots i_{2\nu}} E^{-1}$. Therefore even-rank spinors in the Euclidean complex space $E_{2\nu}^{+}$ are equivalent to certain sets of tensors in $E_{2\nu}^{+}$.

1.5 Semispinors in Even-Dimensional Complex Euclidean Spaces

Let us introduce, in the complex Euclidean space E_n^{+}, $n = 2\nu$, the Levi-Civita pseudotensor

$$\overset{\circ}{\boldsymbol{\varepsilon}} = \overset{\circ}{\varepsilon}{}^{i_1 i_2 \ldots i_n} \boldsymbol{\vartheta}_{i_1} \boldsymbol{\vartheta}_{i_2} \ldots \boldsymbol{\vartheta}_{i_n},$$

determined in an orthonormal basis $\mathfrak{Z}_i$ in E_n^+ by the contravariant components

$$\overset{\circ}{\varepsilon}{}^{i_1 i_2 \dots i_n} = \begin{cases} 1 & \text{if the substitution } \begin{pmatrix} i_1 & i_2 & \dots & i_n \\ 1 & 2 & \dots & n \end{pmatrix} \text{ is even,} \\ -1 & \text{if the substitution } \begin{pmatrix} i_1 & i_2 & \dots & i_n \\ 1 & 2 & \dots & n \end{pmatrix} \text{ is odd,} \\ 0 & \text{if among the indices } i_k \text{ at least two coincide.} \end{cases}$$

By definition, the contravariant components of the Levi-Civita pseudotensor $\overset{\circ}{\varepsilon}{}^{i_1 i_2 \dots i_n}$ are antisymmetric in all indices, and under an orthogonal transformation

$$\mathfrak{Z}'_i = l^j{}_i \mathfrak{Z}_j, \qquad \mathfrak{Z}_i = b^j{}_i \mathfrak{Z}'_j$$

of the vectors of the basis $\mathfrak{Z}_i$ they are transformed as follows:

$$\left(\overset{\circ}{\varepsilon}{}^{i_1 i_2 \dots i_n}\right)' = \frac{1}{\det \|b^m{}_n\|} b^{i_1}{}_{j_1} b^{i_2}{}_{j_2} \cdots b^{i_n}{}_{j_n} \overset{\circ}{\varepsilon}{}^{j_1 j_2 \dots j_n} \equiv \overset{\circ}{\varepsilon}{}^{i_1 i_2 \dots i_n}.$$

Thus the components of the pseudotensor $\overset{\circ}{\varepsilon}{}^{i_1 i_2 \dots i_n}$ are invariant under all orthogonal transformations of bases in the space E_n^+.

The covariant components of the pseudotensor $\overset{\circ}{\varepsilon}_{i_1 i_2 \dots i_n}$ in an orthonormal basis $\mathfrak{Z}_i$ numerically coincide with the contravariant ones, $\overset{\circ}{\varepsilon}{}^{i_1 i_2 \dots i_n}$:

$$\overset{\circ}{\varepsilon}_{i_1 i_2 \dots i_n} = \overset{\circ}{\varepsilon}{}^{i_1 i_2 \dots i_n}.$$

Using the Levi-Civita pseudotensor, definition (1.21) of the components of the second-rank spinor $\overset{\circ}{\gamma}_{2\nu+1}$ may be written in an explicitly invariant form:

$$\overset{\circ}{\gamma}_{2\nu+1} = \frac{\mathrm{i}^{\nu}}{(2\nu)!} \overset{\circ}{\varepsilon}{}^{i_1 i_2 \dots i_{2\nu}} \overset{\circ}{\gamma}_{i_1} \overset{\circ}{\gamma}_{i_2} \cdots \overset{\circ}{\gamma}_{i_{2\nu}}. \tag{1.96}$$

The transformation law (1.91) for the components of $\overset{\circ}{\gamma}_{2\nu+1}$ under transformations of the basis $\mathfrak{Z}_i$ is quite clear from Eq. (1.96).

If the components of the first-rank spinor ψ in the space $E_{2\nu}^+$ are not arbitrary but are related by

$$\psi = \overset{\circ}{\gamma}_{2\nu+1} \psi \quad \text{or} \quad \psi = -\overset{\circ}{\gamma}_{2\nu+1} \psi, \tag{1.97}$$

then such a spinor is called a semispinor in the space $E_{2\nu}^+$.

From the invariance of $\overset{\circ}{\gamma}_{2\nu+1}$ under proper orthogonal transformations of the basis $\boldsymbol{\vartheta}_i$ it follows that Eqs. (1.97) are also invariant under proper orthogonal transformations of bases in $E^+_{2\nu}$. Therefore the sets ψ, defined by Eqs. (1.97), form subspaces in the space S_{2^ν} which are invariant with respect to the proper orthogonal group of transformations of bases in $E^+_{2\nu}$.

Let ψ be the components of an arbitrary spinor in the space $E^+_{2\nu}$, and let us introduce two spinors with components $\psi_{(I)}$ and $\psi_{(II)}$ specified in the same basis as ψ:

$$\psi_{(I)} = \frac{1}{2}\left(I + \overset{\circ}{\gamma}_{2\nu+1}\right)\psi, \qquad \psi_{(II)} = \frac{1}{2}\left(I - \overset{\circ}{\gamma}_{2\nu+1}\right)\psi. \tag{1.98}$$

From definitions (1.98) and from the equality

$$\overset{\circ}{\gamma}_{2\nu+1}\overset{\circ}{\gamma}_{2\nu+1} = I,$$

which holds by virtue of definition (1.21), it follows that the components $\psi_{(I)}$ and $\psi_{(II)}$ satisfy the equations

$$\psi_{(I)} = \overset{\circ}{\gamma}_{2\nu+1}\psi_{(I)}, \qquad \psi_{(II)} = -\overset{\circ}{\gamma}_{2\nu+1}\psi_{(II)}.$$

Thus the components $\psi_{(I)}$ and $\psi_{(II)}$ determine semispinors in the space $E^+_{2\nu}$.

Let us introduce, in the spinor space S_{2^ν}, $\nu > 1$, a special basis, to be denoted $\overset{*}{\boldsymbol{\varepsilon}}_A$, in which the spintensors $\overset{\circ}{\gamma}_i$ are represented by the following matrices:

$$\overset{\circ}{\gamma}_{2\nu} = \begin{Vmatrix} 0 & I \\ I & 0 \end{Vmatrix}, \qquad \overset{\circ}{\gamma}_{\alpha} = \begin{Vmatrix} 0 & -\mathrm{i}\overset{\circ}{\sigma}_{\alpha} \\ \mathrm{i}\overset{\circ}{\sigma}_{\alpha} & 0 \end{Vmatrix}, \qquad \alpha = 1, 2, \ldots, 2\nu - 1. \tag{1.99}$$

Here, 0 is the zero matrix of order $2^{\nu-1}$ and I is the unit matrix of order $2^{\nu-1}$. The matrices $\overset{\circ}{\sigma}_{\alpha}$, also being of order $2^{\nu-1}$ with $\alpha = 1, 2, \ldots, 2(\nu - 1)$, satisfy the equations

$$\overset{\circ}{\sigma}_{\alpha}\overset{\circ}{\sigma}_{\beta} + \overset{\circ}{\sigma}_{\beta}\overset{\circ}{\sigma}_{\alpha} = 2\delta_{\alpha\beta}I,$$

and the matrix $\overset{\circ}{\sigma}_{2\nu-1}$ is defined by the equality

$$\overset{\circ}{\sigma}_{2\nu-1} = \mathrm{i}^{\nu-1}\overset{\circ}{\sigma}_{1}\overset{\circ}{\sigma}_{2}\cdots\overset{\circ}{\sigma}_{2(\nu-1)}.$$

In this case, we have for $\overset{\circ}{\gamma}_{2\nu+1}$:

$$\overset{\circ}{\gamma}_{2\nu+1} = \begin{Vmatrix} I & 0 \\ 0 & -I \end{Vmatrix}.$$

Let us suppose that the matrices $\overset{\circ}{\sigma}_\alpha$ in definitions (1.99) are Hermitian and that they are symmetric for $\alpha = 1, 2, \ldots, \nu - 1$ and antisymmetric for $\alpha = \nu, \nu + 1, \ldots, 2(\nu - 1)$:

$$\overset{\circ}{\sigma}{}^T_1 = \overset{\circ}{\sigma}_1, \qquad \overset{\circ}{\sigma}{}^T_2 = \overset{\circ}{\sigma}_2, \qquad \ldots, \qquad \overset{\circ}{\sigma}{}^T_{\nu-1} = \overset{\circ}{\sigma}_{\nu-1},$$
$$\overset{\circ}{\sigma}{}^T_\nu = -\overset{\circ}{\sigma}_\nu, \qquad \overset{\circ}{\sigma}{}^T_{\nu+1} = -\overset{\circ}{\sigma}_{\nu+1}, \qquad \ldots, \qquad \overset{\circ}{\sigma}{}^T_{2(\nu-1)} = -\overset{\circ}{\sigma}_{2(\nu-1)}.$$

In this case, the metric spinor E in the Euclidean space $E^+_{2\nu}$ for even ν may be defined in the spinbasis $\overset{*}{\boldsymbol{\varepsilon}}_A$ by the matrix of covariant components

$$E = \|e_{BA}\| = \begin{Vmatrix} \varepsilon & 0 \\ 0 & -\varepsilon \end{Vmatrix}, \tag{1.100}$$

where the matrix ε of order $2^{\nu-1}$, satisfies the equation

$$\overset{\circ}{\sigma}{}^T_\alpha = -\varepsilon \overset{\circ}{\sigma}_\alpha \varepsilon^{-1}. \tag{1.101}$$

For odd ν, we define the metric spinor in the spinbasis $\overset{*}{\boldsymbol{\varepsilon}}_A$ in the following way:

$$E = \|e_{BA}\| = \begin{Vmatrix} 0 & -\varepsilon \\ \varepsilon & 0 \end{Vmatrix}, \tag{1.102}$$

where ε satisfies the equation

$$\overset{\circ}{\sigma}{}^T_\alpha = \varepsilon \overset{\circ}{\sigma}_\alpha \varepsilon^{-1}. \tag{1.103}$$

It is not difficult to verify that, with the metric spinor E specified in this way, equality (1.44), serving as a definition of E, really holds. The matrix ε, satisfying Eqs. (1.101) for even ν and Eqs. (1.103) for odd ν, may be written explicitly in the form

$$\varepsilon = \lambda \overset{\circ}{\sigma}_\nu \overset{\circ}{\sigma}_{\nu+1} \cdots \overset{\circ}{\sigma}_{2(\nu-1)}.$$

Here, λ is an arbitrary nonzero complex number.

The infinitesimal operators of the spinor representation, defined by Eq. (1.68), have the following form in the spinbasis $\overset{*}{\boldsymbol{\varepsilon}}_A$:

$$A^{\alpha,2\nu} = \frac{1}{2}\begin{Vmatrix} -\mathrm{i}\overset{\circ}{\sigma}_\alpha & 0 \\ 0 & \mathrm{i}\overset{\circ}{\sigma}_\alpha \end{Vmatrix}, \qquad A^{\alpha\beta} = \frac{1}{2}\begin{Vmatrix} \overset{\circ}{\sigma}{}^{[\alpha}\overset{\circ}{\sigma}{}^{\beta]} & 0 \\ 0 & \overset{\circ}{\sigma}{}^{[\alpha}\overset{\circ}{\sigma}{}^{\beta]} \end{Vmatrix},$$
$$\alpha, \beta = 1, 2, \ldots, 2\nu - 1. \tag{1.104}$$

From expressions (1.104) for the infinitesimal operators A^{ij} it follows that the group of spinor transformations $\{\pm S\}$, corresponding to the proper orthogonal group of transformations of the bases $\mathbf{Э}_i$ in the space $E^+_{2\nu}$, is defined in the chosen special spinbasis $\overset{*}{\boldsymbol{\varepsilon}}_A$ by the matrices S of the form

$$S = \begin{Vmatrix} A & 0 \\ 0 & D \end{Vmatrix}, \tag{1.105}$$

where A and D are some matrices of the order $2^{\nu-1}$.

Evidently, the sets of matrix pairs $\{\pm D\}$, $\{\pm A\}$ corresponding to the group $SO^+_{2\nu}$ of proper orthogonal transformations of bases of the space $E^+_{2\nu}$ form groups which realize representations of the $SO^+_{2\nu}$ group.

It follows from relation (1.105) that the spinor representation of the $SO^+_{2\nu}$ group is reducible and splits into two different representations.

Using the matrices $\overset{\circ}{\sigma}_\alpha$, Eqs. (1.59), which define the spinor representation of the $SO^+_{2\nu}$ group, may be written in the form

$$\begin{aligned}
-\mathrm{i}l^\alpha{}_{2\nu}\overset{\circ}{\sigma}_\alpha + l^{2\nu}{}_{2\nu}I &= A^{-1}D,\\
\mathrm{i}l^\alpha{}_{2\nu}\overset{\circ}{\sigma}_\alpha + l^{2\nu}{}_{2\nu}I &= D^{-1}A,\\
l^\alpha{}_\beta\overset{\circ}{\sigma}_\alpha + \mathrm{i}l^{2\nu}{}_\beta I &= A^{-1}\overset{\circ}{\sigma}_\beta D,\\
l^\alpha{}_\beta\overset{\circ}{\sigma}_\alpha - \mathrm{i}l^{2\nu}{}_\beta I &= D^{-1}\overset{\circ}{\sigma}_\beta A.
\end{aligned} \tag{1.106}$$

Due to orthogonality of the transformation $l^j{}_i$, in Eqs. (1.106), only the following ones are independent:

$$\begin{aligned}
-\mathrm{i}l^\alpha{}_{2\nu}\overset{\circ}{\sigma}_\alpha + l^{2\nu}{}_{2\nu}I &= A^{-1}D,\\
l^\alpha{}_\beta\overset{\circ}{\sigma}_\alpha + \mathrm{i}l^{2\nu}{}_\beta I &= A^{-1}\overset{\circ}{\sigma}_\beta D.
\end{aligned} \tag{1.107}$$

Let us also write the normalization conditions (1.60) in the spinbasis $\overset{*}{\boldsymbol{\varepsilon}}_A$ for even ν:

$$\varepsilon = A^T\varepsilon A, \quad \varepsilon = D^T\varepsilon D. \tag{1.108}$$

For odd ν we have

$$\varepsilon = A^T\varepsilon D. \tag{1.109}$$

Note that it follows from the first equation (1.107) that, for proper orthogonal transformations of the basis $\mathbf{Э}_i$ of the Euclidean space $E^+_{2\nu}$, leaving the basis vector

$\mathfrak{Z}_{2\nu}$ invariable, if the equalities

$$l^{2\nu}{}_{2\nu} = 1, \quad l^{\alpha}{}_{2\nu} = 0, \tag{1.110}$$

are valid, the matrices D and A coincide, $D = A$.

In this case, Eqs. (1.107) pass over to the following equations:

$$l^{\alpha}{}_{\beta}\overset{\circ}{\sigma}_{\alpha} = A^{-1}\overset{\circ}{\sigma}_{\beta}A. \tag{1.111}$$

It is easy to see that, in the spinbasis $\overset{*}{\boldsymbol{\varepsilon}}_A$ under consideration, from the condition $\psi = \overset{\circ}{\gamma}_{2\nu+1}\psi$ it follows

$$\psi^{1+2^{\nu-1}} = \psi^{2+2^{\nu-1}} = \cdots = \psi^{2^{\nu}} = 0,$$

while from the condition $\psi = -\overset{\circ}{\gamma}_{2\nu+1}\psi$ it follows

$$\psi^{1} = \psi^{2} = \cdots = \psi^{2^{\nu-1}} = 0.$$

Let us denote the column of contravariant components $\psi^1, \psi^2, \ldots, \psi^{2^{\nu-1}}$ by the symbol φ and the column of contravariant components $\psi^{1+2^{\nu-1}}, \psi^{2+2^{\nu-1}}, \ldots, \psi^{2^{\nu}}$ by the symbol χ:

$$\varphi = \begin{Vmatrix} \psi^1 \\ \psi^2 \\ \vdots \\ \psi^{2^{\nu-1}} \end{Vmatrix}, \quad \chi = \begin{Vmatrix} \psi^{1+2^{\nu-1}} \\ \psi^{2+2^{\nu-1}} \\ \vdots \\ \psi^{2^{\nu}} \end{Vmatrix}.$$

Then, for the contravariant components of the semispinors $\psi_{(I)}$ and $\psi_{(II)}$, in the spinbasis $\overset{*}{\boldsymbol{\varepsilon}}_A$, we can write

$$\psi_{(I)} = \begin{Vmatrix} \varphi \\ 0 \end{Vmatrix}, \quad \psi_{(II)} = \begin{Vmatrix} 0 \\ \chi \end{Vmatrix}.$$

Here 0 is a column of $2^{\nu-1}$ zeros.

According to definition (1.100), for covariant components of the semispinors $\boldsymbol{\psi}_{(I)}, \boldsymbol{\psi}_{(II)}$ in the spinbasis $\overset{*}{\boldsymbol{\varepsilon}}_A$, for even ν we have

$$\widetilde{\psi}_{(I)} = (\varphi^T\varepsilon^T,\ 0^T), \quad \widetilde{\psi}_{(II)} = (0^T,\ -\chi^T\varepsilon^T).$$

For odd ν, the covariant components of the semispinors $\boldsymbol{\psi}_{(I)}$ and $\boldsymbol{\psi}_{(II)}$, according to the definition (1.102), are given by

$$\widetilde{\psi}_{(I)} = (0^T,\ \varphi^T\varepsilon^T), \quad \widetilde{\psi}_{(II)} = (-\chi^T\varepsilon^T,\ 0^T).$$

Relation (1.105) implies that, under proper orthogonal transformations of the orthonormal basis $\mathfrak{Э}_i$ in the space $E^+_{2\nu}$, the covariant and contravariant components of the spinor, ψ^A and ψ_A $(A = 1, 2, \ldots, 2^{\nu-1})$ and its components ψ^A and ψ_A $(A = 1 + 2^{\nu-1}, 2 + 2^{\nu-1}, \ldots, 2^\nu)$, being calculated in the spinbasis $\overset{*}{\boldsymbol{\varepsilon}}_A$, are transformed separately. Therefore, restricting ourselves to considering only proper orthogonal transformations of the bases $\mathfrak{Э}_i$ in $E^+_{2\nu}$, we can define the covariant and contravariant components of the semispinors $\boldsymbol{\psi}_{(I)}$ and $\boldsymbol{\psi}_{(II)}$ by only $2^{\nu-1}$ components.

Under reflection transformations of bases in the space $E^+_{2\nu}$, the components of semispinors satisfying the equation $\psi = \pm\overset{\circ}{\gamma}_{2\nu+1}\psi$, pass over to components satisfying the equation $\psi = \mp\overset{\circ}{\gamma}_{2\nu+1}\psi$:

$$\psi'_{(I)} = \frac{1}{2}\{(I + \overset{\circ}{\gamma}_{2\nu+1})\psi\}' = \frac{1}{2}(I - \overset{\circ}{\gamma}_{2\nu+1})\psi',$$

$$\psi'_{(II)} = \frac{1}{2}\{(I - \overset{\circ}{\gamma}_{2\nu+1})\psi\}' = \frac{1}{2}(I + \overset{\circ}{\gamma}_{2\nu+1})\psi'.$$

1.6 Spinors in Even-Dimensional Real Euclidean and Pseudo-Euclidean Spaces $E^q_{2\nu}$

1.6.1 The Pseudo-Orthogonal Group of Transformations of Orthonormal Bases in Pseudo-Euclidean Spaces $E^q_{2\nu}$

Consider a 2ν-dimensional pseudo-Euclidean vector space $E^q_{2\nu}$ of index q, i.e., a vector space in which a scalar product of the vectors of an orthonormal basis $\mathfrak{Э}_i$ is specified, being defined in the following way[8]:

$$\begin{aligned} g_{ij} &= (\mathfrak{Э}_i, \mathfrak{Э}_j) = -1 \quad \text{for} \quad i = j = 1, 2, \ldots, q, \\ g_{ij} &= (\mathfrak{Э}_i, \mathfrak{Э}_j) = +1 \quad \text{for} \quad i = j = q+1, q+2, \ldots, 2\nu, \\ g_{ij} &= (\mathfrak{Э}_i, \mathfrak{Э}_j) = 0 \quad \text{for} \quad i \neq j. \end{aligned} \tag{1.112}$$

[8]The space $E^q_{2\nu}$ may be singled out in the complex Euclidean vector space $E^+_{2\nu}$ as a linear real shell of a basis of the space $E^+_{2\nu}$ of the following form:

$$\mathrm{i}\overset{\circ}{\mathfrak{Э}}_1, \mathrm{i}\overset{\circ}{\mathfrak{Э}}_2, \ldots, \mathrm{i}\overset{\circ}{\mathfrak{Э}}_q, \overset{\circ}{\mathfrak{Э}}_{q+1}, \ldots, \overset{\circ}{\mathfrak{Э}}_{2\nu},$$

where $\overset{\circ}{\mathfrak{Э}}_i$ is an orthonormal basis in $E^+_{2\nu}$.

Thus the scalar square of a vector with components a^i in an orthonormal basis of the space $E^q_{2\nu}$ has the form

$$a^2 = -(a^1)^2 - \cdots - (a^q)^2 + (a^{q+1})^2 + \cdots + (a^{2\nu})^2.$$

If $q = 0$, the scalar square of a vector is defined as a sum of squares of its components in an orthonormal basis. In this case, the space $E^0_{2\nu}$ is called a real Euclidean vector space.

Linear real transformations of an orthonormal basis

$$\mathcal{G}'_i = l^j{}_i \mathcal{G}_j, \tag{1.113}$$

leaving invariant the scalar products g_{ij} of the vectors $\mathcal{G}_i$, are called pseudo-orthogonal transformations. As follows from the definition, the pseudo-orthogonality condition for the transformation (1.113) is written in the form

$$g_{ij} = l^m{}_i l^n{}_j g_{mn}. \tag{1.114}$$

Let us introduce notations for the principal minors of the matrix of coefficients of the transformation $l^j{}_i$:

$$\Delta_1 = \begin{Vmatrix} l^1{}_1 & l^1{}_2 & \ldots & l^1{}_q \\ l^2{}_1 & l^2{}_2 & \ldots & l^2{}_q \\ \vdots & \vdots & & \vdots \\ l^q{}_1 & l^q{}_2 & \ldots & l^q{}_q \end{Vmatrix}, \quad \Delta_2 = \begin{Vmatrix} l^{q+1}{}_{q+1} & l^{q+1}{}_{q+2} & \ldots & l^{q+1}{}_{2\nu} \\ l^{q+2}{}_{q+1} & l^{q+2}{}_{q+2} & \ldots & l^{q+2}{}_{2\nu} \\ \vdots & \vdots & & \vdots \\ l^{2\nu}{}_{q+1} & l^{2\nu}{}_{q+2} & \ldots & l^{2\nu}{}_{2\nu} \end{Vmatrix}. \tag{1.115}$$

As known, for any pseudo-orthogonal transformations, the determinants Δ_1 and Δ_2 are nonzero, and therefore for continuous pseudo-orthogonal transformations these determinants preserve their sign. In accordance with four possible sign combinations of Δ_1 and Δ_2, the pseudo-orthogonal transformation group $O^q_{2\nu}$ may be split into four connected components[9]:

1. The first connected component is defined by the conditions

$$\Delta_1 > 0, \qquad \Delta_2 > 0.$$

The set of all pseudo-orthogonal transformations belonging to the first connected component forms a group called the proper pseudo-orthogonal group. For instance, the identical transformation is a representative of the first connected component.

[9]Recall that a connected component of a continuous group is, by definition, such a connected part of the group that its any extension is not connected.

2. The second connected component of the pseudo-orthogonal group $O^q_{2\nu}$ is defined in the following way:

$$\Delta_1 > 0, \qquad \Delta_2 < 0.$$

An example of a representative of the second connected component is the transformation

$$L_2 : \quad \mathcal{E}'_{2\nu} = -\mathcal{E}_{2\nu}, \quad \mathcal{E}'_\alpha = \mathcal{E}_\alpha, \quad \alpha = 1, 2, \ldots, 2\nu - 1. \tag{1.116}$$

Any transformation from the second connected component may be represented as a product $L_2 L$, where L is some proper pseudo-orthogonal transformation.

3. The third connected component of the pseudo-orthogonal group $O^q_{2\nu}$ is defined by the condition

$$\Delta_1 < 0, \quad \Delta_2 > 0.$$

As an example of a pseudo-orthogonal transformation from the third connected component, one can take the transformation

$$L_3 : \quad \mathcal{E}'_1 = -\mathcal{E}_1, \quad \mathcal{E}'_\alpha = \mathcal{E}_\alpha, \quad \alpha = 2, 3, \ldots, 2\nu. \tag{1.117}$$

Any pseudo-orthogonal transformation from the third connected component may be represented as a product $L_3 L$, where L is some proper pseudo-orthogonal transformation.

4. For the fourth connected component of the pseudo-orthogonal group we have

$$\Delta_1 < 0, \qquad \Delta_2 < 0.$$

As a representative of the fourth connected component, one can take, for example, the following transformation:

$$L_4 : \quad \mathcal{E}'_1 = -\mathcal{E}_1, \quad \mathcal{E}'_{2\nu} = -\mathcal{E}_{2\nu}, \quad \mathcal{E}'_\alpha = \mathcal{E}_\alpha, \quad \alpha = 2, 3, \ldots, 2\nu - 1. \tag{1.118}$$

Any pseudo-orthogonal transformation of the basis $\mathcal{E}_i$ from the fourth connected component may be represented in the form $L_4 L$, where L is a proper pseudo-orthogonal transformation.

It follows from Eqs. (1.114) that the determinant of the full matrix of coefficients $l^j{}_i$ is equal to $+1$ or -1:

$$\Delta = \det \|l^j{}_i\| = \pm 1.$$

Therefore the determinant Δ for continuous pseudo-orthogonal transformations (1.113) is always equal to $+1$ or -1; thus it preserves its value in any of the connected components of the pseudo-orthogonal group. Evidently, $\Delta > 0$ for the first and fourth connected components and $\Delta < 0$ for the second and third connected components.

1.6.2 Algebra of γ-Matrices

With the aid of the matrices $\overset{\circ}{\gamma}_j$, satisfying Eq. (1.1), let us introduce the set of matrices γ_j:

$$\begin{aligned} \gamma_j &= \mathrm{i}\overset{\circ}{\gamma}_j \quad \text{for} \quad j = 1, 2, \ldots, q, \\ \gamma_j &= \overset{\circ}{\gamma}_j \quad \text{for} \quad j = q+1, q+2, \ldots, 2\nu. \end{aligned} \tag{1.119}$$

Due to Eq. (1.1) and definitions (1.119), the matrices γ_j satisfy the equation

$$\gamma_i \gamma_j + \gamma_j \gamma_i = 2 g_{ij} I, \tag{1.120}$$

where the components g_{ij} are defined by equalities (1.112).

If the matrices $\overset{\circ}{\gamma}_j$ in definitions (1.119) satisfy Eqs. (1.36) and (1.37), then, evidently, the matrices γ_j for $j = 1, 2, \ldots, q$ are anti-Hermitian, while for $j = q+1$, $q+2, \ldots, 2\nu$ they are Hermitian:

$$\begin{aligned} \dot{\gamma}_1 &= -\gamma_1^T, \quad \dot{\gamma}_2 = -\gamma_2^T, \quad \ldots, \quad \dot{\gamma}_q = -\gamma_q^T, \\ \dot{\gamma}_{q+1} &= \gamma_{q+1}^T, \quad \dot{\gamma}_{q+2} = \gamma_{q+2}^T, \quad \ldots, \quad \dot{\gamma}_{2\nu} = \gamma_{2\nu}^T. \end{aligned} \tag{1.121}$$

Besides, in this case,

$$\begin{aligned} \gamma_1^T &= \gamma_1, \quad \gamma_2^T = \gamma_2, \quad \ldots, \quad \gamma_\nu^T = \gamma_\nu, \\ \gamma_{\nu+1}^T &= -\gamma_{\nu+1}, \quad \gamma_{\nu+2}^T = -\gamma_{\nu+2}, \quad \ldots, \quad \gamma_{2\nu}^T = -\gamma_{2\nu}. \end{aligned} \tag{1.122}$$

All properties of the matrices $\overset{\circ}{\gamma}_j$ mentioned in Sect. 1.1 are in an obvious way extended to the matrices γ_j. In particular, Eqs. (1.7) take the form

$$\begin{aligned} &\operatorname{tr}\left(\gamma_{i_1 i_2 \ldots i_k} \gamma^{j_1 j_2 \ldots j_m}\right) = 0, \quad \text{if} \quad k \neq m, \\ &\operatorname{tr}\left(\gamma_{i_1 i_2 \ldots i_k} \gamma^{j_1 j_2 \ldots j_k}\right) = (-1)^{\frac{1}{2}k(k-1)} k!\, 2^\nu \delta^{j_1}_{[i_1} \delta^{j_2}_{i_2} \cdots \delta^{j_k}_{i_k]}. \end{aligned} \tag{1.123}$$

Here,

$$\gamma_{i_1 i_2 \ldots i_k} = \gamma_{[i_1} \gamma_{i_2} \cdots \gamma_{i_k]}, \qquad \gamma^{j_1 j_2 \ldots j_k} = \gamma^{[j_1} \gamma^{j_2} \cdots \gamma^{j_k]}.$$

The matrices γ^j with an upper index are defined using the components g^{ij}, whose matrix is inverse to the matrix $g_{ij} = (\mathbf{\ni}_i, \mathbf{\ni}_j)$:

$$\gamma^j = g^{ij}\gamma_i, \quad \|g^{ij}\| = \|g_{ij}\|^{-1}.$$

Thus, by virtue of definition (1.112), we have

$$\gamma^1 = -\gamma_1, \quad \gamma^2 = -\gamma_2, \quad \ldots, \quad \gamma^q = -\gamma_q,$$
$$\gamma^{q+1} = \gamma_{q+1}, \quad \ldots, \quad \gamma^{2\nu} = \gamma_{2\nu}.$$

As in the complex space $E_{2\nu}^+$, in the real space $E_{2\nu}^q$ we define the matrix E by the following equality:

$$\gamma_i^T = -E\gamma_i E^{-1}. \tag{1.124}$$

For even ν, the matrix E satisfying Eq. (1.124) may be represented in the form

$$E = \lambda\gamma_1\gamma_2\cdots\gamma_\nu, \tag{1.125}$$

where λ is an arbitrary nonzero complex number. If ν is odd, the matrix E may be determined in the following way:

$$E = \lambda\gamma_{\nu+1}\gamma_{\nu+2}\cdots\gamma_{2\nu}. \tag{1.126}$$

Evidently, the symmetry properties of the matrices γ_i are the same as those of the matrices $\overset{\circ}{\gamma}_i$. Therefore, due to Eqs. (1.53) and (1.55), the following symmetry properties are valid:

$$E^T = (-1)^{\frac{1}{2}\nu(\nu+1)}E,$$
$$\left(E\gamma_{i_1 i_2 \ldots i_k}\right)^T = (-1)^{\frac{1}{2}[\nu(\nu+1)+k(k+1)]}E\gamma_{i_1 i_2 \ldots i_k}.$$

Consider the set of matrices $\dot{\gamma}_i^T$ which are Hermitian conjugates of γ_i. The matrices $-\dot{\gamma}_i^T$ satisfy Eq. (1.120), therefore, due to Pauli's theorem, there is such a matrix β that

$$\dot{\gamma}_i^T = -\beta\gamma_i\beta^{-1}, \tag{1.127}$$

and this β is defined up to multiplication by an arbitrary nonzero complex number.

It is easy to show that, due to Eqs. (1.127), the following equations also hold:

$$\dot{\gamma}_{i_1 i_2 \ldots i_k}^T = (-1)^{\frac{1}{2}k(k+1)}\beta\gamma_{i_1 i_2 \ldots i_k}\beta^{-1}. \tag{1.128}$$

Indeed, let us multiply Eq. (1.127) from the right by $\beta\gamma_{i_2}\ldots\gamma_{i_k}$:

$$\dot{\gamma}_i^T\beta\gamma_{i_2}\ldots\gamma_{i_k} = -\beta\gamma_i\gamma_{i_2}\ldots\gamma_{i_k}.$$

Transposing the matrices β and γ_{i_m} in the left-hand side of this equation using (1.127), we obtain the equation

$$(-1)^{\frac{1}{2}k(k+1)}\left(\dot{\gamma}_i\dot{\gamma}_{i_2}\cdots\dot{\gamma}_{i_k}\right)^T\beta = \beta\gamma_i\gamma_{i_2}\cdots\gamma_{i_k},$$

from which follows (1.128).

Considering the Hermitian conjugate of Eq. (1.127) and multiplying the resulting equation from the left by $\dot{\beta}^T$ and from the right by $(\dot{\beta}^T)^{-1}$, we obtain

$$\dot{\gamma}_i^T = -\dot{\beta}^T\gamma_i\left(\dot{\beta}^T\right)^{-1}. \tag{1.129}$$

Comparing Eqs. (1.127) and (1.129), we find:

$$\beta\gamma_i\beta^{-1} = \dot{\beta}^T\gamma_i\left(\dot{\beta}^T\right)^{-1}.$$

From the latter equation it follows that the matrix $\beta^{-1}\dot{\beta}^T$ commutes with all γ_i:

$$\beta^{-1}\dot{\beta}^T\gamma_i = \gamma_i\beta^{-1}\dot{\beta}^T$$

and consequently the matrix $\beta^{-1}\dot{\beta}^T$ is proportional to the unit matrix,

$$\beta^{-1}\dot{\beta}^T = \mu I. \tag{1.130}$$

Multiplying the Hermitian conjugate of Eq. (1.130) from the left by β^{-1} and from the right by $\dot{\beta}^T$, we obtain

$$\beta^{-1}\dot{\beta}^T = \frac{1}{\dot{\mu}}I = \mu I.$$

Hence it follows that the coefficient μ is equal to unity by absolute value, $\mu = \exp \mathrm{i}\theta$, where θ is an arbitrary real number. Normalizing β by multiplying it by $\exp\left(-\frac{\mathrm{i}}{2}\theta\right)$, we find that the normalized matrix β is Hermitian,

$$\dot{\beta}^T = \beta. \tag{1.131}$$

With such a normalization, the matrix β is defined up to multiplying by an arbitrary nonzero real number.

In what follows, we will suppose that the matrix β is defined in such a way that the Hermitianity condition (1.131) holds.

Multiplying Eq. (1.128) from the right by $\dot{\beta}^T$ and taking into account the Hermitianity of β, we obtain

$$\left[\left(\beta\gamma_{i_1 i_2 \ldots i_k}\right)^{\cdot}\right]^T = (-1)^{\frac{1}{2}k(k+1)}\beta\gamma_{i_1 i_2 \ldots i_k}. \tag{1.132}$$

Thus the product $\beta\gamma_{i_1 i_2 \ldots i_k}$ is either Hermitian or anti-Hermitian, depending on the value of k.

A direct inspection shows that if the matrices γ_i satisfy conditions (1.121) and (1.122), then, for even $q < 2\nu$, the matrix β may be defined by the equality

$$\beta = \mathrm{i}^{\frac{1}{2}q(q+1)-\nu}\gamma_{[q+1}\gamma_{q+2}\cdots\gamma_{2\nu]}. \tag{1.133}$$

If $q = 2\nu$ and relations (1.121) are valid, the matrix β may be defined as the unit matrix, $\beta = I$.

For odd q, if the conditions (1.121) and (1.122) are valid, the matrix β may be defined in the following way:

$$\beta = \mathrm{i}^{\frac{1}{2}q(q+1)}\gamma_{[1}\gamma_2\cdots\gamma_{q]}. \tag{1.134}$$

Let us denote the product $E^{-1}\beta^T$ by the symbol Π,

$$\Pi = E^{-1}\beta^T, \tag{1.135}$$

and calculate a product of the matrix Π by its complex conjugate matrix $\dot{\Pi}$. To do so, we transpose equation (1.127) and substitute γ_i^T in the resulting relation using (1.124):

$$\dot{\gamma}_i = \left(E^{-1}\beta^T\right)^{-1}\gamma_i E^{-1}\beta^T.$$

The latter equation implies that the matrix Π connects the matrices γ_i and the complex conjugate matrices $\dot{\gamma}_i$:

$$\dot{\gamma}_i = \Pi^{-1}\gamma_i\Pi. \tag{1.136}$$

In a similar way, transposing equations (1.128), we find

$$\dot{\gamma}_{i_1 i_2 \ldots i_k} = \Pi^{-1}\gamma_{i_1 i_2 \ldots i_k}\Pi. \tag{1.137}$$

Considering a complex conjugate of Eq. (1.136) and multiplying the result from the left by $\dot{\Pi}$ and from the right by $\dot{\Pi}^{-1}$, we obtain

$$\dot{\Pi}\gamma_i\dot{\Pi}^{-1} = \dot{\gamma}_i.$$

Thus

$$\dot{\Pi}\gamma_i\dot{\Pi}^{-1} = \Pi^{-1}\gamma_i\Pi.$$

Hence it follows

$$\Pi\dot{\Pi}\gamma_i = \gamma_i\Pi\dot{\Pi}.$$

Since the matrix $\Pi\dot{\Pi}$ commutes with all γ_i, the matrix $\Pi\dot{\Pi}$ is a multiple of the unit matrix,

$$\Pi\dot{\Pi} = \eta I. \tag{1.138}$$

Let us multiply Eq. (1.138) from the left by Π^{-1} and from the right by Π. We obtain

$$\dot{\Pi}\Pi = \eta I. \tag{1.139}$$

Complex conjugation of Eq. (1.139) gives

$$\Pi\dot{\Pi} = \dot{\eta} I. \tag{1.140}$$

Comparing Eqs. (1.138) and (1.140), we find that the number η in Eq. (1.138) is real, $\eta = \dot{\eta}$.

Since the Hermitian matrix β is defined by Eqs. (1.127) and (1.131) up to multiplying by an arbitrary nonzero real number, we can normalize β, multiplying it by $|\eta|^{1/2}$, to obtain that the matrix Π satisfies the equation $\Pi\dot{\Pi} = \pm I$. A direct inspection using definitions (1.125) and (1.126) for E and definitions (1.133) and (1.134) for β shows that

$$\Pi\dot{\Pi} = (-1)^{\frac{1}{2}(\nu-q)(\nu-q-1)} I. \tag{1.141}$$

With the above normalizations, the matrices β and Π are defined up to multiplying by -1 (for fixed E). If E is re-defined by $E \to E\rho \exp \mathrm{i}\varphi$, where ρ and φ are arbitrary real numbers, we have

$$\beta \to \pm\rho\beta, \qquad \Pi \to \pm\Pi \exp(-\mathrm{i}\varphi).$$

1.6.3 Real and Imaginary Representation of the Matrices $\boldsymbol{\gamma}_i$

In pseudo-Euclidean spaces $E^q_{2\nu}$ of any dimension 2ν (but with certain values of the index q), there exist pure real and pure imaginary representations for $\boldsymbol{\gamma}$ matrices. The reality condition for $\boldsymbol{\gamma}$ matrices $\gamma_i = \dot{\gamma}_i$ in the space $E^q_{2\nu}$, by virtue of equality (1.136), is written in the form

$$\gamma_i = \Pi^{-1}\gamma_i \Pi, \quad \text{or} \quad \gamma_i \Pi = \Pi \gamma_i.$$

Hence it follows that the matrix Π is proportional to the unit matrix, $\Pi = \lambda I$ (λ is an arbitrary nonzero complex number). Taking into account equality (1.141), we find

$$\dot{\lambda}\lambda = (-1)^{\frac{1}{2}(\nu-q)(\nu-q-1)}.$$

Evidently, this equality may only hold under the condition

$$(-1)^{\frac{1}{2}(\nu-q)(\nu-q-1)} = 1. \tag{1.142}$$

Thus the matrices γ_i can be real in the spaces $E^q_{2\nu}$ with the dimension and index satisfying Eq. (1.142). From Eqs. (1.142) it follows $\nu - q \pm 4k = 0, 1$ ($k = 0, 1, 2, \dots$). This relation holds for any given dimension 2ν for the corresponding value of the index q.

From the condition that the $\boldsymbol{\gamma}$ matrices are imaginary, $\dot{\gamma}_i = -\gamma_i$, follows the equality $\gamma_i \Pi = -\Pi \gamma_i$, which has the solution

$$\Pi = \lambda \gamma_{2\nu+1}, \quad \gamma_{2\nu+1} = \mathrm{i}^{\nu-q}\gamma_1\gamma_2\cdots\gamma_{2\nu}.$$

A direct calculation shows that the matrix $\gamma_{2\nu+1}$, in the case of imaginary γ_i, satisfies the equation $\dot{\gamma}_{2\nu+1}\gamma_{2\nu+1} = (-1)^{\nu-q} I$. Therefore in this case, taking into account equality (1.141), we find

$$\dot{\lambda}\lambda = (-1)^{\frac{1}{2}(\nu-q)(\nu-q+1)}.$$

Thus the matrices γ_i can be pure imaginary only in spaces with the dimension and index satisfying the condition

$$(-1)^{\frac{1}{2}(\nu-q)(\nu-q+1)} = 1,$$

which implies $\nu - q \pm 4k = 0, 3$ ($k = 0, 1, 2, \dots$).

1.6.4 Spinor Representation of the Group of Pseudo-Orthogonal Transformations of Bases of the Space $E^q_{2\nu}$

Due to the existence of four connected components of the pseudo-orthogonal group $O^q_{2\nu}$, the number of different spinor representation of the group $O^q_{2\nu}$ is larger than in the case of the complex orthogonal group $O^+_{2\nu}$. Besides, the reality condition for the coefficients of the pseudo-orthogonal transformation $l^j{}_i$ leads to some additional properties of spinor transformations S which are absent in the complex Euclidean space $E^+_{2\nu}$.

The spinor representations of the $O^q_{2\nu}$ group may be conveniently split into several classes [81, 84].

I. The first class of spinor representations $O^q_{2\nu} \to \{\pm S\}$ can be defined by the equations

$$l^j{}_i\gamma_j = S^{-1}\gamma_i S \tag{1.143}$$

and one of the normalization conditions

$$\begin{aligned} &a. \qquad S^T E S = E, \\ &b. \qquad S^T E S = E \ \operatorname{sign} \Delta, \\ &c. \qquad S^T E S = E \ \operatorname{sign} \Delta_1, \\ &d. \qquad S^T E S = E \ \operatorname{sign} \Delta_2. \end{aligned} \tag{1.144}$$

For all normalization conditions (1.144), as long as Eqs. (1.143) are valid, the following equations hold as well:

$$S^{-1}\gamma_{2\nu+1}S = \gamma_{2\nu+1} \operatorname{sign} \Delta, \qquad \gamma_{2\nu+1} = \mathrm{i}^{\nu-q}\gamma_1\gamma_2\ldots\gamma_{2\nu} \tag{1.145}$$

and

$$\dot{S}^T \beta S = \beta \operatorname{sign} \Delta_1 \tag{1.146a}$$

for an odd index q of the pseudo-Euclidean space $E^q_{2\nu}$ or

$$\dot{S}^T \beta S = \beta \operatorname{sign} \Delta_2 \tag{1.146b}$$

for an even index q.

Equations (1.145) may be proved in the same way as Eqs. (1.72) in the complex Euclidean space $E^+_{2\nu}$.

Equations (1.146a) and (1.146b) realize a connection between the spinor transformations S and the Hermitian conjugate transformations $\dot{S}^T$ and follow from the reality of the coefficients $l^j{}_i$. To prove Eqs. (1.146a) and (1.146b), we take the Hermitian conjugate of Eq. (1.143) and, in the result obtained, we replace $\dot{\gamma}^T_i$ according to the formula (1.127). Due to reality of $l^j{}_i$, we obtain

$$l^j{}_i\beta\gamma_j\beta^{-1} = \dot{S}^T\beta\gamma_i\beta^{-1}(\dot{S}^T)^{-1}.$$

Whence we find

$$l^j{}_i\gamma_j = \beta^{-1}\dot{S}^T\beta\gamma_i\big(\beta^{-1}\dot{S}^T\beta\big)^{-1}. \tag{1.147}$$

Comparing Eqs. (1.143) and (1.147), we obtain

$$\beta^{-1}\dot{S}^T\beta\gamma_i\big(\beta^{-1}\dot{S}^T\beta\big)^{-1} = S^{-1}\gamma_i S.$$

Hence it follows

$$S\beta^{-1}\dot{S}^T\beta\gamma_i = \gamma_i S\beta^{-1}\dot{S}^T\beta.$$

Thus the matrix $S\beta^{-1}\dot{S}^T\beta$ commutes with all γ_i and is consequently proportional to the unit matrix,

$$S\beta^{-1}\dot{S}^T\beta = \eta I.$$

It is convenient to write the latter equation in the form

$$\dot{S}^T\beta S = \eta\beta. \tag{1.148}$$

Let us find the number η in Eq. (1.148). Writing the Hermitian conjugate of Eq. (1.148) taking into account the Hermitianity of the matrices β, we find

$$\dot{S}^T\beta S = \dot{\eta}\beta = \eta\beta.$$

So the coefficient η is real, $\eta = \dot{\eta}$. It is easy to see that Eqs. (1.148) imply

$$\det S \cdot \det \dot{S} = \eta^{2^\nu}.$$

Evidently, for any normalization condition in (1.144), written with the aid of the matrix E, the equality $\big(\det S\big)^2 = 1$ holds. Therefore the determinant $\det S$ is real, $\det S = \det \dot{S}$, and for η we have $\eta^{2^\nu} = 1$. So, since η is real, we obtain $\eta = \pm 1$ and consequently

$$\dot{S}^T\beta S = \pm\beta. \tag{1.149}$$

To determine the sign in Eq. (1.149), let us calculate the trace of the Hermitian matrix $\dot{S}^T S$. We have, using the equality (1.148),

$$\operatorname{tr}\big(\dot{S}^T S\big) = \operatorname{tr}\big(\eta\beta S^{-1}\beta^{-1}S\big). \tag{1.150}$$

If the index q of the pseudo-Euclidean space $E^q_{2\nu}$ is odd, then, substituting the matrix β in Eq. (1.150) according to Eq. (1.134), we find

$$\operatorname{tr}\left(\dot{S}^T S\right) = \operatorname{tr}\left(\eta\gamma_{[1}\gamma_2\cdots\gamma_{q]}S^{-1}\gamma_{[q}^{-1}\cdots\gamma_2^{-1}\gamma_{1]}^{-1}S\right)$$
$$= \operatorname{tr}\left[(-1)^{\frac{1}{2}q(q-1)}\eta\gamma^{[1}\gamma^2\cdots\gamma^{q]}S^{-1}\gamma_{[1}\gamma_2\cdots\gamma_{q]}S\right].$$

It is easy to see that, due to Eqs. (1.143), the following relations hold:

$$l^{j_1}{}_{i_1}l^{j_2}{}_{i_2}\cdots l^{j_q}{}_{i_q}\gamma_{j_1}\gamma_{j_2}\cdots\gamma_{j_q} = S^{-1}\gamma_{i_1}\gamma_{i_2}\cdots\gamma_{i_q}S,$$

and using them, the expression for $\operatorname{tr}\left(\dot{S}^T S\right)$ may be transformed to

$$\operatorname{tr}\left(\dot{S}^T S\right) = \operatorname{tr}\left[(-1)^{\frac{1}{2}q(q-1)}\eta\gamma^{[1}\gamma^2\cdots\gamma^{q]}l^{j_1}{}_1 l^{j_2}{}_2\cdots l^{j_q}{}_q\gamma_{[j_1}\gamma_{j_2}\cdots\gamma_{j_q]}\right]$$
$$= \operatorname{tr}\left[\eta\delta^1_{[j_1}\delta^2_{j_2}\cdots\delta^q_{j_q]}q!\,l^{j_1}{}_1 l^{j_2}{}_2\cdots l^{j_q}{}_q I\right] = 2^\nu\eta\Delta_1.$$

Here, determinant Δ_1 is defined by equality (1.115). The eigenvalues of a matrix formed as a product of any matrix S by its Hermitian conjugate matrix $\dot{S}^T$, are positive, therefore

$$\operatorname{tr}\left(\dot{S}^T S\right) > 0.$$

Consequently, for an odd index q, in Eq. (1.149) we take the "+" sign if $\Delta_1 > 0$ and the "−" sign if $\Delta_1 < 0$:

$$\dot{S}^T\beta S = \beta\,\operatorname{sign}\Delta_1.$$

If the index q is even, then, using definition (1.133), we find for the trace of the matrix $\dot{S}^T S$:

$$\operatorname{tr}\left(\dot{S}^T S\right) = 2^\nu\eta\Delta_2,$$

where the determinant Δ_2 is defined by the equality (1.115). Therefore, for even q, in Eq. (1.149) we take the "+" sign if $\Delta_2 > 0$ and the "−" sign if $\Delta_2 < 0$:

$$\dot{S}^T\beta S = \beta\,\operatorname{sign}\Delta_2.$$

Thus relations (1.146a) and (1.146b) have been proved.

If the index q is odd, then, due to Eqs. (1.146a) and (1.144), the following equation also hold (equation (a) in (1.144) corresponds to equation (a) in (1.151), equation (b) in (1.144) corresponds to equation (b) in (1.151) etc.)

$$\begin{aligned} a.\quad & S^{-1}\Pi\dot{S} = \Pi \operatorname{sign} \Delta_1, \\ b.\quad & S^{-1}\Pi\dot{S} = \Pi \operatorname{sign} \Delta_2, \\ c.\quad & S^{-1}\Pi\dot{S} = \Pi, \\ d.\quad & S^{-1}\Pi\dot{S} = \Pi \operatorname{sign} \Delta. \end{aligned} \tag{1.151}$$

For an even index q, due to Eqs. (1.146b) and (1.144), the following equations hold:

$$\begin{aligned} a.\quad & S^{-1}\Pi\dot{S} = \Pi \operatorname{sign} \Delta_2, \\ b.\quad & S^{-1}\Pi\dot{S} = \Pi \operatorname{sign} \Delta_1, \\ c.\quad & S^{-1}\Pi\dot{S} = \Pi \operatorname{sign} \Delta, \\ d.\quad & S^{-1}\Pi\dot{S} = \Pi. \end{aligned} \tag{1.152}$$

Assuming that the symmetry conditions (1.122) for the matrices γ_i are satisfied, let us write down the spinor transformations S_2, S_3, S_4, corresponding to the reflection transformations L_2, L_3, L_4 defined by equalities (1.116), (1.117) and (1.118) for different normalizations in (1.144):

	a	b	c	d
S_2	$\mathrm{i}^{\nu+1}\gamma_{2\nu}\gamma_{2\nu+1}$	$\mathrm{i}^{\nu}\gamma_{2\nu}\gamma_{2\nu+1}$	$\mathrm{i}^{\nu+1}\gamma_{2\nu}\gamma_{2\nu+1}$	$\mathrm{i}^{\nu}\gamma_{2\nu}\gamma_{2\nu+1}$
S_3	$\mathrm{i}^{\nu}\gamma_1\gamma_{2\nu+1}$	$\mathrm{i}^{\nu+1}\gamma_1\gamma_{2\nu+1}$	$\mathrm{i}^{\nu+1}\gamma_1\gamma_{2\nu+1}$	$\mathrm{i}^{\nu}\gamma_1\gamma_{2\nu+1}$
S_4	$\mathrm{i}\gamma_1\gamma_{2\nu}$	$\mathrm{i}\gamma_1\gamma_{2\nu}$	$\gamma_1\gamma_{2\nu}$	$\gamma_1\gamma_{2\nu}$

For all normalizations (1.144), the spinor transformations S_2, S_3 and S_4 anticommute with each other:

$$S_2S_3 + S_3S_2 = 0, \qquad S_2S_4 + S_4S_2 = 0,$$
$$S_3S_4 + S_4S_3 = 0.$$

Let us also write down the spinor transformations S_J corresponding to the full reflection transformation $\mathcal{Э}'_i = -\mathcal{Э}_i$, $i = 1, 2, \ldots, 2\nu$:

	a	b	c	d
S_J	$\mathrm{i}^{\nu}\gamma_{2\nu+1}$	$\mathrm{i}^{\nu}\gamma_{2\nu+1}$	$\mathrm{i}^{\nu-q}\gamma_{2\nu+1}$	$\mathrm{i}^{\nu-q}\gamma_{2\nu+1}$

II. The second class of spinor representations is defined by the equation

$$l^j{}_i\gamma_j = S^{-1}\gamma_i S \operatorname{sign} \Delta \tag{1.153}$$

and one of the normalization conditions (1.144).

For all normalization conditions (1.144) and for any index q of the space $E^q_{2\nu}$, due to Eqs. (1.153) and (1.144), the following equations hold:

$$\dot{S}^T \beta S = \beta \operatorname{sign} \Delta_2,$$
$$S^{-1}\gamma_{2\nu+1}S = \gamma_{2\nu+1} \operatorname{sign} \Delta. \tag{1.154}$$

Due to Eqs. (1.154) and (1.144), we also have the equations

$$\begin{aligned} a.\quad & S^{-1}\Pi\dot{S} = \Pi \operatorname{sign} \Delta_2,\\ b.\quad & S^{-1}\Pi\dot{S} = \Pi \operatorname{sign} \Delta_1,\\ c.\quad & S^{-1}\Pi\dot{S} = \Pi \operatorname{sign} \Delta,\\ d.\quad & S^{-1}\Pi\dot{S} = \Pi. \end{aligned}$$

If the matrices of components of the spintensor γ_i satisfy the symmetry conditions (1.122), then the spinor transformations S_2, S_3, S_4, corresponding to the transformations (1.116), (1.117), (1.118) of the basis $\boldsymbol{\partial}_i$ have the following form for the class of spinor representations under consideration:

	a	b	c	d
S_2	$\mathrm{i}\gamma_{2\nu}$	$\gamma_{2\nu}$	$\mathrm{i}\gamma_{2\nu}$	$\gamma_{2\nu}$
S_3	γ_1	$\mathrm{i}\gamma_1$	$\mathrm{i}\gamma_1$	γ_1
S_4	$\mathrm{i}\gamma_1\gamma_{2\nu}$	$\mathrm{i}\gamma_1\gamma_{2\nu}$	$\gamma_1\gamma_{2\nu}$	$\gamma_1\gamma_{2\nu}$

(1.155)

For all normalization conditions (1.144), the spinor transformations S_2, S_3, S_4, defined according to (1.155), anti-commute with each other.

The transformation of the full reflection of the basis vectors $\boldsymbol{\partial}'_i = -\boldsymbol{\partial}_i$ is in correspondence with the spinor transformation S_J defined in the following way:

	a	b	c	d
S_J	$\mathrm{i}^{\nu}\gamma_{2\nu+1}$	$\mathrm{i}^{\nu}\gamma_{2\nu+1}$	$\mathrm{i}^{\nu-q}\gamma_{2\nu+1}$	$\mathrm{i}^{\nu-q}\gamma_{2\nu+1}$

The spinor representations $O^q_{2\nu} \to \{\pm S_{II}\}$, defined by equalities (1.153) and (1.144), are equivalent to the spinor representations $O^q_{2\nu} \to \{\pm S_I\}$ defined by equalities (1.143) and (1.144) and are connected with them by the relationship

$$\pm S_{II} = A(\pm S_I)A^{-1},$$

where

$$A = I + \mathrm{i}\gamma_{2\nu+1}.$$

III. The third class of spinor representations of the pseudo-orthogonal group is defined by the equations

$$l^j{}_i \gamma_j = S^{-1} \gamma_i S \operatorname{sign} \Delta_1 \tag{1.156}$$

and one of the normalization conditions (1.144).

Due to Eqs. (1.156), for any normalization in (1.144), Eq. (1.145) holds along with the equation

$$\dot{S}^T \beta S = \beta \tag{1.157a}$$

for odd q and the equation

$$\dot{S}^T \beta S = \beta \operatorname{sign} \Delta_2 \tag{1.157b}$$

for even q.

For the class of spinor representations under consideration, for an odd index q, the four normalizations in (1.144) are in correspondence with the following four relations connecting the spinor transformations S and $\dot{S}$:

$$a. \qquad S^{-1} \Pi \dot{S} = \Pi,$$

$$b. \qquad S^{-1} \Pi \dot{S} = \Pi \operatorname{sign} \Delta,$$

$$c. \qquad S^{-1} \Pi \dot{S} = \Pi \operatorname{sign} \Delta_1,$$

$$d. \qquad S^{-1} \Pi \dot{S} = \Pi \operatorname{sign} \Delta_2.$$

For an odd index q, the normalization conditions (1.144) correspond to Eqs. (1.152).

Let us write out the spinor transformations S_2, S_3, S_4, corresponding to the reflection transformations (1.116), (1.117) and (1.118) in a spinbasis in which the symmetry conditions (1.122) are fulfilled:

	a	b	c	d
S_2	$\mathrm{i}^{\nu+1} \gamma_{2\nu} \gamma_{2\nu+1}$	$\mathrm{i}^{\nu} \gamma_{2\nu} \gamma_{2\nu+1}$	$\mathrm{i}^{\nu+1} \gamma_{2\nu} \gamma_{2\nu+1}$	$\mathrm{i}^{\nu} \gamma_{2\nu} \gamma_{2\nu+1}$
S_3	γ_1	$\mathrm{i} \gamma_1$	$\mathrm{i} \gamma_1$	γ_1
S_4	$\mathrm{i}^{\nu+1} \gamma_1 \gamma_{2\nu} \gamma_{2\nu+1}$	$\mathrm{i}^{\nu+1} \gamma_1 \gamma_{2\nu} \gamma_{2\nu+1}$	$\mathrm{i}^{\nu} \gamma_1 \gamma_{2\nu} \gamma_{2\nu+1}$	$\mathrm{i}^{\nu} \gamma_1 \gamma_{2\nu} \gamma_{2\nu+1}$

For all normalization conditions (1.144), the spinor transformations S_2, S_3, S_4 commute with each other:

$$S_2 S_3 = S_3 S_2, \qquad S_2 S_4 = S_4 S_2, \qquad S_3 S_4 = S_4 S_3.$$

For the transformation of the full reflection, the spinor transformations S_J for an even index q have the form

	a	b	c	d
S_J	$\mathrm{i}^{\nu}\gamma_{2\nu+1}$	$\mathrm{i}^{\nu}\gamma_{2\nu+1}$	$\mathrm{i}^{\nu-q}\gamma_{2\nu+1}$	$\mathrm{i}^{\nu-q}\gamma_{2\nu+1}$

and for an odd index q

	a	b	c	d
S_J	I	I	$\mathrm{i}I$	$\mathrm{i}I$

Hence it is evident that, for an odd index q, the identical transformation and the full reflection transformation $E^q_{2\nu}$ are in correspondence with the same spinor transformations (for normalizations (a) and (b) in (1.144)). Thus these spinor representations are not exact.

IV. The fourth class of spinor representations of the pseudo-Euclidean group $O^q_{2\nu}$ may be specified by the equation

$$l^j{}_i\gamma_j = S^{-1}\gamma_i S \operatorname{sign} \Delta_2 \tag{1.158}$$

and one of the normalization conditions (1.144).

Under any normalization condition (1.144), due to Eqs. (1.158), Eq. (1.145) is valid as well as the equation

$$\dot{S}^T\beta S = \beta \operatorname{sign} \Delta$$

for an odd index q of the pseudo-Euclidean space $E^q_{2\nu}$, or the equation

$$\dot{S}^T\beta S = \beta \operatorname{sign} \Delta_2$$

for an even index q.

For an odd index q, the following relations between S and $\dot{S}$ correspond to the normalization conditions (1.144):

$$a. \quad S^{-1}\Pi\dot{S} = \Pi \operatorname{sign} \Delta,$$

$$b. \quad S^{-1}\Pi\dot{S} = \Pi,$$

$$c. \quad S^{-1}\Pi\dot{S} = \Pi \operatorname{sign} \Delta_2,$$

$$d. \quad S^{-1}\Pi\dot{S} = \Pi \operatorname{sign} \Delta_1.$$

For an even index q, Eqs. (1.152) are valid.

If the matrices γ_i satisfy relations (1.122), then for the spinor transformations corresponding to the reflection transformations (1.116), (1.117) and (1.118) we have

	a	b	c	d
S_2	$\mathrm{i}\gamma_{2\nu}$	$\gamma_{2\nu}$	$\mathrm{i}\gamma_{2\nu}$	$\gamma_{2\nu}$
S_3	$\mathrm{i}^{\nu}\gamma_1\gamma_{2\nu+1}$	$\mathrm{i}^{\nu+1}\gamma_1\gamma_{2\nu+1}$	$\mathrm{i}^{\nu+1}\gamma_1\gamma_{2\nu+1}$	$\mathrm{i}^{\nu}\gamma_1\gamma_{2\nu+1}$
S_4	$\mathrm{i}^{\nu+1}\gamma_1\gamma_{2\nu}\gamma_{2\nu+1}$	$\mathrm{i}^{\nu+1}\gamma_1\gamma_{2\nu}\gamma_{2\nu+1}$	$\mathrm{i}^{\nu}\gamma_1\gamma_{2\nu}\gamma_{2\nu+1}$	$\mathrm{i}^{\nu}\gamma_1\gamma_{2\nu}\gamma_{2\nu+1}$

Under any normalization condition (1.144), the spinor transformations S_2, S_3, S_4 commute with each other. The transformation of the full reflection is in correspondence with the same spinor transformations S_J as for class III spinor representations.

The spinor representations $O^q_{2\nu} \to \{\pm S_{IV}\}$, defined by equalities (1.158) and (1.144), are equivalent to the representations $O^q_{2\nu} \to \{\pm S_{III}\}$ defined by equalities (1.156) and (1.144) and are connected with them by a relation of the form $\pm S_{IV} = A(\pm S_{III})A^{-1}$, where $A = I + \mathrm{i}\gamma_{2\nu+1}$.

A proof of the relations connecting the spinor transformations S and $\dot{S}$ for the second, third and fourth classes of spinor representations is entirely similar to the corresponding proof for spinor representations of the first class.

The spinor representation of the group of proper pseudo-orthogonal transformations is the same in all classes considered above and is defined by the equations

$$l^j{}_i\gamma_j = S^{-1}\gamma_i S, \qquad S^T E S = E.$$

Evidently, all spintensors E, β, Π and $\gamma_{2\nu+1}$ are invariant under the proper pseudo-orthogonal transformations of the basis $\mathfrak{Э}_i$.

For the small proper pseudo-orthogonal transformations

$$l^j{}_i = \delta^j_i + \delta\varepsilon_i{}^j, \qquad \delta\varepsilon_{ij} = -\delta\varepsilon_{ji},$$

the spinor transformations are determined by the formula

$$S = I + \frac{1}{4}\gamma^{ij}\delta\varepsilon_{ij}, \qquad \gamma^{ij} = \frac{1}{2}(\gamma^i\gamma^j - \gamma^j\gamma^i), \tag{1.159}$$

obtained in the same way as in the space $E^+_{2\nu}$.

Since for both real and imaginary representations of the matrices γ^i the infinitesimal operators $\frac{1}{4}\gamma^{ij}$ in Eq. (1.159) are real, for such representations of the matrices γ_i, the spinor representations are realized by real matrices S, and in this case one can consider spinors with real components.

1.6.5 Spinors in the Space $E^q_{2\nu}$

Spinors in the pseudo-Euclidean space $E^q_{2\nu}$ are defined as invariant geometric objects whose components are transformed according to a spinor representation of the pseudo-orthogonal group $O^q_{2\nu} \rightarrow \{\pm S\}$. Along with the spinor whose components are transformed with the matrices S, we will also consider an object in the space $E^q_{2\nu}$ whose components are transformed with the aid of the complex conjugate matrices $\dot{S}$. We will mark the indices of components of such objects by a dot and call them dotted indices. Thus, under a transformation of the orthonormal basis $\mathfrak{Э}_i$ of the space $E^q_{2\nu}$, the components of a first-rank spinor $\boldsymbol{\psi}$ are transformed according to the formulae (we omit the $\pm$ sign)

$$(\psi^A)' = S^A{}_B \psi^B, \qquad (\psi_B)' = Z^A{}_B \psi_A,$$

while the components of a spinor with dotted indices are transformed according to formulas

$$(\psi^{\dot{A}})' = \dot{S}^A{}_B \psi^{\dot{B}}, \qquad (\psi_{\dot{B}})' = \dot{Z}^A{}_B \psi_{\dot{A}}.$$

It is clear that the complex conjugate components of the spinor $\dot{\psi}^A$, $\dot{\psi}_A$, are transformed as components of a spinor with dotted indices $\psi^{\dot{A}}$, $\psi_{\dot{A}}$. Spinors with any number of usual and dotted indices are also defined in an obvious way.

The normalization conditions for the spinor transformations (1.144) and Eqs. (1.146), (1.151), (1.152) (or similar equations for the other classes of spinor representations) mean that the matrices E, β and Π define, in the pseudo-Euclidean space $E^q_{2\nu}$, the components of spintensors with the following structure of indices:

$$E = \|e_{BA}\|, \quad \beta = \|\beta_{\dot{B}A}\|, \quad \Pi = \|\Pi^A{}_{\dot{B}}\|.$$

It is clear that the components of the spinors E, β and Π are invariant under continuous transformations of the orthonormal basis $\mathfrak{Э}_i$ of the space $E^q_{2\nu}$; at some reflection transformations, components of the spinors E, β, Π may change their sign depending on the particular class of spinor representations.

Evidently, the matrices $E\gamma^{i_1 i_2 \ldots i_k}$ and $\beta\gamma^{i_1 i_2 \ldots i_k}$ also form the components of spintensors which are invariant at least under the continuous transformations of bases of the space $E^q_{2\nu}$:

$$E\gamma^{i_1 i_2 \ldots i_k} = \|\gamma_{BA}^{i_1 i_2 \ldots i_k}\|, \quad \beta\gamma^{i_1 i_2 \ldots i_k} = \|\gamma_{\dot{B}A}^{i_1 i_2 \ldots i_k}\|.$$

Using Eqs. (1.151) and (1.152) (or similar equations for other classes of spinor representations) and the transformation law for the components of spinors with the dotted indices, it is easy to obtain that the components ψ^{+B}, determined by the equality

$$\psi^{+B} = \Pi^B{}_{\dot{A}} \dot{\psi}^A, \tag{1.160}$$

are transformed due to transformations of the orthonormal basis $\mathbf{\ni}_i$ of the space $E^q_{2\nu}$ as contravariant components of a first-rank spinor. The spinor $\boldsymbol{\psi}^+ = \pm\psi^{+A}\boldsymbol{\varepsilon}_A$ with the contravariant components ψ^{+A} is called conjugate with respect to the spinor $\boldsymbol{\psi} = \pm\psi^A\boldsymbol{\varepsilon}_A$ with the contravariant components ψ^A, calculated in the same basis as ψ^{+A}.[10]

It is easy to calculate the conjugate spinor of a conjugate spinor $\boldsymbol{\psi}^+$. We have

$$(\psi^{+A})^+ = \Pi^A{}_{\dot{B}} \dot{\psi}^{+B} = \Pi^A{}_{\dot{B}} (\Pi^B{}_{\dot{C}} \dot{\psi}^C)^{\cdot} = \Pi^A{}_{\dot{B}} (\Pi^B{}_{\dot{C}})^{\cdot} \psi^C.$$

Taking into account relation (1.141), we find

$$(\psi^{+A})^+ = (-1)^{\frac{1}{2}(\nu-q)(\nu-q-1)}\psi^A.$$

Since the components ψ^A and $-\psi^A$ define the same spinor, the latter equality implies that the conjugate spinor of a conjugate spinor $\boldsymbol{\psi}$ coincides with the spinor $\boldsymbol{\psi}$.

Covariant components of the conjugate spinor ψ^+_A are determined using the metric spinor $E = \|e_{AB}\|$:

$$\psi^+_B = e_{BA}\psi^{+A} = e_{BA}\Pi^A{}_{\dot{C}}\dot{\psi}^C. \tag{1.161}$$

Bearing in mind definition (1.135), we can also rewrite Eq. (1.161) for the covariant components of a conjugate spinor in the form

$$\psi^+_B = \beta_{\dot{A}B}\dot{\psi}^A. \tag{1.162}$$

Denoting the row of the covariant components ψ^+_A by the symbol ψ^+, we can write definition (1.162) in a matrix form:

$$\psi^+ = \dot{\psi}^T\beta.$$

[10]The conjugate spinor may also be defined with the aid of the spinor β defined, instead of Eq. (1.127), by the equation

$$\dot{\gamma}_i^T = \beta\gamma_i\beta^{-1}. \tag{*}$$

In physical applications, in the four-dimensional space E^1_4 with the metric signature $(+, +, +, -)$, definition (1.127) is used; in the four-dimensional space E^3_4, with the metric signature $(-, -, -, +)$ (where the matrices γ_i are related to the matrices γ_i of the space E^1_4 by the factor i) one uses definition (*). It is easy to see that the matrices β, defined in the same space by Eqs. (1.127) and (*), differ by the factor $\gamma_{2\nu+1}$.

1.6.6 Connection Between Second-Rank Spinors and Tensors in an Even-Dimensional Pseudo-Euclidean Space $E^q_{2\nu}$

In the real space $E^q_{2\nu}$, the expansion of the components of the second-rank spinor ψ^{BA} in invariant spintensors E^{-1} and $\gamma_{i_1 i_2 \cdots i_k} E^{-1}$ has the form

$$\psi^{BA} = \frac{1}{2^\nu}\left[(-1)^{\frac{1}{2}\nu(\nu+1)} F e^{BA} + \sum_{k=1}^{2\nu} \frac{1}{k!} F^{i_1 i_2 \cdots i_k} \gamma^{BA}_{i_1 i_2 \cdots i_k}\right], \tag{1.163}$$

where the quantities F, $F^{i_1 i_2 \ldots i_k}$ are defined by the relations

$$\begin{aligned} F &= e_{BA} \psi^{BA}, \\ F^{i_1 i_2 \cdots i_k} &= (-1)^k \gamma^{i_1 i_2 \cdots i_k}_{BA} \psi^{BA}. \end{aligned} \tag{1.164}$$

Formulae (1.164) for F, $F^{i_1 i_2 \ldots i_k}$ are obtained by contracting equality (1.163) with components e_{BA} and $\gamma^{i_1 i_2 \cdots i_k}_{BA}$ with respect to the indices B, A and takes into account relations (1.123).

At continuous transformations of the basis $\mathfrak{Э}_i$ in the space $E^q_{2\nu}$, the quantity F is invariant, while the quantities $F^{i_1 i_2 \cdots i_k}$ are transformed as components of an antisymmetric tensor of rank k. Transformations of the quantities F and $F^{i_1 i_2 \cdots i_k}$ under reflections of the vectors of the basis $\mathfrak{Э}_i$ depends on the adopted normalization of spinor transformations.

Due to completeness and linear independence of the set of matrices formed by components of the spintensors I, $\gamma_i, \ldots$ and $\gamma_{i_1 i_2 \ldots i_k}$ $(i_1 < i_2 < \cdots < i_k)$ and due to non-degeneracy of the matrix β, the set of matrices formed by components of the spintensors

$$\beta^{-1}, \quad \gamma_i \beta^{-1}, \quad \ldots, \quad \gamma_{i_1 i_2 \cdots i_{2\nu}} \beta^{-1} \qquad (i_1 < i_2 < \cdots < i_k)$$

is also complete and linearly independent. Therefore the components $\psi^{\dot{B}A}$ of a second-rank spinor with one dotted index in the pseudo-Euclidean space $E^q_{2\nu}$ may be expanded with respect to the set of invariant spintensors with components

$$\beta^{-1} = \|\beta^{A\dot{B}}\|, \qquad \gamma_{i_1 i_2 \cdots i_k} \beta^{-1} = \|\gamma^{A\dot{B}}_{i_1 i_2 \cdots i_k}\|.$$

The corresponding expansion may be written in the form

$$\psi^{\dot{B}A} = \frac{1}{2^\nu}\left(H\beta^{A\dot{B}} + \sum_{k=1}^{2\nu} \frac{\mathrm{i}^{\frac{1}{2}k(k-3)}}{k!} H^{i_1 i_2 \cdots i_k} \gamma^{A\dot{B}}_{i_1 i_2 \cdots i_k}\right). \tag{1.165}$$

Contracting equality (1.165) with components of the invariant spintensors $\beta_{\dot{B}A}$ and $\gamma_{\dot{B}A}^{\ i_1 i_2 \cdots i_k} = \beta_{\dot{B}C}\gamma^{C}{}_{A}{}^{i_1 i_2 \cdots i_k}$, we find for the components of the tensors H and $H^{i_1 i_2 \cdots i_k}$:

$$
\begin{aligned}
H &= \beta_{\dot{B}A}\psi^{\dot{B}A}, \\
H^{i_1 i_2 \cdots i_k} &= \mathrm{i}^{\frac{1}{2}k(k+1)}\gamma_{\dot{B}A}^{\ i_1 i_2 \cdots i_k}\psi^{\dot{B}A}.
\end{aligned} \tag{1.166}
$$

In the derivation of Eqs. (1.166), one should take into account the identities

$$
\begin{aligned}
\beta_{\dot{B}A}\beta^{A\dot{B}} &= 2^{\nu}, \\
\beta_{\dot{B}A}\gamma_{i_1 i_2 \cdots i_k}^{A\dot{B}} &= \mathrm{tr}\left(\beta\gamma_{i_1 i_2 \cdots i_k}\beta^{-1}\right) = \mathrm{tr}\left(\gamma_{i_1 i_2 \cdots i_k}\right) = 0, \\
\gamma_{\dot{B}A}^{\ i_1 i_2 \cdots i_k}\gamma_{j_1 j_2 \cdots j_m}^{A\dot{B}} &= \mathrm{tr}\left(\gamma^{i_1 i_2 \cdots i_k}\gamma_{j_1 j_2 \cdots j_m}\right) = 0, \quad \text{if} \quad k \neq m, \\
\gamma_{\dot{B}A}^{\ i_1 i_2 \cdots i_k}\gamma_{j_1 j_2 \cdots j_k}^{A\dot{B}} &= \mathrm{tr}\left(\beta\gamma^{i_1 i_2 \cdots i_k}\gamma_{j_1 j_2 \cdots j_k}\beta^{-1}\right) = \mathrm{tr}\left(\gamma^{i_1 i_2 \cdots i_k}\gamma_{j_1 j_2 \cdots j_k}\right) \\
&= (-1)^{\frac{1}{2}k(k-1)}2^{\nu}k\,!\,\delta_{[j_1}^{[i_1}\delta_{j_2}^{i_2}\cdots\delta_{j_k]}^{i_k]},
\end{aligned}
$$

which follow from the definition of the components of the invariant spintensors $\beta_{\dot{B}A}$, $\beta^{A\dot{B}}$, $\gamma_{\dot{B}A}^{\ i_1 i_2 \cdots i_k}$, $\gamma_{j_1 j_2 \cdots j_k}^{A\dot{B}}$ and from relations (1.123).

If the matrix of components of the second-rank spinor $\psi^{\dot{B}A}$ is Hermitian,

$$
\left(\psi^{\dot{B}A}\right)^{\cdot} = \psi^{\dot{A}B},
$$

then, from the Hermitian properties (1.131) and (1.132) of the matrices of components of the spintensors $\beta_{\dot{B}A}$ and $\gamma_{\dot{B}A}^{\ i_1 i_2 \cdots i_k}$ it follows that the components of the tensors H, $H^{i_1 i_2 \cdots i_k}$ are real.

1.7 Semispinors in Even-Dimensional Real Spaces

Consider an even-dimensional real vector pseudo-Euclidean space $E_{2\nu}^{q}$ referred to an orthonormal basis $\mathfrak{Э}_i$. Semispinors in the space $E_{2\nu}^{q}$ are defined by the equalities

$$
\psi = \gamma_{2\nu+1}\psi \quad \text{or} \quad \psi = -\gamma_{2\nu+1}\psi, \tag{1.167}
$$

where $\gamma_{2\nu+1}$ is defined by the second equation in (1.145).

It is easy to see that the matrix $\gamma_{2\nu+1}$ coincides with the matrix $\overset{\circ}{\gamma}_{2\nu+1}$ defined by Eq. (1.96). Therefore Eqs. (1.167) are identical to Eqs. (1.97) which define semispinors in the complex Euclidean space $E_{2\nu}^{+}$.

Taking the Hermitian conjugate of Eqs. (1.167) and multiplying the result by the matrix of components of the invariant spinor β, using the equality

$$\dot{\gamma}^T_{2\nu+1} = (-1)^q \beta \gamma_{2\nu+1} \beta^{-1}$$

that follows from (1.128), we obtain that the covariant components of the conjugate semispinor ψ^+ in the space $E^q_{2\nu}$ satisfy the equation

$$(-1)^q \psi^+ = \pm \psi^+ \gamma_{2\nu+1}. \tag{1.168}$$

Here, the upper sign corresponds to the first equation in (1.167) and the lower sign to the second equation in (1.167).

Using the components ψ of an arbitrary spinor in the space $E^q_{2\nu}$, one can define two semispinors with components $\psi_{(I)}$ nd $\psi_{(II)}$, calculated in the same basis as ψ:

$$\psi_{(I)} = \frac{1}{2}(I + \gamma_{2\nu+1})\psi, \qquad \psi_{(II)} = \frac{1}{2}(I - \gamma_{2\nu+1})\psi.$$

For the covariant components of the conjugate semispinors $\psi^+_{(I)}$ and $\psi^+_{(II)}$, we have

$$\psi^+_{(I)} = \frac{1}{2}\psi^+\big[I + (-1)^q \gamma_{2\nu+1}\big], \qquad \psi^+_{(II)} = \frac{1}{2}\psi^+\big[I - (-1)^q \gamma_{2\nu+1}\big].$$

Let us introduce, in the spinor space, the spinbasis $\overset{*}{\boldsymbol{\varepsilon}}_A$ in which the components of the spintensors γ_i are defined by the matrices (we assume $q < 2\nu$)

$$\gamma_{2\nu} = \begin{Vmatrix} 0 & I \\ I & 0 \end{Vmatrix}, \qquad \gamma_\alpha = \begin{Vmatrix} 0 & -\mathrm{i}\sigma_\alpha \\ \mathrm{i}\sigma_\alpha & 0 \end{Vmatrix}, \qquad \alpha = 1, 2, \ldots, 2\nu - 1,$$

where σ_α ($\alpha = 1, 2, \ldots, 2\nu - 1$) are matrices of the order $2^{\nu-1}$ related to the matrices $\overset{\circ}{\sigma}_\alpha$ in (1.99) by the equalities

$$\sigma_1 = \mathrm{i}\overset{\circ}{\sigma}_1, \quad \sigma_2 = \mathrm{i}\overset{\circ}{\sigma}_2, \quad \ldots, \quad \sigma_q = \mathrm{i}\overset{\circ}{\sigma}_q,$$

$$\sigma_{q+1} = \overset{\circ}{\sigma}_{q+1}, \quad \sigma_{q+2} = \overset{\circ}{\sigma}_{q+2}, \quad \ldots, \quad \sigma_{2\nu-1} = \overset{\circ}{\sigma}_{2\nu-1}.$$

Evidently, the matrices σ_α satisfy the equations

$$\sigma_\alpha \sigma_\beta + \sigma_\beta \sigma_\alpha = 2 g_{\alpha\beta} I.$$

The components of the metric spinor E in the spinbasis $\overset{*}{\boldsymbol{\varepsilon}}_A$ coincide with those of the metric spinor E defined by relations (1.100)–(1.103). The components of the invariant spinor β in the spinbasis $\overset{*}{\boldsymbol{\varepsilon}}_A$ for even q are specified, according to definition (1.127), by the Hermitian matrix

$$\beta = \begin{Vmatrix} \widetilde{\beta} & 0 \\ 0 & -\widetilde{\beta} \end{Vmatrix}, \tag{1.169}$$

where $\widetilde{\beta}$ satisfies the equations

$$\dot{\sigma}_\alpha^T = \widetilde{\beta}\sigma_\alpha\widetilde{\beta}^{-1}, \qquad \alpha = 1, 2, \ldots, 2\nu - 1.$$

For odd q, the Hermitian matrix β has the form

$$\beta = \begin{Vmatrix} 0 & -\mathrm{i}\widetilde{\beta} \\ \mathrm{i}\widetilde{\beta} & 0 \end{Vmatrix}, \tag{1.170}$$

where $\widetilde{\beta}$ satisfies the equations

$$\dot{\sigma}_\alpha^T = -\widetilde{\beta}\sigma_\alpha\widetilde{\beta}^{-1}, \qquad \alpha = 1, 2, \ldots, 2\nu - 1.$$

From the Hermitianity of the matrix β it follows that the matrix $\widetilde{\beta}$ is also Hermitian. Using definitions (1.133) and (1.134) for β, it is easy to find that for even and odd q the matrix $\widetilde{\beta}$ may be defined as

$$\widetilde{\beta} = \mathrm{i}^{\frac{1}{2}q(q+1)}\sigma_{[1}\sigma_2\cdots\sigma_{q]}.$$

Let us further consider only such orthogonal transformations of bases in the space $E_{2\nu}^q$ which may be obtained in a continuous way from the identical transformation. At such transformations of bases in the space $E_{2\nu}^q$, the corresponding spinor transformations in the spinbasis $\overset{*}{\boldsymbol{\varepsilon}}_A$ have the form

$$S = \begin{Vmatrix} A & 0 \\ 0 & D \end{Vmatrix}. \tag{1.171}$$

Let us write down Eqs. (1.143) defining the spinor transformations S in the spinbasis $\overset{*}{\boldsymbol{\varepsilon}}_A$:

$$\begin{aligned} -\mathrm{i}l^\alpha{}_{2\nu}\sigma_\alpha + l^{2\nu}{}_{2\nu}I &= A^{-1}D, \\ l^\alpha{}_\beta\sigma_\alpha + \mathrm{i}l^{2\nu}{}_\beta I &= A^{-1}\sigma_\beta D. \end{aligned} \tag{1.172}$$

For the class of transformations $l^j{}_i$ under consideration, all normalizations (1.144) are equivalent. Let us write out the normalization conditions (1.144) for even ν:

$$\varepsilon = A^T\varepsilon A, \quad \varepsilon = D^T\varepsilon D.$$

For odd ν:

$$\varepsilon = A^T \varepsilon D.$$

Equations (1.146), connected with the real character of the coefficients $l^j{}_i$, for the considered transformations of bases in the space $E^q_{2\nu}$ have the form

$$\dot{S}^T \beta S = \beta.$$

These equations, in the spinbasis $\overset{*}{\boldsymbol{\varepsilon}}_A$, taking into account the above definition of β, may be written for even q in the following way:

$$\dot{A}^T \widetilde{\beta} A = \widetilde{\beta}, \qquad \dot{D}^T \widetilde{\beta} D = \widetilde{\beta}.$$

For odd q:

$$\dot{A}^T \widetilde{\beta} D = \widetilde{\beta}.$$

It is easy to see that, in the spinbasis $\overset{*}{\boldsymbol{\varepsilon}}_A$, the components of the second-rank spinor $\gamma_{2\nu+1}$ are represented by the matrix

$$\gamma_{2\nu+1} = \begin{Vmatrix} I & 0 \\ 0 & -I \end{Vmatrix}.$$

Using this expression for $\gamma_{2\nu+1}$, from definitions (1.167) we find that, in the spinbasis $\overset{*}{\boldsymbol{\varepsilon}}_A$, the contravariant components of the semispinors $\boldsymbol{\psi}_{(I)}$ and $\boldsymbol{\psi}_{(II)}$ are determined by the equalities

$$\psi_{(I)} = \begin{Vmatrix} \varphi \\ 0 \end{Vmatrix}, \qquad \psi_{(II)} = \begin{Vmatrix} 0 \\ \chi \end{Vmatrix},$$

in which φ is a column of the contravariant components of the spinor ψ^A ($A = 1, 2, \ldots, 2^{\nu-1}$), χ is a column of the contravariant components of ψ^A ($A = 1 + 2^{\nu-1}, 2 + 2^{\nu-1}, \ldots, 2^\nu$), and 0 is a column of $2^{\nu-1}$ zeros.

According to definition (1.169), for the covariant components of the conjugate semispinors $\overset{+}{\psi}_{(I)}$ and $\overset{+}{\psi}_{(II)}$ in the space $E^q_{2\nu}$, calculated in the spinbasis $\overset{*}{\boldsymbol{\varepsilon}}_A$, for even q we have

$$\overset{+}{\psi}_{(I)} = \dot{\psi}^T_{(I)} \beta = (\dot{\varphi}^T \widetilde{\beta},\ 0^T),$$
$$\overset{+}{\psi}_{(II)} = \dot{\psi}^T_{(II)} \beta = (0^T,\ -\dot{\chi}^T \widetilde{\beta}).$$

With definition (1.170), for odd q, for the components of $\psi^+_{(I)}$ and $\psi^+_{(II)}$ we find

$$\psi^+_{(I)} = \left(0^T, \ -\mathrm{i}\dot{\varphi}^T\widetilde{\beta}\right),$$
$$\psi^+_{(II)} = \left(\mathrm{i}\dot{\chi}^T\widetilde{\beta}, \ 0^T\right).$$

From relation (1.171) it follows that the spinor components φ and χ, under the considered transformations of the basis $\mathbf{\ni}_i$ in the space $E^q_{2\nu}$, are transformed separately:

$$\varphi' = A\varphi, \quad \chi' = D\chi.$$

If $l^{2\nu}{}_{2\nu} = 1$ and $l^{\alpha}{}_{2\nu} = 0$, it follows from (1.172) that $D = A$.

1.8 Spinors in Odd-Dimensional Euclidean Spaces

1.8.1 Spinor Representation of the Proper Complex Orthogonal Group

Consider the odd-dimensional complex Euclidean space $E^+_{2\nu-1}$ referred to the orthonormal vector basis $\mathbf{\ni}_i$. Let $\|l^j{}_i\|$ be a proper orthogonal transformation of vectors of the basis $\mathbf{\ni}_i$:

$$\mathbf{\ni}'_i = l^j{}_i\mathbf{\ni}_j, \qquad \det\|l^j{}_i\| = 1. \tag{1.173}$$

The spinor representation $SO^+_{2\nu-1} \to \{\pm S\}$ of the proper orthogonal group $SO^+_{2\nu-1}$ of transformations of the bases $\mathbf{\ni}_i$ in the space $E^+_{2\nu-1}$ is specified by the group $\{\pm S\}$ defined by the equations

$$l^j{}_i\overset{\circ}{\gamma}_j = S^{-1}\overset{\circ}{\gamma}_i S, \qquad S^T E S = E, \tag{1.174}$$

in which the indices i, j take all integer values from 1 to $2\nu - 1$, and the matrix $\overset{\circ}{\gamma}_{2\nu-1}$ is defined by the equality

$$\overset{\circ}{\gamma}_{2\nu-1} = \mathrm{i}^{\nu-1}\overset{\circ}{\gamma}_1\overset{\circ}{\gamma}_2\cdots\overset{\circ}{\gamma}_{2(\nu-1)},$$

while $\overset{\circ}{\gamma}_\alpha$ ($\alpha = 1, 2, \ldots, 2(\nu-1)$) are matrices of order $2^{\nu-1}$ satisfying the equation

$$\overset{\circ}{\gamma}_\alpha\overset{\circ}{\gamma}_\beta + \overset{\circ}{\gamma}_\beta\overset{\circ}{\gamma}_\alpha = 2\delta_{\alpha\beta}I.$$

The matrix E in Eqs. (1.174) is defined by the equations

$$(-1)^{\nu+1}\overset{\circ}{\gamma}{}_i^T = E\overset{\circ}{\gamma}_i E^{-1}. \tag{1.175}$$

The solvability of Eqs. (1.174) with respect to S follows from the subsequent reasoning.

A first-rank spinor in the odd-dimensional complex Euclidean space $E^+_{2\nu-1}$ is defined as an invariant geometric object of the form $\boldsymbol{\psi} = \pm\psi^A \boldsymbol{\varepsilon}_A$, where the pairs of contravariant components $\pm\psi^A$ and the spinbasis $\pm\{\boldsymbol{\varepsilon}_A\}$ are related to some orthonormal basis $\mathfrak{Э}_i$ of the space $E^+_{2\nu-1}$, and is transformed under the orthogonal transformation (1.173) of the bases $\mathfrak{Э}_i$ according to the relations

$$(\pm\psi^B)' = \pm S^B{}_A\psi^A, \qquad (\pm\boldsymbol{\varepsilon}_B)' = \pm Z^A{}_B\boldsymbol{\varepsilon}_A,$$

in which the spinor transformations $S = \|S^B{}_A\|$ and $S^{-1} = \|Z^A{}_B\|$ are defined by Eqs. (1.174).

The spinor index juggling is carried out using the metric spinor whose component matrix E is defined by Eqs. (1.175).[11]

If, in Eq. (1.175), the matrices $\overset{\circ}{\gamma}_i$ for $i = 1, 2, \ldots, \nu-1$ are taken to be symmetric and for $i = \nu, \nu+1, \ldots, 2(\nu-1)$ to be antisymmetric, then, for any ν, the metric spinor E, defined by Eq. (1.175), may be specified by the matrix

$$E = \overset{\circ}{\gamma}_\nu \overset{\circ}{\gamma}_{\nu+1} \cdots \overset{\circ}{\gamma}_{2(\nu-1)}. \tag{1.176}$$

A direct inspection using the above symmetry properties of $\overset{\circ}{\gamma}_i$ in definition (1.176) shows that the components matrix of the spinor E possesses the following symmetry properties:

$$E^T = (-1)^{\frac{1}{2}\nu(\nu-1)}E. \tag{1.177}$$

Performing calculations similar to those of Sect. 1.1 (see page 18), one can find that, due to Eqs. (1.175), the following equations also hold:

$$\left(\overset{\circ}{\gamma}_{i_1 i_2 \ldots i_k}\right)^T = (-1)^{k\nu+\frac{1}{2}k(k+1)} E\overset{\circ}{\gamma}_{i_1 i_2 \ldots i_k} E^{-1}, \tag{1.178}$$

[11] One could express the idea to define the metric spinor in the space $E^+_{2\nu-1}$ as in the space $E^+_{2\nu}$, by Eqs. (1.44), in which $i = 1, 2, \ldots, 2\nu-1$. However, for odd ν, Eq. (1.44) for matrices of order $2^{\nu-1}$ with $i = 1, 2, \ldots, 2\nu-1$ has no solution since the matrix E of order $2^{\nu-1}$ is defined by Eq. (1.44) for $i = 1, 2, \ldots, 2(\nu-1)$ up to a factor and is connected with $\overset{\circ}{\gamma}_{2\nu-1}$ by the equation $\overset{\circ}{\gamma}{}^T_{2\nu-1} = E\overset{\circ}{\gamma}_{2\nu-1}E^{-1}$, which is easily obtained by contracting equations (1.54) for $k = 2(\nu-1)$ with components of the Levi-Civita pseudotensor $\varepsilon^{i_1 i_2 \ldots i_k}$ with respect to the indices $i_1, i_2, \ldots, i_k$.

in which $\overset{\circ}{\gamma}_{i_1 i_2 \ldots i_k} = \overset{\circ}{\gamma}_{[i_1} \overset{\circ}{\gamma}_{i_2} \cdots \overset{\circ}{\gamma}_{i_k]}$. Equations (1.178) imply the symmetry properties of the spintensors $E\overset{\circ}{\gamma}_{i_1 i_2 \ldots i_k}$:

$$\left(E\overset{\circ}{\gamma}_{i_1 i_2 \ldots i_k}\right)^T = (-1)^{k\nu + \frac{1}{2}[\nu(\nu-1)+k(k+1)]} E\overset{\circ}{\gamma}_{i_1 i_2 \ldots i_k}. \tag{1.179}$$

The above symmetry properties of the matrices E and $E\overset{\circ}{\gamma}_{i_1 i_2 \ldots i_k}$ are independent of the specific choice of the matrices $\overset{\circ}{\gamma}_i$.

The above definition of the spinor representation and spinors in the odd-dimensional complex Euclidean space $E^+_{2\nu-1}$ may be obtained in the following way. The space $E^+_{2\nu-1}$ may be considered as a subspace in the complex Euclidean space $E^+_{2\nu}$, orthogonal to the basis vector $\mathcal{Э}_{2\nu}$ in $E^+_{2\nu}$. Then the group of orthogonal transformations of bases $\mathcal{Э}_i$ in the space $E^+_{2\nu-1}$ is isomorphic to the subgroup of the group of orthogonal transformations of bases in the space $E^+_{2\nu}$, which is singled out by the conditions

$$l^{2\nu}{}_{2\nu} = 1, \quad l^{\alpha}{}_{2\nu} = 0.$$

At proper orthogonal transformations of bases $\mathcal{Э}_i$ of the Euclidean space $E^+_{2\nu}$, for which all these conditions are valid, the spinor transformations S, calculated in the spinbasis $\overset{*}{\boldsymbol{\varepsilon}}_A$ according to Eqs. (1.105) and (1.110), have the form

$$S = \begin{Vmatrix} A & 0 \\ 0 & A \end{Vmatrix},$$

where the matrix A satisfies Eqs. (1.111) and the normalization conditions (1.108). Equations (1.111) and (1.108) for $D = A$, up to notations ($\overset{\circ}{\gamma}_i \rightarrow \overset{\circ}{\sigma}_i$, $E \rightarrow \varepsilon$, $S \rightarrow A$), coincide with Eqs. (1.174). Thus the components of a first-rank spinor in the space $E^+_{2\nu-1}$ are transformed as nonzero components of semispinors in the space $E^+_{2\nu}$ in the spinbasis $\overset{*}{\boldsymbol{\varepsilon}}_A$ under proper orthogonal transformations of the subspace orthogonal to the basis vector $\mathcal{Э}_{2\nu}$ in $E^+_{2\nu-1}$. This implies the solvability of Eqs. (1.174).

1.8.2 Spinor Representation of the Full Orthogonal Group

The spinor representation of the full orthogonal group $O^+_{2\nu-1}$ of transformations of the bases $\mathcal{Э}_i$ of the odd-dimensional complex Euclidean space $E^+_{2\nu-1}$ may be defined by the equations[12]

$$l^j{}_i \overset{\circ}{\gamma}_j = S^{-1} \overset{\circ}{\gamma}_i S \operatorname{sign} \Delta, \qquad S^T E S = E, \tag{1.180}$$

or by the equations

$$l^j{}_i \overset{\circ}{\gamma}_j = S^{-1} \overset{\circ}{\gamma}_i S \operatorname{sign} \Delta, \qquad S^T E S = E \operatorname{sign} \Delta. \tag{1.181}$$

Under the proper orthogonal transformations $l^j{}_i$, Eqs. (1.180) and (1.181) are identical to Eqs. (1.174). Let us calculate the spinor transformations S, corresponding to the full reflection transformation $\mathcal{Э}'_i = -\mathcal{Э}_i$ for $i = 1, 2, \ldots, 2\nu - 1$. Equations (1.180) for S give in this case

$$\overset{\circ}{\gamma}_i = S^{-1} \overset{\circ}{\gamma}_i S, \qquad S^T E S = E. \tag{1.182}$$

It is easy to see that Eqs. (1.182) have the solution

$$S = \pm I. \tag{1.183}$$

Since any improper orthogonal transformation may be represented in the form of a product of some proper orthogonal transformation and a full reflection transformation, from (1.183) it follows that the set of spinor transformations defined by Eqs. (1.180) that corresponds to all improper orthogonal transformations coincides with the group of spinor transformations, corresponding to the proper orthogonal group $SO^+_{2\nu-1}$. Thus the spinor representation of the full orthogonal group $O^+_{2\nu-1}$, defined by Eqs. (1.180), is not exact.

Equations (1.181), which, for the full reflection transformation $\mathcal{Э}'_i = -\mathcal{Э}_i$, are written in the form

$$\overset{\circ}{\gamma}_i = S^{-1} \overset{\circ}{\gamma}_i S, \qquad S^T E S = -E,$$

define the spinor transformation $S = \pm \mathrm{i} I$.

[12]The equations $l^j{}_i \overset{\circ}{\gamma}_j = S^{-1} \overset{\circ}{\gamma}_i S$ have no solution for S on the full orthogonal group $O^+_{2\nu-1}$. Indeed, for instance, for the reflection transformation of a single vector from the basis $\mathcal{Э}'_{2\nu-1} = -\mathcal{Э}_{2\nu-1}$, these equations give

$$S \overset{\circ}{\gamma}_{2\nu-1} = -\overset{\circ}{\gamma}_{2\nu-1} S, \qquad S \overset{\circ}{\gamma}_\alpha = \overset{\circ}{\gamma}_\alpha S, \qquad \alpha = 1, 2, \ldots, 2(\nu - 1). \tag{*}$$

From the second equation in (*) it follows that S is proportional to the unit matrix $S = \lambda I$, which, for $\lambda \neq 0$, contradicts the first equation.

1.8.3 Connection Between Second-Rank Spinors and Tensors in an Odd-Dimensional Space $E^{+}_{2\nu-1}$

According to Eqs. (1.24) and (1.25), a second-rank spinor in the space $E^{+}_{2\nu-1}$ is equivalent to a system of tensors, consisting of a vector and odd-rank antisymmetric tensors

$$\boldsymbol{F} = \left\{F^{i},\ F^{i_1 i_2 i_3}, \ldots,\ F^{i_1 i_2 i_3 \ldots i_{2\nu-1}}\right\},$$

or to a system of tensors consisting of a scalar and even-rank anti-symmetric tensors:

$$\boldsymbol{F} = \left\{F,\ F^{i_1 i_2}, \ldots,\ F^{i_1 i_2 \ldots i_{2(\nu-1)}}\right\}.$$

Raising the index A in Eqs. (1.24) and (1.25) with the aid of the metric spinor E, one can write down a relationship between the contravariant components of the second-rank spinor ψ^{BA} and the even-rank tensors $\boldsymbol{F}$ in the form

$$\psi^{BA} = \frac{1}{2^{\nu-1}}\left[(-1)^{\frac{1}{2}\nu(\nu-1)} F e^{BA} + \sum_{k=1}^{\nu-1} \frac{1}{(2k)!} F^{i_1 i_2 \cdots i_{2k}} \overset{\circ}{\gamma}{}^{BA}_{i_1 i_2 \cdots i_{2k}}\right], \tag{1.184}$$

where

$$F = e_{BA}\psi^{BA} = \psi^{A}{}_{A},$$

$$F^{i_1 i_2 \cdots i_{2k}} = \overset{\circ}{\gamma}{}^{i_1 i_2 \cdots i_{2k}}_{BA}\psi^{BA} = (-1)^k \overset{\circ}{\gamma}{}^{A}{}_{B}{}^{i_1 i_2 \cdots i_{2k}} \psi^{B}{}_{A}.$$

A relationship between the components of a second-rank spinor and the odd-rank tensors $\boldsymbol{F}$ has the form

$$\psi^{BA} = \frac{1}{2^{\nu-1}} \sum_{k=0}^{\nu-1} \frac{1}{(2k+1)!} F^{i_1 i_2 \cdots i_{2k+1}} \overset{\circ}{\gamma}{}^{BA}_{i_1 i_2 \cdots i_{2k+1}}. \tag{1.185}$$

Here,

$$F^{i_1 i_2 \cdots i_{2k+1}} = (-1)^{\nu+1} \overset{\circ}{\gamma}{}^{i_1 i_2 \cdots i_{2k+1}}_{BA} \psi^{BA} = (-1)^k \overset{\circ}{\gamma}{}^{A}{}_{B}{}^{i_1 i_2 \cdots i_{2k+1}} \psi^{B}{}_{A}.$$

1.8.4 Spinors in Odd-Dimensional Pseudo-Euclidean Spaces

Let us now consider an odd-dimensional real pseudo-Euclidean space $E^{q}_{2\nu-1}$ of index q ($q < 2\nu - 1$), referred to an orthonormal basis $\mathbf{\Im}_i$, in which the first q

vectors are imaginary-unit, while the remaining $2\nu - q - 1$ vectors are unit:

$$(\mathfrak{I}_i, \mathfrak{I}_i) = -1 \quad \text{for} \quad i = 1, 2, \ldots, q,$$
$$(\mathfrak{I}_i, \mathfrak{I}_i) = 1 \quad \text{for} \quad i = q+1, q+2, \ldots, 2\nu - 1,$$
$$(\mathfrak{I}_i, \mathfrak{I}_j) = 0 \quad \text{for} \quad i \neq j.$$

A spinor representation of the group $SO^q_{2\nu-1}$ of proper pseudo-orthogonal transformations

$$\mathfrak{I}'_i = l^j{}_i \mathfrak{I}_j, \quad \det \|l^j{}_i\| = 1,$$

is defined by the equations

$$l^j{}_i \gamma_j = S^{-1} \gamma_i S, \qquad S^T E S = E,$$

in which the matrix of metric spinor components E is defined by Eqs. (1.175), while the matrices γ_j are connected with $\overset{\circ}{\gamma}_j$ by the relations

$$\gamma_j = \mathrm{i}\overset{\circ}{\gamma}_j \quad \text{for} \quad j = 1, 2, \ldots, q,$$
$$\gamma_j = \overset{\circ}{\gamma}_j \quad \text{for} \quad j = q+1, q+2, \ldots, 2\nu - 1.$$

The reality condition for the coefficients of the pseudo-orthogonal transformation $l^j{}_i$ leads to the existence, in $E^q_{2\nu-1}$, of a spinor β which is invariant with respect to proper pseudo-orthogonal transformations of the bases $\mathfrak{I}_i$:

$$\dot{S}^T \beta S = \beta,$$

The matrix of its components is determined by the equation

$$\dot{\gamma}_i^T = (-1)^q \beta \gamma_i \beta^{-1}.$$

In the spinbasis in which the matrices γ_j satisfy the conditions

$$\gamma_j^T = \gamma_j \quad \text{for} \quad j = 1, 2, \ldots, \nu - 1,$$
$$\gamma_j^T = -\gamma_j \quad \text{for} \quad j = \nu, \nu+1, \ldots, 2(\nu - 1),$$
$$\dot{\gamma}_j = -\gamma_j^T \quad \text{for} \quad j = 1, 2, \ldots, q,$$
$$\dot{\gamma}_j = \gamma_j^T \quad \text{for} \quad j = q+1, q+2, \ldots, 2(\nu - 1),$$

the component matrix of β may be determined from the relation

$$\beta = \mathrm{i}^{\frac{1}{2}q(q+1)}\gamma_{[1}\gamma_2\cdots\gamma_{q]}.$$

In space with a definite metric $E^0_{2\nu-1}$, if all matrices γ_i are Hermitian, $\dot{\gamma}_i^T = \gamma_i$, the equality $\gamma_i\beta = \beta\gamma_i$ holds, and β may be defined as the unit matrix, $\beta = I$.

Using the invariant spinor β in the space $E^q_{2\nu-1}$, the conjugate spinors are defined by their covariant components ψ^+_A, whose row is denoted by the symbol ψ^+:

$$\psi^+ = \dot{\psi}^T\beta. \tag{1.186}$$

From the results of Sect. 1.7 it follows that the components of conjugate spinors introduced in this way are transformed in $E^q_{2\nu-1}$ as nonzero components of conjugate semispinors in $E^q_{2\nu}$ in the spinbasis $\overset{*}{\boldsymbol{\varepsilon}}_A$ under proper orthogonal transformations of the basis in the subspace orthogonal to the basis vector $\boldsymbol{\vartheta}_{2\nu}$ in $E^q_{2\nu}$.

In the space $E^q_{2\nu-1}$, one can consider spinors with the components ψ^A, ψ_A, transformed with the aid of the matrices S and S^{-1}, and spinors defined by the components $\psi^{\dot{A}}$, $\psi_{\dot{A}}$ with dotted indices, transformed with the aid of the complex-conjugate matrices $\dot{S}$, $\dot{S}^{-1}$. A second-rank spinor in the space $E^q_{2\nu-1}$, with components $\psi^{\dot{B}A}$, having one dotted index, may be expanded in the system of invariant spintensors $\gamma_{i_1i_2\ldots i_{2k}}\beta^{-1}$ with an even number of indices:

$$\psi^{\dot{B}A} = \frac{1}{2^{\nu-1}}\left(H\beta^{A\dot{B}} + \sum_{k=1}^{\nu-1}\frac{\mathrm{i}^{k(2k+1)}}{(2k)\,!}H^{i_1i_2\cdots i_{2k}}\gamma^{A\dot{B}}_{i_1i_2\cdots i_{2k}}\right), \tag{1.187}$$

or in the system of invariant spintensors $\gamma_{i_1i_2\ldots i_{2k+1}}\beta^{-1}$ with an odd number of indices:

$$\psi^{\dot{B}A} = \frac{1}{2^{\nu-1}}\sum_{k=0}^{\nu-1}\frac{\mathrm{i}^{(k-1)(2k+1)}}{(2k+1)\,!}H^{i_1i_2\cdots i_{2k+1}}\gamma^{A\dot{B}}_{i_1i_2\cdots i_{2k+1}}. \tag{1.188}$$

The components of antisymmetric tensors $H^{i_1i_2\ldots i_n}$ and the invariant H in Eqs. (1.187) and (1.188) are defined by the relations

$$H = \beta_{\dot{B}A}\psi^{\dot{B}A}, \qquad H^{i_1i_2\ldots i_n} = \mathrm{i}^{\frac{1}{2}n(n+1)}\gamma^{i_1i_2\ldots i_n}_{\dot{B}A}\psi^{\dot{B}A},$$

obtained in the same way as in the corresponding formulae for the space $E^q_{2\nu}$.

1.9 Representation of Spinors by Complex Tensors

1.9.1 Equivalence of Geometric Objects in Euclidean Spaces

Let $O_n^+ \to S = \|S^{\mathcal{B}}{}_{\mathcal{A}}\|$ be a certain representation of the orthogonal group O_n^+ of transformations of orthonormal bases $\mathcal{Э}_i$ of the Euclidean space E_n^+, acting in the space of the variables $a^1, a^2, \ldots, a^N$. In the Euclidean space E_n^+, a geometric object is specified if, in each orthonormal basis $\mathcal{Э}_i$ of the space E_n^+, a system of components $a^1, a^2, \ldots, a^N$ is specified which are transformed with the aid of the group S in a transition from one orthonormal basis $\mathcal{Э}_i$ to another orthonormal basis $\mathcal{Э}'_i$ in the space E_n^+.

This means that if, in an orthonormal basis $\mathcal{Э}_i$ of the space E_n^+, the components $a^{\mathcal{B}}$ ($\mathcal{B} = 1, 2, \ldots, \mathrm{N}$) are specified, then, in the basis $\mathcal{Э}'_i$, obtained from $\mathcal{Э}_i$ by the orthogonal transformation

$$\mathcal{Э}'_i = l^j{}_i \mathcal{Э}_j, \quad \|l^j{}_i\| \in O_n^+,$$

the geometric object is defined by the components $(a^{\mathcal{B}})'$, calculated by the formula

$$(a^{\mathcal{B}})' = s^{\mathcal{B}}{}_{\mathcal{C}} a^{\mathcal{C}},$$

where the coefficients $\|S^{\mathcal{B}}{}_{\mathcal{C}}\|$ correspond to the transformation $\|l^j{}_i\|$ under the mapping $O_n^+ \to S$.

Definition Two geometric objects A_1 and A_2 in the Euclidean space E_n^+ are equivalent, $A_1 \sim A_2$, if, in each orthonormal basis $\mathcal{Э}_i$ of the space E_n^+, there exists a one-to-one correspondence $A_1 = f(A_2)$ between the components of the objects A_1 and A_2, which is invariant under the choice of the basis $\mathcal{Э}_i$.

Geometric objects and their equivalence are defined in pseudo-Euclidean spaces in a similar way.

Evidently, the above definition of equivalent geometric objects possesses the properties of reflexivity,

$$A \sim A,$$

symmetry,

$$A_1 \sim A_2 \Rightarrow A_2 \sim A_1,$$

and transitivity,

$$A_1 \sim A_2, \;\; A_2 \sim A_3 \Rightarrow A_1 \sim A_3.$$

Thus the definition introduced possesses all properties of the equivalence relation.

According to the definition, knowledge of the components of a geometric object in any orthonormal basis $\mathbf{\ni}_i$ in E_n^+ makes it possible to calculate the components of an equivalent geometric object. Therefore it makes quite the same effect whether one uses the components of an object A_1 or those of an object A_2 equivalent to A_1. Meanwhile, the components of two equivalent geometric objects, according to the above definition, may be transformed by essentially different, mutually non-equivalent representations of the transformation group of the bases $\mathbf{\ni}_i$.

As a simple example of equivalent geometric objects, one can take the second-rank spinor $\boldsymbol{\Psi}$ and the systems of tensors $\boldsymbol{F}$ in Euclidean spaces, whose connection between the components (which is one-to-one and, in this example, linear) is given by the equalities (1.94) and (1.95).

A classical and simple example of equivalent geometric objects in the sense of the above definition is given by an axial vector $a_i \mathbf{\ni}^i$ and an antisymmetric second-rank tensor in three-dimensional space $a^{ij} \mathbf{\ni}_i \mathbf{\ni}_j$ $(a^{ij} = -a^{ji})$, whose components in an orthonormal basis $\mathbf{\ni}_i$ are connected by the one-to-one relations

$$a_1 = a^{23}, \quad a_2 = a^{31}, \quad a_3 = a^{12},$$

which are invariant with respect to the choice of an orthonormal basis $\mathbf{\ni}_i$. Unlike the previous example, in this case, the components of the equivalent objects a_i, a^{ij} are transformed by essentially different (non-equivalent) representations $O_3 \to O_3$ and $O_3 \to O_3 \times O_3$.

1.9.2 Connection Between First- and Second-Rank Spinors

Consider an arbitrary square complex matrix $\|\psi^{BA}\|$ of order r. Evidently, if the matrix elements ψ^{BA} are represented in the form of products of complex numbers ψ^A,

$$\psi^{BA} = \psi^B \psi^A, \tag{1.189}$$

then ψ^{BA} satisfy the equations

$$\psi^{AB} = \psi^{BA}, \qquad \psi^{AB}\psi^{CD} = \psi^{BC}\psi^{DA} = \psi^{BD}\psi^{AC}. \tag{1.190}$$

Among Eqs. (1.190), the following $r(r-1)$ equations may be taken as independent ones:

$$\psi^{AB} = \psi^{BA}, \qquad \psi^{AA}\psi^{BC} = \psi^{AB}\psi^{AC} \qquad (\psi^{AA} \neq 0; \quad B, C \neq A).$$

Indeed, all Eqs. (1.190) may be obtained from these ones. Assuming $\psi^{EE} \neq 0$, we have

$$\psi^{AB}\psi^{CD} = \frac{\psi^{EA}\psi^{EB}}{\psi^{EE}}\frac{\psi^{EC}\psi^{ED}}{\psi^{EE}} = \frac{\psi^{EA}\psi^{EC}}{\psi^{EE}}\frac{\psi^{EB}\psi^{ED}}{\psi^{EE}} = \psi^{AC}\psi^{BD},$$

$$\psi^{AB}\psi^{CD} = \frac{\psi^{EA}\psi^{EB}}{\psi^{EE}}\frac{\psi^{EC}\psi^{ED}}{\psi^{EE}} = \frac{\psi^{EA}\psi^{ED}}{\psi^{EE}}\frac{\psi^{EB}\psi^{EC}}{\psi^{EE}} = \psi^{DA}\psi^{BC}.$$

Reversely, if the matrix elements ψ^{BA} satisfy Eqs. (1.190), then there exists a set of r complex numbers ψ^{A}, defined up to a common sign, such that $\psi^{BA} = \psi^{B}\psi^{A}$. Indeed, if all matrix elements $\|\psi^{BA}\|$ are equal to zero, $\psi^{BA} = 0$, we put $\psi^{A} = 0$. If there is at least one nonzero matrix element $\|\psi^{BA}\|$, so that $\psi^{BA} \neq 0$, then we put

$$\psi^{A} = \frac{\psi^{BA}\eta_B}{\pm\sqrt{\psi^{CD}\eta_C\eta_D}}, \tag{1.191}$$

where η_C $(C = 1, 2,\ldots, r)$ are arbitrary, generally complex numbers satisfying the condition $\psi^{CD}\eta_C\eta_D \neq 0$. In Eq. (1.191), summing over the indices B, C, D is performed from 1 to r.

It is easy to see that if ψ^{BA} in Eq. (1.191) satisfy Eqs. (1.190), then the quantities ψ^{A} defined by Eq. (1.191) satisfy (1.189).

Due to Eqs. (1.189), definition (1.191) of the quantities ψ^{A} is independent of the choice of the numbers η_C. Indeed, if the inequalities $\psi^{CD}\eta_C\eta_D \neq 0$ and $\psi^{CD}\eta^*_C\eta^*_D \neq 0$ hold, then, using Eqs. (1.190), we obtain

$$\psi^{A} = \frac{\psi^{BA}\eta_B}{\pm\sqrt{\psi^{CD}\eta_C\eta_D}} = \frac{\psi^{MN}\eta^*_M\eta^*_N \cdot \psi^{BA}\eta_B}{\pm\sqrt{\left(\psi^{MN}\eta^*_M\eta^*_N\right)^2\psi^{CD}\eta_C\eta_D}}$$
$$= \frac{\psi^{NA}\eta^*_N \cdot \psi^{BM}\eta_B\eta^*_M}{\pm\sqrt{\psi^{MN}\eta^*_M\eta^*_N\left(\psi^{CD}\eta_C\eta^*_D\right)^2}} = \frac{\psi^{NA}\eta^*_N}{\pm\sqrt{\psi^{MN}\eta^*_M\eta^*_N}}.$$

If, in Eq. (1.191), we define the numbers η_C by the equality $\eta_C = \delta_{BC}$, where B is a fixed number and δ_{BC} are the Kronecker delta (i.e., all numbers η_C except one are zero, while the number η_B labeled B is equal to unity) and $\psi^{BB} \neq 0$, then Eq. (1.191) is rewritten in a simpler form,

$$\psi^{A} = \frac{\psi^{BA}}{\pm\sqrt{\psi^{BB}}}, \tag{1.192}$$

without summing over the index B. Due to independence of definition (1.191) of the choice of η_C, Eqs. (1.191) and (1.192) determine the same dependence ψ^{A} on ψ^{BA}.

It is easy to show that if the components ψ^{BA} in the equality (1.191) are transformed by the law

$$\psi^{BA} \to \left(\psi^{BA}\right)' = S^B{}_C S^A{}_M \psi^{CM}, \tag{1.193}$$

where $S^B{}_C$ define an arbitrary complex square nondegenerate matrix of order r, then the quantities ψ^A, defined according to (1.191), are transformed in the following way:

$$\psi^A \to \left(\psi^A\right)' = \pm S^A{}_B \psi^B. \tag{1.194}$$

Indeed, since the right-hand side of Eq. (1.191) is insensitive to the choice of the numbers η_B, in (1.191) one can put $\eta'_B = Z^A{}_B \eta_A$, where $Z^A{}_B$ are elements of the inverse matrix S^{-1}:

$$\psi^A = \frac{\psi^{BA}\eta_B}{\pm\sqrt{\psi^{CD}\eta_C\eta_D}} = \frac{\psi^{BA}\eta'_B}{\pm\sqrt{\psi^{CD}\eta'_C\eta'_D}}.$$

Using this equality, we find for the transformed quantities ψ^A:

$$(\psi^A)' = \frac{(\psi^{BA})'\eta'_B}{\pm\sqrt{(\psi^{CD})'\eta'_C\eta'_D}} = \frac{S^A{}_M \psi^{BM}\eta_B}{\pm\sqrt{\psi^{CD}\eta_C\eta_D}} = \pm S^A{}_M \psi^M.$$

Thus the geometric object, whose components ψ^{BA} satisfy Eqs. (1.190) and are transformed with the aid of the group $S \times S$ by the law (1.193), is equivalent to a two-valued object with components $\{\pm\psi^A\}$, transformed with the aid of the factor group $\{\pm S\} = S/\pm I$ by the law (1.194).

In particular, it follows from the above-said that, in Euclidean spaces, a second-rank spinor with components ψ^{BA}, satisfying Eqs. (1.190), is equivalent to a first-rank spinor with the components $\pm\psi^A$.[13]

1.9.3 *Equivalence of a First-Rank Spinor ψ to a Set of Complex Tensors C*

If, in an even-dimensional complex Euclidean space $E^+_{2\nu}$, the components of a second-rank spinor ψ^{BA} are not arbitrary but satisfy Eqs. (1.190) (and consequently are represented in the form of products of components of a first-rank spinor, $\psi^B\psi^A$),

[13] By definition, the components of the second-rank spinor ψ^{BA} are always defined with a sign, although they can be products of components of first-rank spinors defined up to a common sign.

then we will write Eq. (1.94) in the form

$$\psi^{BA} = \psi^B \psi^A = \frac{1}{2^\nu}\left(Ce^{BA} + \sum_{k=1}^{2\nu} \frac{(-1)^k}{k\,!} C^{i_1 i_2 \cdots i_k} \overset{\circ}{\gamma}{}^{BA}_{i_1 i_2 \cdots i_k}\right). \tag{1.195}$$

The same for the components of a spinor $\psi^{BA} = \psi^B \psi^A$ in an odd-dimensional complex Euclidean space $E^+_{2\nu+1}$:

$$\psi^{BA} = \psi^B \psi^A = \frac{1}{2^\nu}\left(Ce^{BA} + \sum_{k=1}^{\nu} \frac{1}{(2k)\,!} C^{i_1 i_2 \cdots i_{2k}} \overset{\circ}{\gamma}{}^{BA}_{i_1 i_2 \cdots i_{2k}}\right), \tag{1.196}$$

or

$$\psi^{BA} = \psi^B \psi^A = \frac{(-1)^\nu}{2^\nu} \sum_{k=o}^{\nu} \frac{1}{(2k+1)\,!} C^{i_1 i_2 \cdots i_{2k+1}} \overset{\circ}{\gamma}{}^{BA}_{i_1 i_2 \cdots i_{2k+1}}. \tag{1.197}$$

In equalities (1.195)–(1.197), the components of the tensors $\boldsymbol{C}$ are defined in the following way:

$$\begin{aligned} C &= e_{BA} \psi^B \psi^A = \psi^T E \psi, \\ C^{i_1 i_2 \cdots i_k} &= \overset{\circ}{\gamma}{}_{BA}^{i_1 i_2 \cdots i_k} \psi^B \psi^A = \psi^T E \overset{\circ}{\gamma}{}^{i_1 i_2 \cdots i_k} \psi. \end{aligned} \tag{1.198}$$

From the symmetry properties (1.56) of the components of E and $E\overset{\circ}{\gamma}{}^{i_1 i_2 \cdots i_k}$, for the tensors $\boldsymbol{C}$ in an even-dimensional space $E^+_{2\nu}$ it follows:

$$\begin{aligned} C &= 0, \quad \text{if} \quad \frac{1}{2}\nu(\nu+1) \quad \text{is odd}, \\ C^{i_1 i_2 \cdots i_k} &= 0, \quad \text{if} \quad \frac{1}{2}[\nu(\nu+1) + k(k+1)] \quad \text{is odd}. \end{aligned}$$

For the tensors $\boldsymbol{C}$ in an odd-dimensional space $E^+_{2\nu+1}$, on the basis of the symmetry properties (1.177) and (1.179), we have (the identities (1.177) and (1.179) are written for the space $E^+_{2\nu-1}$, therefore, in application to the present case, one should replace in them $\nu \to \nu + 1$)

$$\begin{aligned} C &= 0, \quad \text{if} \quad \frac{1}{2}\nu(\nu+1) \quad \text{is odd}, \\ C^{i_1 i_2 \cdots i_k} &= 0, \quad \text{if} \quad \frac{1}{2}[\nu(\nu+1) + k(k-1)] + k\nu \quad \text{is odd}. \end{aligned}$$

By virtue of definitions (1.198), the components of the tensors $\boldsymbol{C}$ satisfy $2^\nu(2^\nu - 1)$ independent bilinear equations. To obtain such equations in the even-dimensional space $E^+_{2\nu}$, we multiply Eqs. (1.195) written for the indices B, A and C, D:

$$2^{2\nu}\psi^{BA}\psi^{CD} = \sum_{m=0}^{2\nu}\sum_{k=0}^{2\nu}\frac{(-1)^{k+m}}{m\,!k\,!}C^{i_1 i_2\cdots i_k}C^{j_1 j_2\cdots j_m}\overset{\circ}{\gamma}{}^{BA}_{i_1 i_2\cdots i_k}\overset{\circ}{\gamma}{}^{CD}_{j_1 j_2\cdots j_m}. \tag{1.199}$$

Contracting equation (1.199) with respect to the indices A, B, C, D with components of the spintensors $\overset{\circ}{\gamma}{}^{n_1\ldots n_l}_{AC}\,\overset{\circ}{\gamma}{}^{s_1\ldots s_q}_{DB}$, we obtain:

$$2^{2\nu}C^{n_1\cdots n_l}C^{s_1\cdots s_q} = \sum_{m=0}^{2\nu}\sum_{k=0}^{2\nu}\frac{(-1)^{k+m}}{m\,!k\,!}C^{i_1\cdots i_k}C^{j_1\cdots j_m}$$
$$\times\ \mathrm{tr}\left(\overset{\circ}{\gamma}_{i_1\cdots i_k}\overset{\circ}{\gamma}{}^{n_1\cdots n_l}\overset{\circ}{\gamma}_{j_1\cdots j_m}\overset{\circ}{\gamma}{}^{s_1\cdots s_q}\right). \tag{1.200}$$

Here and henceforth, to simplify the notation, we omit the tensor indices with zero number, so that $C^{i_0} = C$, and it is supposed that $\overset{\circ}{\gamma}{}^{k_0} = I$ and $\overset{\circ}{\gamma}_{i_0} = I$. Let us note that Eqs. (1.200), connecting the components of the tensors $\boldsymbol{C}$, may also be obtained by contracting the identities (1.20) with the spinor components $\psi_M\psi_E\psi^D\psi^A$ with respect to the indices M, E, D, A.

In the same way one obtains bilinear equations connecting the components of even-rank tensors $\boldsymbol{C}$ in odd-dimensional spaces $E^+_{2\nu+1}$:

$$2^{2\nu}C^{n_1\cdots n_l}C^{s_1\cdots s_q} = \sum_{m=0}^{\nu}\sum_{k=0}^{\nu}\frac{1}{(2m)\,!\,(2k)\,!}C^{i_1\cdots i_{2k}}C^{j_1\cdots j_{2m}}$$
$$\times\ \mathrm{tr}\left(\overset{\circ}{\gamma}_{i_1\cdots i_{2k}}\overset{\circ}{\gamma}{}^{n_1\cdots n_l}\overset{\circ}{\gamma}_{j_1\cdots j_{2m}}\overset{\circ}{\gamma}{}^{s_1\cdots s_q}\right) \tag{1.201}$$

and equations for the components of odd-rank tensors $\boldsymbol{C}$ in odd-dimensional spaces $E^+_{2\nu+1}$:

$$2^{2\nu}C^{n_1\cdots n_l}C^{s_1\cdots s_q} = \sum_{m=0}^{\nu}\sum_{k=0}^{\nu}\frac{1}{(2m+1)\,!\,(2k+1)\,!}$$
$$\times\ C^{i_1\cdots i_{2k+1}}C^{j_1\cdots j_{2m+1}}\,\mathrm{tr}\left(\overset{\circ}{\gamma}_{i_1\cdots i_{2k+1}}\overset{\circ}{\gamma}{}^{n_1\cdots n_l}\overset{\circ}{\gamma}_{j_1\cdots j_{2m+1}}\overset{\circ}{\gamma}{}^{s_1\cdots s_q}\right). \tag{1.202}$$

Evidently, the set of tensors $\boldsymbol{C}$ in the even-dimensional space $E^+_{2\nu}$, satisfying Eqs. (1.200) (or Eqs. (1.201) or (1.202) in the odd-dimensional space $E^+_{2\nu+1}$), is equivalent to a second-rank spinor with components ψ^{BA} in the space $E^+_{2\nu}$ (or $E^+_{2\nu+1}$), satisfying Eqs. (1.190).

Thus a second-rank spinor with components ψ^{BA} satisfying Eqs. (1.190) is, on the one hand, equivalent to the set of tensors $\boldsymbol{C}$, and, on the other hand, it is equivalent to the first-rank spinor $\boldsymbol{\psi} = \pm\psi^A \boldsymbol{\varepsilon}_A$. Therefore, due to transitivity of the equivalence relation, the following theorem is valid.

Theorem **([75])** *The first-rank spinor* $\boldsymbol{\psi} = \pm\psi^A \boldsymbol{\varepsilon}_A$ *in the complex Euclidean space* $E^+_{2\nu}$ *(or in the space* $E^+_{2\nu+1}$*) with the components* ψ^A*, defined up to a common sign, is equivalent to a set of complex tensors* $\boldsymbol{C}$*, satisfying* $2^\nu(2^\nu - 1)$ *independent bilinear equations in* (1.200) *(or Eqs.* (1.201), (1.202) *in* $E^+_{2\nu+1}$*). The one-to-one relation between the components of the tensors* $\boldsymbol{C}$ *and those of the spinor* $\pm\psi^A$*, invariant under the choice of the orthonormal basis of Euclidean space, is performed, in the even-dimensional space* $E^+_{2\nu}$*, by Eqs.* (1.191), (1.195) *and* (1.198)*, and, in the odd-dimensional space* $E^+_{2\nu+1}$*, by Eqs.* (1.191), (1.196) *and* (1.198) *or by* (1.191), (1.197) *and* (1.198)*.*

In real pseudo-Euclidean spaces $E^q_{2\nu}$, the components of the tensors $\boldsymbol{C}$ corresponding to a second-rank spinor with the components ψ^{BA}, satisfying Eqs. (1.190), will be defined in an orthonormal basis $\boldsymbol{\partial}_i$ in the space $E^q_{2\nu}$ by Eqs. (1.198), in which the spintensors $\overset{\circ}{\gamma}{}^{i_1 i_2 \cdots i_m}$ are replaced by the spintensors $\gamma^{i_1 i_2 \cdots i_m} = \gamma^{[i_1}\gamma^{i_2} \ldots \gamma^{i_m]}$:

$$C = \psi^T E \psi,$$
$$C^{i_1 i_2 \cdots i_m} = \psi^T E \gamma^{i_1 i_2 \cdots i_m} \psi. \tag{1.203}$$

In this case, Eqs. (1.195)–(1.197) and (1.200)–(1.202) remain valid if the spintensors $\overset{\circ}{\gamma}{}^{i_1 i_2 \cdots i_m}$ are also replaced in them by the spintensors $\gamma^{i_1 i_2 \cdots i_m}$.

1.10 Representation of Spinors by Real Tensors

Consider an r-dimensional complex square matrix $\|\psi^{\dot{B}A}\|$. Suppose that the components $\psi^{\dot{B}A}$ can be presented in the form $\psi^{\dot{B}A} = \dot{\psi}^B \psi^A$. Then $\psi^{\dot{B}A}$ satisfy the equations

$$\left(\psi^{\dot{B}A}\right)^{\boldsymbol{\cdot}} = \psi^{\dot{A}B}, \qquad \psi^{\dot{A}B}\psi^{\dot{C}D} = \psi^{\dot{A}D}\psi^{\dot{C}B}, \tag{1.204}$$

and the inequality $\psi^{\dot{A}A} \geqslant 0$ is also valid. Among Eqs. (1.204), the following $(r-1)^2$ equations (counting complex equations twice) are independent:

$$\left(\psi^{\dot{B}A}\right)^{\boldsymbol{\cdot}} = \psi^{\dot{A}B}, \qquad \psi^{\dot{A}A}\psi^{\dot{C}D} = \psi^{\dot{A}D}\psi^{\dot{C}A}. \tag{1.205}$$

Indeed, all Eqs. (1.204) follow from (1.205):

$$\psi^{\dot{A}B}\psi^{\dot{C}D} = \frac{\psi^{\dot{A}E}\psi^{\dot{E}B}}{\psi^{\dot{E}E}}\frac{\psi^{\dot{C}E}\psi^{\dot{E}D}}{\psi^{\dot{E}E}} = \frac{\psi^{\dot{A}D}\psi^{\dot{E}E}}{\psi^{\dot{E}E}}\frac{\psi^{\dot{C}B}\psi^{\dot{E}E}}{\psi^{\dot{E}E}} = \psi^{\dot{A}D}\psi^{\dot{C}B}.$$

If Eqs. (1.205) hold, then, from the condition $\psi^{\dot{A}A} \geqslant 0$, for some particular value of the index A, it follows that, for any complex numbers η_C ($C = 1, 2, \ldots r$), the following condition is valid $\psi^{\dot{C}D}\dot{\eta}_C\eta_D \geqslant 0$. Indeed, multiplying the second equation (1.205) by the numbers η_D and $\dot{\eta}_C$ and then summing over the indices C and D from 1 to r, we obtain

$$\psi^{\dot{A}A}\psi^{\dot{C}D}\dot{\eta}_C\eta_D = \psi^{\dot{A}D}\eta_D\psi^{\dot{C}A}\dot{\eta}_C.$$

Substituting, in the right-hand side of this equality, the quantities $\psi^{\dot{C}A}$ according to the first equation (1.205), we find:

$$\psi^{\dot{A}A}\psi^{\dot{C}D}\dot{\eta}_C\eta_D = \psi^{\dot{A}D}\eta_D(\psi^{\dot{A}C}\eta_C)^{\cdot} \geqslant 0.$$

Hence it is evident that if $\psi^{\dot{A}A} \geqslant 0$, then also $\psi^{\dot{C}D}\dot{\eta}_C\eta_D \geqslant 0$.

Evidently, if the components ψ^A define the matrix $\|\psi^{\dot{B}A}\|$, then the components $\psi^A \exp(\mathrm{i}\varphi)$ (φ is an arbitrary real number), and only these components, define the same matrix $\|\psi^{\dot{B}A}\|$.

Conversely, if some components $\psi^{\dot{B}A}$ ($\dot{B}, A = 1, 2, \ldots, r$) satisfy Eqs. (1.204), and also the inequality $\psi^{\dot{A}A} \geqslant 0$ holds, then there exists a set of components ψ^A, defined up to a common factor $\exp(\mathrm{i}\varphi)$, such that $\psi^{\dot{B}A} = \dot{\psi}^B\psi^A$.

Indeed, if all elements of the matrix $\|\psi^{\dot{B}A}\|$ are zero, we put $\psi^A = 0$.

If there is at least one nonzero element of the matrix $\|\psi^{\dot{B}A}\|$, i.e., $\psi^{\dot{B}A} \neq 0$, then we put

$$\psi^A = \frac{\psi^{\dot{B}A}\dot{\eta}_B}{\sqrt{\psi^{\dot{C}D}\dot{\eta}_C\eta_D}}, \tag{1.206}$$

where η_C ($C = 1, 2, \ldots, r$) are arbitrary complex numbers satisfying the condition $\psi^{\dot{C}D}\dot{\eta}_C\eta_D \neq 0$.

Due to Eqs. (1.204), the numbers ψ^A calculated by the formula (1.206), corresponding to any different numbers η_C, differ by the factor $\exp(\mathrm{i}\varphi)$, where φ is a real number. Indeed, assuming that $\psi^{\dot{C}D}\dot{\eta}_C\eta_D \neq 0$ and $\psi^{\dot{M}N}\dot{\eta}^*_M\eta^*_N \neq 0$ ($\eta_C \neq \eta^*_C$), we obtain

$$\begin{aligned}\psi^A &= \frac{\psi^{\dot{B}A}\dot{\eta}_B}{\sqrt{\psi^{\dot{C}D}\dot{\eta}_C\eta_D}} = \frac{\psi^{\dot{M}N}\dot{\eta}^*_M\eta^*_N\psi^{\dot{B}A}\dot{\eta}_B}{\sqrt{(\psi^{\dot{M}N}\dot{\eta}^*_M\eta^*_N)^2\psi^{\dot{C}D}\dot{\eta}_C\eta_D}} \\ &= \frac{\psi^{\dot{M}A}\dot{\eta}^*_M}{\sqrt{\psi^{\dot{M}N}\dot{\eta}^*_M\eta^*_N}}\frac{\psi^{\dot{B}N}\dot{\eta}_B\eta^*_N}{\sqrt{\psi^{\dot{M}N}\dot{\eta}^*_M\eta^*_N\psi^{\dot{C}D}\dot{\eta}_C\eta_D}}.\end{aligned} \tag{1.207}$$

Since the equality

$$\bmod \frac{\psi^{\dot{B}N}\dot{\eta}_B\eta_N^*}{\sqrt{\psi^{\dot{M}N}\dot{\eta}_M^*\eta_N^*\psi^{\dot{C}D}\dot{\eta}_C\eta_D}} = 1,$$

is valid, one can put

$$\frac{\psi^{\dot{B}N}\dot{\eta}_B\eta_N^*}{\sqrt{\psi^{\dot{M}N}\dot{\eta}_M^*\eta_N^*\psi^{\dot{C}D}\dot{\eta}_C\eta_D}} = \exp(\mathrm{i}\varphi),$$

and equality (1.207) can be continued:

$$\psi^A = \frac{\psi^{\dot{M}A}\dot{\eta}_M^*}{\sqrt{\psi^{\dot{M}N}\dot{\eta}_M^*\eta_N^*}}\exp(\mathrm{i}\varphi).$$

Evidently, the set $\{\psi^A\}$, corresponding to all numbers η_C in Eq. (1.206), has the form $\psi^A \exp \mathrm{i}\varphi$, where φ is an arbitrary real number. If, in Eq. (1.206), we put $\eta_C = \delta_{BC}\exp(\mathrm{i}\varphi)$, then Eq. (1.206) takes a form more convenient for a practical calculation of the components ψ^A, corresponding to given $\psi^{\dot{B}A}$:

$$\psi^A = \frac{\psi^{\dot{B}A}}{\sqrt{\psi^{\dot{B}B}}}\exp(\mathrm{i}\varphi).$$

Here, no summing is performed over the index B.

Evidently, in the transformation $\psi^{\dot{B}A} \to \dot{S}^B{}_C S^A{}_M \psi^{\dot{C}M}$, where the coefficients $S^A{}_M$ define a nondegenerate matrix, the whole set $\{\psi^A \exp(\mathrm{i}\varphi)\}$, corresponding to given $\psi^{\dot{B}A}$ and all φ by formula (1.206), is transformed as follows:

$$\{\psi^A \exp \mathrm{i}\varphi\} \to S^A{}_B\{\psi^B \exp(\mathrm{i}\varphi)\}.$$

Thus specifying the components of the object $\psi^{\dot{B}A}$ transformed with the aid of the group $\dot{S} \times S$, satisfying Eqs. (1.204) and the condition $\psi^{\dot{A}A} \geqslant 0$, and specifying the argument φ_0 of one of the components ψ^A entirely determine the components of the object ψ^A, transformed with the aid of the group S.

If the components $\psi^{\dot{B}A}$ define, in the pseudo-Euclidean space $E^q_{2\nu}$, a second-rank spinor with one dotted index and satisfy Eqs. (1.204) and the condition $\psi^{\dot{A}A} \geqslant 0$, then we will write Eq. (1.165) in the form

$$\psi^{\dot{B}A} = \frac{1}{2^\nu}\left(D\beta^{A\dot{B}} + \sum_{k=1}^{2\nu}\frac{1}{k!}\mathrm{i}^{\frac{1}{2}k(k-3)}D^{i_1 i_2\cdots i_k}\gamma^{A\dot{B}}_{i_1 i_2\cdots i_k}\right). \tag{1.208}$$

Also, in the odd-dimensional space $E^{(q)}_{2\nu+1}$, for the components $\psi^{\dot{B}A}$ we have

$$\psi^{\dot{B}A} = \frac{1}{2^\nu}\left(D\beta^{A\dot{B}} + \sum_{k=1}^{\nu} \frac{\mathrm{i}^{k(2k+1)}}{(2k)\,!} D^{i_1 i_2 \cdots i_{2k}} \gamma^{A\dot{B}}_{i_1 i_2 \cdots i_{2k}}\right), \tag{1.209}$$

or

$$\psi^{\dot{B}A} = \frac{1}{2^\nu} \sum_{k=0}^{\nu} \frac{\mathrm{i}^{(k-1)(2k+1)}}{(2k+1)\,!} D^{i_1 i_2 \cdots i_{2k+1}} \gamma^{A\dot{B}}_{i_1 i_2 \cdots i_{2k+1}}. \tag{1.210}$$

In Eqs. (1.208)–(1.210), the real components of the tensors $\boldsymbol{D}$ are defined by the relations

$$D = \beta_{\dot{B}A} \dot{\psi}^B \psi^A,$$

$$D^{i_1 i_2 \cdots i_k} = \mathrm{i}^{\frac{1}{2}k(k+1)} \gamma^{i_1 i_2 \cdots i_k}_{\dot{B}A} \dot{\psi}^B \psi^A. \tag{1.211}$$

It is also convenient to write down definitions (1.211) of the components of the tensors $\boldsymbol{D}$ in terms of the components of the conjugate spinor $\psi^+ = \|\psi^+_A\|$:

$$D = e_{BA} \psi^{+A} \psi^B = \psi^+ \psi, \tag{1.212}$$

$$D^{i_1 i_2 \ldots i_k} = \mathrm{i}^{\frac{1}{2}k(k+1)} \gamma^B{}_A{}^{i_1 i_2 \cdots i_k} \psi^+_B \psi^A = \mathrm{i}^{\frac{1}{2}k(k+1)} \psi^+ \gamma^{i_1 i_2 \cdots i_k} \psi.$$

If the contravariant components of the second-rank spinor $\psi^{\dot{B}A}$ satisfy Eqs. (1.204), then the corresponding components of the tensors $\boldsymbol{D}$ satisfy a set of real equations, which, in the space $E^q_{2\nu}$, may be obtained by multiplying Eqs. (1.208), written for the indices A, B and C, D, and by contracting the resulting equation with components of the spintensors $\gamma^{n_1 \cdots n_p} \gamma^{s_1 \cdots s_r}$:

$$2^{2\nu} D^{n_1 \cdots n_p} D^{s_1 \cdots s_r} = \sum_{k=0}^{2\nu} \sum_{m=0}^{2\nu} \frac{\mathrm{i}^{g_1}}{m\,!\,k\,!} D^{i_1 \cdots i_k} D^{j_1 \cdots j_m}$$
$$\times \operatorname{tr}\left(\gamma_{i_1 \cdots i_k} \gamma^{n_1 \cdots n_p} \gamma_{j_1 \cdots j_m} \gamma^{s_1 \cdots s_r}\right). \tag{1.213}$$

Here,

$$g_1 = \frac{1}{2}[p(p+1) + r(r+1) + m(m-3) + k(k-3)].$$

In a similar way one can also obtain bilinear equations for the components of the tensors $\boldsymbol{D}$ in odd-dimensional spaces $E^q_{2\nu+1}$.

It is clear from the above-said that specifying the real tensors $\boldsymbol{D}$ satisfying Eqs. (1.213) and the argument φ_0 of one of the components ψ^A determines a spinor entirely. Hence it follows that spinor equations may be equivalently written in terms of the components of the tensors $\boldsymbol{D}$ and φ_0. Excluding φ_0 from these equations, one can obtain a closed set of equations in terms of the components of real tensors $\boldsymbol{D}$.

The components of the tensors $\boldsymbol{C}$ and $\boldsymbol{D}$ in the pseudo-Euclidean space $E^q_{2\nu}$ are also connected by crossed equations, which are obtained by multiplying (1.195), written for the indices A and B, by Eq. (1.208), written for the indices C and D, and by contracting the result obtained with the components of the spintensors $\boldsymbol{\gamma}$:

$$2^{2\nu} C^{s_1\cdots s_p} D^{n_1\cdots n_l} = \sum_{k=0}^{2\nu}\sum_{m=0}^{2\nu} \frac{\mathrm{i}^{g_2}}{m\,!k\,!} C^{i_1\cdots i_k} D^{j_1\cdots j_m} \times \operatorname{tr}\left(\gamma_{i_1\cdots i_k}\gamma^{s_1\cdots s_p}\gamma_{j_1\cdots j_m}\gamma^{n_1\cdots n_l}\right).$$

Here, the coefficient g_2 is defined by the relation

$$g_2 = \frac{1}{2}[m(m-3) + l(l+1)] + k(k+3).$$

Equations of another type are obtained by multiplying Eqs. (1.195) by complex-conjugate equations (1.195) and subsequent contraction with components of the spintensors $\boldsymbol{\gamma}$:

$$2^{2\nu} D^{s_1\cdots s_p} D^{n_1\cdots n_l} = \sum_{k=0}^{2\nu}\sum_{m=0}^{2\nu} \frac{\mathrm{i}^{g_3}}{m\,!k\,!} C^{i_1\cdots i_k} \dot{C}^{j_1\cdots j_m} \times \operatorname{tr}\left(\gamma_{i_1\cdots i_k}\gamma^{n_1\cdots n_l}\gamma_{j_1\cdots j_m}\gamma^{s_1\cdots s_p}\right). \tag{1.214}$$

Here,

$$g_3 = \frac{1}{2}[p(p+1) + 3l(l+1)] + \nu(\nu+1) + 2(k+m) + q(q+1) + 2\nu(\nu-q).$$

Equations (1.214) determine the components of the real tensors $\boldsymbol{D}$ in terms of the components of the complex tensors $\boldsymbol{C}$.

To calculate the components of the spinor corresponding to the tensors $\boldsymbol{C}$ and $\boldsymbol{D}$, besides Eqs. (1.191) and (1.206), one can also use the equation

$$\psi^A = \frac{\psi^{\dot{B}A}}{\pm\sqrt{\left(\psi^{BB}\right)^{\cdot}}}, \quad \text{or} \quad \psi^A = \frac{\psi^{\dot{B}A}\dot{\eta}_B}{\sqrt{\left(\psi^{CD}\eta_C\eta_D\right)^{\cdot}}} \tag{1.215}$$

and relations (1.195)–(1.197) and (1.208)–(1.210), determining the components of the second-rank spinor ψ^{BA}, $\psi^{\dot{B}A}$ in terms of the components of the tensors $\boldsymbol{C}$ and $\boldsymbol{D}$.

Due to Eqs. (1.190) and (1.204), the right-hand side of the first formula in (1.215) does not depend on the fixed value of the index B (on the quantities η_C in the second formula (1.215)).

1.11 Tensor Representation of Semispinors in Euclidean Spaces

Let us first consider semispinors in the even-dimensional complex Euclidean space $E^+_{2\nu}$. Evidently, relations (1.191), (1.195) and (1.198) which realize the one-to-one connection between the components of the tensors $\boldsymbol{C}$ and those of the spinor ψ^A, are also valid in the case that the components ψ^A define a semispinor in the space $E^+_{2\nu}$. However, the tensors $\boldsymbol{C}$, corresponding to semispinors, have a specific form, satisfying some additional linear equations.

To find these equations, let us replace in Eqs. (1.198) the components of the spinor ψ^A according to Eq. (1.97)[14] (to simplify the formulae, we do the calculations in a matrix form):

$$C^{i_1 i_2 \cdots i_k} = \psi^T E \overset{\circ}{\gamma}{}^{i_1 i_2 \cdots i_k} \psi = \pm \psi^T E \overset{\circ}{\gamma}{}^{i_1 i_2 \cdots i_k} \overset{\circ}{\gamma}_{2\nu+1} \psi. \tag{1.216}$$

To calculate the products of the matrices $\overset{\circ}{\gamma}{}^{i_1 i_2 \cdots i_k} \overset{\circ}{\gamma}_{2\nu+1}$ in this equation, we contract the identities (1.16e) and (1.16f) with components of the Levi-Civita pseudotensor with respect to the indices $i_1 i_2 \cdots i_{2\nu+1}$. We obtain:

$$\overset{\circ}{\gamma}{}^{i_1 i_2 \cdots i_k} \overset{\circ}{\gamma}_{2\nu+1} = (-1)^{\frac{1}{2}k(k-1)} \frac{\mathrm{i}^\nu}{(2\nu-k)\,!} \overset{\circ}{\varepsilon}{}^{i_1 \ldots i_k i_{k+1} \ldots i_{2\nu}} \overset{\circ}{\gamma}_{i_{k+1} \ldots i_{2\nu}},$$

$$\overset{\circ}{\gamma}_{2\nu+1} \overset{\circ}{\gamma}{}^{i_1 i_2 \cdots i_k} = (-1)^{\frac{1}{2}k(k+1)} \frac{\mathrm{i}^\nu}{(2\nu-k)\,!} \overset{\circ}{\varepsilon}{}^{i_1 \ldots i_k i_{k+1} \ldots i_{2\nu}} \overset{\circ}{\gamma}_{i_{k+1} \ldots i_{2\nu}}. \tag{1.217}$$

Substituting, in Eq. (1.216), the product of the spintensors $\boldsymbol{\gamma}$ according to the first equality in (1.217), we continue Eq. (1.216):

$$\begin{aligned} C^{i_1 i_2 \cdots i_k} &= \pm(-1)^{\frac{1}{2}k(k-1)} \frac{\mathrm{i}^\nu}{(2\nu-k)\,!} \overset{\circ}{\varepsilon}{}^{i_1 \ldots i_k i_{k+1} \ldots i_{2\nu}} \psi^T E \overset{\circ}{\gamma}_{i_{k+1} \ldots i_{2\nu}} \psi \\ &= \pm(-1)^{\frac{1}{2}k(k-1)} \frac{\mathrm{i}^\nu}{(2\nu-k)\,!} \overset{\circ}{\varepsilon}{}^{i_1 \ldots i_k i_{k+1} \ldots i_{2\nu}} C_{i_{k+1} \ldots i_{2\nu}}. \end{aligned}$$

[14]Here and in the subsequent equations, the upper sign corresponds to the first equation (1.97) and the lower sign to the second equation (1.97).

Thus the components of the tensors $\boldsymbol{C}$ corresponding to semispinors in the complex Euclidean space $E_{2\nu}^{+}$ satisfy the linear equation

$$C^{i_1 i_2 \cdots i_k} = \pm \frac{\mathrm{i}^{\nu+k(k+1)}}{(2\nu-k)\,!} \overset{\circ}{\varepsilon}^{i_1 \ldots i_k i_{k+1} \ldots i_{2\nu}} C_{i_{k+1} \ldots i_{2\nu}}. \tag{1.218}$$

In the left-hand side of this equation, the components $C^{i_1 i_2 \cdots i_k}$ are equal to zero if the number $1/2[\nu(\nu+1)+k(k+1)]$ is odd. The rank of the tensor with components $C_{i_{k+1} \ldots i_{2\nu}}$ is equal to $2\nu - k$, therefore the components $C_{i_{k+1} \ldots i_{2\nu}}$ in the right-hand side of Eq. (1.218) are zero if the number $1/2[\nu(\nu+1)+(2\nu-k)(2\nu-k+1)] = 1/2[\nu(\nu-1)+k(k-1)]$ is odd. Thus for the tensors $\boldsymbol{C}$, defined by semispinors in the space $E_{2\nu}^{+}$, we have

$$C^{i_1 i_2 \cdots i_k} = 0, \quad \text{if one of the numbers} \quad \frac{1}{2}[\nu(\nu+1)+k(k+1)],$$
$$\frac{1}{2}[\nu(\nu-1)+k(k-1)] \quad \text{is odd.}$$

Now consider semispinors in the real pseudo-Euclidean space $E_{2\nu}^{q}$. The complex tensors $\boldsymbol{C}$, defined by semispinors in the space $E_{2\nu}^{q}$, satisfy the same equations as those for semispinors in the complex space $E_{2\nu}^{+}$. The real tensors $\boldsymbol{D}$, defined by semispinors in the space $E_{2\nu}^{q}$, also satisfy additional linear equations. Let us find these equations.

Substituting, in definition (1.212) for the component of the tensors $\boldsymbol{D}$, the components of the spinor ψ according to Eq. (1.167) and the components of the conjugate spinor ψ^{+} according to Eq. (1.168), we obtain:

$$D^{i_1 i_2 \ldots i_k} = \mathrm{i}^{\frac{1}{2}k(k+1)} \psi^{+} \gamma^{i_1 i_2 \cdots i_k} \psi = (-1)^q \, \mathrm{i}^{\frac{1}{2}k(k+1)} \psi^{+} \gamma_{2\nu+1} \gamma^{i_1 i_2 \cdots i_k} \gamma_{2\nu+1} \psi.$$

It is easy to see that, from the identities (1.217) and from the definition of the component matrix of the spintensor $\overset{\circ}{\gamma}_{2\nu+1} = \gamma_{2\nu+1}$, it follows

$$\gamma_{2\nu+1} \gamma^{i_1 i_2 \cdots i_k} \gamma_{2\nu+1} = (-1)^k \gamma^{i_1 i_2 \cdots i_k}.$$

Taking into account the latter identity, we find for the components of the tensor $\boldsymbol{D}$:

$$D^{i_1 i_2 \ldots i_k} = (-1)^{k+q} \mathrm{i}^{\frac{1}{2}k(k+1)} \psi^{+} \gamma^{i_1 i_2 \cdots i_k} \psi = (-1)^{k+q} \, D^{i_1 i_2 \ldots i_k}.$$

Hence it follows that the components of the real tensors $D^{i_1 i_2 \ldots i_k}$ defined by semispinors are equal to zero if the number $k+q$ is odd.

Let us now substitute the components of the spinor ψ in the definition (1.212) according to Eq. (1.167):

$$D^{i_1 i_2 \ldots i_k} = \pm \mathrm{i}^{\frac{1}{2}k(k+1)} \psi^+ \gamma^{i_1 i_2 \cdots i_k} \gamma_{2\nu+1} \psi.$$

Taking into account the first identity (1.217), we obtain for $D^{i_1 i_2 \ldots i_k}$:

$$\begin{aligned} D^{i_1 i_2 \ldots i_k} &= \pm(-1)^{\frac{1}{2}k(k-1)} \frac{\mathrm{i}^{\nu-q+\frac{1}{2}k(k+1)}}{(2\nu-k)\,!} \varepsilon^{i_1 \ldots i_k i_{k+1} \ldots i_{2\nu}} \psi^+ \gamma_{i_{k+1} \ldots i_{2\nu}} \psi \\ &= \pm \frac{1}{(2\nu-k)\,!} (-1)^{\frac{1}{2}(k^2-q)+\nu(k+1)} \varepsilon^{i_1 \ldots i_k i_{k+1} \ldots i_{2\nu}} D_{i_{k+1} \ldots i_{2\nu}}. \end{aligned}$$

Thus the components of the real tensors $\boldsymbol{D}$ corresponding to semispinors satisfy the following additional linear equations:

$$\begin{aligned} D^{i_1 i_2 \ldots i_k} &= 0, \quad \text{if the number} \quad k+q \quad \text{is odd,} \\ D^{i_1 i_2 \ldots i_k} &= \pm \frac{1}{(2\nu-k)\,!} (-1)^{\frac{1}{2}(k^2-q)+\nu(k+1)} \varepsilon^{i_1 \ldots i_k i_{k+1} \ldots i_{2\nu}} D_{i_{k+1} \ldots i_{2\nu}}. \end{aligned}$$

1.12 Representation of Two Spinors by Sets of Tensors

If the components ψ^{BA} of the complex square matrix $\|\psi^{BA}\|$ of order r are represented as products, $\psi^{BA} = \chi^B \psi^A$, then it is easy to see that ψ^{BA} satisfy the equations

$$\psi^{AB} \psi^{CD} = \psi^{BC} \psi^{DA}, \tag{1.219}$$

among which there are $(r-1)^2$ independent equations. If $\psi^{AA} \neq 0$, then the following equations in (1.219) are independent:

$$\psi^{AA} \psi^{CD} = \psi^{AC} \psi^{DA} \quad \text{for} \quad C \neq A, \quad A \neq D.$$

Conversely, if Eqs. (1.219) hold, there is a set of components χ^B, defined up to simultaneous multiplication of all components χ^B by an arbitrary nonzero complex number α, and a set of components ψ^A, defined up to simultaneous multiplication of all components ψ^A by $1/\alpha$, such that the equalities $\psi^{BA} = \chi^B \psi^A$ hold. The components ψ^A and χ^B may be defined by the equalities

$$\psi^A = \alpha\frac{\psi^{BA}\eta_B}{\sqrt{\psi^{CD}\eta_C\mu_D}},$$
$$\chi^B = \frac{1}{\alpha}\frac{\psi^{BA}\mu_A}{\sqrt{\psi^{CD}\eta_C\mu_D}}, \tag{1.220}$$

where α is an arbitrary nonzero complex number; η_C, μ_D $(C,\ D = 1,\ 2, \ldots, r)$ are arbitrary, in general complex numbers satisfying the condition

$$\eta_C\mu_D\psi^{CD} \neq 0. \tag{1.221}$$

In relations (1.220) and (1.221), summing from 1 to r by coinciding indices is assumed.

It is easy to show that if the quantities ψ^{BA} satisfy Eqs. (1.219), then the sets $\{\psi^A\}$ and $\{\chi^B\}$ as a whole (corresponding to all numbers α) do not depend on the choice of the numbers η_C and μ_D.

If ψ^A and χ^B are fixed, the corresponding number α is determined by the relation

$$\alpha = \frac{\psi^A\mu_A}{\sqrt{\psi^{CD}\eta_C\mu_D}}.$$

Assuming that, in Eqs. (1.220), the numbers η_C and μ_D are given by the equalities $\eta_C = \delta_{CM}$ and $\mu_D = \delta_{DN}$, Eqs. (1.220) may be written in the form

$$\psi^A = \alpha\frac{\psi^{MA}}{\sqrt{\psi^{MN}}}, \quad \chi^B = \frac{1}{\alpha}\frac{\psi^{BN}}{\sqrt{\psi^{MN}}}. \tag{1.222}$$

The sets $\{\psi^A\}$ and $\{\chi^B\}$, defined by the equalities (1.222), do not depend on the values of the fixed indices M and N if the quantities ψ^{BA} satisfy Eqs. (1.219).

If $r = 2^\nu$, then, for the quantities ψ^{BA} satisfying Eqs. (1.219), according to (1.94), in the even-dimensional space $E^+_{2\nu}$ one can write

$$\psi^{BA} = \frac{1}{2^\nu}\left[(-1)^{\frac{1}{2}\nu(\nu+1)}Ke^{BA} + \sum_{m=1}^{2\nu}\frac{1}{m\,!}K^{i_1i_2\cdots i_m}\overset{\circ}{\gamma}{}^{BA}_{i_1i_2\cdots i_m}\right]. \tag{1.223}$$

In Eqs. (1.223), the components $K^{i_1i_2\cdots i_m}$, which are antisymmetric over all indices $i_1i_2\cdots i_m$, are defined by the equalities

$$K = e_{BA}\chi^B\psi^A = \chi^A\psi_A,$$
$$K^{i_1i_2\cdots i_m} = (-1)^m\overset{\circ}{\gamma}{}_{BA}^{i_1i_2\cdots i_m}\chi^B\psi^A = (-1)^{\frac{1}{2}m(m-1)}\overset{\circ}{\gamma}{}^A{}_B{}^{i_1i_2\cdots i_m}\chi^B\psi_A. \tag{1.224}$$

By virtue of definitions (1.224), the components of the tensors $\boldsymbol{K}$ satisfy the following bilinear equations:

$$2^{2\nu} K^{n_1\cdots n_p} K^{s_1\cdots s_q} = (-1)^{\frac{1}{2}[p(p-1)+q(q-1)]} \sum_{k=0}^{2\nu} \sum_{m=0}^{2\nu} \frac{1}{m\,!\,k\,!} K^{i_1\cdots i_k} K^{j_1\cdots j_m}$$
$$\times \operatorname{tr}\left(\gamma_{i_1\cdots i_k} \gamma^{n_1\cdots n_p} \gamma_{j_1\cdots j_m} \gamma^{s_1\cdots s_q}\right). \qquad (1.225)$$

Similar relations can be easily written also in odd-dimensional spaces.

Evidently, specifying the set of components of K satisfying Eqs. (1.225), is equivalent to specifying the quantities ψ^{BA} satisfying Eq. (1.219). It means that specifying the set of components of K satisfying Eqs. (1.225), determines the set of components χ^B and ψ^A up to multiplying all components χ^B and ψ^A by an arbitrary complex number according to Eqs. (1.220).

If the components χ^B and ψ^A define two spinors in Euclidean space, then the quantities $K^{i_1 i_2\cdots i_m}$ are the components of an antisymmetric tensor of rank m, while K is an invariant (at least under proper transformations of orthonormal bases of Euclidean space). Evidently, if the relative sign of the components of the spinors χ and ψ is not fixed, the quantities K are determined up to a common sign.

Chapter 2
Spinor Fields in a Riemannian Space

2.1 Riemannian Space

2.1.1 Basic Definitions

Let us recall the basic elementary data concerning Riemannian spaces. The material to be presented in this section is of auxiliary nature, and proofs of the relations appearing here can be found in known textbooks on Riemannian geometry (see, e.g., [16, 55, 60]).

Consider a Riemannian space V_n of dimension n with the metric tensor $\boldsymbol{g}$, referred to the coordinate system x with the variables x^i $(i = 1, 2, \ldots, n)$. We denote by g_{ij} and g^{ij} the covariant and contravariant components of the metric tensor of the space V_n specified in the coordinate system x. Let us introduce, at each point of V_n, the Euclidean tangent vector space E_n, and let us choose in E_n a local covariant vector basis $\mathcal{Э}_i$, tangent to the coordinate lines of x^i.

Let L be a curve in the Riemannian space V_n, defined by the parametric equations $x^i = x^i(s)$, where $x^i(s)$ are continuously differentiable functions whose derivatives do not vanish simultaneously,

$$\left(\frac{dx^1}{ds}\right)^2 + \left(\frac{dx^2}{ds}\right)^2 + \cdots + \left(\frac{dx^n}{ds}\right)^2 \neq 0$$

(this inequality is the condition that L does not contain singular points). As is well-known, parallel transport of vectors of the basis $\mathcal{Э}_i$ along the curve L may be defined in a Riemannian space V_n by an equality of the form

$$d\mathcal{Э}_i = \Gamma^k_{ij}\mathcal{Э}_k dx^j, \tag{2.1}$$

where $d\mathcal{Э}_i$ is the differential of the basis vectors in the course of their parallel transport from a point with the coordinates x^i to a point with the coordinates $x^i + dx^i$

V. A. Zhelnorovich, *Theory of Spinors and Its Application in Physics and Mechanics*, https://doi.org/10.1007/978-3-030-27836-6_2

on L; Γ^k_{ij} are the connection coefficients (Christoffel symbols), defined in terms of the metric tensor components,

$$\Gamma^k_{ij} = \frac{1}{2} g^{ks}(-\partial_s g_{ij} + \partial_i g_{sj} + \partial_j g_{si}), \tag{2.2}$$

where $\partial_i = \partial/\partial x^i$ is the symbol of a partial derivative with respect to the variables x^i.

In a Riemannian space, the differential $d\mathbf{\Im}_i$, defined by relation (2.1), is not holonomic; in curvilinear coordinate systems in Euclidean spaces, the basis vectors $\mathbf{\Im}_i$ may be presented as derivatives of the radius vector, $\mathbf{\Im}_i = \partial_i \boldsymbol{r}$. Therefore, in this case, the differential $d\mathbf{\Im}_i$ is holonomic and may be presented in the form $d\mathbf{\Im}_i = dx^j \partial_j \mathbf{\Im}_i = dx^j \partial_j \partial_i \boldsymbol{r}$.

The coefficients Γ^k_{ij}, defined by equality (2.2), are symmetric in their lower indices: $\Gamma^k_{ij} = \Gamma^k_{ji}$.

Let us note the following relations for contractions of the Christoffel symbols with the metric tensor components which directly follow from definition (2.2):

$$\Gamma^j_{ij} = \frac{1}{2g} \partial_i g,$$

$$g^{ij} \Gamma^k_{ij} = -\frac{1}{\sqrt{|g|}} \partial_j \left(\sqrt{|g|}\, g^{jk}\right). \tag{2.3}$$

Here, g is the determinant of the covariant components of the metric tensor, $g = \det \|g_{ij}\|$.

When the variables of the coordinate system are transformed, $x^i \to y^i(x^j)$, the Christoffel symbols are transformed in the following way:

$$\left(\Gamma^k_{ij}\right)' = \frac{\partial y^k}{\partial x^m} \frac{\partial x^l}{\partial y^i} \frac{\partial x^n}{\partial y^j} \Gamma^m_{ln} + \frac{\partial y^k}{\partial x^m} \frac{\partial x^m}{\partial y^i \partial y^j}.$$

Thus the transformation law for the Christoffel symbols is not tensor.

With the aid of the Christoffel symbols Γ^k_{ij}, one defines covariant derivatives of tensor fields in V_n. For a tensor field with components $\mu^{j_1 \dots j_q}_{i_1 \dots i_p}$, with p covariant indices and q contravariant indices, the covariant derivative is defined by the equality

$$\begin{aligned} \nabla_k \mu^{j_1 \dots j_q}_{i_1 \dots i_p} &= \partial_k \mu^{j_1 \dots j_q}_{i_1 \dots i_p} \\ &+ \Gamma^{j_1}_{ks} \mu^{s j_2 \dots j_q}_{i_1 \dots i_p} + \sum_{\nu=2}^{q-1} \Gamma^{j_\nu}_{ks} \mu^{j_1 \dots j_{\nu-1} s j_{\nu+1} \dots j_q}_{i_1 \dots i_p} + \Gamma^{j_q}_{ks} \mu^{j_1 \dots j_{q-1} s}_{i_1 \dots i_p} \\ &- \Gamma^s_{k i_1} \mu^{j_1 \dots j_q}_{s i_2 \dots i_p} - \sum_{\nu=2}^{p-1} \Gamma^s_{k i_\nu} \mu^{j_1 \dots j_q}_{i_1 \dots i_{\nu-1} s i_{\nu+1} \dots i_p} - \Gamma^s_{k i_p} \mu^{j_1 \dots j_q}_{i_1 \dots i_{p-1} s}. \end{aligned} \tag{2.4}$$

In particular, for contravariant and covariant components of a vector we have

$$\nabla_k \mu^j = \partial_k \mu^j + \Gamma^j_{ks} \mu^s,$$
$$\nabla_k \mu_j = \partial_k \mu_j - \Gamma^s_{kj} \mu_s.$$

Denoting the components of a tensor of arbitrary rank by the symbol $\mu^{\mathcal{A}}$ ($\mathcal{A} = 1, 2, \ldots, N$ is a generalized index), we can rewrite Eq. (2.4) for the covariant derivative of $\mu^{\mathcal{A}}$ in the form

$$\nabla_k \mu^{\mathcal{A}} = \partial_k \mu^{\mathcal{A}} + F^{\mathcal{A}i}_{\mathcal{B}j} \mu^{\mathcal{B}} \Gamma^j_{ki}. \tag{2.5}$$

The components of the tensor $F^{\mathcal{A}i}_{\mathcal{B}j}$, appearing in this relation, represent a sum of different products of Kronecker deltas. A definition of the components $F^{\mathcal{A}i}_{\mathcal{B}j}$ is evident from a comparison between Eqs. (2.4) and (2.5).

Unlike partial derivatives, an alternated product of covariant derivatives of a tensor field in a Riemannian space is, in general, not zero. The following relation is valid:

$$\nabla_i \nabla_j \mu^{\mathcal{A}} - \nabla_j \nabla_i \mu^{\mathcal{A}} = -F^{\mathcal{A}k}_{\mathcal{B}s} \mu^{\mathcal{B}} R_{ijk}{}^s, \tag{2.6}$$

where the components $F^{\mathcal{A}i}_{\mathcal{B}j}$ are the same as in Eqs. (2.5), while the components $R_{ijk}{}^s$ are defined by the equality

$$R_{ijk}{}^s = \partial_j \Gamma^s_{ik} + \Gamma^s_{jp} \Gamma^p_{ik} - \partial_i \Gamma^s_{jk} - \Gamma^s_{ip} \Gamma^p_{jk}. \tag{2.7}$$

For covariant and contravariant components of a vector, Eq. (2.6) takes the form

$$\begin{aligned} (\nabla_i \nabla_j - \nabla_j \nabla_i) \mu^s &= -R_{ijk}{}^s \mu^k, \\ (\nabla_i \nabla_j - \nabla_j \nabla_i) \mu_k &= R_{ijk}{}^s \mu_s. \end{aligned} \tag{2.8}$$

The components $R_{ijk}{}^s$, related to a basis $\mathbf{Э}_i$, define in Riemannian space a fourth-rank tensor called the curvature tensor of the Riemannian space. The components R_{ik} and R defined by

$$R_{ik} = R_{ijk}{}^j, \qquad R = g^{ik} R_{ik} \tag{2.9}$$

define the Ricci tensor and the scalar curvature of the Riemannian space, respectively.[1]

[1]Some authors use another definition of the Ricci tensor components, $R_{jk} = R_{ijk}{}^i$, which differs in sign from definition (2.9) adopted here.

Directly from definition (2.7), it follows that the covariant components of the curvature tensor $R_{ijkm} = g_{sm} R_{ijk}{}^{s}$ possess the following simple symmetry properties:

$$R_{ijkm} = R_{kmij}, \quad R_{[ijk]m} = 0,$$
$$R_{ijkm} = -R_{ijmk}, \quad R_{ijkm} = -R_{jikm}.$$

It is easy to count that, in a Riemannian space V_n, the number of independent components of the curvature tensor is equal to $\frac{1}{12}n^2(n^2 - 1)$; in the case of a four-dimensional Riemannian space V_4, important for applications, the number of independent components of the curvature tensor is 20.

Due to definition (2.7), the curvature tensor components also satisfy the Bianchi differential identity

$$\nabla_m R_{ijk}{}^{s} + \nabla_i R_{jmk}{}^{s} + \nabla_j R_{mik}{}^{s} = 0. \tag{2.10}$$

A contraction of the Bianchi identity (2.10) with the tensor components $\delta^i_s g^{km}$ with respect to the indices s, i, k, m leads to the following important identity:

$$\nabla_j \left(R_i{}^j - \frac{1}{2} R \delta_i^j \right) = 0.$$

2.1.2 Lie Derivatives

Consider a domain V_0 of a Riemannian space V_n, and let the differentiable functions

$$x^i = f^i(x_0^j), \quad \det \left\| \frac{\partial x^i}{\partial x_0^j} \right\| \neq 0, \tag{2.11}$$

realize a one-to-one, continuous correspondence between points of the domain V_0 and points of some domain V in the space V_n. Under the transformation (2.11), each point M_0 of the domain V_0 is put into correspondence with a point M of the domain V, and the coordinate lines x^i in V_0 pass into certain lines x'^i in the domain V, to be considered as coordinate lines of the coordinate system x' with the covariant vector basis $\mathbf{\Im}'_i$ in V. We thus have, in the domain V, two coordinate systems, x and x'.

By definition of the coordinate system x', the coordinates of point M with respect to x' are the coordinates of the preimage of M (i.e., point M_0) with respect to the system x. Let x^i be the coordinates of point M in system x, x'^i the coordinates of point M in system x', and consequently, according to (2.11), the coordinates x^i and x'^i of point M are connected by the equality $x^i = f^i(x'^j)$, while the bases $\mathbf{\Im}_i$ and $\mathbf{\Im}'_i$ of the coordinate systems x and x' are connected by the equalities

$$\mathbf{\Im}'_i = \frac{\partial x^j}{\partial x'^i} \mathbf{\Im}_j, \qquad \mathbf{\Im}_i = \frac{\partial x'^j}{\partial x^i} \mathbf{\Im}'_j.$$

Consider, in a domain V_0 of the space V_n, a tensor field $\boldsymbol{\mu}(M_0)$, specified in a coordinate system x by the components $\mu^{\mathcal{A}}(M_0)$:

$$\boldsymbol{\mu}(M_0) = \mu^{\mathcal{A}}(M_0)\boldsymbol{\ni}_{\mathcal{A}}(M_0).$$

Here, $\boldsymbol{\ni}_{\mathcal{A}}$ are polyadic products of the basis vectors $\boldsymbol{\ni}_i$, corresponding to the structure of indices in the components of the tensor $\boldsymbol{\mu}$; M_0 is an arbitrary point in V_0. Let us define, in the domain V, the tensor field $\widetilde{\boldsymbol{\mu}}(M)$ with components $\widetilde{\mu}'^{\mathcal{A}}(M)$, calculated in the basis $\boldsymbol{\ni}'_i$ and numerically equal to the components $\mu^{\mathcal{A}}(M_0)$ at point M_0 whose image is point M under the transformation (2.11). Thus,

$$\widetilde{\boldsymbol{\mu}}(M) = \widetilde{\mu}'^{\mathcal{A}}(M)\boldsymbol{\ni}'_{\mathcal{A}}(M), \qquad \widetilde{\mu}'^{\mathcal{A}}(M) = \mu^{\mathcal{A}}(M_0).$$

The coordinate system x' in the domain V, with the variables x'^i, is called a dragged coordinate system, and the tensor field $\widetilde{\boldsymbol{\mu}}(M)$ is called a dragged field under the transformation (2.11).

The dragged tensor field $\widetilde{\boldsymbol{\mu}}(M)$ may also be defined in the basis $\boldsymbol{\ni}_i$ of the coordinate system x:

$$\widetilde{\boldsymbol{\mu}}(M) = \left[\widetilde{\mu}'^{\mathcal{A}}(M)\frac{\partial x^{\mathcal{B}}}{\partial x'^{\mathcal{A}}}\right]\boldsymbol{\ni}_{\mathcal{B}}(M) = \left[\mu^{\mathcal{A}}(M_0)\frac{\partial x^{\mathcal{B}}}{\partial x'^{\mathcal{A}}}\right]\boldsymbol{\ni}_{\mathcal{B}}(M).$$

Here, $\partial x^{\mathcal{B}}/\partial x'^{\mathcal{A}}$ are the transformation coefficients for the components of the tensor $\mu^{\mathcal{A}}$ under the transformation of the coordinate system $x \to x'$. If the domains V and V_0 intersect, then, at each point of the intersection $V_0 \cap V$ of V_0 and V, two tensors, $\boldsymbol{\mu}(M)$ and $\widetilde{\boldsymbol{\mu}}(M)$, are defined. Let us calculate the difference $-L_u\boldsymbol{\mu}\,dt = \widetilde{\boldsymbol{\mu}}(M) - \boldsymbol{\mu}(M)$ at some point $M \in V_0 \cap V$ under a small transformation (2.11), which we will write in the form

$$x^i = x'^i + u^i dt, \qquad x'^i = x^i - u^i dt, \tag{2.12}$$

where dt is an arbitrary (small) parameter, $\boldsymbol{u}(x^i) = u^j\boldsymbol{\ni}_j$ is a certain specified vector field. We have

$$-L_u\boldsymbol{\mu}\,dt = \widetilde{\boldsymbol{\mu}}(M) - \boldsymbol{\mu}(M) = \left[\mu^{\mathcal{A}}(M_0)\frac{\partial x^{\mathcal{B}}}{\partial x'^{\mathcal{A}}} - \mu^{\mathcal{B}}(M)\right]\boldsymbol{\ni}_{\mathcal{B}}(M). \tag{2.13}$$

It is easy to verify that the coefficients $\partial x^{\mathcal{B}}/\partial x'^{\mathcal{A}}$ of the transformation (2.12) may be written in the form

$$\frac{\partial x^{\mathcal{B}}}{\partial x'^{\mathcal{A}}} = \delta^{\mathcal{B}}_{\mathcal{A}} + F^{\mathcal{B}j}_{\mathcal{A}i}\,\partial_j u^i\,dt, \tag{2.14}$$

where the quantities $F^{\mathcal{B}j}_{\mathcal{A}i}$ are the same as in Eq. (2.5). Using the notation (2.14), expression (2.13) for the tensor $-L_u\boldsymbol{\mu}\,dt$ may be written as

$$- L_u\boldsymbol{\mu}\,dt = \left[\mu^{\mathcal{B}}(M_0) + F^{\mathcal{B}j}_{\mathcal{A}i}\,\mu^{\mathcal{A}}(M_0)\partial_j u^i\,dt - \mu^{\mathcal{B}}(M)\right]\boldsymbol{\ni}_{\mathcal{B}}(M). \tag{2.15}$$

Since for the transformation (2.12), up to first-order quantities with respect to dt, the functions $\mu^{\mathcal{A}}(M_0)$ and $\mu^{\mathcal{A}}(M)$ are related by

$$\mu^{\mathcal{A}}(M_0) = \mu^{\mathcal{A}}(M) - u^i\,\partial_i\mu^{\mathcal{A}}\,dt,$$

Equation (2.15), up to first-order small quantities, may be transformed to

$$-L_u\boldsymbol{\mu}\,dt = \left(-u^i\,\partial_i\mu^{\mathcal{B}} + F^{\mathcal{B}j}_{\mathcal{A}i}\,\mu^{\mathcal{A}}\partial_j u^i\right)\boldsymbol{\ni}_{\mathcal{B}}\,dt.$$

The tensor $L_u\boldsymbol{\mu}\,dt$ is called the Lie differential of the tensor $\boldsymbol{\mu}$ with respect to the vector field $\boldsymbol{u}$. It is obvious that, geometrically, the quantity $-L_u\boldsymbol{\mu}\,dt$ is an increment of the tensor field $\boldsymbol{\mu}(M)$ at some point M due to its dragging along $\boldsymbol{u}\,dt$.

The tensor

$$L_u\boldsymbol{\mu} = \left(u^i\,\partial_i\mu^{\mathcal{B}} - F^{\mathcal{B}j}_{\mathcal{A}i}\,\mu^{\mathcal{A}}\partial_j u^i\right)\boldsymbol{\ni}_{\mathcal{B}} \tag{2.16}$$

is called the Lie derivative of the tensor field $\boldsymbol{\mu}(M)$ with respect to the vector field $\boldsymbol{u}(M)$. In the Riemannian space V_n, the Christoffel symbols are symmetric, $\Gamma^s_{ij} = \Gamma^s_{ji}$, therefore, due to definition (2.5), we can also write expression (2.16) for the Lie derivative in the form

$$L_u\boldsymbol{\mu} = \left(u^i\nabla_i\mu^{\mathcal{B}} - F^{\mathcal{B}j}_{\mathcal{A}i}\,\mu^{\mathcal{A}}\nabla_j u^i\right)\boldsymbol{\ni}_{\mathcal{B}}.$$

In particular, from definition (2.16) it follows that the Lie derivative of a scalar field μ with respect to the vector field $\boldsymbol{u}$ coincides with the usual directional derivative along the direction of the vector $\boldsymbol{u}$:

$$L_u\mu = u^i\,\partial_i\mu. \tag{2.17}$$

Lie derivatives for covariant and contravariant components of a vector field in V_n are determined as

$$\begin{aligned} L_u\mu^j &= u^i\,\partial_i\mu^j - \mu^i\,\partial_i u^j \equiv u^i\nabla_i\mu^j - \mu^i\nabla_i u^j, \\ L_u\mu_j &= u^i\,\partial_i\mu_j + \mu_i\partial_j u^i \equiv u^i\nabla_i\mu_j + \mu_i\nabla_j u^i. \end{aligned} \tag{2.18}$$

Let us also present an expression for the Lie derivative of Christoffel symbols Γ^k_{ij} of the Riemannian space,

$$L_u \Gamma^k_{ij} = u^s \partial_s \Gamma^k_{ij} - \Gamma^s_{ij} \partial_s u^k + \Gamma^k_{sj} \partial_i u^s + \Gamma^k_{si} \partial_j u^s + \partial_i \partial_j u^k$$

and for the covariant components of the metric tensor in a Riemannian space:

$$L_u g_{ij} = u^s \partial_s g_{ij} + g_{is} \partial_j u^s + g_{js} \partial_i u^s \equiv \nabla_i u_j + \nabla_j u_i .$$

If the metric tensor in the space V_n satisfies the equation

$$L_u g_{ij} = \nabla_i u_j + \nabla_j u_i = 0, \tag{2.19}$$

it is said that the space V_n admits a group of motions. Equation (2.19) is called the Killing equation, while the vector $\boldsymbol{u}$ is in this case called a Killing vector. Geometrically, the validity of Eq. (2.19) means that the Riemannian space V_n under consideration possesses a symmetry (isometry), specified by the Killing vector $\boldsymbol{u}$.

2.2 Nonholonomic Systems of Orthonormal Bases in a Riemannian Space

In the Euclidean (or pseudo-Euclidean) space E_n, tangent to V_n at a certain point x^i, in addition to the local basis $\boldsymbol{\vartheta}_i$, let us choose a local orthonormal basis $\boldsymbol{e}_a$, $a = 1, 2, \ldots, n$. We will denote the indices of tensor components, specified in orthonormal bases $\boldsymbol{e}_a$, by the first letters of the Latin alphabet, a, b, c, d, e, f; those of tensor components specified in the local bases $\boldsymbol{\vartheta}_i$, will be denoted by the Latin letters $i, j, k, \ldots$. Let us denote by the symbol g_{ab} the components of the metric tensor of the space E_n in the basis $\boldsymbol{e}_a$. Thus

$$g_{ab} = \pm 1 \quad \text{for} \quad a = b,$$
$$g_{ab} = 0 \quad \text{for} \quad a \neq b.$$

The basis $\boldsymbol{\vartheta}_i$ is connected with the orthonormal basis $\boldsymbol{e}_a$ by the relations

$$\boldsymbol{\vartheta}_i = h_i{}^a \boldsymbol{e}_a, \qquad \boldsymbol{e}_a = h^i{}_a \boldsymbol{\vartheta}_i, \tag{2.20}$$

in which the coefficients $h_i{}^a$ and $h^i{}_a$ are usually called the scale factors (or Lamé coefficients). Taking into account that, for the scalar products of the basis vectors $\boldsymbol{\vartheta}_i$ and $\boldsymbol{e}_a$, the following equalities hold,

$$\left(\boldsymbol{\vartheta}_i, \boldsymbol{\vartheta}_j\right) = g_{ij}, \quad \left(\boldsymbol{e}_a, \boldsymbol{e}_b\right) = g_{ab}, \tag{2.21}$$

we find, multiplying Eqs. (2.20) scalarly by the vectors $\boldsymbol{\ni}^i$ and $\boldsymbol{e}^a$ that

$$h_i{}^a = \left(\boldsymbol{\ni}_i, \boldsymbol{e}^a\right), \quad h^i{}_a = \left(\boldsymbol{\ni}^i, \boldsymbol{e}_a\right). \tag{2.22}$$

Replacing, in the first formula in (2.21), the vectors $\boldsymbol{\ni}_i$ and, in the second formula in (2.21), the vectors $\boldsymbol{e}_a$ using Eqs. (2.20), we obtain equalities connecting the components of the metric tensor:

$$g_{ij} = h_i{}^a h_j{}^b g_{ab}, \quad g_{ab} = h^i{}_a h^j{}_b g_{ij}. \tag{2.23}$$

It is also easy to see that the following equations are valid:

$$h_i{}^a h^j{}_a = \delta_i^j, \quad h_i{}^a h^i{}_b = \delta_b^a.$$

They show that the matrices of the scale factors $\|h_i{}^a\|$ and $\|h^j{}_b\|$ are mutually reciprocal.

Let us write the first equation in (2.23) in a matrix form:

$$\|g_{ij}\| = \|h_i{}^a\| \|g_{ab}\| \|h_j{}^b\|^T. \tag{2.24}$$

Since $\det \|h_i{}^a\| = \det \|h_j{}^b\|^T$ and $\det \|g_{ab}\| = (-1)^q$, Eq. (2.24) implies

$$\det \|g_{ij}\| = \left(\det \|h_i{}^a\|\right)^2 \det \|g_{ab}\| = (-1)^q \left(\det \|h_i{}^a\|\right)^2,$$

where q is the index of the tangent space E_n.

In the general case, in a Riemannian (or Euclidean) space, there are no coordinate systems for whose coordinate lines the vectors $\boldsymbol{e}_a(x^i)$ would be tangent vectors. In this sense, the systems of orthonormal bases $\boldsymbol{e}_a(x^i)$ are nonholonomic.

In what follows, we will suppose that the coefficients $h_i{}^a$ form in the space V_n at least twice continuously differentiable fields.

2.2.1 Ricci Rotation Coefficients

Parallel transport of the vectors of orthonormal bases $\boldsymbol{e}_a(x^i)$ in a Riemannian space may be defined using the relation

$$d\boldsymbol{e}_a = dx^i \Delta_{i,a}{}^b \boldsymbol{e}_b, \tag{2.25}$$

where $\Delta_{i,a}{}^b$ are certain coefficients called the Ricci rotation coefficients. Multiplying Eq. (2.25) scalarly by the basis vector $\boldsymbol{e}_c$, we obtain the equality

$$dx^i \Delta_{i,ac} = \left(\boldsymbol{e}_c, d\boldsymbol{e}_a\right), \tag{2.26}$$

in which $\Delta_{i,ac} = g_{bc}\Delta_{i,a}{}^{b}$.

In Eq. (2.26), let us permute the indices a and c:

$$dx^i \Delta_{i,ca} = (\boldsymbol{e}_a, d\boldsymbol{e}_c) \tag{2.27}$$

and add Eqs. (2.26) and (2.27). As a result, we obtain the equation

$$dx^i (\Delta_{i,ac} + \Delta_{i,ca}) = (\boldsymbol{e}_a, d\boldsymbol{e}_c) + (\boldsymbol{e}_c, d\boldsymbol{e}_a) = d(\boldsymbol{e}_a, \boldsymbol{e}_c). \tag{2.28}$$

Taking into account equalities (2.21) and the constancy of the metric tensor components g_{ac}, we continue Eq. (2.28):

$$dx^i (\Delta_{i,ac} + \Delta_{i,ca}) = dg_{ac} = 0. \tag{2.29}$$

From Eq. (2.29) it follows that the Ricci rotation coefficients $\Delta_{i,ac}$ are antisymmetric in the indices a, c: $\Delta_{i,ac} = -\Delta_{i,ca}$.

Evidently, the quantities $dx^i \Delta_{i,ac}$ geometrically determine (up to first-order small quantities) a transformation from the orthonormal basis $\boldsymbol{e}_a(x^i)$, parallel-transported from a point x^i to a point $x^i + dx^i$, to the basis $\boldsymbol{e}_a(x^i + dx^i)$. The antisymmetric nature of the components $\Delta_{i,ac}$ with respect to the indices a and c is a consequence of the orthogonality of such transformation.

Consider the transformation properties of the Ricci symbols $\Delta_{i,ab}$ under transformations of the orthonormal bases $\boldsymbol{e}_a$ and under transformations of the holonomic coordinates with the variables x^i. Let the orthonormal bases $\boldsymbol{e}_a(x^i)$ be in correspondence with the Ricci rotation coefficients $\Delta_{i,ab}$, and the bases $\boldsymbol{e}'_a(x^i)$, obtained from $\boldsymbol{e}_a(x^i)$ by the smooth orthogonal transformation

$$\boldsymbol{e}'_a = l^b{}_a \boldsymbol{e}_b, \tag{2.30}$$

be in correspondence with the Ricci rotation coefficients $\Delta'_{i,ab}$, determined by the relation

$$dx^i \Delta'_{i,ab} = (\boldsymbol{e}'_b, d\boldsymbol{e}'_a).$$

Let us replace the vectors $\boldsymbol{e}'_a$ in this equality using Eq. (2.30):

$$\begin{aligned} dx^i \Delta'_{i,ab} &= l^c{}_a l^d{}_b (\boldsymbol{e}_d, d\boldsymbol{e}_c) + (\boldsymbol{e}_d, \boldsymbol{e}_c) l^d{}_b dl^c{}_a \\ &= \left[l^c{}_a l^d{}_b (\boldsymbol{e}_d, \boldsymbol{e}_f) \Delta_{i,c}{}^f + (\boldsymbol{e}_d, \boldsymbol{e}_c) l^d{}_b \partial_i l^c{}_a \right] dx^i. \end{aligned}$$

Using relations (2.21) and (2.26), we find

$$dx^i \Delta'_{i,ab} = \left(l^c{}_a l^d{}_b \Delta_{i,cd} + g_{cd} l^d{}_b \partial_i l^c{}_a \right) dx^i.$$

This leads, due to arbitrariness of the quantities dx^i, to the transformation formula for the rotation coefficients:

$$\Delta'_{i,ab} = l^c{}_a l^d{}_b \Delta_{i,cd} + g_{cd} l^d{}_b \partial_i l^c{}_a. \tag{2.31}$$

Thus the transformation of the Ricci rotation coefficients $\Delta_{i,ab}$ under orthogonal transformations of the orthonormal bases $\boldsymbol{e}_a$ is not tensor.

Since under any smooth transformation of the of the coordinates x^i

$$x^i \to y^i = y^i(x^j) \tag{2.32}$$

the vectors of an orthonormal basis $\boldsymbol{e}_a$ remain unchanged, for the Ricci rotation coefficients $\Delta'_{i,ab}$ in the coordinate system with the variables y^i we have

$$dy^i \Delta'_{i,ab} = \left(\boldsymbol{e}'_b, d\boldsymbol{e}'_a\right) = \left(\boldsymbol{e}_b, d\boldsymbol{e}_a\right) = dx^j \Delta_{j,ab}.$$

This leads to the transformation formula for the Ricci rotation coefficients

$$\Delta'_{i,ab} = \frac{\partial x^j}{\partial y^i} \Delta_{j,ab}. \tag{2.33}$$

Thus, under the transformation of variables of a holonomic coordinate system (2.32), the Ricci rotation coefficients $\Delta_{i,ab}$, corresponding to the orthonormal bases $\boldsymbol{e}_a(x^i)$, are transformed as covariant components of a vector.

The Ricci rotation coefficients may be expressed in terms of the scale factors. To obtain the corresponding expression, we substitute in equality (2.26) the vectors $\boldsymbol{e}_a$ in terms of the vectors $\boldsymbol{\ni}_i$ according to Eq. (2.20):

$$dx^i \Delta_{i,ac} = \left(\boldsymbol{e}_c, d\left\{h^k{}_a \boldsymbol{\ni}_k\right\}\right) = \left(\boldsymbol{e}_c, \boldsymbol{\ni}_k\right) dh^k{}_a + h^k{}_a\left(\boldsymbol{e}_c, d\boldsymbol{\ni}_k\right). \tag{2.34}$$

Using Eqs. (2.1) and taking into account the relations (2.22), we continue equality (2.34):

$$dx^i \Delta_{i,ac} = \left[\left(\boldsymbol{e}_c, \boldsymbol{\ni}_k\right)\partial_i h^k{}_a + h^k{}_a \Gamma^j_{ik}\left(\boldsymbol{e}_c, \boldsymbol{\ni}_j\right)\right] dx^i = h_{jc} h^k{}_a\left(\Gamma^j_{ik} + h_k{}^b \partial_i h^j{}_b\right) dx^i.$$

Hence it follows

$$\Delta_{i,ac} = h_{jc} h^k{}_a\left(\Gamma^j_{ik} + h_k{}^b \partial_i h^j{}_b\right). \tag{2.35}$$

It is also useful to write (2.35) in an explicitly invariant and symmetric form:

$$\Delta_{i,ac} = h_{jc} \nabla'_i h^j{}_a \equiv \frac{1}{2}\left(h_{jc} \nabla'_i h^j{}_a - h_{ja} \nabla'_i h^j{}_c\right).$$

Here, ∇'_i is the symbol of a covariant derivative acting only upon the indices that refer to the coordinate system with the variables x^i. Thus

$$\nabla'_i h^j{}_a = \partial_i h^j{}_a + \Gamma^j_{ik} h^k{}_a.$$

Equation (2.35) gives an expression of the Ricci rotation coefficients in terms of the Christoffel symbols and the scale factors. Substituting, in this expression, the Christoffel symbols Γ^k_{ij} in terms of the metric tensor components according to the equality (2.2), and, in the expression thus obtained, the components g_{ij} in terms of the scale factors according to (2.23), we can obtain the following expression of the Ricci rotation symbols $\Delta_{i,ac}$ in terms of the scale factors:

$$\Delta_{i,ac} = \frac{1}{2}\left[h^j{}_c\left(\partial_i h_{ja} - \partial_j h_{ia}\right) - h^j{}_a\left(\partial_i h_{jc} - \partial_j h_{ic}\right) + h_i{}^b h^j{}_a h^s{}_c\left(\partial_j h_{sb} - \partial_s h_{jb}\right)\right].$$

Along with the Ricci rotation coefficients $\Delta_{i,ac}$, the symbols $\Delta_{i,jm}$ are also often used:

$$\Delta_{i,jm} = h_j{}^a h_m{}^c \Delta_{i,ac} = \frac{1}{2}\left[h_i{}^a\left(\partial_j h_{ma} - \partial_m h_{ja}\right) + h_j{}^a\left(\partial_i h_{ma} - \partial_m h_{ia}\right) - h_m{}^a\left(\partial_i h_{ja} - \partial_j h_{ia}\right)\right] \tag{2.36}$$

or

$$\Delta_{i,jm} = \frac{1}{2}\left(h_j{}^a \nabla'_i h_{ma} - h_m{}^a \nabla'_i h_{ja}\right) \tag{2.37}$$

as well as the symbols $\Delta_{a,bc}$:

$$\Delta_{a,bc} = h^i{}_a \Delta_{i,bc} = \frac{1}{2}\left(h_{jc}\nabla'_a h^j{}_b - h_{jb}\nabla'_a h^j{}_c\right),$$

where $\nabla'_a = h^i{}_a \nabla'_i$, or

$$\Delta_{a,bc} = \frac{1}{2}\left[h^j{}_a(\partial_b h_{jc} - \partial_c h_{jb}) + h^j{}_c(\partial_a h_{jb} + \partial_b h_{ja}) - h^j{}_b(\partial_a h_{jc} + \partial_c h_{ja})\right], \tag{2.38}$$

where $\partial_a = h^i{}_a \partial_i$.

Using Eqs. (2.3), it is easy to show that, due to definitions (2.38), the Ricci rotation coefficients $\Delta_{a,bc}$ satisfy the equality

$$\Delta_{b,a}{}^b = \frac{1}{\sqrt{|g|}}\,\partial_i\left(h^i{}_a\sqrt{|g|}\right). \tag{2.39}$$

2.2.2 Covariant Derivatives

Covariant derivatives for the tensor components, specified in an orthonormal bases $\boldsymbol{e}_a$, are defined in the following way:

$$\begin{aligned}\nabla_k \eta_{a_1\ldots a_p}^{c_1\ldots c_q} &= \partial_k \eta_{a_1\ldots a_p}^{c_1\ldots c_q} \\ &+ \Delta_{k,b}{}^{c_1} \eta_{a_1\ldots a_p}^{bc_2\ldots c_q} + \sum_{s=2}^{q-1} \Delta_{k,b}{}^{c_s} \eta_{a_1\ldots a_p}^{c_1\ldots c_{s-1}bc_{s+1}\ldots c_q} + \Delta_{k,b}{}^{c_q} \eta_{a_1\ldots a_p}^{c_1\ldots c_{q-1}b} \\ &- \Delta_{k,a_1}{}^{b} \eta_{ba_2\ldots a_p}^{c_1\ldots c_q} - \sum_{s=2}^{p-1} \Delta_{k,a_s}{}^{b} \eta_{a_1\ldots a_{s-1}ba_{s+1}\ldots a_p}^{c_1\ldots c_q} - \Delta_{k,a_p}{}^{b} \eta_{a_1\ldots a_{p-1}b}^{c_1\ldots c_q}. \end{aligned} \tag{2.40}$$

In particular, for components of a vector field we have

$$\nabla_k \eta^c = \partial_k \eta^c + \Delta_{k,b}{}^c \eta^b,$$

$$\nabla_k \eta_a = \partial_k \eta_a - \Delta_{k,a}{}^b \eta_b.$$

Denoting the components of a tensor of an arbitrary rank, calculated in a basis $\boldsymbol{e}_a$, by the symbol $\eta^{\mathcal{A}}$, we can write an expression for the covariant derivative of $\eta^{\mathcal{A}}$ in the form

$$\nabla_k \eta^{\mathcal{A}} = \partial_k \eta^{\mathcal{A}} + F_{\mathcal{B}b}^{\mathcal{A}a} \eta^{\mathcal{B}} \Delta_{k,a}{}^b. \tag{2.41}$$

A definition of the quantities $F_{\mathcal{B}b}^{\mathcal{A}a}$ in Eq. (2.41) is evident from a comparison of the equalities (2.40) and (2.41).

Contracting equation (2.35) with components $h_s{}^a$ with respect to the index a, we obtain

$$\partial_i h_s{}^c - \Gamma_{is}^j h_j{}^c + \Delta_{i,b}{}^c h_s{}^b = 0. \tag{2.42}$$

The left-hand side of Eq. (2.42) represents a covariant derivative of the components $h_s{}^c$:

$$\nabla_i h_s{}^c = \partial_i h_s{}^c - \Gamma_{is}^j h_j{}^c + \Delta_{i,b}{}^c h_s{}^b.$$

Thus Eq. (2.42) may be written as vanishing of the covariant derivative of the components $h_s{}^c$: $\nabla_i h_s{}^c = 0$. Equation (2.42) may also be written in the form

$$\nabla_i' h_s{}^c = -\Delta_{i,b}{}^c h_s{}^b. \tag{2.43}$$

The tetrad components of the curvature tensor of a Riemannian space are easily expressed in terms of the Ricci rotation coefficients $\Delta_{a,bc}$. To this end, it is sufficient to replace in the equality

$$(\nabla'_i\nabla'_j - \nabla'_j\nabla'_i)h^s{}_a = -R_{ijk}{}^s h^k{}_a$$

(see (2.8)) the derivatives of $h^s{}_a$ according to (2.43). After simple transformations, the following equality is obtained:

$$\begin{aligned} R_{abcd} &= \partial_b \Delta_{a,cd} - \partial_a \Delta_{b,cd} \\ &\quad + \Delta_{f,cd}\left(\Delta_{a,b}{}^f - \Delta_{b,a}{}^f\right) - \Delta_{a,cf}\Delta_{b,d}{}^f + \Delta_{a,df}\Delta_{b,c}{}^f. \end{aligned} \tag{2.44}$$

2.2.3 The Nonholonomity Object

The quantities

$$\Omega_{ab}{}^c = -\Omega_{ba}{}^c = \frac{1}{2}h^i{}_a h^j{}_b\left(\partial_i h_j{}^c - \partial_j h_i{}^c\right) \tag{2.45}$$

form the nonholonomity object. Evidently, if the nonholonomity object vanishes, $\Omega_{ab}{}^c = 0$, then the inequality holds,

$$\partial_i h_j{}^a = \partial_j h_i{}^a,$$

from which it follows that the scale factors $h_i{}^a$ are represented in the form

$$h_i{}^a = \frac{\partial y^a}{\partial x^i}.$$

Hence it follows that vanishing of the nonholonomity object is a necessary and sufficient condition for holonomity of the basis system $\boldsymbol{e}_a(x^i)$.

A direct calculation shows that there holds the equality

$$\partial_a\partial_b - \partial_b\partial_a = -2\Omega_{ab}{}^c\partial_c,$$

in which $\partial_a = h^i{}_a\partial_i$ are the differentiation operators in the directions of the vectors $\boldsymbol{e}_a$.

From definitions (2.45) and (2.38) we find that the nonholonomity object is connected with the Ricci rotation coefficients by the relation

$$\Omega_{abc} = \frac{1}{2}\left(\Delta_{b,ac} - \Delta_{a,bc}\right).$$

The inverse relation has the form

$$\Delta_{a,bc} = -\Omega_{abc} + \Omega_{bca} + \Omega_{acb}.$$

2.2.4 Fermi-Walker Transport

In some physical applications, along with the parallel transport considered above, one also uses the so-called Fermi-Walker transport. Consider, in the Riemannian space V_n, some non-null curve L without singular points, specified by the parametric equations $x^i = x^i(s)$, where s is the arc length along the curve L and $x^i(s)$ are continuously differentiable functions. Let $\boldsymbol{a} = a^i \boldsymbol{\ni}_i$ be a vector field specified at each point on L and tangent to L, such that

$$a_i a^i = \varepsilon = \pm 1 \tag{2.46}$$

(the minus sign in Eq. (2.46) can appear for curves in Riemannian spaces with an indefinite metric). The tensor $\boldsymbol{\mu} = \mu^{\mathcal{A}} \boldsymbol{\ni}_{\mathcal{A}}$ is Fermi-Walker transported along the curve L if the components $\mu^{\mathcal{A}}$ satisfy the equation

$$\frac{D}{ds}\mu^{\mathcal{A}} = F^{\mathcal{A}i}_{\mathcal{B}j}\mu^{\mathcal{B}}\varepsilon\left(a_i\frac{D}{ds}a^j - a^j\frac{D}{ds}a_i\right), \tag{2.47}$$

in which the quantities $F^{\mathcal{A}i}_{\mathcal{B}j}$ are the same as in Eq. (2.5). $D/ds = a^i\nabla_i$ is the symbol of an absolute derivative along L. In particular, for components of a vector $\mu^i \boldsymbol{\ni}_i = \mu_i \boldsymbol{\ni}^i$, Eq. (2.47) is written in the following way:

$$\frac{D}{ds}\mu^i = \varepsilon\mu^j\left(a_j\frac{D}{ds}a^i - a^i\frac{D}{ds}a_j\right),$$

$$\frac{D}{ds}\mu_j = -\varepsilon\mu_i\left(a_j\frac{D}{ds}a^i - a^i\frac{D}{ds}a_j\right). \tag{2.48}$$

It is easy to see that the Fermi-Walker transport conserves the scalar product of vectors. Indeed, if the vectors $\mu^i \boldsymbol{\ni}_i$ and $\eta^i \boldsymbol{\ni}_i$ are Fermi-Walker transported along the curve L, then

$$\frac{D}{ds}(\mu^i\eta_i) = \mu^i\frac{D}{ds}\eta_i + \eta_i\frac{D}{ds}\mu^i = \varepsilon(\eta^i\mu^j + \eta^j\mu^i)\left(a_j\frac{D}{ds}a_i - a_i\frac{D}{ds}a_j\right) = 0.$$

Hence it follows, in particular that an orthonormal basis $\boldsymbol{e}_a$, being Fermi-Walker transported along L, remains orthonormal.

Evidently, the field of the tangent vector $a^i \boldsymbol{\ni}_i$ of the curve L satisfies Eq. (2.48), and thus the tangent vector $a^i \boldsymbol{\ni}_i$ is Fermi-Walker transported along L. In particular, if one of the vectors $\boldsymbol{e}_b$ of an orthonormal basis is taken to be tangent to L, then, being Fermi-Walker transported along L, the basis remains orthonormal, and its vector $\boldsymbol{e}_b$ is tangent to L at all points of L.

2.3 The Spinor as an Invariant Geometric Object in a Riemannian Space

All finite-dimensional representations of the full linear group which includes transformations of bases of the Euclidean vector space E_n are known, and it turns out that, among them, there are no representations that would coincide, on the orthogonal subgroup, with the spinor representation. Hence it follows that the spinor representation of the orthogonal group cannot be extended to a linear finite-dimensional representation of the full linear group.[2]

Therefore the components ψ^A of a spinor in a Euclidean vector space may be introduced and considered, in general, only relative to a set of bases connected by orthogonal transformations. In a similar way, the field of spinor components $\psi^A(x^i)$ in a Euclidean point space may be, in general, introduced only relative to coordinate systems connected by orthogonal transformations. In this connection, in Riemannian spaces V_n, spinor fields are usually introduced at each point x^i as objects with components transformed by means of a spinor representation of the orthogonal group of transformations of the orthonormal bases $\boldsymbol{e}_a$ of the Euclidean (or pseudo-Euclidean) space E_n, tangent to V_n at the point x^i. Then, the parallel transport rule for spinor components may be defined by establishing a correspondence with parallel transport of the tensors defined by the spinor $\boldsymbol{\psi}$. In such a consideration, the spinor at point x^i in Riemannian space is introduced, in essence, in the tangent Euclidean space at this point.

An essential shortcoming of such a definition of spinor fields in the Riemannian space V_n is that the choice of the systems of orthonormal bases at different points of V_n cannot be fixed in a unique way, without additional conditions. The choice of a set of orthonormal bases $\boldsymbol{e}_a$ is not connected with the geometric properties of

[2]However, a certain extension of the representation of orthogonal groups does exist. Thus, for instance, the group of linear transformations of a plane, defined by the matrices

$$M = \begin{Vmatrix} m_1 & -m_2 \\ m_2 & m_1 \end{Vmatrix},$$

has the representation $M \to \{\pm S\}$, where S is defined in the following way:

$$S = \begin{Vmatrix} \sqrt{m_1 + im_2} & 0 \\ 0 & 1/\sqrt{m_1 + im_2} \end{Vmatrix}.$$

This representation passes into the spinor representation of the group of rotations of the plane for $m_1 = \cos\varphi, m_2 = \sin\varphi$ (see Chap. 4).

a Riemannian space and is in general, for a Riemannian space, a supplementary construction. Introduction of a nonholonomic system of bases $\boldsymbol{e}_a$ in a Riemannian space V_n requires that one additionally introduces $\frac{1}{2}n(n-1)$ components of a two-valent tensor defining the bases $\boldsymbol{e}_a$.

The conclusions obtained above in a representation of spinors by systems of tensors lead to another possibility of defining spinor fields in a Riemannian space. Namely, as was shown above, the components ψ^A of a spinor in an orthonormal basis in a Euclidean space are defined by the complex tensor components $\boldsymbol{C}$ with some invariant algebraic relations between them. Thus, in an orthonormal basis, the spinor may be defined as an invariant geometric object (in E_n) not only by the set of components ψ^A but also by the set of components $\boldsymbol{C}$. It is important that, unlike the components ψ^A, the tensor components $\boldsymbol{C}$ may be defined in any non-orthogonal basis of a Euclidean vector space, while the tensor fields $\boldsymbol{C}(x^i)$ may be defined in any curvilinear coordinate system in a Euclidean point space and in a Riemannian space. The tensor fields $\boldsymbol{C}(x^i)$ in a Riemannian space may also be defined in non-holonomic orthonormal bases $\boldsymbol{e}_a$. In such orthonormal bases, one can also define the corresponding spinor components $\psi^A(x^i)$.

Thus the spinor field in a Riemannian space may be defined either by the fields of components $\boldsymbol{C}(x^i)$ in an arbitrary coordinate system or by the field of components $\psi^A(x^i)$ in a nonholonomic system of orthonormal bases $\boldsymbol{e}_a$. However, introduction of the spinor components ψ^A is connected with additionally introducing $\frac{1}{2}n(n-1)$ functions determining the base $\boldsymbol{e}_a$, and such orthonormal sets of bases are external and extraneous for a Riemannian space from the viewpoint of its geometric properties. Therefore, introduction of spinors $\boldsymbol{\psi}$ into V_n with the aid of the tensor components $\boldsymbol{C}$ is naturally and has certain advantages.

2.3.1 Parallel Transport and Covariant Differentiation of Spinors in a Riemannian Space

To establish the parallel transport law for spinors in an n-dimensional Riemannian space V_n, let us adopt, that the invariant spintensors E and γ_a are the same in all spaces E_n and consequently do not depend on the variables x^i[3]:

$$\partial_i E = 0, \qquad \partial_i \gamma_a = 0. \tag{2.49}$$

Consider, in a Riemannian space V_n, some continuously differentiable curve L. By definition, the spinor $\boldsymbol{\psi}$ of any rank is parallel-transported along the curve L if its components in an orthonormal basis $\boldsymbol{e}_a(x^i)$, parallel-transported along L, are constant.

[3]The theory of parallel transport of spinors in Riemannian spaces may also be developed without this restriction (see [4]).

It is easy to show that this definition of the parallel transport for spinors does not depend on the choice of the parallel-transported basis $\boldsymbol{e}_a(x^i)$.

Let $\boldsymbol{e}_a(x^i)$ and $\boldsymbol{e}_a(x^i + dx^i)$ be orthonormal bases at the points x^i and $x^i + dx^i$, respectively, on a curve L in the space V_n. Let us denote the basis $\boldsymbol{e}_a(x^i)$, parallel-transported from the point x^i to the point $x^i + dx^i$, by the symbol $\boldsymbol{e}^{\shortparallel}_a(x^i + dx^i)$. Thus, at the point $x^i + dx^i$, there are two bases: $\boldsymbol{e}_a(x^i + dx^i)$ and $\boldsymbol{e}^{\shortparallel}_a(x^i + dx^i)$.

By definition, the components of the spinor ψ, parallel-transported along L, are the same in the basis $\boldsymbol{e}_a(x^i)$ and in the basis $\boldsymbol{e}^{\shortparallel}_a(x^i + dx^i)$. In the general case, the basis $\boldsymbol{e}^{\shortparallel}_a(x^i + dx^i)$ does not coincide with the basis $\boldsymbol{e}_a(x^i + dx^i)$, therefore the components of the parallel-transported spinor are, in general, different in the bases $\boldsymbol{e}_a(x^i)$ and $\boldsymbol{e}_a(x^i + dx^i)$ on L. Let us calculate the differential for the covariant and contravariant components of a first-rank spinor which is parallel-transported from the point x^i to the point $x^i + dx^i$ on L, calculated in the bases $\boldsymbol{e}_a(x^i)$. According to the equality (2.25), the orthogonal transformation from the basis $\boldsymbol{e}^{\shortparallel}_a(x^i + dx^i)$ to the basis $\boldsymbol{e}_a(x^i + dx^i)$ at the point $x^i + dx^i$ on L is written, up to first-order small quantities, as

$$\boldsymbol{e}^{\shortparallel}_a(x^i + dx^i) = \left(\delta^b_a + dx^j \Delta_{j,a}{}^b\right)\boldsymbol{e}_b(x^i + dx^i), \tag{2.50}$$

where $\Delta_{j,a}{}^b$ are the Ricci rotation coefficients calculated at the point x^i. The transformation S of the spinor components, corresponding to the orthogonal transformation (2.50), according to definition (1.159), may be written as follows:

$$S = \|S^B{}_A\| = I + \frac{1}{4}dx^i \Delta_{i,ab}\gamma^{ab}. \tag{2.51}$$

Here, I is the unit matrix of the order 2^ν. For the inverse transformation S^{-1}, up to first-order small quantities, we have

$$S^{-1} = \|Z^B{}_A\| = I - \frac{1}{4}dx^i \Delta_{i,ab}\gamma^{ab}. \tag{2.52}$$

Let, at a point x^i of the curve L in the basis $\boldsymbol{e}_a(x^i)$, the spinor $\boldsymbol{\psi}$ have the contravariant components $\psi^A(x^i)$ and the covariant components $\psi_A(x^i)$. By definition, the spinor $\boldsymbol{\psi}$, parallel-transported to the point $x^i + dx^i$ on the curve L in the basis $\boldsymbol{e}^{\shortparallel}_a(x^i + dx^i)$, has the same contravariant components $\psi^A(x^i)$ and covariant components $\psi_A(x^i)$. Therefore, to calculate the components of a parallel-transported spinor at a point $x^i + dx^i$ of the curve L in the basis $\boldsymbol{e}_a(x^i + dx^i)$, it is sufficient to subject its components $\psi^A(x^i)$, $\psi_A(x^i)$ to the spinor transformation (2.51), (2.52), corresponding to a transition from the basis $\boldsymbol{e}^{\shortparallel}_a(x^i + dx^i)$ to the basis $\boldsymbol{e}_a(x^i + dx^i)$.

Thus, denoting the contravariant and covariant components of the spinor parallel-transported from the point x^i to the point $x^i + dx^i$ of the curve L in the basis $\boldsymbol{e}_a(x^i + dx^i)$ by the symbols $\psi_{_{\parallel}}^{B}(x^i + dx^i)$ and $\psi^{\parallel}_{B}(x^i + dx^i)$, we can write

$$\begin{aligned} \psi_{_{\parallel}}^{B}(x^i + dx^i) &= S^B{}_A \psi^A(x^i), \\ \psi^{\parallel}_{B}(x^i + dx^i) &= Z^A{}_B \psi_A(x^i). \end{aligned} \tag{2.53}$$

Using expressions (2.51) and (2.52) for S and S^{-1}, we can write Eq. (2.53) for contravariant components of a parallel-transported spinor as follows:

$$\psi_{_{\parallel}}^{B}(x^i + dx^i) = \left(\delta^B_A + dx^i \Gamma^B_{iA}\right)\psi^A(x^i), \tag{2.54}$$

and for covariant components

$$\psi^{\parallel}_{B}(x^i + dx^i) = \left(\delta^A_B - dx^i \Gamma^A_{iB}\right)\psi_A(x^i). \tag{2.55}$$

The coefficients Γ^B_{iA}, called the spinor connection coefficients, are, according to Eqs. (2.51) and (2.52), defined by the formula

$$\|\Gamma^B_{iA}\| = \Gamma_i = \frac{1}{4}\Delta_{i,ab}\gamma^{ab}. \tag{2.56}$$

Just as for tensors, the result of the above-defined parallel transport of spinors between two points of a Riemannian space depends on the curve L connecting these points, along which the parallel transport is carried out.

Consider the transformation of the spinor connection symbols Γ_i under transformations of the coordinate system.

From definition (2.56) and from the transformation law (2.33) for the Ricci rotation coefficients it follows that, under an arbitrary smooth transformation of the variables x^i of a holonomic coordinate system of the Riemannian space, $x^i \to y^i = y^i(x^j)$, the coefficients Γ_i are transformed as covariant components of a vector,

$$\Gamma'_i = \frac{\partial x^j}{\partial y^i}\Gamma_j.$$

Let us also establish the transformation law for the spinor connection symbols Γ_i under arbitrary smooth transformations of the set of orthonormal bases $\boldsymbol{e}_a(x^i)$ obtained in a continuous way from the identical transformation

$$\boldsymbol{e}'_a = l^b{}_a \boldsymbol{e}_b. \tag{2.57}$$

Taking into account the invariance of the spintensor components γ^{ab} under the transformation (2.57) and Eq. (2.31), for Γ_i' we obtain:

$$\Gamma_i' = \frac{1}{4}\Delta'_{i,ab}\gamma^{ab} = \frac{1}{4}(l^c{}_a l^d{}_b \Delta_{i,cd} + g_{cd} l^d{}_b \partial_i l^c{}_a)\gamma^{ab}. \tag{2.58}$$

Let S be the matrix of a spinor transformation corresponding to the orthogonal transformation (2.57). By definition, the matrix S satisfies the equations

$$l^b{}_a \gamma_b = S^{-1}\gamma_a S, \qquad E = S^T E S. \tag{2.59}$$

From definitions (2.59) it follows

$$l^b{}_a \gamma^a = S\gamma^b S^{-1}, \qquad l^c{}_a l^d{}_b \gamma^{ab} = S\gamma^{cd} S^{-1}. \tag{2.60}$$

Taking into account the second relation in (2.60), we can write (2.58) in the form

$$\Gamma_i' = \frac{1}{4}(\Delta_{i,ab} S\gamma^{ab} S^{-1} + g_{cd} l^d{}_b \partial_i l^c{}_a \gamma^{ab}) = S\Gamma_i S^{-1} + \frac{1}{4} g_{cd} l^d{}_b \partial_i l^c{}_a \gamma^{ab}. \tag{2.61}$$

It is helpful to represent Eq. (2.61) in another form. To this end, let us differentiate the first equation in (2.59), multiplied from the left by S, with respect to x^i. Taking into account the constancy of the matrices γ_a, we find

$$(\partial_i l^b{}_a) S\gamma_b + \partial_i S(l^b{}_a \gamma_b) = \gamma_a \partial_i S. \tag{2.62}$$

Multiplying Eq. (2.62) from the right by the matrix S^{-1}, after simple transformations with the aid of the first equations in (2.59) and (2.60), we obtain the equation

$$\gamma_a(\partial_i S \cdot S^{-1}) - (\partial_i S \cdot S^{-1})\gamma_a = (g_{cd} l^c{}_b \partial_i l^d{}_a)\gamma^b. \tag{2.63}$$

Differentiating the second equation in (2.59) with respect to x^i, after a simple transformation, we arrive at the equation

$$E(\partial_i S \cdot S^{-1}) + (\partial_i S \cdot S^{-1})^T E = 0. \tag{2.64}$$

From (2.63) and (2.64) it follows:

$$\partial_i S \cdot S^{-1} = \frac{1}{4} g_{cd} l^c{}_b \partial_i l^d{}_a \gamma^{ab}, \tag{2.65}$$

or

$$\partial_i S = \frac{1}{4} g_{cd} l^c{}_b \partial_i l^d{}_a \gamma^{ab} S, \quad \partial_i S^{-1} = -\frac{1}{4} g_{cd} l^c{}_b \partial_i l^d{}_a S^{-1}\gamma^{ab}.$$

Replacing the second term in (2.61) according to Eq. (2.65), we finally write down the transformation law for the spinor connection symbols under the orthogonal transformation (2.57) of the bases $\boldsymbol{e}_a(x^i)$ in the form

$$\Gamma_i' = S\Gamma_i S^{-1} + \partial_i S \cdot S^{-1}.$$

Consider now a spinor field $\boldsymbol{\psi}(x^i)$ along a curve L in a Riemannian space V_n, defined in a nonholonomic system of orthonormal bases $\boldsymbol{e}_a(x^i)$ by a continuously differentiable field of contravariant components $\psi^A(x^i)$, or by a continuously differentiable field of covariant components $\psi_A(x^i)$. For the covariant and contravariant components of the spinor at the point $x^i + dx^i$, specified in the basis $\boldsymbol{e}_a(x^i + dx^i)$, up to first-order small quantities we have:

$$\begin{aligned} \psi^A(x^i + dx^i) &= \psi^A(x^i) + dx^j \partial_j \psi^A(x^i), \\ \psi_A(x^i + dx^i) &= \psi_A(x^i) + dx^j \partial_j \psi_A(x^i). \end{aligned} \tag{2.66}$$

Let us denote by the symbol $\psi_{_{||}}^A(x^i)$ the components of the spinor $\boldsymbol{\psi}(x^i + dx^i)$ parallel-transported from the point $x^i + dx^i$ to the point x^i in the basis $\boldsymbol{e}_a(x^i)$. Due to Eq. (2.54), for the components $\psi_{_{||}}^A(x^i)$, the following equality holds:

$$\psi_{_{||}}^B(x^i) = \left(\delta_A^B - dx^i \Gamma_{iA}^B\right)\psi^A(x^i + dx^i), \tag{2.67}$$

where the spinor connection coefficients Γ_{iA}^B are defined according to (2.56).

Consider the difference $D\psi^A$ between the contravariant components of the spinor $\psi^A(x^i)$ and the contravariant components of the spinor $\psi_{_{||}}^A(x^i)$, parallel-transported from the point $x^i + dx^i$ to the point x^i. Taking into account the equality (2.67), up to first-order small quantities, we find for the components $D\psi^A$ in the basis $\boldsymbol{e}_a(x^i)$:

$$D\psi^A = \psi_{_{||}}^A(x^i) - \psi^A(x^i) = \psi^A(x^i + dx^i) - dx^j \Gamma_{jB}^A \psi^B(x^i) - \psi^A(x^i).$$

Replacing, in this formula, the components $\psi^A(x^i + dx^i)$ via $\psi^A(x^i)$ according to (2.66), we obtain

$$D\psi^A = dx^i\left(\partial_i \psi^A - \Gamma_{iB}^A \psi^B\right). \tag{2.68}$$

In the same manner, for the difference of covariant components of the spinor $D\psi_A = \psi_A^{||}(x^i) - \psi_A(x^i)$ we can find

$$D\psi_A = dx^i\left(\partial_i \psi_A + \Gamma_{iA}^B \psi_B\right). \tag{2.69}$$

It is convenient to write Eqs. (2.68) and (2.69) in a matrix form:

$$D\psi = dx^i\big(\partial_i\psi - \Gamma_i\psi\big) = dx^i\left(\partial_i\psi - \frac{1}{4}\Delta_{i,ab}\gamma^{ab}\psi\right),$$

$$D\widetilde{\psi} = dx^i\big(\partial_i\widetilde{\psi} + \widetilde{\psi}\,\Gamma_i\big) = dx^i\left(\partial_i\widetilde{\psi} + \frac{1}{4}\Delta_{i,ab}\widetilde{\psi}\gamma^{ab}\right). \tag{2.70}$$

Here, ψ is a column of the contravariant components of the spinor ψ^A, $\widetilde{\psi}$ is a row of the covariant components of ψ_A, and Γ_i are the matrices of spinor connection symbols Γ^B_{iA}.

The first-rank spinor $D\boldsymbol{\psi}$, defined by the covariant components $D\psi_A$ or by the contravariant component $D\psi^A$, is called the covariant (or absolute) differential of the spinor $\boldsymbol{\psi}(x^i)$ at the point x^i.

Similarly to (2.66)–(2.69), one can introduce definitions of covariant differentials for spinor fields of any rank with q contravariant indices and p covariant indices:

$$D\psi^{B_1\ldots B_q}_{A_1\ldots A_p} = dx^i\Bigg(\partial_i\psi^{B_1\ldots B_q}_{A_1\ldots A_p}$$

$$- \Gamma^{B_1}_{iC}\psi^{CB_2\ldots B_q}_{A_1\ldots A_p} - \sum_{\alpha=2}^{q-1}\Gamma^{B_\alpha}_{iC}\psi^{B_1\ldots B_{\alpha-1}CB_{\alpha+1}\ldots B_q}_{A_1\ldots A_p} - \Gamma^{B_q}_{iC}\psi^{B_1\ldots B_{q-1}C}_{A_1\ldots A_p}$$

$$+ \Gamma^{E}_{iA_1}\psi^{B_1\ldots B_q}_{EA_2\ldots A_p} + \sum_{\alpha=2}^{p-1}\Gamma^{E}_{iA_\alpha}\psi^{B_1\ldots B_q}_{A_1\ldots A_{\alpha-1}EA_{\alpha+1}\ldots A_p} + \Gamma^{E}_{iA_p}\psi^{B_1\ldots B_q}_{A_2\ldots A_{p-1}E}\Bigg). \tag{2.71}$$

By the sense of the definition, the covariant differential $D\boldsymbol{\psi}$ of a spinor is a spinor of the same rank as $\boldsymbol{\psi}$.

Evidently, if the spinor $\boldsymbol{\psi}$ is parallel-transported along a curve L, then its covariant differential along L is zero, $D\psi = 0$.

Since, at continuous rotations of orthonormal bases $\boldsymbol{e}_a(x^i)$, the components of the spintensors E and γ_a are invariant, they do not change in parallel transport along L. Therefore the covariant differentials of the spintensors E and γ_a are equal to zero,

$$DE = 0, \quad D\gamma_a = 0. \tag{2.72}$$

Taking into account definition (2.71) and conditions (2.49), one can write Eqs. (2.72) in the form

$$DE = dx^i\big(E\Gamma_i + \Gamma_i^T E\big) = 0,$$

$$D\gamma_a = dx^i\big(-\Delta_{i,a}{}^b\gamma_b - \Gamma_i\gamma_a + \gamma_a\Gamma_i\big) = 0. \tag{2.73}$$

It is easy to verify directly that Eqs. (2.73) are fulfilled identically due to definition (2.56) of the spinor connection symbols Γ_i.

Carrying out a parallel transport of the spinor $\boldsymbol{\psi}$ along the curve L, it is possible to put this spinor, at each point of L, in correspondence to the real tensors $\boldsymbol{D}$ and to the complex tensors $\boldsymbol{C}$ according to Eqs. (1.203), (1.212). Let us show that, with the above-defined parallel transport of spinors, the tensors $\boldsymbol{C}$ and $\boldsymbol{D}$, defined by the parallel-transported spinor, are also a result of parallel transport according to the usual tensor law.

Let, at a point x^i on the curve L, a spinor $\boldsymbol{\psi}(x^i)$ in the basis $\boldsymbol{e}_a(x^i)$ be determined by the contravariant components $\psi(x^i)$, while a spinor parallel-transported to the point $x^i + dx^i$ on L is determined in the basis $\boldsymbol{e}_a(x^i + dx^i)$ by the components $\psi_{\shortparallel} = \big(I + dx^i \Gamma_i\big)\psi$. At the point x^i, the spinor $\boldsymbol{\psi}(x^i)$ is in correspondence with the tensor $\boldsymbol{C}(x^i)$ with components $C(x^i)$ and $C^{a_1 a_2 \ldots a_s}(x^i)$ defined in the basis $\boldsymbol{e}_a(x^i)$:

$$
\begin{aligned}
C(x^i) &= \psi^T(x^i) E \psi(x^i), \\
C^{a_1 a_2 \ldots a_s}(x^i) &= \psi^T(x^i) E \gamma^{a_1 a_2 \ldots a_s} \psi(x^i).
\end{aligned}
$$

At the point $x^i + dx^i$, the parallel-transported spinor is, according to relations (2.54), in correspondence with the tensors $\boldsymbol{C}_{\shortparallel}(x^i + dx^i)$, defined in the basis $\boldsymbol{e}_a(x^i + dx^i)$ by the components

$$
\begin{aligned}
C_{\shortparallel}(x^i + dx^i) &= \big(\psi + dx^j \Gamma_j \psi\big)^T E \big(\psi + dx^s \Gamma_s \psi\big), \\
C_{\shortparallel}^{a_1 a_2 \ldots a_s}(x^i + dx^i) &= \big(\psi + dx^j \Gamma_j \psi\big)^T E \gamma^{a_1 a_2 \ldots a_s} \big(\psi + dx^s \Gamma_s \psi\big).
\end{aligned} \tag{2.74}
$$

From Eqs. (2.74), up to small first-order quantities, we have

$$
\begin{aligned}
C_{\shortparallel}(x^i + dx^i) &= \psi^T E \psi + dx^j \psi^T \big(E \Gamma_j + \Gamma_j^T E\big)\psi, \\
C_{\shortparallel}^{a_1 a_2 \ldots a_s}(x^i + dx^i) &= \psi^T E \gamma^{a_1 a_2 \ldots a_s} \psi \\
&\quad + dx^j \psi^T \big(E \gamma^{a_1 a_2 \ldots a_s} \Gamma_j + \Gamma_j^T E \gamma^{a_1 a_2 \ldots a_s}\big)\psi.
\end{aligned} \tag{2.75}
$$

Replacing in the first equation (2.75) the coefficients Γ_j according to Eq. (2.56) and taking into account the relation

$$
E \gamma^{ab} = -(\gamma^{ab})^T E \tag{2.76}
$$

that follows from (1.54) for $k = 2$, we find for the scalar C:

$$
C_{\shortparallel}(x^i + dx^i) = \psi^T(x^i) E \psi(x^i) = C(x^i).
$$

Thus the scalar C is invariant under the parallel transport we consider. To transform the second equation (2.75), we will use the identity

$$\frac{1}{2}(\gamma^{bc}\gamma^{a_1a_2\ldots a_s} - \gamma^{a_1a_2\ldots a_s}\gamma^{bc}) = (-1)^s s\big(-g^{c\,[a_1}\gamma^{a_2\ldots a_s]\,b} + g^{b\,[a_1}\gamma^{a_2\ldots a_s]\,c}\big), \tag{2.77}$$

which is obtained by adding of the identities (1.16c) and (1.16d). Substituting in the second equation (2.75) the coefficients Γ_j according to the law (2.56) and taking into account relations (2.76) and (2.77), we find for the components of the tensors $\underset{\shortparallel}{C}^{a_1a_2\ldots a_s}(x^i + dx^i)$ at the point $x^i + dx^i$:

$$\begin{aligned}\underset{\shortparallel}{C}^{a_1a_2\ldots a_s}(x^i + dx^i) &= \psi^T(x^i)E\gamma^{a_1a_2\ldots a_s}\psi(x^i) \\ &+ dx^j\psi^T(x^i)E\Big(\Delta_{j,b}{}^{a_1}\gamma^{ba_2\ldots a_s} + \sum_{\nu=2}^{s-1}\Delta_{j,b}{}^{a_\nu}\gamma^{a_1\ldots a_{\nu-1}ba_{\nu+1}\ldots a_s} \\ &+ \Delta_{j,b}{}^{a_s}\gamma^{a_1\ldots a_{s-1}b}\Big)\psi(x^i) = C^{a_1a_2\ldots a_s}(x^i) + dx^j\Big(\Delta_{j,b}{}^{a_1}C^{ba_2\ldots a_s}(x^i) \\ &+ \sum_{\nu=2}^{s-1}\Delta_{j,b}{}^{a_\nu}C^{a_1a_2\ldots a_{\nu-1}ba_{\nu+1}\ldots a_s}(x^i) + \Delta_{j,b}{}^{a_s}C^{a_1a_2\ldots a_{s-1}b}(x^i)\Big).\end{aligned} \tag{2.78}$$

Equation (2.78) defines a usual tensor parallel transpost. In the same manner, one can show that the real tensors $\boldsymbol{D}$, defined by the spinor $\boldsymbol{\psi}$, are also subject to parallel transport according to the tensor law when the spinor is parallel-transported according to (2.54) and (2.55).

It should be noted that the parallel transport of spinors could be defined by precisely this property of correspondence to the parallel transport of the tensors $\boldsymbol{C}$ or $\boldsymbol{D}$ defined by the spinor. Using the real tensors $\boldsymbol{D}$, it easy to find that this condition is written in the form of the set of equations

$$\begin{aligned}E\Gamma_i + \Gamma_i^T E &= 0, \\ E\gamma^{a_1a_2\ldots a_s}\Gamma_i + \Gamma_i^T E\gamma^{a_1a_2\ldots a_s} &= E\Big(\Delta_{j,b}{}^{a_1}\gamma^{ba_2\ldots a_s} \\ &+ \sum_{\nu=2}^{s-1}\Delta_{j,b}{}^{a_\nu}\gamma^{a_1\ldots a_{\nu-1}ba_{\nu+1}\ldots a_s} + \Delta_{j,b}{}^{a_s}\gamma^{a_1\ldots a_{s-1}b}\Big),\end{aligned}$$

which may also be rewritten in the form

$$E\Gamma_i + \Gamma_i^T E = 0, \tag{2.79}$$

$$\gamma^{a_1 a_2 \dots a_s}\Gamma_i - \Gamma_i \gamma^{a_1 a_2 \dots a_s} = \Delta_{j,b}{}^{a_1}\gamma^{b a_2 \dots a_s} + \sum_{\nu=2}^{s-1}\Delta_{j,b}{}^{a_\nu}\gamma^{a_1 \dots a_{\nu-1} b a_{\nu+1} \dots a_s} + \Delta_{j,b}{}^{a_s}\gamma^{a_1 \dots a_{s-1} b}.$$

The set of Eqs. (2.79) has a solution for Γ_i in the form (2.56). Using the complex tensors $\boldsymbol{C}$ for defining the parallel transport for spinors also leads to Eqs. (2.79).

In a four-dimensional pseudo-Riemannian space, the above way of defining the symbols Γ_i, connected with using the real tensors $\boldsymbol{D}$, is realized in [20–24]. However, Refs. [20–24] did not use the first equation in (2.79), therefore in [20–24] the symbols Γ_i are defined by the equations used there only up to an arbitrary spherical matrix.

Equations (2.68) and (2.69) for the covariant differential of the components of a first-rank spinor may be written in the form

$$D\psi^A = dx^i \nabla_i \psi^A, \quad D\psi_A = dx^i \nabla_i \psi_A,$$

where by definition

$$\nabla_i \psi^A = \partial_i \psi^A - \Gamma_{iB}^A \psi^B, \tag{2.80}$$

and

$$\nabla_i \psi_A = \partial_i \psi_A + \Gamma_{iA}^B \psi_B. \tag{2.81}$$

Evidently, the quantities $\nabla_i \psi^A$ and $\nabla_i \psi_A$ form a spintensor having one additional covariant tensor index as compared with ψ^A and ψ_A. Spintensors with components $\nabla_i \psi^A$ and $\nabla_i \psi_A$ are called the covariant derivatives of the spinor fields $\psi^A(x^i)$ and $\psi_A(x^i)$. In a Cartesian coordinate system in a Euclidean space, the connection symbols Γ_i are zero, and the covariant derivatives ∇_i turn into usual partial derivatives.

Equations (2.80) and (2.81) may be conveniently written in a matrix form:

$$\nabla_i \psi = \partial_i \psi - \Gamma_i \psi, \qquad \nabla_i \widetilde{\psi} = \partial_i \widetilde{\psi} + \widetilde{\psi}\Gamma_i.$$

It follows from Eqs. (2.73) that covariant derivatives of the metric spinor E and the spintensors γ_a are equal to zero:

$$\nabla_i E = E\Gamma_i + \Gamma_i^T E = 0,$$

$$\nabla_i \gamma_a = -\Delta_{i,a}{}^b \gamma_b - \Gamma_i \gamma_a + \gamma_a \Gamma_i = 0.$$

Just as for tensor fields, covariant derivatives of spinor fields are not commutative. Rather a simple calculation leads to the following expressions for alternated covariant derivatives of contravariant components of a first-rank spinor field:

$$\nabla_i \nabla_j \psi - \nabla_j \nabla_i \psi = \frac{1}{4} R_{ijks} \gamma^{ks} \psi,$$

and for alternated covariant components of a first-rank spinor field:

$$\nabla_i \nabla_j \widetilde{\psi} - \nabla_j \nabla_i \widetilde{\psi} = -\frac{1}{4} R_{ijks} \widetilde{\psi} \gamma^{ks}.$$

2.4 Fermi-Walker Transport of Spinors

Consider a Riemannian space V_n, referred to a coordinate system x^i, with a covariant vector basis $\boldsymbol{\vartheta}_i$. Let L be some nonnull continuously differentiable curve without singular points in the space V_n, and let an arbitrary orthonormal basis $\check{\boldsymbol{e}}_a(x^i)$, connected with the basis $\boldsymbol{\vartheta}_i$ by the scale factors $\check{\boldsymbol{e}}_a = \check{h}^i{}_a \boldsymbol{\vartheta}_i$, be Fermi-Walker transported along L. By definition of the Fermi-Walker transport for vectors, for the scale factors $\check{h}^i{}_a$ determining the basis $\check{\boldsymbol{e}}_a$, the following condition is valid:

$$\frac{D'}{ds} \check{h}^i{}_a = \varepsilon \check{h}^j{}_a \left(a_j \frac{D}{ds} a^i - a^i \frac{D}{ds} a_j \right), \tag{2.82}$$

where a_j and a^i are components of a unit (or imaginary-unit, depending on the kind of L) vector $\boldsymbol{a}$, tangent to L; $\varepsilon = a_i a^i$ is the sign indicator of the squared vector $\boldsymbol{a}$. For the derivative $D'\check{h}^i{}_a/ds$, we have by definition

$$\frac{D'}{ds} \check{h}^i{}_a = a^j \nabla'_j \check{h}^i{}_a = a^j (\partial_j \check{h}^i{}_a + \Gamma^i_{js} \check{h}^s{}_a),$$

where Γ^i_{js} are the Christoffel symbols corresponding to the coordinate system with the variables x^i. Contracting equation (2.43), written for the bases $\check{\boldsymbol{e}}_a$, with the tensor components $a^i \check{h}^s{}_a$ with respect to the indices i and s, we obtain that for the Ricci rotation coefficients $\check{\Delta}_{j,a}{}^b$, corresponding to the bases $\check{\boldsymbol{e}}_a(x^i)$, the following equation holds:

$$a^i \check{\Delta}_{i,a}{}^b = -\check{h}^i{}_a \frac{D'}{ds} \check{h}_i{}^b = \check{h}_i{}^b \frac{D'}{ds} \check{h}^i{}_a.$$

Therefore, replacing here the derivative D'/Ds according to Eq. (2.82), we obtain

$$a^j \check{\Delta}_{j,a}{}^b = \varepsilon \check{h}^j{}_a \check{h}_i{}^b \left(a_j \frac{D}{ds} a^i - a^i \frac{D}{ds} a_j \right). \tag{2.83}$$

Let us introduce, in the space V_n, an arbitrary smooth system of orthonormal bases $\boldsymbol{e}_a(x^i)$, corresponding to the Ricci rotation symbols $\Delta_{i,ab}$. For simplicity of the subsequent transformations (and without loss of generality), let us further suppose that, at a single selected point x^i on L, the bases $\boldsymbol{e}_a(x^i)$ and $\check{\boldsymbol{e}}_a(x^i)$ coincide. Then, for the bases $\boldsymbol{e}_a(x^i+dx^i)$ and $\check{\boldsymbol{e}}_a(x^i+dx^i)$ at the point x^i+dx^i on L, up to small first-order quantities, we have

$$\begin{aligned} \boldsymbol{e}_a(x^i+dx^i) &= \big(\delta_a^b + dx^j \Delta_{j,a}{}^b\big)\boldsymbol{e}_b(x^i), \\ \check{\boldsymbol{e}}_a(x^i+dx^i) &= \big(\delta_a^b + dx^j \check{\Delta}_{j,a}{}^b\big)\boldsymbol{e}_b(x^i). \end{aligned} \tag{2.84}$$

From the second equation (2.84), also up to small first-order quantities, we obtain

$$\boldsymbol{e}_b(x^i) = \big(\delta_b^c - dx^j \Delta_{j,b}{}^c\big)\check{\boldsymbol{e}}_c(x^i+dx^i). \tag{2.85}$$

Replacing in the first equation (2.84) the vectors $\boldsymbol{e}_b(x^i)$ according to Eq. (2.85), we obtain that the bases $\boldsymbol{e}_a(x^i+dx^i)$ and $\check{\boldsymbol{e}}_a(x^i+dx^i)$ are connected by an orthogonal transformation of the form

$$\begin{aligned} \boldsymbol{e}_a(x^i+dx^i) &= \big(\delta_a^b + dx^j \Delta_{j,a}{}^b\big)\big(\delta_b^c - dx^s \check{\Delta}_{s,b}{}^c\big)\check{\boldsymbol{e}}_c(x^i+dx^i) \\ &= \big[\delta_a^c + dx^j\big(\Delta_{j,a}{}^c - \check{\Delta}_{j,a}{}^c\big)\big]\check{\boldsymbol{e}}_c(x^i+dx^i). \end{aligned} \tag{2.86}$$

By definition, a spinor (of arbitrary rank) is Fermi-Walker transported on the curve L in the Riemannian space V_n if its components are constant in the local orthonormal basis $\check{\boldsymbol{e}}_a(x^i)$, Fermi-Walker transported on L.

Consider, at a point x^i on the curve L in the space V_n, a first-rank spinor $\boldsymbol{\psi}(x^i)$ specified by its contravariant or covariant components $\psi^A(x^i)$ or $\psi_A(x^i)$ in the orthonormal basis $\boldsymbol{e}_a(x^i)$. Let us Fermi-Walker transport the spinor $\boldsymbol{\psi}(x^i)$ from a point x^i to a point x^i+dx^i on the curve L. By definition, the components of the transported spinor at the point x^i+dx^i in the basis $\check{\boldsymbol{e}}_a(x^i+dx^i)$ coincide with its components at the point x^i in the basis $\boldsymbol{e}_a(x^i)$. Therefore, to calculate the components $\psi^A(x^i+dx^i)$ and $\psi_A(x^i+dx^i)$ of the transported spinor in the basis $\boldsymbol{e}_a(x^i+dx^i)$, it is sufficient to subject the components $\psi^A(x^i)$ and $\psi_A(x^i)$ to the spinor transformations S, S^{-1} corresponding to the orthogonal transformation (2.86). For the spinor transformations S and S^{-1}, according to definition (1.159), we have

$$\begin{aligned} S &= I + \frac{1}{4}dx^j\big(\Delta_{j,ab} - \check{\Delta}_{j,ab}\big)\gamma^a\gamma^b, \\ S^{-1} &= I - \frac{1}{4}dx^j\big(\Delta_{j,ab} - \check{\Delta}_{j,ab}\big)\gamma^a\gamma^b, \end{aligned}$$

where $dx^j = a^j ds$. Therefore, for the components $\psi^A(x^i+dx^i)$ and $\psi_A(x^i+dx^i)$ of the spinor $\boldsymbol{\psi}(x^i)$, Fermi-Walker transported to the point x^i+dx^i and calculated in the basis $\boldsymbol{e}_a(x^i + dx^i)$, we have (in a matrix form)

$$\psi(x^i + dx^i) = \psi(x^i) + \frac{1}{4}a^j(\Delta_{j,ab} - \check{\Delta}_{j,ab})\gamma^a\gamma^b\psi(x^i)\,ds,$$
$$\widetilde{\psi}(x^i + dx^i) = \widetilde{\psi}(x^i) - \frac{1}{4}a^j(\Delta_{j,ab} - \check{\Delta}_{j,ab})\widetilde{\psi}(x^i)\gamma^a\gamma^b\,ds. \tag{2.87}$$

Equations (2.87) imply the relations for differentials of the contravariant and covariant components of the spinor in the basis $\boldsymbol{e}_a(x^i)$ when it is Fermi-Walker transported:

$$d\psi = \frac{1}{4}a^j(\Delta_{j,ab} - \check{\Delta}_{j,ab})\gamma^a\gamma^b\psi\,ds,$$
$$d\widetilde{\psi} = -\frac{1}{4}a^j(\Delta_{j,ab} - \check{\Delta}_{j,ab})\widetilde{\psi}\gamma^a\gamma^b\,ds. \tag{2.88}$$

Taking into account definitions (2.70) of the absolute differential and definitions (2.83) for the quantities $a^j\check{\Delta}_{j,a}{}^b$, Eqs. (2.88) may be written in the form

$$D\psi = \frac{1}{4}\varepsilon\left(a_j\frac{D}{ds}a_i - a_i\frac{D}{ds}a_j\right)\gamma^i\gamma^j\psi\,ds,$$
$$D\widetilde{\psi} = -\frac{1}{4}\varepsilon\left(a_j\frac{D}{ds}a_i - a_i\frac{D}{ds}a_j\right)\widetilde{\psi}\gamma^i\gamma^j\,ds. \tag{2.89}$$

It is also easy to obtain Eqs. (2.89) by postulating the correspondence between the transport of spinors and the Fermi-Walker transport of the tensors $\boldsymbol{C}$ and $\boldsymbol{D}$ defined by the spinor.

One can also, in an evident way, obtain equations defining the Fermi-Walker transport for spinors of any rank.

2.5 Lie Differentiation of Spinor Fields

Let V_n be a Riemannian space of dimension n, related to a holonomic coordinate system with the variables x^i and the covariant vector basis $\boldsymbol{\ni}_i$. Let us introduce, in the space V_n, a smooth field of bases $\boldsymbol{e}_a(x^i)$. Consider a motion of the Riemannian space V_n, determined by the Killing vector field $\boldsymbol{u}(x^i) = u^i\boldsymbol{\ni}_i$. By definition, the components of the Killing vector u^i satisfy the equation

$$L_u g_{ij} = \nabla_i u_j + \nabla_j u_i = 0.$$

When the Riemannian space is dragged along the Killing vector $\boldsymbol{u}$, the orthonormal bases $\boldsymbol{e}_a(x^i - u^i dt)$ pass into orthonormal bases, to be denoted by the symbol

$\overset{*}{\boldsymbol{e}}_a(x^i)$. For the Lie derivative of the scale factors, defining the bases $\boldsymbol{e}_a(x^i)$ by formulae (2.20), with respect to the Killing vector $\boldsymbol{u}$, according to definitions (2.18) and (2.42), we have:

$$L_u h^i{}_a = u^s \partial_s h^i{}_a - h^j{}_a \partial_j u^i = u^j \Delta_{j,a}{}^c h^i{}_c - h^j{}_a \nabla_j u^i, \tag{2.90}$$

where $\Delta_{j,a}{}^c$ are the Ricci rotation coefficients corresponding to the orthonormal bases $\boldsymbol{e}_a(x^i)$. Using equality (2.90), we find that the transformation of the orthonormal basis $\overset{*}{\boldsymbol{e}}_a(x^i)$, dragged to the point x^i, to the basis $\boldsymbol{e}_a(x^i)$,

$$\overset{*}{\boldsymbol{e}}_a(x^i) = \left(\delta_a^b + \varepsilon_a{}^b\, dt\right)\boldsymbol{e}_b(x^i), \tag{2.91}$$

is determined by the relation

$$\varepsilon_a{}^b = h_i{}^b L_u h^i{}_a = u^s \Delta_{s,a}{}^b - h^j{}_a h_i{}^b \nabla_j u^i.$$

Let us define, in the space V_n, the spinor field $\boldsymbol{\psi}(x^i)$ with the components specified in the bases $\boldsymbol{e}_a(x^i)$. The spinor transformation $S(dt)$, corresponding to the orthogonal transformation (2.91), according to definition (1.159), has the form

$$S\left(dt\right) = I + \frac{1}{4}\gamma^{ab}\varepsilon_{ab} dt = I + \frac{1}{4}\gamma^{ij}\left(u^s \Delta_{s.ij} - \nabla_i u_j\right)dt. \tag{2.92}$$

By definition, the spinor field $\boldsymbol{\psi}(x^i)$ is dragged by the motion $x^i - u^i dt \to x^i$ if its components, calculated with respect to the dragged bases $\overset{*}{\boldsymbol{e}}_a(x^i)$, are constant. From this definition it follows that, if in the basis $\boldsymbol{e}_a(x^i - u^i dt)$ a first-rank spinor is specified by the contravariant components $\psi^A(x^i - u^i dt)$ or by the covariant components $\psi_A(x^i - u^i dt)$, then in the basis $\overset{*}{\boldsymbol{e}}_a(x^i)$, dragged to the point x^i, the components of the spinor are the same as in the basis $\boldsymbol{e}_a(x^i - u^i dt)$, and in order to obtain the components of the dragged spinor field in the basis $\boldsymbol{e}_a(x^i)$, it is sufficient to subject its components to the spinor transformation (2.92):

$$\overset{*}{\psi}{}^A(x^i) = S^A{}_B \psi^B(x^i - u^i dt) = S^A{}_B\left[\psi^B(x^i) - u^j \partial_j \psi^B\, dt\right],$$

$$\overset{*}{\psi}_B(x^i) = Z^A{}_B \psi_A(x^i - u^i dt) = Z^A{}_B\left[\psi_A(x^i) - u^j \partial_j \psi_A\, dt\right].$$

Here, $\|Z^A{}_B\| = \|S^A{}_B\|^{-1}$.

Let us define the Lie differential of a spinor field at a point with the coordinates x^i by the components $L_u \psi dt$:

$$-L_u \psi dt = \overset{*}{\psi}(x^i) - \psi(x^i).$$

Taking into account equality (2.92) for $L_u \psi^A dt$, we obtain up to small first-order quantities:

$$L_u \psi^A dt = -\left[\delta^A_B + \frac{1}{4}(u^s \Delta_{s,ij} - \nabla_i u_j)\gamma^A{}_B{}^{ij} dt\right]\left[\psi^B(x^i) - u^j \partial_j \psi^B \, dt\right]$$
$$+ \psi^A(x^i) = \left(u^j \nabla_j \psi^A + \frac{1}{4}\gamma^A{}_B{}^{ij}\psi^B \nabla_i u_j\right) dt.$$

where $\|\gamma^A{}_B{}^{ij}\| = \gamma^{ij}$. In the same manner, for the Lie differential $L_u \psi_B dt$ we find:

$$L_u \psi_B dt = \left(u^i \nabla_i \psi_B - \frac{1}{4}\psi_A \gamma^A{}_B{}^{ij} \nabla_i u_j\right) dt.$$

We define Lie derivatives of a spinor field with respect to the vector field $\boldsymbol{u}$ by the equalities

$$L_u \psi^A = u^i \nabla_i \psi^A + \frac{1}{4}\gamma^A{}_B{}^{ij}\psi^B \nabla_i u_j,$$
$$L_u \psi_B = u^i \nabla_i \psi_B - \frac{1}{4}\psi_A \gamma^A{}_B{}^{ij} \nabla_i u_j.$$

It is not hard to generalize the relations obtained to a spinor field of any rank, with components $\psi^{A_1 \dots A_p}_{B_1 \dots B_q}$:

$$L_u \psi^{A_1 \dots A_p}_{B_1 \dots B_q} = u^i \nabla_i \psi^{A_1 \dots A_p}_{B_1 \dots B_q} + \frac{1}{4}\Bigg(- \gamma^{A_1}{}_C{}^{ij} \psi^{C A_2 \dots A_p}_{B_1 \dots B_q}$$
$$- \sum_{\alpha=2}^{p-1} \gamma^{A_\alpha}{}_C{}^{ij} \psi^{A_1 \dots A_{\alpha-1} C A_{\alpha+1} \dots A_p}_{B_1 \dots B_q} - \gamma^{A_p}{}_C{}^{ij} \psi^{A_1 \dots A_{p-1} C}_{B_1 \dots B_q} + \gamma^E{}_{B_1}{}^{ij} \psi^{A_1 \dots A_p}_{E B_2 \dots B_q}$$
$$+ \sum_{\alpha=2}^{q-1} \gamma^E{}_{B_\alpha}{}^{ij} \psi^{A_1 \dots A_p}_{B_1 \dots B_{\alpha-1} E B_{\alpha+1} \dots B_q} + \gamma^E{}_{B_q}{}^{ij} \psi^{A_1 \dots A_p}_{B_1 \dots B_{q-1} E}\Bigg) \nabla_i u_j. \qquad (2.93)$$

If one considers an object with components having tensor and spinor indices, then the Lie derivatives are formed depending on the structure of indices according to Eqs. (2.93) and (2.18).

Due to invariance of the components of the spintensors E and γ_a under continuous orthogonal transformations, the Lie derivatives of E and γ_a with respect to the Killing vector $\boldsymbol{u}$ are equal to zero,

$$L_u E = 0, \qquad L_u \gamma_a = 0. \qquad (2.94)$$

Chapter 3
Spinors in the Four-Dimensional Pseudo-Euclidean Space

3.1 Dirac Matrices and the Spinor Representation of the Lorentz Group

3.1.1 The Lorentz Group

Let us consider the four-dimensional pseudo-Euclidean vector space E_4^1 of index 1 referred to an orthonormal basis Э_i, $i = 1, 2, 3, 4$. The scalar products of the vectors of orthonormal basis $g_{ij} = \left(\text{Э}_i, \text{Э}_j\right)$ in the space E_4^1 are defined by the matrix

$$g_{ij} = \begin{Vmatrix} 1 & 0 & 0 & 0 \\ 0 & 1 & 0 & 0 \\ 0 & 0 & 1 & 0 \\ 0 & 0 & 0 & -1 \end{Vmatrix}.$$

Consider the set of all transformations of basis Э_i:

$$\text{Э}'_i = l^j{}_i \boldsymbol{e}_j, \tag{3.1}$$

defined by the condition of invariance of the scalar products g_{ij}:

$$l^m{}_i l^n{}_j g_{mn} = g_{ij}. \tag{3.2}$$

Transformations (3.1) of an orthonormal basis in the space E_4^1 under condition (3.2) are called the Lorentz transformations.

Equations (3.2) for $i = n = 4$ are written in the form

$$\left(l^4{}_4\right)^2 - \sum_{\alpha=1}^{3} \left(l^\alpha{}_4\right)^2 = 1. \tag{3.3}$$

V. A. Zhelnorovich, *Theory of Spinors and Its Application in Physics and Mechanics*, https://doi.org/10.1007/978-3-030-27836-6_3

From (3.3) it follows $\left(l^4{}_4\right)^2 \geqslant 1$ and therefore coefficient $l^4{}_4$ of the Lorentz transformation always either greater than unity or less than minus unity:

$$l^4{}_4 \geqslant +1 \quad \text{or} \quad l^4{}_4 \leqslant -1.$$

Denote by the symbol L the matrix of the transformation coefficients $l^j{}_i$ and by the symbol $\boldsymbol{g}$ the matrix of the components g_{ij}; then condition (3.2) can be written in the matrix form:

$$L^T \boldsymbol{g} L = \boldsymbol{g}, \tag{3.4}$$

where the symbol "T" means transposition. Since $\det L^T = \det L$, then from Eq. (3.4) it follows $(\det L)^2 = 1$. From this it follows that the determinant of matrix L is equal to $+1$ or -1

$$\det L = \pm 1. \tag{3.5}$$

It is easy to see that if transformations L_1 and L_2 of the bases $\mathfrak{I}_i$ satisfy Eq. (3.4) that the product $L_1 L_2$ also satisfy Eq. (3.4). From Eq. (3.5) it follows that for each transformation L, satisfying Eq. (3.4), there always exists the inverse transformation L^{-1}, and L^{-1} also satisfies Eq. (3.4). Thus the set of all Lorentz transformations forms the group O^1_4. O^1_4 is called the Lorentz group.

The Lorentz group may be split into four connected components [18, 25, 54]:

1. The first connected component $\boldsymbol{L}^{\uparrow}_{+}$ consists of the Lorentz transformations, which are defined by the conditions

$$\det \|l^j{}_i\| = +1, \quad l^4{}_4 \geqslant +1.$$

These transformations do not change the direction of time therefore they are also called the proper orthochronous transformations, or the restricted Lorentz transformation. It is easy to see that the set of all transformations from $\boldsymbol{L}^{\uparrow}_{+}$ forms a group, called the proper orthochronous Lorentz group, or the restricted Lorentz group.

2. The second connected component $\boldsymbol{L}^{\uparrow}_{-}$ of the Lorentz group consists of all transformations of the form $L = PG$, where G is the arbitrary restricted Lorentz transformation and P is the space inversion transformation

$$P = \|l^j{}_i\| = \begin{Vmatrix} -1 & 0 & 0 & 0 \\ 0 & -1 & 0 & 0 \\ 0 & 0 & -1 & 0 \\ 0 & 0 & 0 & 1 \end{Vmatrix}.$$

The transformations $L = PG$ are called improper orthochronous transformations. It is obvious that for the improper orthochronous transformations are carried out the relations

$$\det \|l^j{}_i\| = -1, \quad l^4{}_4 \geqslant +1.$$

The set of all transformations $\boldsymbol{L}^{\uparrow} = \boldsymbol{L}_{+}^{\uparrow} \cup \boldsymbol{L}_{-}^{\uparrow}$ forms a group called the orthochronous Lorentz group.

3. The third connected component $\boldsymbol{L}_{-}^{\downarrow}$ of the Lorentz group consists of all transformations $L = TG$, where G is the arbitrary restricted Lorentz transformation; T is the time-reversal transformation, determined by the following matrix:

$$T = \|l^j{}_i\| = \begin{Vmatrix} 1 & 0 & 0 & 0 \\ 0 & 1 & 0 & 0 \\ 0 & 0 & 1 & 0 \\ 0 & 0 & 0 & -1 \end{Vmatrix}.$$

Transformations of the form of $L = TG$ are called the improper non-orthochronous transformations. For the improper non-orthochronous transformations we have

$$\det \|l^j{}_i\| = -1, \quad l^4{}_4 \leqslant -1.$$

The set of all transformations $\boldsymbol{L}_{+}^{\uparrow} \cup \boldsymbol{L}_{-}^{\downarrow}$ forms a group.

4. The fourth connected component $\boldsymbol{L}_{+}^{\downarrow}$ of the Lorentz group consists of all transformations $L = JG$, where G is the arbitrary restricted Lorentz transformation and J is the transformation of total reflection determined by the matrix

$$J = \|l^j{}_i\| = \begin{Vmatrix} -1 & 0 & 0 & 0 \\ 0 & -1 & 0 & 0 \\ 0 & 0 & -1 & 0 \\ 0 & 0 & 0 & -1 \end{Vmatrix}.$$

Transformations of the form $L = JG$ are called the proper non-orthochronous transformations. The proper non-orthochronous transformations are defined by conditions

$$\det \|l^j{}_i\| = +1, \quad l^4{}_4 \leqslant -1.$$

The set of all transformations $\boldsymbol{L}_{+} = \boldsymbol{L}_{+}^{\uparrow} \cup \boldsymbol{L}_{+}^{\downarrow}$ forms a group called the proper Lorentz group. This group consists of all Lorentz transformations for which the equality $\det \|l^j{}_i\| = +1$ is satisfied.

It is easy to see that the identical transformation I and the transformations of reflections P, T, J form a finite discrete group in which multiplication is defined by the equalities

$$\begin{gathered} I^2 = P^2 = T^2 = J^2 = I, \\ PT = TP = J, \quad PI = IP = P, \quad PJ = JP = T, \\ TI = IT = T, \quad TJ = JT = P, \quad JI = IJ = J. \end{gathered} \tag{3.6}$$

The commutative group consisting of elements I, P, T, J with multiplication (3.6), is called the reflection group.

3.1.2 Algebra of the Four-Dimensional Dirac Matrices

In the space E_4^1 the minimum order of the matrices of the spintensor components $\gamma_i = \|\gamma^B{}_{Ai}\|$ $(i = 1, 2, 3, 4)$, satisfying the equation

$$\gamma_i\gamma_j + \gamma_j\gamma_i = 2g_{ij}I, \tag{3.7}$$

is equal to four. In this case the matrices γ_i are called the Dirac matrices.

The system of sixteen different matrices of the fourth order

$$I, \quad \gamma_i, \quad \gamma_{ij} = \gamma_{[i}\gamma_{j]}, \quad \gamma_{ijk} = \gamma_{[i}\gamma_j\gamma_{k]}, \quad \gamma_{ijks} = \gamma_{[i}\gamma_j\gamma_k\gamma_{s]} \tag{3.8}$$

(for example, for $i < j < k < s$) is full and linearly independent. Instead of matrices γ_{ijk} and γ_{ijks} in the space E_4^1 it is convenient to use the matrices γ^5 and $\overset{*}{\gamma}{}^i$, which are defined in terms of γ_i as follows

$$\gamma^5 = -\frac{1}{24}\varepsilon^{ijks}\gamma_i\gamma_j\gamma_k\gamma_s, \quad \overset{*}{\gamma}{}^i = -\frac{1}{6}\varepsilon^{ijks}\gamma_j\gamma_k\gamma_s, \tag{3.9}$$

Here ε^{ijks} are the antisymmetric in all indices contravariant components of the Levi-Civita pseudotensor[1]

$$\varepsilon^{ijks} = -1, \quad \text{if substitution } \begin{pmatrix} i & j & k & s \\ 1 & 2 & 3 & 4 \end{pmatrix} \text{ is even },$$

$$\varepsilon^{ijks} = 1, \quad \text{if substitution } \begin{pmatrix} i & j & k & s \\ 1 & 2 & 3 & 4 \end{pmatrix} \text{ is odd },$$

$$\varepsilon^{ijks} = 0, \quad \text{if among the indices } i, j, k, s \text{ at least two coincide .}$$

---

[1]By virtue of definition the components of the Levi-Civita pseudotensor satisfy the following relations

$$\varepsilon_{pqmn}\varepsilon^{ijks} = -\begin{vmatrix} \delta^i_p & \delta^i_q & \delta^i_m & \delta^i_n \\ \delta^j_p & \delta^j_q & \delta^j_m & \delta^j_n \\ \delta^k_p & \delta^k_q & \delta^k_m & \delta^k_n \\ \delta^s_p & \delta^s_q & \delta^s_m & \delta^s_n \end{vmatrix}, \quad \varepsilon_{pqmn}\varepsilon^{ijkn} = -\begin{vmatrix} \delta^i_p & \delta^i_q & \delta^i_m \\ \delta^j_p & \delta^j_q & \delta^j_m \\ \delta^k_p & \delta^k_q & \delta^k_m \end{vmatrix},$$

$$\varepsilon_{pqmn}\varepsilon^{ijmn} = -2(\delta^i_p\delta^j_q - \delta^j_p\delta^i_q), \quad \varepsilon_{pjmn}\varepsilon^{ijmn} = -6\delta^i_p,$$

$$\varepsilon_{pqmn}\varepsilon^{pqmn} = -24.$$

Thus $\varepsilon^{1234} = -1$, $\varepsilon_{1234} = 1$.

The matrices γ_{ijk} and γ_{ijks} are expressed in terms of matrices γ^5, $\overset{*}{\gamma}{}^i$ by the relations

$$\gamma_{ijk} = -\varepsilon_{ijks}\overset{*}{\gamma}{}^s, \quad \gamma_{ijks} = \gamma^5\varepsilon_{ijks},$$

which can be obtained by contracting definitions (3.9) with components of the Levi-Civita pseudotensor.

The traces of all γ-matrices are equal to zero:

$$\operatorname{tr}\gamma_i = 0, \quad \operatorname{tr}\gamma_{ij} = 0, \quad \operatorname{tr}\overset{*}{\gamma}{}^i = 0, \quad \operatorname{tr}\gamma^5 = 0.$$

It is easy to calculate also the traces of bilinear products of the γ-matrices

$$\begin{gathered}
\operatorname{tr}\left(\gamma_i\gamma^j\right) = 4\delta_i^j, \quad \operatorname{tr}\left(\overset{*}{\gamma}_i\overset{*}{\gamma}{}^j\right) = 4\delta_i^j, \\
\operatorname{tr}\left(\gamma^{ij}\gamma_{ks}\right) = 4\left(\delta_s^i\delta_k^j - \delta_k^i\delta_s^j\right), \\
\operatorname{tr}\left(\gamma_i\gamma_{js}\right) = 0, \quad \operatorname{tr}\left(\gamma_{js}\gamma_i\right) = 0, \\
\operatorname{tr}\left(\overset{*}{\gamma}_i\gamma_{js}\right) = 0, \quad \operatorname{tr}\left(\gamma_{js}\overset{*}{\gamma}_i\right) = 0, \\
\operatorname{tr}\left(\gamma_i\gamma^5\right) = 0, \quad \operatorname{tr}\left(\gamma^5\gamma_i\right) = 0, \\
\operatorname{tr}\left(\overset{*}{\gamma}_i\gamma^5\right) = 0, \quad \operatorname{tr}\left(\gamma^5\overset{*}{\gamma}_i\right) = 0, \\
\operatorname{tr}\left(\gamma_{ij}\gamma^5\right) = 0, \quad \operatorname{tr}\left(\gamma^5\gamma_{ij}\right) = 0, \\
\operatorname{tr}\left(\overset{*}{\gamma}_i\gamma_j\right) = 0, \quad \operatorname{tr}\left(\gamma_j\overset{*}{\gamma}_i\right) = 0, \\
\operatorname{tr}\left(\gamma^5\gamma^5\right) = -4.
\end{gathered} \tag{3.10}$$

The first relation in (3.10) can be obtained, calculating the trace of Eq. (3.7); the other equations in (3.10) can be obtained, calculating the trace of Eq. (3.7), multiplied beforehand by the matrices γ_i, γ_{ij}, $\overset{*}{\gamma}_i$, and γ^5.

For products of the γ-matrices are valid the following relations, which are contained in identities (1.15) for $\nu = 2$:

$$\begin{aligned}
&\gamma^i\gamma^j = \overset{*}{\gamma}{}^i\overset{*}{\gamma}{}^j = \gamma^{ij} + g^{ij}I, \\
&\gamma^5\overset{*}{\gamma}{}^i = -\overset{*}{\gamma}{}^i\gamma^5 = \gamma^i, \quad \gamma^i\gamma^5 = -\gamma^5\gamma^i = \overset{*}{\gamma}{}^i, \\
&\gamma^i\overset{*}{\gamma}{}^j = -\overset{*}{\gamma}{}^i\gamma^j = \gamma^5 g^{ij} + \frac{1}{2}\varepsilon^{ijks}\gamma_{ks}, \\
&\gamma^{ij}\gamma^5 = \gamma^5\gamma^{ij} = \frac{1}{2}\varepsilon^{ijks}\gamma_{ks}, \quad \gamma^5\gamma^5 = -I,
\end{aligned}$$

$$\begin{aligned}
\overset{*}{\gamma}{}^s \gamma^{ij} &= \varepsilon^{sijk}\gamma_k + g^{is}\overset{*}{\gamma}{}^j - g^{js}\overset{*}{\gamma}{}^i,\\
\gamma^{ij}\overset{*}{\gamma}{}^s &= \varepsilon^{sijk}\gamma_k - g^{is}\overset{*}{\gamma}{}^j + g^{js}\overset{*}{\gamma}{}^i,\\
\gamma^s\gamma^{ij} &= -\varepsilon^{sijk}\overset{*}{\gamma}_k + g^{is}\gamma^j - g^{js}\gamma^i,\\
\gamma^{ij}\gamma^s &= -\varepsilon^{sijk}\overset{*}{\gamma}_k - g^{is}\gamma^j + g^{js}\gamma^i,\\
\gamma^{ij}\gamma^{ks} &= \gamma^5\varepsilon^{ijks} + \left(\delta^i_m\delta^j_n - \delta^j_m\delta^i_n\right)\left(\gamma^{ms}g^{nk} - \gamma^{mk}g^{ns}\right)\\
&\quad + \left(g^{is}g^{jk} - g^{ik}g^{js}\right) I. \qquad (3.11)
\end{aligned}$$

Let us note that the form of Eqs. (3.11) does not depend on the order of matrices γ_i, satisfying Eq. (3.7).

The invariant metric spinor $E = \|e_{BA}\|$ in the pseudo-Euclidean space E^1_4 is defined by the equation[2]

$$\gamma_i^T = -E\gamma_i E^{-1}. \qquad (3.12)$$

up to multiplying by an arbitrary nonzero complex number.

According to Eqs. (1.53) and (1.55) the matrices E, $E\gamma_i$, $E\gamma_{ij}$, $E\gamma_{ijk}$, and $E\gamma_{ijks}$ have the following symmetry properties

$$\begin{aligned}
E^T = -E, \quad (E\gamma_i)^T = E\gamma_i, \quad \left(E\gamma_{ij}\right)^T = E\gamma_{ij},\\
\left(E\gamma_{ijk}\right)^T = -E\gamma_{ijk}, \quad \left(E\gamma_{ijks}\right)^T = -E\gamma_{ijks}. \qquad (3.13)
\end{aligned}$$

From definitions (3.9) and Eqs. (3.13) it follows also

$$\left(E\overset{*}{\gamma}{}^i\right)^T = -E\overset{*}{\gamma}{}^i, \quad \left(E\gamma^5\right)^T = -E\gamma^5. \qquad (3.14)$$

From Eqs. (3.13) it is seen that the matrix of the metric spinor components E in the space E^1_4 is antisymmetric. Therefore we must bear in mind that for contraction of the components of two spinors the equality takes place

$$\psi_A\xi^A = -\psi^A\xi_A,$$

and for contraction of the metric spinor components the following equalities hold

$$e^{AB}e^{CD}e_{BD} = -e^{AC}, \quad e_{AB}e_{CD}e^{BD} = -e_{AC}.$$

[2]In physical literature the matrix of the components of the second rank spinor $E^{-1} = \|e^{BA}\|$ is usually denoted by symbol C (in the Dirac theory the matrix C is operator of the charge conjugation).

Since the components of the metric spinor e_{BA} are antisymmetric, we see that the spinor components

$$\varepsilon^{ABCD} = -e^{AB}e^{CD} + e^{AC}e^{BD} - e^{AD}e^{BC},$$
$$\varepsilon_{ABCD} = -e_{AB}e_{CD} + e_{AC}e_{BD} - e_{AD}e_{BC} \tag{3.15}$$

are antisymmetric in all indices. It is easy to verify that the components of spinors ε_{ABCD} and ε^{ABCD} are connected by the relations

$$\varepsilon^{ABCD} = e^{AA'}e^{BB'}e^{CC'}e^{DD'}\varepsilon_{A'B'C'D'},$$
$$\varepsilon_{ABCD} = e_{AA'}e_{BB'}e_{CC'}e_{DD'}\varepsilon^{A'B'C'D'}.$$

By virtue of definition (3.15) for contractions of the components ε_{ABCD} and ε^{ABCD} are carried out the equalities

$$\varepsilon_{ABCD}\varepsilon^{EFQH} = \begin{vmatrix} \delta^E_A & \delta^E_B & \delta^E_C & \delta^E_D \\ \delta^F_A & \delta^F_B & \delta^F_C & \delta^F_D \\ \delta^Q_A & \delta^Q_B & \delta^Q_C & \delta^Q_D \\ \delta^H_A & \delta^H_B & \delta^H_C & \delta^H_D \end{vmatrix}, \quad \varepsilon_{ABCD}\varepsilon^{EFQD} = \begin{vmatrix} \delta^E_A & \delta^E_B & \delta^E_C \\ \delta^F_A & \delta^F_B & \delta^F_C \\ \delta^Q_A & \delta^Q_B & \delta^Q_C \end{vmatrix},$$

$$\varepsilon_{ABCD}\varepsilon^{EFCD} = 2(\delta^E_A\delta^F_B - \delta^F_A\delta^E_B), \quad \varepsilon_{AFCD}\varepsilon^{EFCD} = 6\delta^E_A,$$

$$\varepsilon_{ABCD}\varepsilon^{ABCD} = 24.$$

As it was already noted, the components of the metric spinor e_{AB} are determined by Eq. (3.12) up to multiplying by an arbitrary nonzero complex number $\lambda \neq 0$. Therefore it is always possible to define e_{AB} so that the equality was carried out

$$\varepsilon_{1234} = -e_{12}e_{34} + e_{13}e_{24} - e_{14}e_{23} = 1. \tag{3.16}$$

In this case the components ε_{ABCD} (and ε^{ABCD}) are the Levi-Civita symbols, which determine an invariant spinor of the fourth rank in the space E^1_4. This condition can be accepted as the normalization condition for definition of the components of the metric spinor e_{AB}; obviously that as a result of such normalization the components of the metric spinor are determined up to simultaneous multiplication of all components e_{AB} only by -1.

The invariant spinor of the second rank $\beta_{\dot{B}A}$ in pseudo-Euclidean space E^1_4 is defined by the equations

$$\dot{\gamma}^T_i = -\beta\gamma_i\beta^{-1}, \quad \dot{\beta}^T = \beta. \tag{3.17}$$

The following relations are also valid (see (1.128))

$$\dot{\gamma}_{ij}^{T} = -\beta\gamma_{ij}\beta^{-1}, \quad \dot{\gamma}_{ijk}^{T} = \beta\gamma_{ijk}\beta^{-1}, \quad \dot{\gamma}_{ijks}^{T} = \beta\gamma_{ijks}\beta^{-1}. \tag{3.18a}$$

It is obvious that from the last two equations in (3.18a) it follows

$$(\overset{*}{\gamma}{}_{i}^{T})^{\cdot} = \beta\overset{*}{\gamma}_{i}\beta^{-1}, \quad (\dot{\gamma}^{5})^{T} = \beta\gamma^{5}\beta^{-1}. \tag{3.18b}$$

From Eqs. (3.17) and (3.18) we see that matrices of the spintensor components $\beta\gamma_i$, $\beta\gamma_{ij}$, $\beta\gamma_{ijk}$, $\beta\gamma_{ijks}$, $\beta\overset{*}{\gamma}{}^{i}$, $\beta\gamma^5$ have the following symmetry properties:

$$\begin{aligned}
(\beta\gamma_i)^T &= -(\beta\gamma_i)^{\cdot}, & (\beta\gamma_{ij})^T &= -(\beta\gamma_{ij})^{\cdot},\\
(\beta\gamma_{ijk})^T &= (\beta\gamma_{ijk})^{\cdot}, & (\beta\gamma_{ijks})^T &= (\beta\gamma_{ijks})^{\cdot},\\
(\beta\overset{*}{\gamma}{}^{i})^T &= (\beta\overset{*}{\gamma}{}^{i})^{\cdot}, & (\beta\gamma^5)^T &= (\beta\gamma^5)^{\cdot}.
\end{aligned} \tag{3.19}$$

The symmetry properties (3.13) and (3.19) have invariant character and do not depend on a specific expression of the matrices γ_i.

The matrix of the components of the invariant spinor $\Pi = \|\Pi^{B}{}_{\dot{A}}\| = E^{-1}\beta^{T}$ in four-dimensional pseudo-Euclidean space E_4^1, according to equalities (1.136), (1.137), and (1.141), satisfies the equations

$$\begin{aligned}
\Pi\dot{\Pi} &= I,\\
\dot{\gamma}_i = \Pi^{-1}\gamma_i\Pi, &\quad \dot{\gamma}_{ij} = \Pi^{-1}\gamma_{ij}\Pi,\\
(\overset{*}{\gamma}_i)^{\cdot} = \Pi^{-1}\overset{*}{\gamma}_i\Pi, &\quad \dot{\gamma}^5 = \Pi^{-1}\gamma^5\Pi.
\end{aligned} \tag{3.20}$$

The Pauli identity (1.19) (with the covariant spinor indices) in the space E_4^1 takes the form

$$4e_{DE}e_{BA} = -e_{DA}e_{EB} + \gamma_{iDA}\gamma^{i}_{EB} - \frac{1}{2}\gamma_{ijDA}\gamma^{ij}_{EB} - \overset{*}{\gamma}_{iDA}\overset{*}{\gamma}{}^{i}_{EB} + \overset{5}{\gamma}_{DA}\overset{5}{\gamma}_{EB}. \tag{3.21}$$

The expressions for bilinear products of any spintensors $\boldsymbol{\gamma}$ are given in Appendix C.

Many different sets of the specific numerical matrices γ_i, satisfying Eqs. (3.7), are known and in different problems it is convenient to use the different presentations for γ_i. In particular, if γ_i are chosen in such a way that γ_2, $i\gamma_4$ are Hermitian and symmetric, while γ_1, γ_3 are Hermitian and antisymmetric:

$$\begin{aligned}
\gamma_1^T &= \dot{\gamma}_1, & \gamma_2^T &= \dot{\gamma}_2, & \gamma_3^T &= \dot{\gamma}_3, & \gamma_4^T &= -\dot{\gamma}_4,\\
\gamma_1^T &= -\gamma_1, & \gamma_2^T &= \gamma_2, & \gamma_3^T &= -\gamma_3, & \gamma_4^T &= \gamma_4,
\end{aligned} \tag{3.22}$$

then from definitions (3.12) and (3.17) it follows that the invariant spinors E and β can be defined by the equalities

$$E = \|e_{BA}\| = -\mathrm{i}\gamma_2\gamma_4, \quad \beta = \|\beta_{\dot{B}A}\| = \mathrm{i}\gamma^4 = -\mathrm{i}\gamma_4. \tag{3.23}$$

As matrices γ_i, for example, can be taken the following matrices (the spinor representation of the matrices γ_i):

$$\gamma_1 = \begin{Vmatrix} 0 & 0 & 0 & \mathrm{i} \\ 0 & 0 & \mathrm{i} & 0 \\ 0 & -\mathrm{i} & 0 & 0 \\ -\mathrm{i} & 0 & 0 & 0 \end{Vmatrix}, \quad \gamma_2 = \begin{Vmatrix} 0 & 0 & 0 & 1 \\ 0 & 0 & -1 & 0 \\ 0 & -1 & 0 & 0 \\ 1 & 0 & 0 & 0 \end{Vmatrix},$$

$$\gamma_3 = \begin{Vmatrix} 0 & 0 & \mathrm{i} & 0 \\ 0 & 0 & 0 & -\mathrm{i} \\ -\mathrm{i} & 0 & 0 & 0 \\ 0 & \mathrm{i} & 0 & 0 \end{Vmatrix}, \quad \gamma_4 = \begin{Vmatrix} 0 & 0 & \mathrm{i} & 0 \\ 0 & 0 & 0 & \mathrm{i} \\ \mathrm{i} & 0 & 0 & 0 \\ 0 & \mathrm{i} & 0 & 0 \end{Vmatrix}. \tag{3.24}$$

In this case the components of invariant spinors of the second rank E, γ^5, β, and Π can be defined by the matrices

$$E = \begin{Vmatrix} 0 & 1 & 0 & 0 \\ -1 & 0 & 0 & 0 \\ 0 & 0 & 0 & -1 \\ 0 & 0 & 1 & 0 \end{Vmatrix}, \quad \gamma^5 = \begin{Vmatrix} -\mathrm{i} & 0 & 0 & 0 \\ 0 & -\mathrm{i} & 0 & 0 \\ 0 & 0 & \mathrm{i} & 0 \\ 0 & 0 & 0 & \mathrm{i} \end{Vmatrix},$$

$$\beta = \begin{Vmatrix} 0 & 0 & 1 & 0 \\ 0 & 0 & 0 & 1 \\ 1 & 0 & 0 & 0 \\ 0 & 1 & 0 & 0 \end{Vmatrix}, \quad \Pi = \begin{Vmatrix} 0 & 0 & 0 & -1 \\ 0 & 0 & 1 & 0 \\ 0 & 1 & 0 & 0 \\ -1 & 0 & 0 & 0 \end{Vmatrix}. \tag{3.25}$$

Sometimes the other set of the matrices γ_i is used (standard representation or the Pauli representation)

$$\gamma_1 = \begin{Vmatrix} 0 & 0 & 0 & -\mathrm{i} \\ 0 & 0 & -\mathrm{i} & 0 \\ 0 & \mathrm{i} & 0 & 0 \\ \mathrm{i} & 0 & 0 & 0 \end{Vmatrix}, \quad \gamma_2 = \begin{Vmatrix} 0 & 0 & 0 & -1 \\ 0 & 0 & 1 & 0 \\ 0 & 1 & 0 & 0 \\ -1 & 0 & 0 & 0 \end{Vmatrix},$$

$$\gamma_3 = \begin{Vmatrix} 0 & 0 & -\mathrm{i} & 0 \\ 0 & 0 & 0 & \mathrm{i} \\ \mathrm{i} & 0 & 0 & 0 \\ 0 & -\mathrm{i} & 0 & 0 \end{Vmatrix}, \quad \gamma_4 = \begin{Vmatrix} \mathrm{i} & 0 & 0 & 0 \\ 0 & \mathrm{i} & 0 & 0 \\ 0 & 0 & -\mathrm{i} & 0 \\ 0 & 0 & 0 & -\mathrm{i} \end{Vmatrix}. \tag{3.26}$$

In this case for the matrices E, γ^5, β, and Π we have

$$E = \begin{Vmatrix} 0 & 0 & 0 & 1 \\ 0 & 0 & -1 & 0 \\ 0 & 1 & 0 & 0 \\ -1 & 0 & 0 & 0 \end{Vmatrix}, \quad \gamma^5 = \begin{Vmatrix} 0 & 0 & -i & 0 \\ 0 & 0 & 0 & -i \\ -i & 0 & 0 & 0 \\ 0 & -i & 0 & 0 \end{Vmatrix},$$

$$\beta = \begin{Vmatrix} 1 & 0 & 0 & 0 \\ 0 & 1 & 0 & 0 \\ 0 & 0 & -1 & 0 \\ 0 & 0 & 0 & -1 \end{Vmatrix}, \quad \Pi = \begin{Vmatrix} 0 & 0 & 0 & 1 \\ 0 & 0 & -1 & 0 \\ 0 & -1 & 0 & 0 \\ 1 & 0 & 0 & 0 \end{Vmatrix}. \tag{3.27}$$

In the four-dimensional pseudo-Euclidean space E_4^1 there is the representation in which all matrices γ_i are real (the Majorana representation)[3]

$$\gamma_1 = \begin{Vmatrix} 1 & 0 & 0 & 0 \\ 0 & 1 & 0 & 0 \\ 0 & 0 & -1 & 0 \\ 0 & 0 & 0 & -1 \end{Vmatrix}, \quad \gamma_2 = \begin{Vmatrix} 0 & 0 & 0 & -1 \\ 0 & 0 & 1 & 0 \\ 0 & 1 & 0 & 0 \\ -1 & 0 & 0 & 0 \end{Vmatrix},$$

$$\gamma_3 = \begin{Vmatrix} 0 & 0 & 1 & 0 \\ 0 & 0 & 0 & 1 \\ 1 & 0 & 0 & 0 \\ 0 & 1 & 0 & 0 \end{Vmatrix}, \quad \gamma_4 = \begin{Vmatrix} 0 & 0 & 0 & -1 \\ 0 & 0 & -1 & 0 \\ 0 & 1 & 0 & 0 \\ 1 & 0 & 0 & 0 \end{Vmatrix}. \tag{3.28}$$

In this case spintensors E, β, Π and γ^5 can be defined as follows

$$E = -\beta = \begin{Vmatrix} 0 & 0 & 0 & -i \\ 0 & 0 & -i & 0 \\ 0 & i & 0 & 0 \\ i & 0 & 0 & 0 \end{Vmatrix}, \quad \gamma^5 = \begin{Vmatrix} 0 & 0 & 1 & 0 \\ 0 & 0 & 0 & -1 \\ -1 & 0 & 0 & 0 \\ 0 & 1 & 0 & 0 \end{Vmatrix},$$

$$\Pi = I = \begin{Vmatrix} 1 & 0 & 0 & 0 \\ 0 & 1 & 0 & 0 \\ 0 & 0 & 1 & 0 \\ 0 & 0 & 0 & 1 \end{Vmatrix}. \tag{3.29}$$

---

[3]Real representations of the $\boldsymbol{\gamma}$-matrices exist in four-dimensional pseudo-Euclidean spaces E_4^1, E_4^2. In the space E_4^3 (and in E_4^2) there are purely imaginary representations $\boldsymbol{\gamma}$-matrices, obtained from real $\boldsymbol{\gamma}$-matrices in spaces E_4^1, E_4^2 by multiplication by imaginary unit $i = \sqrt{-1}$. Therefore in the spaces E_4^1, E_4^2 and E_4^3 there are the real spinor representations of corresponding pseudo-orthogonal groups. In four-dimensional pseudo-Euclidean space E_4^2 with the metric signature $(+, +, -, -)$ there are also real semispinors. In spaces E_4^0, E_4^4 do not exist real and imaginary representations for γ_i and therefore in these spaces there are no real spinor representations (see in connection with it also Chap. 1).

For all definitions (3.25), (3.27), and (3.29) of the metric spinor E the normalization condition (3.16) is fulfilled.

3.1.3 The Spinor Representations of the Lorentz Group

According to the results of Sect. 1.6.4 the spinor representations of the Lorentz group O_4^1 may be split into four classes.

I. The first class of spinor representations of the Lorentz group can be defined by the equations

$$l^j{}_i \gamma_j = S^{-1} \gamma_i S \tag{3.30}$$

and by one of the normalization condition

$$\begin{aligned}
&a. \quad S^T E S = E, \\
&b. \quad S^T E S = E \operatorname{sign} \Delta, \qquad \Delta = \det \|l^j{}_i\|, \\
&c. \quad S^T E S = E \operatorname{sign} (\Delta l^4{}_4), \\
&d. \quad S^T E S = E \operatorname{sign} l^4{}_4.
\end{aligned} \tag{3.31}$$

In this case

$$\dot{S}^T \beta S = \beta \operatorname{sign} l^4{}_4. \tag{3.32}$$

If matrices γ_i satisfy conditions (3.22), then the spinor transformations S_P, S_T, S_J, corresponding to reflection transformations P, T, J, are defined by the table

	a	b	c	d
S_P	γ_4	$i\gamma_4$	$i\gamma_4$	γ_4
S_T	$i\overset{*}{\gamma}_4$	$\overset{*}{\gamma}_4$	$i\overset{*}{\gamma}_4$	$\overset{*}{\gamma}_4$
S_J	$i\gamma^5$	$i\gamma^5$	γ^5	γ^5

(3.33)

Direct check shows that the spinor transformations S_P, S_T, S_J for all normalization conditions in (3.31) anti-commute with one another:

$$S_P S_T = -S_T S_P, \quad S_P S_J = -S_J S_P, \quad S_T S_J = -S_J S_T.$$

II. The second class of spinor representations of the Lorentz group is defined by the equations

$$l^j{}_i \gamma_j = S^{-1} \gamma_i S \operatorname{sign} \Delta \tag{3.34}$$

and by one of the normalization conditions in (3.31). The spinor transformations S and the Hermitian conjugate transformations $\dot{S}^T$ for this class of representations are connected by the relation

$$\dot{S}^T \beta S = \beta \operatorname{sign}(\Delta l^4{}_4). \tag{3.35}$$

If Eqs. (3.22) are valid, then the spinor transformations S_P, S_T, S_J, corresponding to the reflections P, T, J, are defined by the table

	a	b	c	d
S_P	$\mathrm{i}\overset{*}{\gamma}_4$	$\overset{*}{\gamma}_4$	$\overset{*}{\gamma}_4$	$\mathrm{i}\overset{*}{\gamma}_4$
S_T	γ_4	$\mathrm{i}\gamma_4$	γ_4	$\mathrm{i}\gamma_4$
S_J	$\mathrm{i}\gamma^5$	$\mathrm{i}\gamma^5$	γ^5	γ^5

(3.36)

The spinor transformations S_P, S_T, S_J determined by table (3.36) anti-commute among themselves.

It is easy to show that all spinor representations $O_4^1 \rightarrow \{\pm S\}$, defined by Eqs. (3.34) and (3.31), are equivalent to the spinor representations, defined by Eqs. (3.30) and (3.31).

III. Let us define the third class of spinor representations of the Lorentz group by the equations

$$l^j{}_i \gamma_j = S^{-1} \gamma_i S \operatorname{sign}\left(\Delta l^4{}_4\right) \tag{3.37}$$

and by one of the normalization conditions (3.31). For the considered spinor representations we have

$$\dot{S}^T \beta S = \beta \operatorname{sign} \Delta. \tag{3.38}$$

Let us write out also the spinor transformations S_P, S_T, S_J, corresponding to the reflection transformations P, T, J:

	a	b	c	d
S_P	$\mathrm{i}\overset{*}{\gamma}_4$	$\overset{*}{\gamma}_4$	$\overset{*}{\gamma}_4$	$\mathrm{i}\overset{*}{\gamma}_4$
S_T	$\mathrm{i}\overset{*}{\gamma}_4$	$\overset{*}{\gamma}_4$	$\mathrm{i}\overset{*}{\gamma}_4$	$\overset{*}{\gamma}_4$
S_J	I	I	$\mathrm{i}I$	$\mathrm{i}I$

(3.39)

Formulas (3.39) are obtained under assumption (3.22). The spinor transformations S_P, S_T, S_J defined by table (3.39) commute among themselves

$$S_P S_T = S_T S_P, \quad S_P S_J = S_J S_P, \quad S_T S_J = S_J S_T.$$

IV. The fourth class of spinor representations of the Lorentz group is defined by the equations

$$l^j{}_i\gamma_j = S^{-1}\gamma_i S \operatorname{sign} l^4{}_4 \tag{3.40}$$

and by one of the normalization conditions (3.31). Under any condition of the normalization (3.31) due to Eqs. (3.40) it is fulfilled also the equation

$$\dot{S}^T \beta S = \beta. \tag{3.41}$$

If Eqs. (3.22) are valid, then the spinor transformations S_P, S_T, S_J corresponding to the reflection transformations P, T, J are defined as follows

	a	b	c	d
S_P	γ_4	$i\gamma_4$	$i\gamma_4$	γ_4
S_T	γ_4	$i\gamma_4$	γ_4	$i\gamma_4$
S_J	I	I	iI	iI

(3.42)

Under any condition of the normalization, the spinor transformations S_P, S_T, S_J defined by table (3.42) commute with each other.

All spinor representations $O_4^1 \to \{\pm S\}$, defined by Eqs. (3.40) and (3.31), are equivalent to the spinor representations defined by Eqs. (3.37) and (3.31) respectively.

The spinor representations described above contain eight nonequivalent representations; for example, eight representations in I and III, I and IV, II and III, II and IV classes are nonequivalent.

It is obvious that the spinor representation of the restricted Lorentz group is the same in all classes of the spinor representations considered above and is defined by the equations

$$l^j{}_i\gamma_j = S^{-1}\gamma_i S, \quad S^T E S = E. \tag{3.43}$$

Similar to the derivation of Eq. (1.69) in the complex Euclidean spaces, we obtain that the spinor transformation corresponding to the small restricted Lorentz transformation

$$l^j{}_i = \delta_i^j + \delta\varepsilon_i{}^j,$$

has the form

$$S = I + \frac{1}{4}\gamma^{ij}\delta\varepsilon_{ij}. \tag{3.44}$$

It is easy to find the spinor transformations S in an explicit form for some finite restricted Lorentz transformations. In particular, for restricted Lorentz transformation — rotation through an angle φ in the plane passing through the basis vectors $\mathcal{E}_\alpha$ and $\mathcal{E}_\beta$ ($\alpha, \beta = 1, 2, 3; \alpha \neq \beta$)

$$\begin{aligned} \mathcal{E}'_\alpha &= \quad \mathcal{E}_\alpha \cos\varphi + \mathcal{E}_\beta \sin\varphi, \\ \mathcal{E}'_\beta &= -\mathcal{E}_\alpha \sin\varphi + \mathcal{E}_\beta \cos\varphi, \end{aligned}$$

we have

$$S = I\cos\frac{\varphi}{2} + \gamma^{\alpha\beta}\sin\frac{\varphi}{2}. \tag{3.45}$$

It is seen from formula (3.45) that under continuous rotation of basis $\mathcal{E}_\alpha$ through the angle 2π, the matrix S changes a sign:

$$S(\varphi + 2\pi) = -S(\varphi).$$

But the rotation through 2π coincides with the identical transformation; thus, to the transformation of rotation through the angle φ formula (3.45) puts in correspondence two matrices S and $-S$.

For the Lorentz transformation (boost or hyperbolic rotation through the angle φ in the plane passing through the basis vectors $\mathcal{E}_\alpha$ and $\mathcal{E}_4$)

$$\begin{aligned} \mathcal{E}'_\alpha &= \quad \mathcal{E}_\alpha \cosh\varphi - \mathcal{E}_4 \sinh\varphi, \\ \mathcal{E}'_4 &= -\mathcal{E}_\alpha \sinh\varphi + \mathcal{E}_4 \cosh\varphi, \end{aligned}$$

we have

$$S = I\cosh\frac{\varphi}{2} - \gamma^{4\alpha}\sinh\frac{\varphi}{2}.$$

3.1.4 Spinors in the Pseudo-Euclidean Space E_4^1

Invariant geometric object $\boldsymbol{\psi} = \pm\psi^A \boldsymbol{\varepsilon}_A$, where the pairs of contravariant components $\pm\psi^A$ and spinbases $\pm\{\boldsymbol{\varepsilon}_A\}$ are referred to an orthonormal basis $\mathcal{E}_i$ in the pseudo-Euclidean space E_4^1 and under the Lorentz transformation of bases $\mathcal{E}_i$ are transformed according to the spinor representation of the Lorentz group, is called a first-rank spinor in the pseudo-Euclidean space E_4^1.

The covariant components ψ_A of the spinor $\boldsymbol{\psi}$ are defined by means of the invariant metric spinor $E = \|e_{AB}\|$:

$$\psi_A = e_{AB}\psi^B. \tag{3.46}$$

Note that the contraction in the right-hand side of Eq. (3.46) is performed over the second index of e_{AB}.

The conjugate spinor $\boldsymbol{\psi}^+$ in the space E_4^1 is defined by the covariant components ψ_A^+ or contravariant components ψ^{+A}, which are expressed in terms of the complex conjugate components of spinor $\dot{\psi}^B$ by the equalities

$$\psi_A^+ = \beta_{\dot{B}A}\dot{\psi}^B, \quad \psi^{+A} = \Pi^A{}_{\dot{B}}\dot{\psi}^B. \tag{3.47}$$

The components of the second rank invariant spinors $\Pi = \|\Pi^A{}_{\dot{B}}\|$ and $\beta = \|\beta_{\dot{B}A}\|$ in Eqs. (3.47) are connected by the relations

$$\Pi^A{}_{\dot{B}} = e^{AC}\beta_{\dot{B}C}, \quad \beta_{\dot{B}A} = e_{AC}\Pi^C{}_{\dot{B}}$$

or, in the matrix form

$$\Pi = E^{-1}\beta^T, \quad \beta = (E\Pi)^T$$

From Eqs. (3.31), (3.32), (3.35), (3.38), and (3.41) it follows that components $\beta_{\dot{B}A}$, $\Pi^A{}_{\dot{B}}$ determine the spinors of the second rank with one dotted index, which are invariant, in any case, under the continuous Lorentz transformations.

Definitions (3.46) and (3.47) of the covariant components ψ_A^+, ψ_A, and the contravariant components ψ^{+A} can be written in the matrix form

$$\psi^+ = \dot{\psi}^T\beta, \quad \widetilde{\psi} = (E\psi)^T = -\psi^T E, \quad \bar{\psi} = \Pi\dot{\psi}. \tag{3.48}$$

Here ψ^+ and $\widetilde{\psi}$ denote the row of the covariant components ψ_A^+ and ψ_A respectively; $\bar{\psi}$ is the column of the contravariant components of conjugate spinor ψ^{+A}.

Since $\Pi\dot{\Pi} = I$ (see (3.20)), we obtain that in the space E_4^1 the conjugation of a conjugate spinor gives

$$(\psi^+)^+ = \psi.$$

3.2 Tensor Representation of Spinors in the Pseudo-Euclidean Space E_4^1

3.2.1 Representation of Spinors in the Space E_4^1 by Complex Tensors

In the pseudo-Euclidean space E_4^1 an expansion of the contravariant components of the second-rank spinor in invariant spintensors $\boldsymbol{\gamma}$ has the form

$$\psi^{BA} = \frac{1}{4}\Big(- Fe^{BA} + F^j\gamma_j^{BA} + \frac{1}{2}F^{js}\gamma_{js}^{BA} + \overset{*}{F}{}^j\overset{*}{\gamma}{}_j^{BA} + \overset{*}{F}\gamma^{5BA}\Big), \tag{3.49}$$

where the tensor components $\boldsymbol{F}$ are expressed in terms of the spinor components ψ^{BA} as follows

$$F = e_{BA}\psi^{BA}, \quad F^j = -\gamma^j_{BA}\psi^{BA}, \quad F^{js} = \gamma^{js}_{BA}\psi^{BA},$$
$$\overset{*}{F}{}^j = \overset{*}{\gamma}{}^j_{BA}\psi^{BA}, \quad \overset{*}{F} = -\gamma^5_{BA}\psi^{BA}.$$

If the components of a spinor ψ^{BA} are symmetric $\psi^{BA} = \psi^{AB} = \psi^{(BA)}$, then by virtue of the symmetry properties of spintensors (3.14), only the terms with F^j and F^{js} remain in formula (3.49):

$$\psi^{(BA)} = \frac{1}{4}\left(F^j\gamma_j^{BA} + \frac{1}{2}F^{js}\gamma_{js}^{BA}\right).$$

If the components of spinor ψ^{BA} are antisymmetric $\psi^{BA} = -\psi^{AB} = \psi^{[BA]}$, then formula (3.49) takes the form

$$\psi^{[BA]} = \frac{1}{4}\Big(- Fe^{BA} + \overset{*}{F}{}^j\overset{*}{\gamma}{}_j^{BA} + \overset{*}{F}\gamma^{5BA}\Big).$$

From the symmetry properties (3.14) and (3.13) it follows that the complex tensors $\boldsymbol{C}$, defined by the spinor $\boldsymbol{\psi}$, in the pseudo-Euclidean space E_4^1 consist only of the vector and antisymmetric second-rank tensor

$$\boldsymbol{C} = \{C^i \boldsymbol{\ni}_i, C^{ij}\boldsymbol{\ni}_i\boldsymbol{\ni}_j\}.$$

The components C^i, C^{ij} in an orthonormal basis $\boldsymbol{\ni}_i$ are defined by the equalities

$$C^i = \gamma^i_{BA}\psi^B\psi^A, \quad C^{ij} = \gamma^{ij}_{BA}\psi^B\psi^A \tag{3.50}$$

or, in the matrix notations

$$C^i = \psi^T E\gamma^i\psi, \quad C^{ij} = \psi^T E\gamma^{ij}\psi. \tag{3.51}$$

Here ψ is the column of the contravariant components of the first-rank spinor ψ^A. Contracting identities (C.1)[4] with the spinor components $\psi^B\psi^A\psi^D\psi^E$ with respect to the indices B, A, D, E and using definitions (3.50), we obtain that the components C^i, C^{ij} satisfy the following invariant algebraic equations [37]

$$C_{ij}C^{ij} = 0, \tag{3.52a}$$

$$\varepsilon_{ijks}C^{ij}C^{ks} = 0, \tag{3.52b}$$

[4]See Appendix C.

$$C^i C^j + C^i{}_s C^{sj} = 0. \tag{3.52c}$$

Among Eqs. (3.52) there are the six independent equations. For example, Eqs. (3.52a), (3.52b) and four equations in (3.52c) for $i = j$ are independent.

By virtue of Eqs. (3.52) also the equations hold

$$C_i C^i = 0, \quad C_i C^{ij} = 0, \quad \varepsilon_{ijks} C^j C^{ks} = 0, \tag{3.53}$$

which can be obtained by contracting identities (C.1) with the spinor components $\psi^A \psi^B \psi^D \psi^E$.

According to the general formulas (1.191) and (1.195) the components of spinor ψ^A are defined by the tensor components C^i, C^{ij} as follows

$$\psi^A = \frac{\psi^{BA}\eta_B}{\pm\sqrt{\psi^{CD}\eta_C\eta_D}}, \quad \psi^{BA} = \frac{1}{4}\left(-C^i \gamma_i^{BA} + \frac{1}{2} C^{ij} \gamma_{ij}^{BA}\right). \tag{3.54}$$

Here η_C $(C = 1, 2, 3, 4)$ are arbitrary complex numbers satisfying the condition $\psi^{CD}\eta_C\eta_D \neq 0$; the components of the spintensors γ_i^{BA} and γ_{ij}^{BA} are defined by the relations

$$\gamma_i^{BA} = e^{AC}\gamma^B{}_{Ci}, \quad \gamma_{ij}^{BA} = e^{AC}\gamma^B{}_{Cij},$$
$$\|\gamma^B{}_{Ci}\| = \gamma_i, \quad \|\gamma^B{}_{Cij}\| = \gamma_{[i}\gamma_{j]}.$$

The right-hand side of the first formula in (3.54) does not depend on the choice of the components η_C, for which $\psi^{CD}\eta_C\eta_D \neq 0$.

Instead of the first formula in (3.54) we can use the simpler relation

$$\psi^A = \frac{\psi^{BA}}{\pm\sqrt{\psi^{BB}}},$$

where we do not sum over B. The right-hand side of this formula does not depend on the value of the index B, for which $\psi^{BB} \neq 0$.

The second formula in (3.54) for the symmetric components ψ^{BA} can be written in the matrix form

$$\|\psi^{BA}\| = \frac{1}{4}\left(C^i \gamma_i - \frac{1}{2} C^{ij} \gamma_i \gamma_j\right) E^{-1}.$$

Equations (3.50) and (3.54), which realize one-to-one connection between the spinor $\boldsymbol{\psi}$ and the tensors $\boldsymbol{C}$, are invariant with respect to the choice of orthonormal basis $\boldsymbol{\vartheta}_i$ in the space E_4^1. Thus, it is valid the following theorem.[5]

Theorem ([74, 75]) *The spinor of the first rank* $\boldsymbol{\psi}$ *in the pseudo-Euclidean space* E_4^1 *with components* ψ^A *defined up to a common sign, is equivalent to the complex vector and the complex antisymmetric tensor of the second rank, whose components* C^i, C^{ij} *satisfy the six independent algebraic equations in* (3.52).

It follows from Eqs. (3.52) that the components C^{ij} determine an arbitrary antisymmetric tensor with zero invariants, while the vector components C^i are defined by C^{ij} up to the common sign. By means of definitions (3.51) and equation $\gamma^5\gamma^i\gamma^5 = \gamma^i$, following from the third equation in (3.11), it is easy to find that if the tensor components C^i, C^{ij} determine the components of a first-rank spinor ψ, then the tensor components $-C^i$, C^{ij} determine the spinor components $\mathrm{i}\gamma^5\psi$. Therefore if only components of an antisymmetric tensor C^{ij} with zero invariants

$$C_{ij}C^{ij} = 0, \quad \varepsilon_{ijks}C^{ij}C^{ks} = 0 \tag{3.55}$$

are given, then they determine two spinors of the first rank with components ψ and $i\gamma^5\psi$.

If spintensors E and γ_i are defined by matrices (3.24) and (3.25), then the tensor components C^i, C^{ij}, according to definitions (3.51), are defined by the equalities

$$\begin{aligned}
C^1 &= 2\mathrm{i}(\psi^1\psi^3 - \psi^2\psi^4),\\
C^2 &= -2(\psi^1\psi^3 + \psi^2\psi^4),\\
C^3 &= -2\mathrm{i}(\psi^1\psi^4 + \psi^2\psi^3),\\
C^4 &= 2\mathrm{i}(-\psi^1\psi^4 + \psi^2\psi^3),\\
C^{14} &= \psi^1\psi^1 - \psi^2\psi^2 + \psi^3\psi^3 - \psi^4\psi^4,\\
C^{24} &= \mathrm{i}(\psi^1\psi^1 + \psi^2\psi^2 + \psi^3\psi^3 + \psi^4\psi^4),
\end{aligned}$$

---

[5]A connection between the first-rank spinors and complex tensors $\boldsymbol{C}$ in the space E_4^1 for the first time was considered, apparently, by Whittaker [73]. The Whittaker formulas define each component of spinor ψ^A in terms of the tensor components $\boldsymbol{C}$ up to the sign and therefore do not carry out a one-to-one connection between the spinors and the tensors $\boldsymbol{C}$. In subsequent papers of various authors some particular cases have been considered (real and two-components spinors in the four-dimensional and three-dimensional spaces), but the explicit formulas, that realize one-to-one connection between spinors and tensors, have not been obtained. After a number of attempts to establish such connection it was appeared opinion on not reducibility of the first-rank spinor to tensors and "elementary nature" of the spinor. The one-to-one invariant connection between spinors and various systems of tensors has been established in [74, 75] in the spaces of any dimension. Various aspects of the connection between tensors and spinors were considered, for example, in [15, 35, 57, 67–70]. A geometric illustration of two-component spinors in the four-dimensional space is given in [50].

$$
\begin{aligned}
C^{34} &= -2(\psi^1\psi^2 + \psi^3\psi^4),\\
C^{23} &= \mathrm{i}(\psi^1\psi^1 - \psi^2\psi^2 - \psi^3\psi^3 + \psi^4\psi^4),\\
C^{31} &= -\psi^1\psi^1 - \psi^2\psi^2 + \psi^3\psi^3 + \psi^4\psi^4,\\
C^{12} &= 2\mathrm{i}(-\psi^1\psi^2 + \psi^3\psi^4).
\end{aligned} \tag{3.56}
$$

In this case the relations expressing the symmetric components of the second rank spinor $\psi^{BA} = \psi^B\psi^A$ in terms of the components of tensors C^i, C^{ij}, have the form

$$
\begin{aligned}
\psi^{11} &= \frac{1}{4}(-C^{31} - \mathrm{i}C^{23} + C^{14} - \mathrm{i}C^{24}),\\
\psi^{22} &= \frac{1}{4}(-C^{31} + \mathrm{i}C^{23} - C^{14} - \mathrm{i}C^{24}),\\
\psi^{33} &= \frac{1}{4}(C^{31} + \mathrm{i}C^{23} + C^{14} - \mathrm{i}C^{24}),\\
\psi^{44} &= \frac{1}{4}(C^{31} - \mathrm{i}C^{23} - C^{14} - \mathrm{i}C^{24}),
\end{aligned}
$$

$$
\begin{aligned}
\psi^{14} &= \frac{\mathrm{i}}{4}(C^3 + C^4), & \psi^{23} &= \frac{\mathrm{i}}{4}(C^3 - C^4),\\
\psi^{31} &= -\frac{1}{4}(C^2 + \mathrm{i}C^1), & \psi^{24} &= \frac{1}{4}(-C^2 + \mathrm{i}C^1),\\
\psi^{12} &= \frac{1}{4}(-C^{34} + \mathrm{i}C^{12}), & \psi^{34} &= -\frac{1}{4}(C^{34} + \mathrm{i}C^{12}).
\end{aligned}
$$

3.2.2 Representation of Spinors in Pseudo-Euclidean Space E_4^1 by Real Tensors

The system of real tensors $\boldsymbol{D}$ defined by the first-rank spinor $\boldsymbol{\psi}$, in pseudo-Euclidean space of E_4^1 consists of scalar Ω, vector $\boldsymbol{j} = j^i\,\boldsymbol{\ni}_i$ and antisymmetric tensors of the second, third and fourth ranks $\boldsymbol{M} = M^{ij}\,\boldsymbol{\ni}_i\boldsymbol{\ni}_j$, $\boldsymbol{S} = S^{ijk}\,\boldsymbol{\ni}_i\boldsymbol{\ni}_j\boldsymbol{\ni}_k$, and $\boldsymbol{N} = N^{ijks}\,\boldsymbol{\ni}_i\boldsymbol{\ni}_j\boldsymbol{\ni}_k\boldsymbol{\ni}_s$, whose components are defined by the following relations

$$
\begin{aligned}
\Omega &= -e_{AB}\psi^{+A}\psi^B, & j^i &= -\mathrm{i}\gamma^i_{AB}\psi^{+A}\psi^B,\\
M^{ij} &= -\mathrm{i}\gamma^{ij}_{AB}\psi^{+A}\psi^B, & S^{ijk} &= -\gamma^{ijk}_{AB}\psi^{+A}\psi^B,\\
& & N^{ijks} &= -\gamma^{ijks}_{AB}\psi^{+A}\psi^B.
\end{aligned} \tag{3.57}
$$

Let us write down definition (3.57) also in the matrix form

$$\begin{aligned} \Omega &= \psi^+\psi, \quad j^i = \mathrm{i}\psi^+\gamma^i\psi, \\ M^{ij} &= \mathrm{i}\psi^+\gamma^{ij}\psi, \quad S^{ijk} = \psi^+\gamma^{ijk}\psi, \\ N^{ijks} &= \psi^+\gamma^{ijks}\psi. \end{aligned} \tag{3.58}$$

Here $\psi^+ = \dot{\psi}^T\beta$ is the row of the covariant components of the conjugate spinor ψ^+_A.

Instead of the components of tensors S^{ijk}, N^{ijks} it is convenient to use the components of the pseudo-vector S_i and pseudo-scalar N, which are connected with S^{ijk}, N^{ijks} by the relations

$$S_i = -\frac{1}{6}\varepsilon_{ijks}S^{jks}, \qquad S^{jks} = S_i\varepsilon^{ijks},$$

$$N = -\frac{1}{24}\varepsilon_{ijks}N^{ijks}, \quad N^{ijks} = N\varepsilon^{ijks}.$$

The pseudo-scalar N and the components of the pseudo-vector S_i are expressed in terms of the spinor components ψ as follows

$$S_i = \psi^+\overset{*}{\gamma}_i\psi, \quad N = \psi^+\gamma^5\psi. \tag{3.59}$$

From definitions (3.57) and (3.59) it follows that the components of tensors $\boldsymbol{D}$ satisfy the Pauli–Fierz invariant algebraic identities, which can be obtained by contracting identities (C.1) with components of spinor $\psi^{+B}\psi^A\psi^{+D}\psi^E$ with respect to the indices B, A, D, E:

$$a.\quad j_i j^i = -\Omega^2 - N^2,$$

$$b.\quad S_i S^i = \Omega^2 + N^2,$$

$$c.\quad S_i j^i = 0,$$

$$d.\quad \frac{1}{2}M_{ij}M^{ij} = \Omega^2 - N^2,$$

$$e.\quad \frac{1}{4}\varepsilon_{ijks}M^{ij}M^{ks} = 2\Omega N,$$

$$f.\quad \Omega j^i = \frac{1}{2}\varepsilon^{ijks}S_j M_{ks},$$

$$g.\quad N j^i = -S_j M^{ij},$$

$$h.\quad \Omega S_i = \frac{1}{2}\varepsilon_{ijks}j^j M^{ks},$$

$$
\begin{aligned}
&i. \quad NS_i = -j^n M_{in},\\
&j. \quad j^i j^j = S^i S^j + M^i{}_s M^{js} - \Omega^2 g^{ij},\\
&k. \quad \Omega M_{ij} + \frac{1}{2} N \varepsilon_{ijks} M^{ks} = -\varepsilon_{ijks} j^k S^s,\\
&l. \quad M^{ij} M^{ks} = \left(\Omega^2 + N^2\right)\left(g^{ik} g^{js} - g^{is} g^{jk}\right) - \frac{1}{4}\varepsilon^{ijpq}\varepsilon^{ksmn} M_{pq} M_{mn}\\
&\quad + g^{ik}\left(j^s j^j - S^s S^j\right) - g^{jk}\left(j^s j^i - S^s S^i\right) - g^{is}\left(j^j j^k - S^j S^k\right)\\
&\qquad + g^{js}\left(j^i j^k - S^i S^k\right),\\
&m. \quad M^i{}_j M^{sj} - \frac{1}{4} g^{is} M_{jq} M^{jq} = \frac{1}{2} g^{is}\left(\Omega^2 + N^2\right) + j^i j^s - S^i S^s.
\end{aligned}
\tag{3.60}
$$

A connection between the components of the spinor $\boldsymbol{\psi}$ and the components of tensors $\boldsymbol{D}$ in the space E_4^1 is given by the following formulas [74, 75]

$$
\psi^A = \frac{\psi^{\dot{B}A}\dot{\eta}_B}{\sqrt{\psi^{\dot{C}D}\dot{\eta}_C \eta_D}},
\tag{3.61}
$$

$$
\psi^{\dot{B}A} = \frac{1}{4}\left(\Omega\beta^{A\dot{B}} - \mathrm{i} j^s \gamma_s^{A\dot{B}} + \frac{\mathrm{i}}{2} M^{js}\gamma_{js}^{A\dot{B}} - S^i \overset{*}{\gamma}{}_i^{A\dot{B}} + N\gamma^{5A\dot{B}}\right),
$$

where η_C ($C = 1, 2, 3, 4$) are arbitrary complex numbers, satisfying the condition $\psi^{\dot{C}D}\dot{\eta}_C\eta_D \neq 0$; the components of spintensors $\boldsymbol{\gamma}$ are defined as follows

$$
\begin{aligned}
\gamma_s^{A\dot{B}} &= \gamma^A{}_{Cs}\beta^{C\dot{B}}, & \gamma_{js}^{A\dot{B}} &= \gamma^A{}_{Cjs}\beta^{C\dot{B}},\\
\overset{*}{\gamma}{}_i^{A\dot{B}} &= \overset{*}{\gamma}{}^A{}_{Ci}\beta^{C\dot{B}}, & \gamma^{5A\dot{B}} &= \gamma^{5A}{}_C\beta^{C\dot{B}}.
\end{aligned}
$$

Here the components of the invariant spinor $\beta^{-1} = \|\beta^{A\dot{B}}\|$ are defined by Eqs. (3.17).

The second formula (3.61) in the matrix notations has the form

$$
\|\psi^{\dot{B}A}\|^T = \frac{1}{4}\left(\Omega I - \mathrm{i} j^s \gamma_s + \frac{\mathrm{i}}{2} M^{js}\gamma_{js} - S^i \overset{*}{\gamma}_i + N\gamma^5\right)\beta^{-1}.
$$

Instead of the first formula in (3.61) it is convenient to use the relation

$$
\psi^A = \frac{\psi^{\dot{B}A}}{\sqrt{\psi^{\dot{B}B}}}\exp i\varphi,
$$

where φ is an arbitrary real number.

The tensors $\boldsymbol{C}$, $\boldsymbol{D}$ are connected by the following cross equations, which are obtained by contraction of the identities (C.1) with the spinor components $\psi^{+B}\psi^{A}\psi^{D}\psi^{E}$ with respect to the indices B, A, D, E:

$$
\begin{aligned}
&a. \quad C_i S^i = 0,\\
&b. \quad C_i j^i = 0,\\
&c. \quad C_{ij} M^{ij} = 0,\\
&d. \quad \varepsilon_{ijks} C^{ij} M^{ks} = 0,\\
&e. \quad \Omega C^i = \frac{1}{2}\varepsilon^{ijks} S_j C_{ks} = \mathrm{i} C_j M^{ij} = \mathrm{i} C^{si} j_s,\\
&f. \quad N C^i = -\frac{\mathrm{i}}{2}\varepsilon^{ijks} j_j C_{ks} = \frac{\mathrm{i}}{2}\varepsilon^{ijks} C_j M_{ks} = -C^{ij} S_j,\\
&g. \quad \Omega C^{ij} + \frac{1}{2}\varepsilon^{ijks} N C_{ks} = -\varepsilon^{ijks} C_k S_s,\\
&h. \quad \varepsilon^{ijks} C_k S_s = \mathrm{i}\left(C^i j^j - C^j j^i\right),\\
&i. \quad \Omega C^{ij} - \frac{1}{2}\varepsilon^{ijks} N C_{ks} = \mathrm{i}\left(C^i{}_s M^{js} - C^j{}_s M^{is}\right),\\
&j. \quad C^i j^j + C^j j^i = C^{is} M^j{}_s + C^{js} M^i{}_s.
\end{aligned}
\tag{3.62}
$$

From definitions (3.23), (3.47), and (3.58) it follows that for any nonzero spinor $\boldsymbol{\psi}$ the vector component j^4 is positive

$$j^4 = \dot{\psi}^1\psi^1 + \dot{\psi}^2\psi^2 + \dot{\psi}^3\psi^3 + \dot{\psi}^4\psi^4 > 0.$$

The condition $j^4 > 0$ is not connected with a specific choice of the Dirac matrices γ_i.

Since the spinor $\boldsymbol{\psi}$ is fully defined by the complex tensors with components C^i and C^{ij}, it is clear that the tensors with the real components Ω, j^i, M^{ij}, S^i, N must be fully defined in terms of C^i and C^{ij}. The respective equations have the form (for unique definition Ω, j^i, M^{ij}, S^i, N in terms of C^i and C^{ij} we must take into account that $j^4 > 0$):

$$
\begin{aligned}
&a. \quad 4\Omega^2 = \dot{C}_i C^i - \frac{1}{2}\dot{C}_{ij} C^{ij},\\
&b. \quad 4\Omega j^i = \mathrm{i}\left(-\dot{C}_j C^{ij} + C_j \dot{C}^{ij}\right),\\
&c. \quad 4\Omega M^{ij} = \mathrm{i}\delta^{ij}_{ks}\left(\dot{C}^k C^s - \dot{C}^{nk} C_n{}^s\right),
\end{aligned}
$$

$$
\begin{aligned}
&d.\quad 4\Omega S^i = -\frac{1}{2}\varepsilon^{ijks}\left(\dot{C}_j C_{ks} + C_j \dot{C}_{ks}\right),\\
&e.\quad 4\Omega N = -\frac{1}{4}\varepsilon_{ijks}\dot{C}^{ij}C^{ks},\\
&f.\quad 4Nj^i = \frac{\mathrm{i}}{2}\varepsilon^{ijks}\left(-\dot{C}_j C_{ks} + C_j \dot{C}_{ks}\right),\\
&g.\quad 2\varepsilon^{ijks}NM_{ks} = \mathrm{i}\delta^{ij}_{ks}\left(\dot{C}^k C^s + \dot{C}^{nk}C_n{}^s\right),\\
&h.\quad 4NS^i = \dot{C}_j C^{ij} + C_j \dot{C}^{ij}, \qquad\qquad (3.63)\\
&i.\quad 4N^2 = \dot{C}_i C^i + \frac{1}{2}\dot{C}_{ij}C^{ij},\\
&j.\quad 4j^i j^s = \dot{C}^i C^s + C^i \dot{C}^s + \dot{C}^{ij}C^s{}_j + C^{ij}\dot{C}^s{}_j\\
&\qquad\qquad - g^{si}\left(\dot{C}_j C^j + \frac{1}{2}\dot{C}_{jn}C^{jn}\right),\\
&k.\quad 4j^n M^{qj} = \delta^{qj}_{ks}\left[\frac{1}{2}\dot{C}^{ks}C^n + \frac{1}{2}C^{ks}\dot{C}^n - \dot{C}^k C^{sn} - C^k \dot{C}^{sn}\right.\\
&\qquad\qquad \left. + g^{ns}\left(\dot{C}^{mk}C_m + C^{mk}\dot{C}_m\right)\right],\\
&l.\quad 4j^q S^s = \frac{\mathrm{i}}{4}g^{qs}\varepsilon_{jkmn}\dot{C}^{jk}C^{mn} + \mathrm{i}\varepsilon^{qsjn}\dot{C}_j C_n\\
&\qquad\qquad - \frac{\mathrm{i}}{2}\left(\varepsilon^{qjkm}\dot{C}^s{}_j + \varepsilon^{sjkm}\dot{C}^q{}_j\right)C_{km},\\
&m.\quad 4M^{ij}M^s{}_j = -g^{si}\left(\dot{C}_j C^j + \frac{1}{2}\dot{C}_{jm}C^{jm}\right) + 2\left(\dot{C}^i C^s + \dot{C}^s C^i\right),\\
&n.\quad 4S^i S^j = -\dot{C}^i C^j - \dot{C}^j C^i + C^{in}\dot{C}^j{}_n + C^{jn}\dot{C}^i{}_n\\
&\qquad\qquad + g^{ij}\left(\dot{C}_k C^k - \frac{1}{2}\dot{C}_{ns}C^{ns}\right),\\
&o.\quad 4S^m M^{ij} = \mathrm{i}\left[\frac{1}{2}\varepsilon^{impq}\left(\dot{C}_{pq}C^j - \dot{C}^j C_{pq}\right)\right.\\
&\qquad\qquad \left. - \frac{1}{2}\varepsilon^{jmpq}\left(\dot{C}_{pq}C^i - \dot{C}^i C_{pq}\right) + \varepsilon^{ijks}\left(\dot{C}_k{}^m C_s - \dot{C}_s C_k{}^m\right)\right].
\end{aligned}
$$

Here $\delta^{ij}_{ks} = \delta^i_k\delta^j_s - \delta^i_s\delta^j_k$, the complex conjugate components $\dot{C}^i$, $\dot{C}^{ij}$ are expressed in terms of the components of the conjugate spinor

$$
\dot{C}^i = -\gamma^i_{BA}\psi^{+B}\psi^{+A}, \quad \dot{C}^{ij} = -\gamma^{ij}_{BA}\psi^{+B}\psi^{+A}.
$$

Relations (3.63) are obtained by contraction of identities (C.1) with the spinor components $\psi^B \psi^{+A} \psi^{+D} \psi^E$.

If the matrices γ_i and β are defined by equalities (3.24) and (3.25), then the components of the real tensors $\boldsymbol{D}$ are expressed in terms of the components of the spinor ψ^A and the complex conjugate spinor $\dot{\psi}^A$ by the equalities

$$
\begin{aligned}
\Omega &= \dot{\psi}^1\psi^3 + \dot{\psi}^2\psi^4 + \dot{\psi}^3\psi^1 + \dot{\psi}^4\psi^2,\\
j^1 &= \dot{\psi}^1\psi^2 + \dot{\psi}^2\psi^1 - \dot{\psi}^3\psi^4 - \dot{\psi}^4\psi^3,\\
j^2 &= \mathrm{i}(-\dot{\psi}^1\psi^2 + \dot{\psi}^2\psi^1 + \dot{\psi}^3\psi^4 - \dot{\psi}^4\psi^3),\\
j^3 &= \dot{\psi}^1\psi^1 - \dot{\psi}^2\psi^2 - \dot{\psi}^3\psi^3 + \dot{\psi}^4\psi^4,\\
j^4 &= \dot{\psi}^1\psi^1 + \dot{\psi}^2\psi^2 + \dot{\psi}^3\psi^3 + \dot{\psi}^4\psi^4,\\
M^{23} &= -\dot{\psi}^1\psi^4 - \dot{\psi}^2\psi^3 - \dot{\psi}^3\psi^2 - \dot{\psi}^4\psi^1,\\
M^{31} &= \mathrm{i}(\dot{\psi}^1\psi^4 - \dot{\psi}^2\psi^3 + \dot{\psi}^3\psi^2 - \dot{\psi}^4\psi^1),\\
M^{12} &= -\dot{\psi}^1\psi^3 + \dot{\psi}^2\psi^4 - \dot{\psi}^3\psi^1 + \dot{\psi}^4\psi^2,\\
M^{14} &= \mathrm{i}(-\dot{\psi}^1\psi^4 - \dot{\psi}^2\psi^3 + \dot{\psi}^3\psi^2 + \dot{\psi}^4\psi^1),\\
M^{24} &= -\dot{\psi}^1\psi^4 + \dot{\psi}^2\psi^3 + \dot{\psi}^3\psi^2 - \dot{\psi}^4\psi^1,\\
M^{34} &= \mathrm{i}(-\dot{\psi}^1\psi^3 + \dot{\psi}^2\psi^4 + \dot{\psi}^3\psi^1 - \dot{\psi}^4\psi^2),\\
S^1 &= -\dot{\psi}^1\psi^2 - \dot{\psi}^2\psi^1 - \dot{\psi}^3\psi^4 - \dot{\psi}^4\psi^3,\\
S^2 &= \mathrm{i}(\dot{\psi}^1\psi^2 - \dot{\psi}^2\psi^1 + \dot{\psi}^3\psi^4 - \dot{\psi}^4\psi^3),\\
S^3 &= -\dot{\psi}^1\psi^1 + \dot{\psi}^2\psi^2 - \dot{\psi}^3\psi^3 + \dot{\psi}^4\psi^4,\\
S^4 &= -\dot{\psi}^1\psi^1 - \dot{\psi}^2\psi^2 + \dot{\psi}^3\psi^3 + \dot{\psi}^4\psi^4,\\
N &= \mathrm{i}(\dot{\psi}^1\psi^3 + \dot{\psi}^2\psi^4 - \dot{\psi}^3\psi^1 - \dot{\psi}^4\psi^2).
\end{aligned}
\tag{3.64}
$$

The components $\psi^{\dot{B}A} = \dot{\psi}^B\psi^A$ in this case are expressed in terms of the real tensor components $\boldsymbol{D}$ as follows:

$$
\begin{aligned}
\psi^{\dot{1}1} &= \frac{1}{4}(j^4 + j^3 - S^4 - S^3),\\
\psi^{\dot{2}2} &= \frac{1}{4}(j^4 - j^3 - S^4 + S^3),\\
\psi^{\dot{3}3} &= \frac{1}{4}(j^4 - j^3 + S^4 - S^3),\\
\psi^{\dot{4}4} &= \frac{1}{4}(j^4 + j^3 + S^4 + S^3),
\end{aligned}
$$

$$
\begin{aligned}
\psi^{\dot{1}2} &= \frac{1}{4}(j^1 + \mathrm{i}j^2 - S^1 - \mathrm{i}S^2),\\
\psi^{\dot{3}4} &= \frac{1}{4}(-j^1 - \mathrm{i}j^2 - S^1 - \mathrm{i}S^2),\\
\psi^{\dot{1}3} &= \frac{1}{4}(\Omega - \mathrm{i}N - M^{12} + \mathrm{i}M^{34}),\\
\psi^{\dot{2}4} &= \frac{1}{4}(\Omega - \mathrm{i}N + M^{12} - \mathrm{i}M^{34}),\\
\psi^{\dot{1}4} &= \frac{1}{4}(-M^{24} + \mathrm{i}M^{14} - M^{23} - \mathrm{i}M^{31}),\\
\psi^{\dot{2}3} &= \frac{1}{4}(M^{24} + \mathrm{i}M^{14} - M^{23} + \mathrm{i}M^{31}).
\end{aligned} \tag{3.65}
$$

Along with the tensors $\boldsymbol{C}$ and $\boldsymbol{D}$, defined by equalities (3.50) and (3.58), in the sequel we shall use the real tensors with components ρ, u^i, μ^{ij}, η defined with the aid of the tensor components $\boldsymbol{D}$:

$$
\begin{aligned}
\rho &= +\sqrt{\Omega^2 + N^2},\\
\rho \exp \mathrm{i}\eta &= \Omega + \mathrm{i}N,\\
\rho u^i &= j^i,\\
\rho \mu^{ij} &= \Omega M^{ij} + \frac{1}{2}\varepsilon^{ijks} N M_{ks}
\end{aligned} \tag{3.66}
$$

and the complex tensors $\boldsymbol{Z}$ with components Z^i, Z^{ij} defined with the aid of the tensor components $\boldsymbol{C}$ and $\boldsymbol{D}$:

$$
\rho Z^i = C^i, \quad \rho Z^{ij} = \Omega C^{ij} + \frac{1}{2}\varepsilon^{ijks} N C_{ks}. \tag{3.67}
$$

Definitions (3.66) and (3.67) make a sense if $\Omega^2 + N^2 \neq 0$.

By virtue of the equation (k) in (3.60) the tensor components $\rho\mu^{ij}$ are expressed also in terms of the vector components j_i and S_i:

$$
\rho \mu^{ij} = -\varepsilon^{ijks} j_k S_s. \tag{3.68}
$$

It is easy to see that the tensor components C^{ij} are expressed in terms of Z^{ij} and Ω, N:

$$
C^{ij} = \frac{1}{\rho}\left(\Omega Z^{ij} - \frac{1}{2}\varepsilon^{ijks} N Z_{ks}\right). \tag{3.69}
$$

From Eqs. (3.60) and definitions (3.66) it follows that the components of the real tensors ρ, u^i, μ^{ij} satisfy the following algebraic equations

$$u_i u^i = -1, \quad u_j \mu^{ij} = 0,$$
$$\varepsilon_{ijks} \mu^{ij} \mu^{ks} = 0, \quad \frac{1}{2} \mu_{ij} \mu^{ij} = \rho^2,$$
$$S^j \mu_{ij} = \rho^2 u_i, \quad \frac{1}{2} \varepsilon_{ijks} u^j \mu^{ks} = S_i. \tag{3.70}$$

From Eqs. (3.52), (3.53), and definitions (3.67) it follows that the components of the complex tensors Z^i, Z^{ij} satisfy the invariant algebraic equations of the form

$$Z_i Z^i = 0, \quad Z_i \dot{Z}^i = 2, \quad Z_{ij} Z^{ij} = 0, \quad Z^{[ij} Z^{ks]} = 0,$$
$$Z_i Z^{ij} = 0, \quad Z^{[i} Z^{jk]} = 0, \quad \rho^2 Z^i Z^j + Z^{is} Z_s{}^j = 0.$$

Using Eqs. (3.63), it is possible to obtain an expression of the real tensor components ρ, u_i, S_i, μ_{ij} in terms of the components of the complex tensors Z^i, Z^{ij}:

$$\rho^2 = -\frac{1}{4} Z_{ij} \dot{Z}^{ij}, \quad \rho u_i = -\frac{\mathrm{i}}{2} Z_{ij} \dot{Z}^j, \quad S_i = -\frac{1}{4} \varepsilon_{ijks} Z^{ks} \dot{Z}^j,$$
$$\rho \mu_{ij} = \frac{\mathrm{i}}{2} \big(Z_{js} \dot{Z}^s{}_i - Z_{is} \dot{Z}^s{}_j \big) = \frac{\mathrm{i}}{2} \rho^2 \big(- Z_i \dot{Z}_j + Z_j \dot{Z}_i \big).$$

The components of tensors ρ, u_i, S_i, μ_{ij} and the components of tensors Z^i, Z^{ij} are also connected by the equations

$$S_i Z^{ij} = 0, \quad u^{[i} Z^{jk]} = 0, \quad Z^{[i} \mu^{jk]} = 0,$$
$$\rho^2 Z^i = \frac{1}{2} \varepsilon^{ijks} S_j Z_{ks} = \mathrm{i} \rho u_s Z^{si} = \mathrm{i} \rho Z_j \mu^{ij}$$

and

$$Z^{ij} = -\varepsilon^{ijks} Z_k S_s = \mathrm{i} \big(- Z^i j^j + Z^j j^i \big).$$

3.2.3 *Representation of Two Spinors by Systems of Tensors*

Tensors $\boldsymbol{K}$ determined by the first-rank spinors $\boldsymbol{\psi}$ and $\boldsymbol{\chi}$ in the space E_4^1, consist of the following tensors $\boldsymbol{K} = \{K, K^i, K^{ij}, K^{ijk}, K^{ijks}\}$.

The tensor components $\boldsymbol{K}$ may be determined by the equalities

$$K = e_{AB} \chi^A \psi^B, \quad K^j = \mathrm{i} \gamma^j_{AB} \chi^A \psi^B, \quad K^{js} = \mathrm{i} \gamma^{js}_{AB} \chi^A \psi^B,$$
$$K^{ijk} = \gamma^{ijk}_{AB} \chi^A \psi^B, \quad K^{ijks} = \gamma^{ijks}_{AB} \chi^A \psi^B. \tag{3.71}$$

The complex conjugate components of tensors $\dot{\boldsymbol{K}}$ are expressed in terms of the conjugate spinor components

$$\dot{K} = -e_{AB}\chi^{+A}\psi^{+B}, \quad \dot{K}^j = \mathrm{i}\gamma^j_{AB}\chi^{+A}\psi^{+B}, \quad \dot{K}^{js} = \mathrm{i}\gamma^{js}_{AB}\chi^{+A}\psi^{+B},$$
$$\dot{K}^{ijk} = -\gamma^{ijk}_{AB}\chi^{+A}\psi^{+B}, \quad \dot{K}^{ijks} = -\gamma^{ijks}_{AB}\chi^{+A}\psi^{+B}.$$

Instead of the components of tensors K^{ijk}, K^{ijks} it is convenient to use also the dual components $\overset{*}{K}_i$, $\overset{*}{K}$:

$$\overset{*}{K}_i = -\frac{1}{6}\varepsilon_{ijks}K^{jks} = \overset{*}{\gamma}_{ABi}\chi^A\psi^B, \qquad K^{jks} = \overset{*}{K}_i\varepsilon^{ijks},$$
$$\overset{*}{K} = -\frac{1}{24}\varepsilon_{ijks}K^{ijks} = \gamma^5_{AB}\chi^A\psi^B, \quad K^{ijks} = \overset{*}{K}\varepsilon^{ijks}. \tag{3.72}$$

An expansion of the spinor components $\chi^A\psi^B$ in system invariant spintensors $\boldsymbol{\gamma}$ has the form

$$\chi^A\psi^B = \frac{1}{4}\Big(- Ke^{AB} + \mathrm{i}K^j\gamma^{AB}_j - \frac{\mathrm{i}}{2}K^{ij}\gamma^{AB}_{ij} + \overset{*}{K}{}^j\overset{*}{\gamma}{}^{AB}_j - \overset{*}{K}\gamma^{5AB}\Big).$$

The relations expressing spinor $\boldsymbol{\psi}$ in terms of the tensors $\boldsymbol{K}$, $\boldsymbol{D}'$ and spinor χ^+, have the form

$$4\Omega'\psi^B = \Big(- Ke^{AB} + \mathrm{i}K^j\gamma^{AB}_j - \frac{\mathrm{i}}{2}K^{ij}\gamma^{AB}_{ij} + \overset{*}{K}{}^j\overset{*}{\gamma}{}^{AB}_j - \overset{*}{K}\gamma^{5AB}\Big)\chi^+_A,$$

$$4j'^i\psi^B = \Big[(\mathrm{i}Kg^{ij} + K^{ij})\gamma^{CB}_j + \frac{1}{2}(- K^s g^{ik} + K^k g^{is} + \mathrm{i}\overset{*}{K}_j\varepsilon^{ijks})\gamma^{CB}_{ks}$$
$$+ \Big(- \mathrm{i}\overset{*}{K}g^{ij} - \frac{1}{2}\varepsilon^{ijks}K_{ks}\Big)\overset{*}{\gamma}{}^{CB}_j + \mathrm{i}\overset{*}{K}{}^i\gamma^{5CB} + K^i e^{CB}\Big]\chi^+_C,$$

$$4M'^{ij}\psi^B = \Big\{(\mathrm{i}\overset{*}{K}_k\varepsilon^{ijks} - K^j g^{is} + K^i g^{js})\gamma^{CB}_s + \Big[- \frac{\mathrm{i}}{2}\overset{*}{K}\varepsilon^{ijks}$$
$$+ \frac{\mathrm{i}}{2}K(g^{ik}g^{js} - g^{is}g^{jk}) + K^{js}g^{ik} - K^{is}g^{jk}\Big]\gamma^{CB}_{ks} + (K_k\varepsilon^{ijks}$$
$$+ \mathrm{i}\overset{*}{K}{}^j g^{is} - \mathrm{i}\overset{*}{K}{}^i g^{js})\overset{*}{\gamma}{}^{CB}_s + K^{ij}e^{CB} + \frac{1}{2}K_{ks}\varepsilon^{ijks}\gamma^{5CB}\Big\}\chi^+_C,$$

$$4S'^i\psi^B = \Big[- \overset{*}{K}{}^i e^{CB} + \Big(\overset{*}{K}g^{ij} - \frac{\mathrm{i}}{2}\varepsilon^{ijks}K_{ks}\Big)\gamma^{CB}_j + \frac{1}{2}(- \mathrm{i}K_j\varepsilon^{ijks}$$
$$+ \overset{*}{K}{}^s g^{ik} - \overset{*}{K}{}^k g^{is})\gamma^{CB}_{ks} + (Kg^{ij} - \mathrm{i}K^{ij})\overset{*}{\gamma}{}^{CB}_j - \mathrm{i}K^i\gamma^{5CB}\Big]\chi^+_C,$$

$$4N'\psi^B = \Big(-\overset{*}{K} e^{CB} + \overset{*}{K}{}^j \gamma_j^{CB} - \frac{\mathrm{i}}{4}\varepsilon^{ijks} K_{ij}\gamma_{ks}^{CB} - \mathrm{i}K^j \overset{*}{\gamma}{}_j^{CB} + K\gamma^{5CB}\Big)\chi_C^+. \tag{3.73}$$

Let us give also an expression of the spinor components $\boldsymbol{\psi}$ in terms of the components of tensors $\boldsymbol{K}$, $\boldsymbol{C}'$ and spinor $\boldsymbol{\chi}$:

$$\begin{aligned}
C'^j\psi &= -(\overset{*}{K}{}^j\gamma^5 + \mathrm{i}K_s\gamma^{sj})\chi = \Big(-\mathrm{i}K^j I + \frac{1}{2}\varepsilon^{jspq}\overset{*}{K}_s\gamma_{pq}\Big)\chi \\
&= \Big(K\gamma^j - \frac{\mathrm{i}}{2}\varepsilon^{jpqs}K_{pq}\overset{*}{\gamma}_s\Big)\chi = \Big(\overset{*}{K}\overset{*}{\gamma}{}^j + \mathrm{i}K^{sj}\gamma_s\Big)\chi, \\
C'_{ij}\psi &= \Big[-\mathrm{i}K_{ij}I + \frac{1}{2}\varepsilon_{ijks}\Big(-\overset{*}{K}\gamma^{ks} + \overset{*}{K}{}^k\gamma^s - \overset{*}{K}{}^s\gamma^k\Big)\Big]\chi \\
&= \Big[K\gamma_{ij} + \frac{\mathrm{i}}{2}\varepsilon_{ijks}\Big(K^{ks}\gamma^5 + K^k\overset{*}{\gamma}{}^s - K^s\overset{*}{\gamma}{}^k\Big)\Big]\chi.
\end{aligned} \tag{3.74}$$

By virtue of definitions (3.71) and (3.72) the tensor components $\boldsymbol{K}$ satisfy the following equations

$$\begin{gathered}
K_iK^i = -\overset{*}{K}_i\overset{*}{K}{}^i = -K^2 - \overset{*}{K}{}^2, \quad K_i\overset{*}{K}{}^i = 0, \\
\frac{1}{2}K_{ij}K^{ij} = K^2 - \overset{*}{K}{}^2, \quad \frac{1}{4}\varepsilon_{ijks}K^{ij}K^{ks} = 2K\overset{*}{K}, \\
KK^i = \frac{1}{2}\varepsilon^{ijks}\overset{*}{K}_jK_{ks}, \quad K\overset{*}{K}{}^i = \frac{1}{2}\varepsilon^{ijks}K_jK_{ks}, \\
\overset{*}{K}K^i = -\overset{*}{K}_jK^{ij}, \quad \overset{*}{K}\overset{*}{K}{}^i = -K_jK^{ij}, \\
K^iK^j = \overset{*}{K}{}^i\overset{*}{K}{}^j + K^i{}_sK^{js} - K^2g^{ij}, \\
\overset{*}{K}K^{ij} - \frac{1}{2}\varepsilon^{ijks}KK_{ks} = -K^i\overset{*}{K}{}^j + K^j\overset{*}{K}{}^i, \\
K^{ij}K^{mn} = (K^2 + \overset{*}{K}{}^2)(g^{im}g^{jn} - g^{in}g^{jm}) - \frac{1}{4}\varepsilon^{ijks}\varepsilon^{mnpq}K_{ks}K_{pq} \\
+g^{im}(K^jK^n - \overset{*}{K}{}^j\overset{*}{K}{}^n) - g^{jm}(K^iK^n - \overset{*}{K}{}^i\overset{*}{K}{}^n) \\
-g^{in}(K^jK^m - \overset{*}{K}{}^j\overset{*}{K}{}^m) + g^{jn}(K^iK^m - \overset{*}{K}{}^i\overset{*}{K}{}^m).
\end{gathered} \tag{3.75}$$

The tensors $\boldsymbol{K}$ and $\boldsymbol{C}$ are connected by the following equations

$$\begin{gathered}
K_iC^i = 0, \quad \overset{*}{K}_iC^i = 0, \\
K_{ij}C^{ij} = 0, \quad \varepsilon_{ijks}K^{ij}C^{ks} = 0,
\end{gathered}$$

$$
\begin{aligned}
KC^i &= \frac{1}{2}\varepsilon^{ijks}\overset{*}{K}_j C_{ks} = \mathrm{i}K^{ij}C_j = -\mathrm{i}K_j C^{ij},\\
\overset{*}{K}C^i &= -\overset{*}{K}_j C^{ij} = \frac{\mathrm{i}}{2}\varepsilon^{ijks}K_{ks}C_j = -\frac{1}{2}\varepsilon^{ijks}K_j C_{ks},\\
KC^{ij} + \frac{1}{2}\varepsilon^{ijks}\overset{*}{K}C_{ks} &= -\varepsilon^{ijks}C_k\overset{*}{K}_s,\\
\varepsilon^{ijks}C_k\overset{*}{K}_s &= \mathrm{i}\big(C^iK^j - C^jK^i\big),\\
KC^{ij} - \frac{1}{2}\varepsilon^{ijks}\overset{*}{K}C_{ks} &= \mathrm{i}\big(C^i{}_sK^{js} - C^j{}_sK^{is}\big),\\
K^iC^j + K^jC^i &= K^i{}_sC^{js} + K^j{}_sC^{is}.
\end{aligned}
\tag{3.76}
$$

Let us give also the relations expressing the tensors $\boldsymbol{D}$ and $\boldsymbol{D}'$ in terms of the tensors $\boldsymbol{K}$:

$$
\begin{aligned}
a.\quad & 4\Omega'\Omega = -\dot{K}K + \dot{K}_iK^i - \frac{1}{2}\dot{K}_{ij}K^{ij} - \big(\overset{*}{K}_i\big)^{\cdot}\overset{*}{K}^i + \big(\overset{*}{K}\big)^{\cdot}\overset{*}{K},\\
b.\quad & 4\Omega'j^i = -\dot{K}^iK - \dot{K}K^i - \mathrm{i}\dot{K}^{si}K_s + \mathrm{i}\dot{K}_sK^{si}\\
& \quad - \frac{1}{2}\varepsilon^{ijks}\big[\dot{K}_{ks}\overset{*}{K}_j + \big(\overset{*}{K}_j\big)^{\cdot}K_{ks}\big] + \mathrm{i}\big(\overset{*}{K}^i\big)^{\cdot}\overset{*}{K} - \mathrm{i}\big(\overset{*}{K}\big)^{\cdot}\overset{*}{K}^i,\\
c.\quad & 4\Omega'M^{ij} = \delta^{ij}_{ks}\Big[-\frac{1}{2}\dot{K}^{ks}K - \frac{1}{2}\dot{K}K^{ks} + \mathrm{i}\dot{K}^kK^s - \mathrm{i}\big(\overset{*}{K}^k\big)^{\cdot}\overset{*}{K}^s\\
& \quad - \mathrm{i}\dot{K}^{nk}K_n{}^s\Big] + \varepsilon^{ijks}\Big[\big(\overset{*}{K}_s\big)^{\cdot}K_k + \dot{K}_k\overset{*}{K}_s + \frac{1}{2}\dot{K}_{ks}\overset{*}{K} + \frac{1}{2}\big(\overset{*}{K}\big)^{\cdot}K_{ks}\Big],\\
d.\quad & 4\Omega'S^i = -\big(\overset{*}{K}^i\big)^{\cdot}K - \dot{K}\overset{*}{K}^i - \mathrm{i}\big(\overset{*}{K}\big)^{\cdot}K^i + \mathrm{i}\dot{K}^i\overset{*}{K}\\
& \quad + \mathrm{i}\dot{K}^{ij}\overset{*}{K}_j - \mathrm{i}\big(\overset{*}{K}_j\big)^{\cdot}K^{ij} - \frac{1}{2}\varepsilon^{ijks}\big(\dot{K}_sK_{jk} + \dot{K}_{jk}K_s\big),\\
e.\quad & 4\Omega'N = -\big(\overset{*}{K}\big)^{\cdot}K - \dot{K}\overset{*}{K} - \mathrm{i}\big(\overset{*}{K}_i\big)^{\cdot}K^i + \mathrm{i}\dot{K}_i\overset{*}{K}^i - \frac{1}{4}\varepsilon^{ijks}\dot{K}_{ij}K_{ks},\\
f.\quad & 4N'\Omega = -\big(\overset{*}{K}\big)^{\cdot}K - \dot{K}\overset{*}{K} + \mathrm{i}\big(\overset{*}{K}_i\big)^{\cdot}K^i - \mathrm{i}\dot{K}_i\overset{*}{K}^i - \frac{1}{4}\varepsilon^{ijks}\dot{K}_{ij}K_{ks},\\
g.\quad & 4N'j^i = -\mathrm{i}\big(\overset{*}{K}^i\big)^{\cdot}K + \mathrm{i}\dot{K}\overset{*}{K}^i - \big(\overset{*}{K}\big)^{\cdot}K^i - \dot{K}^i\overset{*}{K}\\
& \quad + \dot{K}^{ij}\overset{*}{K}_j + \big(\overset{*}{K}_j\big)^{\cdot}K^{ij} + \frac{\mathrm{i}}{2}\varepsilon^{ijks}\big(-\dot{K}_sK_{jk} + \dot{K}_{jk}K_s\big),\\
h.\quad & 4N'M^{ij} = \delta^{ij}_{ks}\Big[-\frac{1}{2}\dot{K}^{ks}\overset{*}{K} - \frac{1}{2}\big(\overset{*}{K}\big)^{\cdot}K^{ks} + \frac{1}{2}\varepsilon^{knqr}\dot{K}^s{}_nK_{qr} + \dot{K}^k\overset{*}{K}^s
\end{aligned}
$$

$$+ (\overset{*}{K}{}^{s})^{\cdot}K^{k}\Big] + \varepsilon^{ijks}\Big[\mathrm{i}(\overset{*}{K}_{k})^{\cdot}\overset{*}{K}_{s} - \mathrm{i}\dot{K}_{k}K_{s} - \frac{1}{2}\dot{K}_{ks}K - \frac{1}{2}\dot{K}K_{ks}\Big],$$

$$i.\quad 4N'S^{i} = -\mathrm{i}\dot{K}^{i}K + \mathrm{i}\dot{K}K^{i} + \dot{K}^{ij}K_{j} + \dot{K}_{j}K^{ij}$$

$$- \frac{\mathrm{i}}{2}\varepsilon^{ijks}\big[(\overset{*}{K}_{s})^{\cdot}K_{jk} - \dot{K}_{jk}\overset{*}{K}_{s}\big] - (\overset{*}{K}{}^{i})^{\cdot}\overset{*}{K} - (\overset{*}{K})^{\cdot}\overset{*}{K}{}^{i},$$

$$j.\quad 4N'N = \dot{K}K + \dot{K}_{i}K^{i} + \frac{1}{2}\dot{K}_{ij}K^{ij} - (\overset{*}{K}_{i})^{\cdot}\overset{*}{K}{}^{i} - (\overset{*}{K})^{\cdot}\overset{*}{K},$$

$$k.\quad 4j'^{i}\Omega = \dot{K}^{i}K + \dot{K}K^{i} - \mathrm{i}\dot{K}^{si}K_{s} + \mathrm{i}\dot{K}_{s}K^{si}$$

$$+ \frac{1}{2}\varepsilon^{ijks}\big[\dot{K}_{ks}\overset{*}{K}_{j} + (\overset{*}{K}_{j})^{\cdot}K_{ks}\big] + \mathrm{i}(\overset{*}{K}{}^{i})^{\cdot}\overset{*}{K} - \mathrm{i}(\overset{*}{K})^{\cdot}\overset{*}{K}{}^{i},$$

$$l.\quad 4j'^{i}j^{j} = -\mathrm{i}\dot{K}^{ij}K + \mathrm{i}\dot{K}K^{ij} + \dot{K}^{i}K^{j} + \dot{K}^{j}K^{i} + (\overset{*}{K}{}^{i})^{\cdot}\overset{*}{K}{}^{j} + (\overset{*}{K}{}^{j})^{\cdot}\overset{*}{K}{}^{i}$$

$$+ \dot{K}^{i}{}_{s}K^{js} + \dot{K}^{js}K^{i}{}_{s} - g^{ij}\Big[\dot{K}K + \dot{K}_{s}K^{s} + \frac{1}{2}\dot{K}_{ks}K^{ks} + (\overset{*}{K}_{s})^{\cdot}\overset{*}{K}{}^{s}$$

$$+ (\overset{*}{K})^{\cdot}\overset{*}{K}\Big] + \varepsilon^{ijks}\Big[\mathrm{i}\dot{K}_{k}\overset{*}{K}_{s} - \mathrm{i}(\overset{*}{K}_{s})^{\cdot}K_{k} + \frac{\mathrm{i}}{2}(\overset{*}{K})^{\cdot}K_{ks} - \frac{\mathrm{i}}{2}\dot{K}_{ks}\overset{*}{K}\Big],$$

$$m.\quad 4j'^{s}M^{ij} = \delta^{ij}_{kn}\Big\{\frac{1}{2}\dot{K}^{kn}K^{s} + \frac{1}{2}\dot{K}^{s}K^{kn} - \dot{K}^{k}K^{ns} - \dot{K}^{ns}K^{k}$$

$$+ g^{ns}\big[\mathrm{i}\dot{K}^{k}K - \mathrm{i}\dot{K}K^{k} - (\overset{*}{K}{}^{k})^{\cdot}\overset{*}{K} - (\overset{*}{K})^{\cdot}\overset{*}{K}{}^{k} + \dot{K}^{mk}K_{m} + \dot{K}_{m}K^{mk}\big]$$

$$+ \frac{1}{2}\varepsilon^{ksmp}\big[\mathrm{i}(\overset{*}{K}{}^{n})^{\cdot}K_{mp} - \mathrm{i}\dot{K}_{mp}\overset{*}{K}{}^{n}\big]\Big\} + \varepsilon^{ijkn}\big\{\delta^{s}_{k}\big[(\overset{*}{K}_{n})^{\cdot}K$$

$$+ \dot{K}\overset{*}{K}_{n} + \mathrm{i}\dot{K}_{n}\overset{*}{K} - \mathrm{i}(\overset{*}{K})^{\cdot}K_{n}\big] + \mathrm{i}(\overset{*}{K}_{n})^{\cdot}K_{k}{}^{s} - \mathrm{i}\dot{K}_{k}{}^{s}\overset{*}{K}_{n}\big\},$$

$$n.\quad 4j'^{i}S^{j} = \dot{K}^{i}\overset{*}{K}{}^{j} + (\overset{*}{K}{}^{j})^{\cdot}K^{i} + (\overset{*}{K}{}^{i})^{\cdot}K^{j} + \dot{K}^{j}\overset{*}{K}{}^{i} - (\overset{*}{K})^{\cdot}K^{ij} - \dot{K}^{ij}\overset{*}{K}$$

$$+ g^{ij}\Big[\mathrm{i}\dot{K}\overset{*}{K} - \mathrm{i}(\overset{*}{K})^{\cdot}K - (\overset{*}{K}_{s})^{\cdot}K^{s} - \dot{K}_{s}\overset{*}{K}{}^{s} + \frac{\mathrm{i}}{4}\varepsilon^{ksmn}\dot{K}_{ks}K_{mn}\Big]$$

$$+ \varepsilon^{ijks}\Big[\frac{1}{2}\dot{K}K_{ks} + \frac{1}{2}\dot{K}_{ks}K + \mathrm{i}\dot{K}_{k}K_{s} + \mathrm{i}(\overset{*}{K}_{k})^{\cdot}\overset{*}{K}_{s}\Big]$$

$$- \frac{\mathrm{i}}{2}\big(\varepsilon^{impq}\dot{K}^{j}{}_{m} + \varepsilon^{jmpq}\dot{K}^{i}{}_{m}\big)K_{pq},$$

$$o.\quad 4j'^{i}N = -\mathrm{i}(\overset{*}{K}{}^{i})^{\cdot}K + \mathrm{i}\dot{K}\overset{*}{K}{}^{i} + (\overset{*}{K})^{\cdot}K^{i} + \dot{K}^{i}\overset{*}{K}$$

$$- \dot{K}^{ij}\overset{*}{K}_{j} - (\overset{*}{K}_{j})^{\cdot}K^{ij} - \frac{\mathrm{i}}{2}\varepsilon^{ijks}\big(\dot{K}_{s}K_{jk} - \dot{K}_{jk}K_{s}\big). \tag{3.77}$$

All Eqs. (3.73)–(3.77) are obtained by contraction of identities (C.1) with components of spinors $\chi^{+D}\chi^E\psi^B$, $\chi^D\chi^E\psi^B$, ...

3.3 Tensor Representation of Semispinors in Pseudo-Euclidean Space E_4^1

3.3.1 Semispinors in Pseudo-Euclidean Space E_4^1

If the four components of the first-rank spinor ψ^A in the pseudo-Euclidean space E_4^1 are connected by the relation

$$\psi = \pm \mathrm{i}\gamma^5\psi, \tag{3.78}$$

in which one can take signs $+$ or $-$, then components ψ^A are defined only by two independent complex parameters. Spinor $\boldsymbol{\psi}$ in this case is called the semispinor in the space E_4^1.

Multiplying the Hermitian conjugate equation (3.78) by the matrix of the components of invariant second-rank spinor β, for the components of the conjugate semispinor we find

$$\psi^+ = \mp \mathrm{i}\psi^+\gamma^5. \tag{3.79}$$

Since Eq. (3.78) is invariant under transformations of the restricted Lorentz group, the set of all spinors with components ψ^A, satisfying Eq. (3.78), forms in the spinor space a subspace that is invariant with respect to the restricted Lorentz group.

To an arbitrary spinor with components ψ in the four-dimensional pseudo-Euclidean space E_4^1 it is possible to put in correspondence two semispinors with the contravariant components $\psi_{(I)}$ and $\psi_{(II)}$ determined in the same basis, as ψ:

$$\psi_{(I)} = \frac{1}{2}\big(I + \mathrm{i}\gamma^5\big)\psi, \quad \psi_{(II)} = \frac{1}{2}\big(I - \mathrm{i}\gamma^5\big)\psi. \tag{3.80}$$

From definitions (3.80) and equality $\gamma^5\gamma^5 = -I$ (see formulas (3.11)) it follows that the components of semispinors $\psi_{(I)}$ and $\psi_{(II)}$ satisfy the equations

$$\psi_{(I)} = \mathrm{i}\gamma^5\psi_{(I)}, \quad \psi_{(II)} = -\mathrm{i}\gamma^5\psi_{(II)}.$$

It follows from definitions (3.80) that the relative sign of the semispinors components $\psi_{(I)}$ and $\psi_{(II)}$ is fixed. It means that the product $\psi_{(I)}\psi_{(II)}$ has the certain sign, though the components $\psi_{(I)}$ and $\psi_{(II)}$ are two-valued.

Let us introduce a spinbasis $\overset{*}{\boldsymbol{\varepsilon}}_A$, in which the components of the invariant spintensors γ_i are determined by the matrices

$$\gamma_1 = \begin{Vmatrix} 0 & 0 & 0 & \mathrm{i} \\ 0 & 0 & \mathrm{i} & 0 \\ 0 & -\mathrm{i} & 0 & 0 \\ -\mathrm{i} & 0 & 0 & 0 \end{Vmatrix}, \quad \gamma_2 = \begin{Vmatrix} 0 & 0 & 0 & 1 \\ 0 & 0 & -1 & 0 \\ 0 & -1 & 0 & 0 \\ 1 & 0 & 0 & 0 \end{Vmatrix},$$

$$\gamma_3 = \begin{Vmatrix} 0 & 0 & \mathrm{i} & 0 \\ 0 & 0 & 0 & -\mathrm{i} \\ -\mathrm{i} & 0 & 0 & 0 \\ 0 & \mathrm{i} & 0 & 0 \end{Vmatrix}, \quad \gamma_4 = \begin{Vmatrix} 0 & 0 & \mathrm{i} & 0 \\ 0 & 0 & 0 & \mathrm{i} \\ \mathrm{i} & 0 & 0 & 0 \\ 0 & \mathrm{i} & 0 & 0 \end{Vmatrix}, \tag{3.81}$$

while the components of the metric spinor E and the invariant spinors γ^5, β, and Π are determined by the matrices

$$E = \begin{Vmatrix} 0 & 1 & 0 & 0 \\ -1 & 0 & 0 & 0 \\ 0 & 0 & 0 & -1 \\ 0 & 0 & 1 & 0 \end{Vmatrix}, \quad \gamma^5 = \begin{Vmatrix} -\mathrm{i} & 0 & 0 & 0 \\ 0 & -\mathrm{i} & 0 & 0 \\ 0 & 0 & \mathrm{i} & 0 \\ 0 & 0 & 0 & \mathrm{i} \end{Vmatrix},$$

$$\beta = \begin{Vmatrix} 0 & 0 & 1 & 0 \\ 0 & 0 & 0 & 1 \\ 1 & 0 & 0 & 0 \\ 0 & 1 & 0 & 0 \end{Vmatrix}, \quad \Pi = \begin{Vmatrix} 0 & 0 & 0 & -1 \\ 0 & 0 & 1 & 0 \\ 0 & 1 & 0 & 0 \\ -1 & 0 & 0 & 0 \end{Vmatrix}. \tag{3.82}$$

According to definitions (3.47) and (3.82), for the covariant components of the conjugate spinor in the spinbasis $\overset{*}{\boldsymbol{\varepsilon}}_A$ we have

$$\psi^+ = \dot{\psi}^T \beta = \|\psi^+_A\| = (\dot{\psi}^3, \dot{\psi}^4, \dot{\psi}^1, \dot{\psi}^2), \tag{3.83}$$

while the covariant components of the spinor ψ_A are defined as follows

$$\widetilde{\psi} = (E\psi)^T = \|\psi_A\| = (\psi^2, -\psi^1, -\psi^4, \psi^3). \tag{3.84}$$

It follows from Eqs. (3.43) that the spinor transformations $\{\pm S\}$, corresponding to the restricted Lorentz transformations of the bases of the space E^1_4, are defined in the chosen special spinbasis $\overset{*}{\boldsymbol{\varepsilon}}_A$ by the matrices

$$S = \pm \begin{Vmatrix} A & 0 \\ 0 & (\dot{A}^{-1})^T \end{Vmatrix}, \tag{3.85}$$

where A is some two-dimensional matrix, 0 is the two-dimensional null matrix

$$A = \begin{Vmatrix} \alpha & \beta \\ \gamma & \delta \end{Vmatrix}, \quad 0 = \begin{Vmatrix} 0 & 0 \\ 0 & 0 \end{Vmatrix}.$$

It follows from Eqs. (3.43) that the matrix A is unimodular

$$\det A = \alpha\delta - \beta\gamma = 1. \tag{3.86}$$

It is obvious that the sets of matrices $\{\pm A\}$ and $\{\pm(\dot{A}^{-1})^T\}$, corresponding to the restricted Lorentz transformation, form groups, which realize two-dimensional representations of the restricted Lorentz group. Thus, the spinor representation of the restricted Lorentz group is reducible.

It is easy to see that in the spinbasis $\overset{*}{\boldsymbol{\varepsilon}}_A$ under consideration, from condition $\psi = \mathrm{i}\gamma^5\psi$ it follows $\psi^3 = \psi^4 = 0$, while from condition $\psi = -\mathrm{i}\gamma^5\psi$ it follows $\psi^1 = \psi^2 = 0$. Therefore for the contravariant components of semispinors ψ_I and ψ_{II} in the spinbasis $\overset{*}{\boldsymbol{\varepsilon}}_A$ we can write

$$\psi_{(I)} = \begin{Vmatrix} \psi^1 \\ \psi^2 \\ 0 \\ 0 \end{Vmatrix}, \quad \psi_{(II)} = \begin{Vmatrix} 0 \\ 0 \\ \psi^3 \\ \psi^4 \end{Vmatrix}. \tag{3.87}$$

According to definitions (3.84) and (3.87), for the covariant components of semispinors $\widetilde{\psi}_I$ and $\widetilde{\psi}_{II}$ in the spinbasis $\overset{*}{\boldsymbol{\varepsilon}}_A$ we have

$$\widetilde{\psi}_{(I)} = (\psi^2, -\psi^1, 0, 0), \quad \widetilde{\psi}_{(II)} = (0, 0, -\psi^4, \psi^3).$$

From definitions (3.83) and (3.87) we find expressions for the covariant components of the conjugate semispinors ψ_I^+ and ψ_{II}^+ in the spinbasis $\overset{*}{\boldsymbol{\varepsilon}}_A$:

$$\psi_{(I)}^+ = (0, 0, \dot{\psi}^1, \dot{\psi}^2), \quad \psi_{(II)}^+ = (\dot{\psi}^3, \dot{\psi}^4, 0, 0).$$

From formula (3.85) it follows that under the restricted Lorentz transformations the components of an arbitrary spinor ψ^1, ψ^2 and the components ψ^3, ψ^4, calculated in the spinbasis $\overset{*}{\boldsymbol{\varepsilon}}_A$, are transformed separately. Therefore, restricting ourselves to considering only restricted orthogonal transformations of the bases $\boldsymbol{Э}_i$ in $E_{2\nu}^+$, we can define the covariant and contravariant components of the semispinors $\psi_{(I)}$ and $\psi_{(II)}$ in spinbasis $\overset{*}{\boldsymbol{\varepsilon}}_A$ by only two nonzero components. In this connection in a calculations with semispinors it is possible (and sometimes simpler) to use two-dimensional matrix notations.

Next, let us denote two nonzero contravariant and covariant components of a semispinor ψ_I, respectively, by symbols ξ^A and $\xi_A (A = 1, 2)$, while the column of contravariant components ξ^A and the row of covariant components ξ_A by symbols ξ and $\widetilde{\xi}$. The nonzero contravariant and covariant components of the semispinor ψ_{II} we denote by symbols $\eta_{\dot{A}}$ and $\eta^{\dot{A}}$ ($\dot{A} = 1, 2$), while the row of the contravariant components $\eta^{\dot{A}}$ and the column of the covariant components $\eta_{\dot{A}}$ by symbols $\widetilde{\eta}$ and η. Thus,

$$\xi = \begin{Vmatrix} \xi^1 \\ \xi^2 \end{Vmatrix} = \begin{Vmatrix} \psi^1 \\ \psi^2 \end{Vmatrix}, \qquad \widetilde{\xi} = \|\xi_1, \xi_2\| = \|\psi^2, -\psi^1\|,$$

$$\eta = \begin{Vmatrix} \eta_{\dot{1}} \\ \eta_{\dot{2}} \end{Vmatrix} = \begin{Vmatrix} \psi^3 \\ \psi^4 \end{Vmatrix}, \qquad \widetilde{\eta} = \|\eta^{\dot{1}}, \eta^{\dot{2}}\| = \| - \psi^4, \psi^3\|. \tag{3.88}$$

By means of the introduced notations the contravariant and covariant components of an arbitrary four-component spinor $\boldsymbol{\psi}$ in the spinbasis $\overset{*}{\boldsymbol{e}}_A$ can be written in the form

$$\psi = \|\psi^A\| = \begin{Vmatrix} \psi^1 \\ \psi^2 \\ \psi^3 \\ \psi^4 \end{Vmatrix} = \begin{Vmatrix} \xi^1 \\ \xi^2 \\ \eta_{\dot{1}} \\ \eta_{\dot{2}} \end{Vmatrix},$$

$$\widetilde{\psi} = \|\psi_A\| = \|\psi^2, -\psi^1, -\psi^4, \psi^3\| = \|\xi_1, \xi_2, \eta^{\dot{1}}, \eta^{\dot{2}}\|. \tag{3.89}$$

In the same spinbasis for the covariant and contravariant components of the conjugate spinor, according to definitions (3.48), (3.82) and (3.83), we have

$$\bar{\psi} = \|\psi^{+A}\| = \begin{Vmatrix} -\dot{\psi}^4 \\ \dot{\psi}^3 \\ \dot{\psi}^2 \\ -\dot{\psi}^1 \end{Vmatrix} = \begin{Vmatrix} -\dot{\eta}_{\dot{2}} \\ \dot{\eta}_{\dot{1}} \\ \dot{\xi}^2 \\ -\dot{\xi}^1 \end{Vmatrix} = \begin{Vmatrix} \dot{\eta}^{\dot{1}} \\ \dot{\eta}^{\dot{2}} \\ \dot{\xi}_1 \\ \dot{\xi}_2 \end{Vmatrix},$$

$$\psi^+ = \|\psi^+_A\| = \|\dot{\psi}^3, \dot{\psi}^4, \dot{\psi}^1, \dot{\psi}^2\| = \|\dot{\eta}_{\dot{1}}, \dot{\eta}_{\dot{2}}, \dot{\xi}^1, \dot{\xi}^2\|.$$

3.3.2 *Two-Component Spinors in the Four-Dimensional Pseudo-Euclidean Space* E_4^1

For the restricted Lorentz transformation of bases $\mathit{Э}_i$ of the space E_4^1 two-component quantities ξ, η in accordance with equalities (3.85) are transformed as follows:

$$\xi' = A\xi, \quad \eta' = \left(\dot{A}^T\right)^{-1}\eta. \tag{3.90}$$

Since groups of the matrices $\{\pm A\}$, $\{\pm(\dot{A}^{-1})^T\}$ realize representations of the restricted Lorentz group, components ξ and η define in the space E_4^1 geometric objects, which will be further called two-component spinors.[6]

From definitions (3.88) it is seen that the contravariant components ξ^A, $\eta^{\dot{A}}$ and the covariant components ξ_A, $\eta_{\dot{A}}$ are connected by the relations[7]

$$\xi_1 = \xi^2, \quad \xi_2 = -\xi^1,$$

$$\eta_{\dot{1}} = \eta^{\dot{2}}, \quad \eta_{\dot{2}} = -\eta^{\dot{1}},$$

which can be written in the form

$$\xi_A = \varepsilon_{AB}\xi^B, \quad \xi^A = \varepsilon^{AB}\xi_B,$$

$$\eta_{\dot{A}} = \varepsilon_{\dot{A}\dot{B}}\eta^{\dot{B}}, \quad \eta^{\dot{A}} = \varepsilon^{\dot{A}\dot{B}}\eta_{\dot{B}},$$

where the components of the metric spinor ε_{AB}, $\varepsilon_{\dot{A}\dot{B}}$, ε^{AB}, and $\varepsilon^{\dot{A}\dot{B}}$ are defined by the matrices

$$\varepsilon = \|\varepsilon_{AB}\| = \|\varepsilon_{\dot{A}\dot{B}}\| = \left\| \begin{matrix} 0 & 1 \\ -1 & 0 \end{matrix} \right\|,$$

$$\varepsilon^{-1} = \|\varepsilon^{AB}\| = \|\varepsilon^{\dot{A}\dot{B}}\| = \left\| \begin{matrix} 0 & -1 \\ 1 & 0 \end{matrix} \right\| = -\varepsilon. \tag{3.91}$$

We note the useful relation

$$\varepsilon_{AB}\varepsilon_{CD} + \varepsilon_{AD}\varepsilon_{BC} + \varepsilon_{AC}\varepsilon_{DB} = 0, \tag{3.92}$$

which is fulfilled by virtue of the fact that in a two-dimensional space the completely antisymmetric components with the number of indices greater than two are equal to zero. Lifting the indices C, D in equality (3.92), we get also

$$\varepsilon_{AB}\varepsilon^{CD} = \delta_A^D\delta_B^C - \delta_A^C\delta_B^D.$$

Let us contract this equality with components $\xi_C\eta_D$, where ξ_C and η_D determine two-component spinors in E_4^1:

$$\varepsilon_{AB}\xi_C\eta^C = \xi_B\eta_A - \xi_A\eta_B.$$

---

[6]Sometimes (especially in physical literature) a four-component spinor in the space E_4^1 is called a bispinor, while ξ, η are called, respectively, undotted and dotted spinors in the space E_4^1.

[7]For two-component spinors one uses also the following connection between contravariant and covariant components of a spinor $\xi_2 = \xi^1$, $\xi_1 = -\xi^2$ that is related to different definition of covariant and contravariant components ε_{AB}, ε^{AB} of the metric spinor. With such definition of the connection between ξ_A and ξ^A in some formulas the sign changes.

From this equality under condition $\xi_C\eta^C = 1$ we find the following representation for the components of the metric spinor ε_{AB}:

$$\varepsilon_{AB} = \xi_B\eta_A - \xi_A\eta_B, \quad \xi_C\eta^C = 1.$$

Further on we introduce four two-dimensional matrices

$$\sigma_1 = \begin{Vmatrix} 0 & 1 \\ 1 & 0 \end{Vmatrix}, \quad \sigma_2 = \begin{Vmatrix} 0 & -\mathrm{i} \\ \mathrm{i} & 0 \end{Vmatrix}, \quad \sigma_3 = \begin{Vmatrix} 1 & 0 \\ 0 & -1 \end{Vmatrix}, \quad \sigma_4 = I = \begin{Vmatrix} 1 & 0 \\ 0 & 1 \end{Vmatrix}. \tag{3.93}$$

Using two-dimensional matrices σ_i and ε, the four-dimensional matrices E, β, Π, γ_i, γ_{ij}, $\overset{*}{\gamma}_i$, and γ^5, determined by equalities (3.8), (3.9), (3.81), and (3.82), can be written in the form:

$$\beta = \begin{Vmatrix} 0 & I \\ I & 0 \end{Vmatrix}, \quad E = \begin{Vmatrix} \varepsilon & 0 \\ 0 & -\varepsilon \end{Vmatrix} = \begin{Vmatrix} \mathrm{i}\sigma_2 & 0 \\ 0 & -\mathrm{i}\sigma_2 \end{Vmatrix}, \quad \Pi = \begin{Vmatrix} 0 & -\mathrm{i}\sigma_2 \\ \mathrm{i}\sigma_2 & 0 \end{Vmatrix},$$

$$\gamma_\alpha = \begin{Vmatrix} 0 & \mathrm{i}\sigma_\alpha \\ -\mathrm{i}\sigma_\alpha & 0 \end{Vmatrix}, \quad \gamma_4 = \begin{Vmatrix} 0 & \mathrm{i}I \\ \mathrm{i}I & 0 \end{Vmatrix}, \quad \gamma^5 = \begin{Vmatrix} -\mathrm{i}I & 0 \\ 0 & \mathrm{i}I \end{Vmatrix},$$

$$\overset{*}{\gamma}_\alpha = \begin{Vmatrix} 0 & -\sigma_\alpha \\ -\sigma_\alpha & 0 \end{Vmatrix}, \quad \overset{*}{\gamma}_4 = \begin{Vmatrix} 0 & -I \\ I & 0 \end{Vmatrix},$$

$$\gamma_{\alpha\beta} = \begin{Vmatrix} \sigma_{[\alpha}\sigma_{\beta]} & 0 \\ 0 & \sigma_{[\alpha}\sigma_{\beta]} \end{Vmatrix}, \quad \gamma_{4\alpha} = \begin{Vmatrix} \sigma_\alpha & 0 \\ 0 & -\sigma_\alpha \end{Vmatrix}. \tag{3.94}$$

By means of the matrices σ_i and ε, Eqs. (3.43), determining spinor transformations, one can write in the form of the two-dimensional matrix equations:

$$A^T\varepsilon A = \varepsilon, \quad l^j{}_i\sigma_j = A^{-1}\sigma_i(\dot{A}^T)^{-1}. \tag{3.95}$$

From Eqs. (3.90) and (3.95) it follows that the contravariant dotted components $\eta^{\dot{A}}$ are transformed as the complex conjugate undotted contravariant components $\dot{\xi}^A$:

$$\tilde{\eta}' = \tilde{\eta}\dot{A}^T.$$

From (3.95) it follows that the matrices σ_j and ε form the invariant spintensor components with the following structure of the indices:

$$\sigma_j = \|\sigma_j^{B\dot{A}}\|, \quad \varepsilon = \|\varepsilon_{BA}\|. \tag{3.96}$$

For the covariant components of the spintensor $\sigma_{\dot{B}Aj} = \varepsilon_{\dot{B}\dot{C}}\varepsilon_{AD}\sigma_j^{D\dot{C}}$ by virtue of equalities (3.91), (3.93), and (3.96) we have

$$\|\sigma_{\dot{B}A4}\| = I, \quad \|\sigma_{\dot{B}A\alpha}\| = -\sigma_\alpha, \quad \alpha = 1, 2, 3. \tag{3.97}$$

Therefore, taking into account the metric signature of the pseudo-Euclidean space E_4^1, we find

$$\sigma^{j}_{\dot{B}A} = -\sigma_{j}^{B\dot{A}}.$$

It is obvious that the matrices σ_i are Hermitian $\dot{\sigma}_i^T = \sigma_i$, or

$$\dot{\sigma}^{j}_{\dot{B}A} = \sigma^{j}_{\dot{A}B}, \quad \dot{\sigma}_{j}^{B\dot{A}} = \sigma_{j}^{A\dot{B}}.$$

Using definitions (3.93), (3.96), and (3.97) it is not difficult to verify that the components $\sigma^{j}_{\dot{B}A}$ and $\sigma_{j}^{B\dot{A}}$ satisfy the equations

$$\begin{aligned}
\sigma_i^{B\dot{C}}\sigma_{\dot{C}Aj} + \sigma_j^{B\dot{C}}\sigma_{\dot{C}Ai} &= -2g_{ij}\delta^B_A, \\
\sigma_{\dot{A}Ci}\sigma_j^{C\dot{B}} + \sigma_{\dot{A}Cj}\sigma_i^{C\dot{B}} &= -2g_{ij}\delta^{\dot{B}}_{\dot{A}}.
\end{aligned} \tag{3.98}$$

Let us now consider the spintensors with components $\sigma^{B}{}_{Aij}$ and $\sigma^{\dot{B}}{}_{\dot{A}ij}$:

$$\begin{aligned}
\sigma^{B}{}_{Aij} &= -\frac{\mathrm{i}}{2}\left(\sigma_i^{B\dot{C}}\sigma_{\dot{C}Aj} - \sigma_j^{B\dot{C}}\sigma_{\dot{C}Ai}\right), \\
\sigma^{\dot{B}}{}_{\dot{A}ij} &= -\frac{\mathrm{i}}{2}\left(\sigma_{\dot{A}Ci}\sigma_j^{C\dot{B}} - \sigma_{\dot{A}Cj}\sigma_i^{C\dot{B}}\right).
\end{aligned} \tag{3.99}$$

Due to definitions (3.93), (3.96), and (3.97) we have

$$\begin{aligned}
\|\sigma^{B}{}_{A\alpha\beta}\| &= \mathrm{i}\sigma_{[\alpha}\sigma_{\beta]} = -\varepsilon_{\alpha\beta\lambda}\sigma^{\lambda}, \\
\|\sigma^{B}{}_{A4\alpha}\| &= -\ \|\sigma^{B}{}_{A\alpha 4}\| = \mathrm{i}\sigma_{\alpha}, \\
\|\sigma^{\dot{B}}{}_{\dot{A}\alpha\beta}\| &= -\ \mathrm{i}\dot{\sigma}_{[\alpha}\dot{\sigma}_{\beta]} = -\varepsilon_{\alpha\beta\lambda}\dot{\sigma}^{\lambda}, \\
\|\sigma^{\dot{B}}{}_{\dot{A}4\alpha}\| &= -\ \|\sigma^{\dot{B}}{}_{\dot{A}\alpha 4}\| = -\mathrm{i}\dot{\sigma}_{\alpha},
\end{aligned} \tag{3.100}$$

where $\sigma^{\alpha} = \sigma_{\alpha}$; $\varepsilon_{\alpha\beta\lambda}$ are the components of the three-dimensional Levi-Civita pseudotensor; the Greek indices α, β, and λ have the values 1, 2, 3. From definitions (3.91), (3.93), and (3.100) it follows that the matrices $\sigma_{ij} = \|\sigma^{B}{}_{Aij}\|$ and ε are connected by the relation

$$\sigma_{ij}^{T} = -\varepsilon\sigma_{ij}\varepsilon^{-1}. \tag{3.101}$$

From equalities (3.100) it is seen that the components of the spintensors $\sigma^{\dot{B}}{}_{\dot{A}ij}$ are complex conjugate with components $\sigma^{B}{}_{Aij}$:

$$\sigma^{\dot{B}}{}_{\dot{A}ij} = \dot{\sigma}^{B}{}_{Aij}. \tag{3.102}$$

Since the components of the metric spinor $\boldsymbol{\varepsilon}$ are real, the same relations are valid for the covariant components of the spintensors $\sigma_{BAij} = \varepsilon_{BC}\sigma^{C}{}_{Aij}$ and $\sigma_{\dot{B}\dot{A}ij} = \varepsilon_{\dot{B}\dot{C}}\sigma^{\dot{C}}{}_{\dot{A}ij}$ and for the contravariant components of the spintensors $\sigma_{ij}^{BA} = \varepsilon^{AC}\sigma^{B}{}_{Cij}$ and $\sigma_{ij}^{\dot{B}\dot{A}} = \varepsilon^{\dot{A}\dot{C}}\sigma^{\dot{B}}{}_{\dot{C}ij}$:

$$\sigma_{\dot{B}\dot{A}ij} = \dot{\sigma}_{BAij}, \quad \sigma_{ij}^{\dot{B}\dot{A}} = \dot{\sigma}_{ij}^{BA}.$$

For the covariant components σ_{BAij} and the contravariant components σ_{ij}^{BA} we have

$$\|\sigma_{BA12}\| = \| - \sigma_{12}^{BA}\| = \| - \mathrm{i}\sigma_{BA34}\| = \|\mathrm{i}\sigma_{34}^{BA}\| = \sigma_1 = \begin{Vmatrix} 0 & 1 \\ 1 & 0 \end{Vmatrix}, \tag{3.103}$$

$$\|\sigma_{BA23}\| = \| - \sigma_{23}^{BA}\| = \| - \mathrm{i}\sigma_{BA14}\| = \|\mathrm{i}\sigma_{14}^{BA}\| = -\sigma_3 = \begin{Vmatrix} -1 & 0 \\ 0 & 1 \end{Vmatrix},$$

$$\|\sigma_{BA31}\| = \|\sigma_{31}^{BA}\| = \| - \mathrm{i}\sigma_{BA24}\| = \| - \mathrm{i}\sigma_{24}^{BA}\| = -\mathrm{i}I = \begin{Vmatrix} -\mathrm{i} & 0 \\ 0 & -\mathrm{i} \end{Vmatrix}.$$

From equalities (3.103) it is seen that the components of the spintensors σ_{ij}^{BA} and σ_{BAij} are symmetric in the spinor indices

$$\sigma_{ij}^{BA} = \sigma_{ij}^{AB}, \quad \sigma_{BAij} = \sigma_{ABij}$$

and satisfy the invariant linear identities

$$\sigma_{BAij} = \frac{\mathrm{i}}{2}\varepsilon_{ijks}\sigma_{BA}^{ks}, \quad \sigma_{\dot{B}\dot{A}ij} = -\frac{\mathrm{i}}{2}\varepsilon_{ijks}\sigma_{\dot{B}\dot{A}}^{ks}. \tag{3.104}$$

For the product of matrices σ_i and σ_{ij}, the following relations are valid, which can be checked directly taking into account definitions (3.96), (3.100) and Eqs. (3.98):

$$\begin{aligned}
\sigma_i^{B\dot{C}}\sigma_{\dot{C}Aj} &= \mathrm{i}\sigma^{B}{}_{Aij} - g_{ij}\delta_A^B, \\
\sigma_{\dot{A}Ci}\sigma_j^{C\dot{B}} &= \mathrm{i}\sigma^{\dot{B}}{}_{\dot{A}ij} - g_{ij}\delta_{\dot{A}}^{\dot{B}}, \\
\sigma^{A}{}_{Bij}\sigma_s^{B\dot{D}} &= \varepsilon_{ijsm}\sigma^{A\dot{D}m} - \mathrm{i}g_{si}\sigma_j^{A\dot{D}} + \mathrm{i}g_{sj}\sigma_i^{A\dot{D}}, \\
\sigma_{\dot{D}Bs}\sigma^{B}{}_{Aij} &= -\varepsilon_{ijsm}\sigma_{\dot{D}A}^{m} + \mathrm{i}g_{si}\sigma_{\dot{D}Aj} - \mathrm{i}g_{sj}\sigma_{\dot{D}Ai}, \\
\sigma^{B}{}_{Aij}\sigma^{A}{}_{Dks} &= \delta_D^B\big(g_{ik}g_{js} - g_{is}g_{jk} + \mathrm{i}\varepsilon_{ijks}\big) \\
&\quad + \mathrm{i}\big(- g_{ik}\sigma^{B}{}_{Djs} + g_{is}\sigma^{B}{}_{Djk} + g_{jk}\sigma^{B}{}_{Dis} - g_{js}\sigma^{B}{}_{Dik}\big).
\end{aligned} \tag{3.105}$$

and

$$\sigma^{i}_{\dot{C}D}\sigma_i^{A\dot{B}} = -2\delta^{\dot{B}}_{\dot{C}}\delta^{A}_{D},$$

$$\frac{1}{4}\sigma^{ij}_{CD}\sigma_{ij}^{AB} + \varepsilon^{AB}\varepsilon_{CD} = -2\delta^{A}_{C}\delta^{B}_{D}. \tag{3.106}$$

Formulae (3.106) are analogous to the Pauli identity (3.21). Identities (3.106) are not difficult to obtain if to use the relations, which are checked directly by means of the definitions of matrices (3.103) and (3.91):

$$\sigma_i^{A\dot{B}}\sigma^{j}_{\dot{B}A} = -2\delta^{j}_{i},$$

$$\varepsilon_{AB}\varepsilon^{AB} = -2, \quad \varepsilon_{AB}\sigma_{ij}^{AB} = 0,$$

$$\sigma_{ABij}\sigma_{ks}^{AB} = -2\big(g_{ik}g_{js} - g_{is}g_{jk} + \mathrm{i}\varepsilon_{ijks}\big). \tag{3.107}$$

The contraction of the first identity in (3.106) with the spintensors components $\boldsymbol{\sigma}$ gives the following identities

$$2\sigma^{j}_{\dot{A}B}\varepsilon_{CD} = \sigma^{j}_{\dot{A}D}\varepsilon_{CB} - \mathrm{i}\sigma_{\dot{A}Di}\sigma^{ij}_{CB},$$

$$2\sigma^{j}_{\dot{A}B}\sigma^{si}_{CD} = \sigma^{j}_{\dot{A}D}\sigma^{si}_{BC} - \sigma^{s}_{\dot{A}D}\sigma^{ij}_{BC} + \sigma^{i}_{\dot{A}D}\sigma^{sj}_{BC}$$
$$+\sigma_{\dot{A}Dq}\big(\sigma^{qs}_{BC}g^{ij} - \sigma^{qi}_{BC}g^{js}\big) + \mathrm{i}\sigma_{\dot{A}Dq}\varepsilon_{BC}\big(g^{qs}g^{ij} - g^{iq}g^{js} + \mathrm{i}\varepsilon^{qjsi}\big),$$

$$2\sigma^{i}_{\dot{A}B}\sigma^{j}_{\dot{C}D} = \sigma^{i}_{\dot{A}D}\sigma^{j}_{\dot{C}B} + \sigma^{j}_{\dot{A}D}\sigma^{i}_{\dot{C}B} - g^{ij}\sigma_{\dot{A}Dn}\sigma^{n}_{\dot{C}B} + \mathrm{i}\varepsilon^{ijmn}\sigma_{\dot{A}Dm}\sigma_{\dot{C}Bn}.$$

In the sequel the following identities will be used also

$$2e_{BC}\sigma^{mn}_{DA} = -\varepsilon_{CD}\sigma^{mn}_{BA} + \sigma^{mn}_{CD}\varepsilon_{BA} + \frac{\mathrm{i}}{2}\big(\sigma^{mj}_{CD}\sigma_{BA}{}^{n}{}_{j} - \sigma^{nj}_{CD}\sigma_{BA}{}^{m}{}_{j}\big),$$

$$2\sigma^{mn}_{BC}e_{DA} = -\varepsilon_{CD}\sigma^{mn}_{BA} + \sigma^{mn}_{CD}\varepsilon_{BA} - \frac{\mathrm{i}}{2}\big(\sigma^{mj}_{CD}\sigma_{BA}{}^{n}{}_{j} - \sigma^{nj}_{CD}\sigma_{BA}{}^{m}{}_{j}\big), \tag{3.108}$$

which can be obtained by the contraction of the second relation in (3.106) with the spintensor components σ_{mn} with respect to the spinor indices.

It is obvious that the system of four matrices $\sigma_i = \|\sigma_i^{B\dot{A}}\|$ and the system of four different matrices $\varepsilon^{-1} = \|\varepsilon_{AB}\|$, $\sigma_{ij} = \|\sigma_{ij}^{AB}\|$(for example, matrices ε^{-1}, σ_{12}, σ_{23}, σ_{31}) are linearly independent and form bases in the space of the second-order complex matrices.

3.3.3 Representation of Semispinors in the Space E_4^1 by Complex and Real Tensors

Using definitions (3.51), (3.58), and (3.59) of the tensors $\boldsymbol{C}$, $\boldsymbol{D}$ and definition (3.80) of semispinors $\psi_{(I)}$, $\psi_{(II)}$, the components of the tensors $\boldsymbol{C}$, $\boldsymbol{D}$ can be expressed in terms of the components of the semispinors $\psi_{(I)}$, $\psi_{(II)}$. For the components of the complex tensors $\boldsymbol{C}$ we have

$$\begin{aligned} C^i &= \psi^T_{(I)} E \gamma^i \psi_{(II)} + \psi^T_{(II)} E \gamma^i \psi_{(I)}, \\ C^{ij} &= \psi^T_{(I)} E \gamma^{ij} \psi_{(I)} + \psi^T_{(II)} E \gamma^{ij} \psi_{(II)}. \end{aligned} \tag{3.109}$$

The components of the real tensors $\boldsymbol{D}$ are expressed in terms of the components of the semispinors by the equalities of the form

$$\begin{aligned} \Omega &= \psi^+_{(I)} \psi_{(II)} + \psi^+_{(II)} \psi_{(I)}, \\ j^s &= \mathrm{i}\big(\psi^+_{(I)} \gamma^s \psi_{(I)} + \psi^+_{(II)} \gamma^s \psi_{(II)}\big), \\ M^{sj} &= \mathrm{i}\big(\psi^+_{(I)} \gamma^{sj} \psi_{(II)} + \psi^+_{(II)} \gamma^{sj} \psi_{(I)}\big), \\ S^i &= \psi^+_{(I)} \overset{*}{\gamma}{}^i \psi_{(I)} + \psi^+_{(II)} \overset{*}{\gamma}{}^i \psi_{(II)}, \\ N &= \psi^+_{(I)} \gamma^5 \psi_{(II)} + \psi^+_{(II)} \gamma^5 \psi_{(I)}. \end{aligned} \tag{3.110}$$

Replacing in definitions (3.109) the matrices E, γ^i, γ^{ij} by formulas (3.94), and the components of the semispinors $\psi_{(I)}$, $\psi_{(II)}$ by formulas (3.87) and (3.88), definitions (3.109) for the components of tensors $\boldsymbol{C}$ can be written by means of the two-dimensional matrix notations

$$\begin{aligned} C^\alpha &= \mathrm{i}\big(-\widetilde{\xi}\sigma^\alpha\eta + \widetilde{\eta}\sigma^\alpha\xi\big), \quad C^4 = \mathrm{i}\big(\widetilde{\xi}\eta + \widetilde{\eta}\xi\big), \\ C^{4\alpha} &= \widetilde{\xi}\sigma^\alpha\xi - \widetilde{\eta}\sigma^\alpha\eta, \quad C^{\alpha\beta} = -\mathrm{i}\varepsilon^{\alpha\beta\lambda}\big(\widetilde{\xi}\sigma_\lambda\xi + \widetilde{\eta}\sigma_\lambda\eta\big), \end{aligned}$$

or in the invariant contractions

$$\begin{aligned} C^j &= -2\mathrm{i}\sigma^j_{\dot{B}A}\eta^{\dot{B}}\xi^A, \\ C^{js} &= -\mathrm{i}\sigma^{js}_{BA}\xi^B\xi^A - \mathrm{i}\sigma^{js}_{\dot{B}\dot{A}}\eta^{\dot{B}}\eta^{\dot{A}}. \end{aligned} \tag{3.111}$$

For the components of the real tensors $\boldsymbol{D}$ in the same way we get

$$\begin{aligned} \Omega &= \dot{\eta}^T\xi + \dot{\xi}^T\eta, \quad N = \mathrm{i}\big(-\dot{\eta}^T\xi + \dot{\xi}^T\eta\big), \\ j^\alpha &= -\dot{\eta}^T\sigma^\alpha\eta + \dot{\xi}^T\sigma^\alpha\xi, \quad j^4 = \dot{\eta}^T\eta + \dot{\xi}^T\xi, \end{aligned}$$

$$M^{\alpha\beta} = -\varepsilon^{\alpha\beta\lambda}(\dot{\eta}^T\sigma_\lambda\xi + \dot{\xi}^T\sigma_\lambda\eta), \quad M^{4\alpha} = \mathrm{i}(-\dot{\eta}^T\sigma^\alpha\xi + \dot{\xi}^T\sigma^\alpha\eta),$$
$$S^\alpha = -\dot{\eta}^T\sigma^\alpha\eta - \dot{\xi}^T\sigma^\alpha\xi, \quad S^4 = \dot{\eta}^T\eta - \dot{\xi}^T\xi,$$

or, in an explicitly invariant form

$$\begin{gathered}
\Omega = \varepsilon_{BA}\xi^B\dot{\eta}^{\dot{A}} + \varepsilon_{\dot{B}\dot{A}}\dot{\xi}^B\eta^{\dot{A}}, \quad N = \mathrm{i}(-\varepsilon_{BA}\xi^B\dot{\eta}^{\dot{A}} + \varepsilon_{\dot{B}\dot{A}}\dot{\xi}^B\eta^{\dot{A}}),\\
j^i = \sigma^i_{\dot{B}A}(-\dot{\xi}^B\xi^A - \eta^{\dot{B}}\dot{\eta}^{\dot{A}}), \quad S^i = \sigma^i_{\dot{B}A}(\dot{\xi}^B\xi^A - \eta^{\dot{B}}\dot{\eta}^{\dot{A}}),\\
M^{js} = -\sigma^{js}_{BA}\xi^B\dot{\eta}^{\dot{A}} - \sigma^{js}_{\dot{B}\dot{A}}\dot{\xi}^B\eta^{\dot{A}}.
\end{gathered} \tag{3.112}$$

If the components of a spinor $\boldsymbol{\psi}$ satisfy Eq. (3.78), then the components of the vector C^i determined by spinor $\boldsymbol{\psi}$, are identically equal to zero, while the components of the antisymmetric tensor C^{ij} satisfy the additional linear equation

$$C^i = 0, \quad C^{ij} = \pm\frac{\mathrm{i}}{2}\varepsilon^{ijks}C_{ks}. \tag{3.113}$$

Equations (3.113) follow directly from definitions (3.111) and identities (3.104), since if the equation $\psi = \pm\mathrm{i}\gamma^5\psi$ is satisfied, then $\xi = 0$ or $\eta = 0$.

Let us give a proof of Eqs. (3.113), which is not connected with the choice of the special basis $\overset{*}{\boldsymbol{\varepsilon}}_A$. Replacing the components of the spinor $\boldsymbol{\psi}$ in definitions (3.50) of the components C^i, C^{ij} by formula (3.78), we find

$$\begin{aligned}
C^i &= \gamma^i_{AB}\psi^A\psi^B = \pm\mathrm{i}\gamma^i_{AB}\gamma^{5B}{}_C\psi^A\psi^C = \pm\mathrm{i}\overset{*}{\gamma}{}^i_{AC}\psi^A\psi^C = 0,\\
C^{ij} &= \gamma^{ij}_{AB}\psi^A\psi^B = \pm\mathrm{i}\gamma^{ij}_{AB}\gamma^{5B}{}_C\psi^A\psi^C\\
&= \pm\frac{\mathrm{i}}{2}\varepsilon^{ijks}\gamma_{ACks}\psi^A\psi^C = \pm\frac{\mathrm{i}}{2}\varepsilon^{ijks}C_{ks}.
\end{aligned} \tag{3.114}$$

Under the transformation of Eqs. (3.114) it is necessary to use relations (3.11).

If the components of complex tensors C^i, C^{ij} satisfy Eqs. (3.113), then all algebraic bilinear equations (3.52) and (3.53) by virtue of Eqs. (3.113) are satisfied identically except the equation

$$C_{ij}C^{ij} = 0. \tag{3.115}$$

Thus, the semispinor $\psi = \pm\mathrm{i}\gamma^5\psi$ in the pseudo-Euclidean space E_4^1 is equivalent to the complex antisymmetric tensor of the second rank with components C^{ij} that satisfy Eqs. (3.113), (3.115) [74, 75].

The one-to-one connection between the components of the semispinor ψ and the tensor C^{ij} is realize by relations

$$C^{ij} = \gamma^{ij}_{BA}\psi^B\psi^A,$$

$$\psi^A = \frac{\psi^{BA}}{\pm\sqrt{\psi^{BB}}}, \qquad \psi^{BA} = \frac{1}{8}C^{ij}\gamma^{BA}_{ij}.$$

Due to the second equation in (3.113), the components of the tensor C^{ij} one can determine by the matrix (the choice of sign in this formula corresponds to the choice of sign in (3.113))

$$C^{ij} = \begin{Vmatrix} 0 & p^3 & -p^2 & \pm\mathrm{i}p^1 \\ -p^3 & 0 & p^1 & \pm\mathrm{i}p^2 \\ p^2 & -p^1 & 0 & \pm\mathrm{i}p^3 \\ \mp\mathrm{i}p^1 & \mp\mathrm{i}p^2 & \mp\mathrm{i}p^3 & 0 \end{Vmatrix}, \tag{3.116}$$

where components p^α determine a three-dimensional complex null vector

$$(p^1)^2 + (p^2)^2 + (p^3)^2 = 0. \tag{3.117}$$

Hence we can say also that a semispinor (or two-component spinor) in the pseudo-Euclidean space E^1_4 is equivalent to the three-dimensional complex null vector with components p^α. The transformation of the components p^α under pseudo-orthogonal transformation of the basis $\mathbf{Э}_i$ in the space E^1_4 is considered in Sect. 3.5 of this chapter.

For the real tensors $\boldsymbol{D}$ determined by semispinors, it is easy to show the validity of the formulas

$$\Omega = N = M^{ij} = 0, \quad j^i = \mp S^i. \tag{3.118}$$

For example, for the invariant Ω we have

$$\Omega = -e_{AB}\big(\pm\,\mathrm{i}\gamma^{5B}{}_C\psi^C\big)\big(\mp\,\mathrm{i}\gamma^{5A}{}_D\psi^{+D}\big) = e_{AB}\psi^{+A}\psi^B = -\Omega.$$

From this it follows $\Omega = 0$.

Thus, the tensors $\boldsymbol{C}$, $\boldsymbol{D}$ corresponding to a semispinor ψ take the following special form

$$\boldsymbol{C} = \{0, C^{ij}\}, \quad \boldsymbol{D} = \{0, j^i, 0, \mp j^i, 0\}.$$

The bilinear equations (3.60), (3.62), (3.63) for tensors $\boldsymbol{C}$, $\boldsymbol{D}$ determined by semispinors are written as follows:

$$j_i j^i = 0, \quad j_i C^{ij} = 0, \quad 2j^i j^s = C^i{}_m \dot{C}^{sm},$$
$$\dot{C}^{ij} C^{ks} = g^{sj} j^i j^k - g^{si} j^j j^k + g^{ik} j^j j^s - g^{kj} j^s j^i$$
$$\mp \frac{\mathrm{i}}{2} j_m \left(j^i \varepsilon^{jksm} - j^j \varepsilon^{iksm} + j^s \varepsilon^{kijm} - j^k \varepsilon^{sijm} \right).$$

In conclusion, we emphasize once again that one four-component spinor $\boldsymbol{\psi}$ determines two two-component spinors $\boldsymbol{\xi}$ and $\boldsymbol{\eta}$ (or two semispinors $\boldsymbol{\psi}_{(I)}$ and $\boldsymbol{\psi}_{(II)}$) with a fixed relative sign. However, one cannot assert (as is sometimes done) that two arbitrary two-component spinors $\boldsymbol{\xi}$, $\boldsymbol{\eta}$ determine one four-component spinor with components (3.89). Indeed, since the components of the spinors $\boldsymbol{\xi}$ and $\boldsymbol{\eta}$ are two-valued, they determine, in any case, two different four-component spinors ψ and $\mathrm{i}\gamma^5\psi$:

$$\psi = \begin{Vmatrix} \xi \\ \eta \end{Vmatrix}, \quad \mathrm{i}\gamma^5\psi = \begin{Vmatrix} \xi \\ -\eta \end{Vmatrix}.$$

From the point of view of the tensor representation of spinors, the components of arbitrary spinors ξ and η determine completely the tensor with components C^{ij} by formula (3.111), but components (3.111) of the vectors C^i are defined by them only up to the sign since the relative sign of the components of arbitrary spinors ξ, η is not defined. As it was already noted (see Sect. 3.2 of this chapter), if the components of tensors $\{C^i, C^{ij}\}$ determine the spinor with components ψ, then the components of tensors $\{-C^i, C^{ij}\}$ determine the spinor with components $\mathrm{i}\gamma^5\psi$.

Thus, the use of the four-component spinor and two arbitrary two-component spinors is not equivalent. This, in essence is a manifestation of the fact that the space of spinors in Euclidean spaces, with components defined up to the common sign, is not linear; in the space of spinors the addition operation is not defined.

3.3.4 Representation of Two Semispinors by Systems of Tensors in the Space E_4^1

If a spinor $\boldsymbol{\chi}$ is arbitrary, while a spinor $\boldsymbol{\psi}$ satisfies Eq. (3.78), then tensors $\boldsymbol{K}$, defined by Eqs. (3.71) and (3.72), satisfy the additional linear equations

$$\overset{*}{K} = \mp \mathrm{i} K, \quad \overset{*}{K}{}^i = \mp K^i, \quad K_{ij} = \pm \frac{\mathrm{i}}{2} \varepsilon_{ijks} K^{ks},$$

which are obtained in the same way, as well as Eq. (3.113). Thus, we have $\boldsymbol{K} = \{K, K^i, K^{ij}, \mp K^i, \mp \mathrm{i} K\}$.

Equations (3.75) in this case pass into the following ones

$$K_i K^i = 0, \quad K_{ij} K^{ij} = 4K^2,$$
$$K K^i = \mathrm{i} K_j K^{ij}, \quad K^{is} K^j{}_s = g^{ij} K^2.$$

Equations (3.77) expressing the components of real tensors $\boldsymbol{D}$ and $\boldsymbol{D}'$ in terms of the components of tensors $\boldsymbol{K}$, in the case under consideration can be written in the form

$$2\Omega' j^i = -K\dot{K}^i - \dot{K}K^i - \mathrm{i}\dot{K}^{si}K_s + \mathrm{i}\dot{K}_s K^{si},$$
$$\mp 2N' j^i = -\mathrm{i}K\dot{K}^i + \mathrm{i}\dot{K}K^i + \dot{K}^{is}K_s + \dot{K}_s K^{is},$$
$$2j'^s j^i = \mathrm{i}\dot{K}K^{si} - \mathrm{i}\dot{K}^{si}K + \dot{K}^s K^i + \dot{K}^i K^s$$
$$+ \dot{K}^{sj}K^i{}_j - g^{si}\left(\dot{K}K + \dot{K}_j K^j\right) \pm \mathrm{i}\varepsilon^{iskj}\dot{K}_k K_j.$$

If a spinor $\boldsymbol{\psi}$ is arbitrary, while $\boldsymbol{\chi}$ is semispinor ($\chi = \pm \mathrm{i}\gamma^5 \chi$), then for the tensors $\boldsymbol{K}$ we have

$$\overset{*}{K} = \mp \mathrm{i} K, \quad \overset{*}{K}{}^i = \pm K^i, \quad K_{ij} = \pm \frac{\mathrm{i}}{2}\varepsilon_{ijks}K^{ks}.$$

Thus, in this case $K = \{K, K^i, K^{ij}, \pm K^i, \mp \mathrm{i} K\}$.

If the tensors $\boldsymbol{K}$ are defined by the semispinors $\psi = \pm \mathrm{i}\gamma^5 \psi$ and $\chi = \pm \mathrm{i}\gamma^5 \chi$, then for tensors $\boldsymbol{K}$ it is possible to write

$$K^i = 0, \quad \overset{*}{K}{}^i = 0, \quad \overset{*}{K} = \mp \mathrm{i} K, \quad K^{ij} = \pm \frac{\mathrm{i}}{2}\varepsilon^{ijks}K_{ks},$$

so $\boldsymbol{K} = \{K, 0, K^{ij}, 0, \mp \mathrm{i} K\}$.

Equations (3.75)–(3.77) for this case can be written as:

$$K_{ij} K^{ij} = 4K^2, \quad K^{is} K^j{}_s = g^{ij} K^2,$$
$$2j'^s j^i = \mathrm{i}\dot{K}K^{si} - \mathrm{i}\dot{K}^{si}K + \dot{K}^{sj}K^i{}_j - g^{si}\dot{K}K,$$
$$K^{is}C^j{}_s + K^{js}C^i{}_s = 0.$$

If the tensors $\boldsymbol{K}$ are defined by semispinors $\psi = \pm \mathrm{i}\gamma^5 \psi$ and $\chi = \mp \mathrm{i}\gamma^5 \chi$, we have

$$K = 0, \quad \overset{*}{K} = 0, \quad K^{ij} = 0, \quad \overset{*}{K}{}^i = \mp K^i. \tag{3.119}$$

Hence, in this case $\boldsymbol{K} = \{0, K^i, 0, \mp K^i, 0\}$.

Equations (3.75)–(3.77) in the presence of relations (3.119) pass into the equations

$$K_i K^i = 0, \quad K_i C^{ij} = 0, \quad \varepsilon_{ijks} K^j C^{ks} = 0,$$
$$2j'^s j^i = \dot{K}^s K^i + \dot{K}^i K^s - g^{si} \dot{K}_j K^j \pm \mathrm{i}\varepsilon^{iskj} \dot{K}_k K_j.$$

3.3.5 *Tensor Representation of Two-Component Spinors in the Pseudo-Euclidean Space* E_4^1

Due to completeness of the system of the matrices $\varepsilon^{-1} = \|\varepsilon^{AB}\|$ and $\sigma^{ij} = \|\sigma_{ij}^{AB}\|$, the components of an arbitrary second-rank spinor ξ^{AB} $(A,\ B = 1,\ 2)$ can be represented in the form

$$\xi^{AB} = \frac{1}{2}\Big(F\varepsilon^{AB} + \frac{1}{4}F^{js}\sigma_{js}^{AB}\Big).$$

Using identities (3.107) we find that the complex scalar F and complex components of the antisymmetric second-rank tensor F^{js} are expressed in terms of ξ^{AB} by the relations

$$F = -\varepsilon_{AB}\xi^{AB}, \quad F^{js} = -\sigma_{AB}^{js}\xi^{AB}.$$

The contraction of the first identity (3.104) with the spinor components χ^{AB} gives the linear equation

$$F^{js} = \frac{\mathrm{i}}{2}\varepsilon^{jsmn}F_{mn}. \tag{3.120}$$

Taking into account expressions (3.91) and (3.103) for ε and σ_{js}, we find

$$F = -\xi^{12} + \xi^{21}, \qquad F^{12} = \mathrm{i}F^{34} = -\xi^{12} - \xi^{21},$$
$$F^{23} = \mathrm{i}F^{14} = \xi^{11} - \xi^{22}, \quad F^{31} = \mathrm{i}F^{24} = \mathrm{i}(\xi^{11} + \xi^{22}).$$

The inverse relations have the form

$$\xi^{11} = \frac{1}{2}(F^{23} - \mathrm{i}F^{31}), \quad \xi^{12} = \frac{1}{2}(-F - F^{12}),$$
$$\xi^{21} = \frac{1}{2}(F - F^{12}), \qquad \xi^{22} = \frac{1}{2}(-F^{23} - \mathrm{i}F^{31}).$$

Due to antisymmetry of the metric spinor components $\varepsilon_{AB} = -\varepsilon_{BA}$ and symmetry of the spintensor components $\sigma_{ij}^{BA} = \sigma_{ij}^{AB}$, the equalities for the antisymmetric and symmetric parts of the components ξ^{AB} are valid

$$\xi^{AB} - \xi^{BA} = F\varepsilon^{AB}, \quad \xi^{AB} + \xi^{BA} = \frac{1}{4}F^{js}\sigma_{js}^{AB}.$$

From this it is seen that any antisymmetric spinor of the second rank with components $\xi^{[AB]}$ is equivalent to a complex scalar F; any spinor of the second rank with symmetric components $\xi^{(AB)}$ is equivalent to a complex antisymmetric tensor of the second rank with components satisfying relation (3.120).

If the components of spinor ξ^{AB} are represented in the form of product of the undotted components of the first-rank spinor $\xi^{AB} = \xi^{A}\xi^{B}$, then an expansion in the invariant spintensors can be written in the form

$$\xi^{AB} = \xi^{A}\xi^{B} = -\frac{\mathrm{i}}{8}C^{js}\sigma_{js}^{AB},$$

where the components of the antisymmetric tensor

$$C^{js} = -\mathrm{i}\sigma_{AB}^{js}\xi^{A}\xi^{B} \tag{3.121}$$

satisfy the equations

$$C_{js}C^{js} = 0, \quad C^{js} = \frac{\mathrm{i}}{2}\varepsilon^{jsmn}C_{mn}. \tag{3.122}$$

The first equation in (3.122) can be obtained by contraction of the second identity in (3.106) with spinor components $\xi_A\xi_B\xi_C\xi_D$ with respect to the indices A, B, C, D.

Similar relations can be written for the spinor components of the second rank with dotted indices

$$\eta^{\dot{A}\dot{B}} = \frac{1}{2}\left(\Phi\varepsilon^{\dot{A}\dot{B}} + \frac{1}{4}\Phi^{js}\sigma_{js}^{\dot{A}\dot{B}}\right),$$

where

$$\Phi = -\varepsilon_{\dot{A}\dot{B}}\eta^{\dot{A}\dot{B}}, \quad \Phi^{js} = -\sigma_{\dot{A}\dot{B}}^{js}\eta^{\dot{A}\dot{B}}.$$

Contracting the second identity (3.104) with components $\eta^{\dot{A}\dot{B}}$, we obtain that the components of the antisymmetric tensor Φ^{js} satisfy the linear equation

$$\Phi^{js} = -\frac{\mathrm{i}}{2}\varepsilon^{jsmn}\Phi_{mn}.$$

Using definitions (3.91) and (3.103) for ε, σ_{js}, we obtain

$$\Phi = -\eta^{\dot{1}\dot{2}} + \eta^{\dot{2}\dot{1}}, \quad \Phi^{12} = -\mathrm{i}\Phi^{34} = -\eta^{\dot{1}\dot{2}} - \eta^{\dot{2}\dot{1}},$$

$$\Phi^{23} = -\mathrm{i}\Phi^{14} = \eta^{\dot{1}\dot{1}} - \eta^{\dot{2}\dot{2}}, \quad \Phi^{31} = -\mathrm{i}\Phi^{24} = -\mathrm{i}(\eta^{\dot{1}\dot{1}} + \eta^{\dot{2}\dot{2}}),$$

$$\eta^{\dot{1}\dot{1}} = \frac{1}{2}(\Phi^{23} + \mathrm{i}\Phi^{31}), \quad \eta^{\dot{1}\dot{2}} = \frac{1}{2}(-\Phi - \Phi^{12}),$$

$$\eta^{\dot{2}\dot{1}} = \frac{1}{2}(\Phi - \Phi^{12}), \quad \eta^{\dot{2}\dot{2}} = \frac{1}{2}(-\Phi^{23} + \mathrm{i}\Phi^{31}).$$

For the product of the dotted components of the first-rank spinor we have

$$\eta^{\dot{A}}\eta^{\dot{B}} = -\frac{\mathrm{i}}{8}C^{js}\sigma_{js}^{\dot{A}\dot{B}}, \quad C^{js} = -\mathrm{i}\sigma_{\dot{A}\dot{B}}^{js}\eta^{\dot{A}}\eta^{\dot{B}}.$$

The components of the antisymmetric tensor C^{js} in this equality satisfy the equations

$$C_{js}C^{js} = 0, \quad C^{js} = -\frac{\mathrm{i}}{2}\varepsilon^{jsmn}C_{mn}.$$

Let us now consider the components of the second rank spinor with one dotted index $\xi^{\dot{B}A}$. Due to completeness of the system of four matrices $\sigma_i = \|\sigma_i^{B\dot{A}}\|$, for the arbitrary components $\xi^{\dot{B}A}$ $(A, \dot{B} = 1, 2)$ one can write

$$\xi^{\dot{B}A} = \frac{1}{2}j^i\sigma_i^{A\dot{B}},$$

where the components of the vector j^i are expressed in terms of $\xi^{\dot{B}A}$ by the equality

$$j^i = -\sigma^i_{\dot{B}A}\xi^{\dot{B}A}.$$

Using definitions (3.96) and (3.93) of the matrices $\sigma_i^{\dot{B}A}$, we find

$$\xi^{\dot{1}1} = \frac{1}{2}(j^4 + j^3), \quad \xi^{\dot{1}2} = \frac{1}{2}(j^1 + \mathrm{i}j^2),$$

$$\xi^{\dot{2}1} = \frac{1}{2}(j^1 - \mathrm{i}j^2), \quad \xi^{\dot{2}2} = \frac{1}{2}(j^4 - j^3).$$

The inverse relations have the form

$$j^1 = \xi^{\dot{1}2} + \xi^{\dot{2}1}, \quad j^2 = \mathrm{i}(-\xi^{\dot{1}2} + \xi^{\dot{2}1}),$$

$$j^3 = \xi^{\dot{1}1} - \xi^{\dot{2}2}, \quad j^4 = \xi^{\dot{1}1} + \xi^{\dot{2}2}.$$

If the matrix of components $\xi^{\dot{B}A}$ is Hermitian $\|\xi^{\dot{B}A}\|^{\cdot} = \|\xi^{\dot{B}A}\|^{T}$, then due to hermiticity of matrices σ_i the vector components j^i are real $\left(j^i\right)^{\cdot} = j^i$.

If the components of a spinor $\xi^{\dot{B}A}$ are represented in the form of product $\xi^{\dot{B}A} = \eta^{\dot{B}}\xi^{A}$, then the vector $\boldsymbol{j} = j^i \boldsymbol{\vartheta}_i$ is null

$$j_i j^i = 0. \tag{3.123}$$

Equation (3.123) can be obtained by contracting the first identity in (3.106) with components of spinor $\eta^{\dot{C}}\xi^{D}\eta_{\dot{B}}\xi_{A}$ with respect to the indices A, $\dot{B}$, $\dot{C}$, D.

It is obvious that the tensor components of any rank $j^{i_1 i_2 \ldots i_n}$ can be represented in the form

$$j^{i_1 i_2 \ldots i_n} = (-1)^n \sigma^{i_1}_{\dot{B}_1 A_1} \sigma^{i_2}_{\dot{B}_2 A_2} \cdots \sigma^{i_n}_{\dot{B}_n A_n} \xi^{\dot{B}_1 A_1 \dot{B}_2 A_2 \ldots \dot{B}_n A_n}$$

with the corresponding choice of the spinor components $\xi^{\dot{B}_1 A_1 \dot{B}_2 A_2 \ldots \dot{B}_n A_n}$. To each tensor index i_s there corresponds the pair of the spinor indices $\dot{B}_s$, A_s.

The inverse relation has the form

$$\xi^{\dot{B}_1 A_1 \dot{B}_2 A_2 \ldots \dot{B}_n A_n} = \frac{1}{2^n} \sigma_{i_1}^{A_1 \dot{B}_1} \sigma_{i_2}^{A_2 \dot{B}_2} \cdots \sigma_{i_n}^{A_n \dot{B}_n} j^{i_1 i_2 \cdots i_n}.$$

In particular, for the components F^{ij} of an arbitrary real antisymmetric tensor of the second rank, the relation is valid

$$F^{ij} = \frac{1}{2} \sigma^{i}_{\dot{B}A} \sigma^{j}_{\dot{D}C} \left(\varepsilon^{AC}\xi^{\dot{B}\dot{D}} + \varepsilon^{\dot{B}\dot{D}}\xi^{AC}\right), \tag{3.124}$$

where the components of the spinor $\xi^{\dot{A}\dot{B}}$ are symmetric in the indices $\dot{A}$, $\dot{B}$ and complex conjugate with components ξ^{AB}.

Let R_{ijks} be components of a fourth rank tensor satisfying the identities

$$R_{ijkm} = R_{kmij}, \quad R_{[ijk]m} = 0,$$
$$R_{ijkm} = -R_{ijmk}, \quad R_{ijkm} = -R_{jikm}.$$

Then for the components R_{ijks} the following representation is valid

$$\begin{aligned} R_{ijks} = \frac{1}{4} \sigma_i^{A\dot{A}} \sigma_j^{B\dot{B}} \sigma_k^{C\dot{C}} \sigma_s^{D\dot{D}} \big[& \Psi_{ABCD}\varepsilon_{\dot{A}\dot{B}}\varepsilon_{\dot{C}\dot{D}} + \Psi_{\dot{A}\dot{B}\dot{C}\dot{D}}\varepsilon_{AB}\varepsilon_{CD} \\ & + \Phi_{AB\dot{C}\dot{D}}\varepsilon_{\dot{A}\dot{B}}\varepsilon_{CD} + \Phi_{CD\dot{A}\dot{B}}\varepsilon_{AB}\varepsilon_{\dot{C}\dot{D}} \\ & + 2\Lambda\left(\varepsilon_{AB}\varepsilon_{CD}\varepsilon_{\dot{A}\dot{D}}\varepsilon_{\dot{B}\dot{C}} + \varepsilon_{AC}\varepsilon_{BD}\varepsilon_{\dot{A}\dot{B}}\varepsilon_{\dot{C}\dot{D}}\right)\big]. \end{aligned} \tag{3.125}$$

Here Λ is a real scalar; the components of the spinor Ψ_{ABCD} are symmetric in all indices; the components $\Psi_{\dot{A}\dot{B}\dot{C}\dot{D}}$ are complex conjugate with Ψ_{ABCD}; the components $\Phi_{AB\dot{C}\dot{D}}$ are symmetric with respect to each pair of the indices A, B and $\dot{C}$, $\dot{D}$ and satisfy the condition $\dot{\Phi}_{AB\dot{C}\dot{D}} = \Phi_{CD\dot{A}\dot{B}}$.

Formulas (3.124) and (3.125) are used[8] in the Newman–Penrose formalism [46, 50]

3.4 Definition of Orthonormal Tetrads in Four-Dimensional Pseudo-Euclidean Space E_4^1 by First-Rank Spinors

3.4.1 The Proper Tetrads Defined by the First-Rank Four-Component Spinors in the Space E_4^1

Let $\boldsymbol{\psi}$ be a four-component spinor of the first-rank in the pseudo-Euclidean space E_4^1, referred to an orthonormal basis $\mathfrak{Э}_i$. Let us consider four vectors with components p^i, q^i, S^i, and j^i determined by the spinor $\boldsymbol{\psi}$

$$
\begin{aligned}
p^i &= \operatorname{Im} C^i = -\frac{\mathrm{i}}{2}\gamma^i_{AB}(\psi^A\psi^B + \psi^{+A}\psi^{+B}),\\
q^i &= \operatorname{Re} C^i = \frac{1}{2}\gamma^i_{AB}(\psi^A\psi^B - \psi^{+A}\psi^{+B}),\\
S^i &= -\overset{*}{\gamma}{}^i_{AB}\psi^{+A}\psi^B, \quad j^i = -\mathrm{i}\gamma^i_{AB}\psi^{+A}\psi^B,
\end{aligned} \tag{3.126}
$$

where ψ^A and ψ^{+A} are the contravariant components of the spinor $\boldsymbol{\psi}$ and conjugate spinor $\boldsymbol{\psi}^+$ calculated in the basis $\mathfrak{Э}_i$.

From the first equation in (3.53), equations (a), (b), (c) in (3.60), and equations (a), (b) in (3.62) it follows that the vector components p^i, q^i, S^i, j^i satisfy the equations

$$
\begin{aligned}
p_i p^i &= q_i q^i = S_i S^i = -j_i j^i = \Omega^2 + N^2,\\
p_i q^i &= p_i S^i = p_i j^i = q_i S^i = q_i j^i = S_i j^i = 0
\end{aligned} \tag{3.127}
$$

and, consequently, vectors with components j^i, q^i, S^i, j^i are mutually orthogonal and have the same modulus $\rho = +\sqrt{\Omega^2 + N^2}$. Therefore, if $\rho \neq 0$, then in the space E_4^1 it is possible to introduce the orthonormal basis (tetrad) $\breve{\boldsymbol{e}}_a$, $a = 1, 2, 3, 4$:

$$
\breve{\boldsymbol{e}}_1 = \pi^i\,\mathfrak{Э}_i, \quad \breve{\boldsymbol{e}}_2 = \xi^i\,\mathfrak{Э}_i, \quad \breve{\boldsymbol{e}}_3 = \sigma^i\,\mathfrak{Э}_i, \quad \breve{\boldsymbol{e}}_4 = u^i\,\mathfrak{Э}_i, \tag{3.128}
$$

[8]We use here the Pauli matrices (3.93), which differ by the factor $1/\sqrt{2}$ from the matrices used in [50]. Therefore formulas (3.124) and (3.125) differ from the corresponding Newman–Penrose formulas by the numerical factor.

where the components π^i, ξ^i, σ^i, u^i of the basis vectors $\breve{\boldsymbol{e}}_a$ are defined by the equalities

$$\pi^i = \frac{1}{\rho}\, p^i, \quad \xi^i = \frac{1}{\rho}\, q^i, \quad \sigma^i = \frac{1}{\rho}\, S^i, \quad u^i = \frac{1}{\rho}\, j^i. \tag{3.129}$$

The orthonormal vector basis $\breve{\boldsymbol{e}}_a$ determined by the first-rank spinor $\boldsymbol{\psi}$ by formulas (3.128), (3.129), and (3.126) we shall call the proper basis (or the proper tetrad) of the spinor $\boldsymbol{\psi}$.

The components of vectors, tensors and spinors calculated in the proper basis $\breve{\boldsymbol{e}}_a$, we shall denote by the symbol $\breve{}$. For example, $\boldsymbol{M} = M^{ij}\boldsymbol{\mathcal{Э}}_i\boldsymbol{\mathcal{Э}}_j = M^{ab}\boldsymbol{e}_a\boldsymbol{e}_b = \breve{M}^{ab}\breve{\boldsymbol{e}}_a\breve{\boldsymbol{e}}_b$. The covariant and contravariant components of the metric tensor of the space E_4^1 calculated in any orthonormal basis are the same $g^{ab} = g_{ab} = \breve{g}^{ab} = \breve{g}_{ab} = \mathrm{diag}(1, 1, 1, -1)$, therefore the sign $\breve{}$ over the components of the metric tensor may always be omitted.

Due to Eqs. (3.129) and (3.127), the components $\pi_i, \xi_i, \sigma_i, u_i$ of the proper tetrad vectors $\breve{\boldsymbol{e}}_a$ satisfy the equalities

$$\begin{gathered} \pi_i\pi^i = \xi_i\xi^i = \sigma_i\sigma^i = -u_iu^i = 1, \\ \pi_i\xi^i = \pi_i\sigma^i = \pi_iu^i = \xi_i\sigma^i = \xi_iu^i = \sigma_iu^i = 0. \end{gathered} \tag{3.130}$$

Let us consider the equations connecting the vector components C^i, j^i, S^i, which are contained in (3.62), (3.63):

$$C^i j^j - C^j j^i = -\mathrm{i}\varepsilon^{ijks} C_k S_s, \quad 2(j^i S^j - j^j S^i) = \mathrm{i}\varepsilon^{ijks} \dot{C}_k C_s.$$

Replacing the components $C^i = q^i + \mathrm{i}p^i$, j^i, S^i in these equations in terms of π^i, ξ^i, σ^i, u^i by formulas (3.129), we obtain that the components of the proper tetrad vectors $\breve{\boldsymbol{e}}_a$ satisfy the equations

$$\begin{gathered} \pi^i\xi^j - \pi^j\xi^i = \varepsilon^{ijks}\sigma_k u_s, \quad \sigma^i u^j - \sigma^j u^i = -\varepsilon^{ijks}\pi_k\xi_s, \\ \xi^i\sigma^j - \xi^j\sigma^i = \varepsilon^{ijks}\pi_k u_s, \quad \pi^i u^j - \pi^j u^i = -\varepsilon^{ijks}\xi_k\sigma_s, \\ \sigma^i\pi^j - \sigma^j\pi^i = \varepsilon^{ijks}\xi_k u_s, \quad \xi^i u^j - \xi^j u^i = -\varepsilon^{ijks}\sigma_k\pi_s, \end{gathered} \tag{3.131}$$

defining, in particular, the orientation of the tetrad $\breve{\boldsymbol{e}}_a$.

A connection between the orthonormal bases $\boldsymbol{\mathcal{Э}}_i$ and $\breve{\boldsymbol{e}}_a$ can be written in the form

$$\breve{\boldsymbol{e}}_a = \breve{h}^i{}_a\boldsymbol{\mathcal{Э}}_i, \quad \boldsymbol{\mathcal{Э}}_i = \breve{h}_i{}^a\breve{\boldsymbol{e}}_a, \tag{3.132}$$

where scale factors $\breve{h}^i{}_a$, $\breve{h}_i{}^a$ in accordance with (3.128) and (3.129) are defined by the matrices

$$\breve{h}^i{}_a = \begin{Vmatrix} \pi^1 & \xi^1 & \sigma^1 & u^1 \\ \pi^2 & \xi^2 & \sigma^2 & u^2 \\ \pi^3 & \xi^3 & \sigma^3 & u^3 \\ \pi^4 & \xi^4 & \sigma^4 & u^4 \end{Vmatrix}, \quad \breve{h}_i{}^a = \begin{Vmatrix} \pi^1 & \xi^1 & \sigma^1 & -u^1 \\ \pi^2 & \xi^2 & \sigma^2 & -u^2 \\ \pi^3 & \xi^3 & \sigma^3 & -u^3 \\ -\pi^4 & -\xi^4 & -\sigma^4 & u^4 \end{Vmatrix}. \tag{3.133}$$

From the equation $u_i u^i = -1$ it follows $|u^4| \geqslant 1$, and from the condition $j^4 \geqslant 0$ and definition $u^4 = j^4/\rho$ it follows $u^4 \geqslant 0$. Therefore $u^4 \geqslant 1$. Besides, the contraction of the first equation in (3.131) with components $\pi_j \xi_i$ gives

$$\det \|\breve{h}^i{}_a\| = \varepsilon_{ijks} \pi^i \xi^j \sigma^k u^s = 1.$$

Thus, matrix (3.133) of the scale factors $\breve{h}^i{}_a$ determines the restricted Lorentz transformation (3.132) from the basis $\boldsymbol{\vartheta}_i$ to the basis $\breve{\boldsymbol{e}}_a$.[9]

Let $g^{ab} = \mathrm{diag}(1, 1, 1, -1)$ be the contravariant components of metric tensor of the pseudo-Euclidean space E_4^1, calculated in the proper basis $\breve{\boldsymbol{e}}_a$. The components g^{ij} of metric tensor of the space E_4^1, calculated in the basis $\boldsymbol{\vartheta}_i$, are connected with g^{ab} by the relation

$$g^{ij} = \breve{h}^i{}_a \breve{h}^j{}_b g^{ab}.$$

Replacing here the quantities $\breve{h}^i{}_a$ by formula (3.133), we obtain an expression of the components g^{ij} of the metric tensor in terms of the vector components π^i, ξ^i, σ^i, and u^i

$$g^{ij} = \pi^i \pi^j + \xi^i \xi^j + \sigma^i \sigma^j - u^i u^j. \tag{3.134}$$

It is obvious that Eqs. (3.134) are equivalent to conditions (3.130).

From definitions (3.128), (3.129), and (3.126) it follows that both a spinor with components ψ and a spinor with components

$$\eta = \left(\alpha I + \mu \gamma^5\right) \psi, \quad \eta^+ = \psi^+ \left(\alpha I + \mu \gamma^5\right), \tag{3.135}$$

where α and μ are arbitrary real numbers that are not simultaneously equal to zero $\alpha^2 + \mu^2 \neq 0$, correspond one and the same proper basis $\breve{\boldsymbol{e}}_a$.

---

[9]The orthonormal vector tetrads defined by the spinor field of the first-rank in the space E_4^1, were introduced by Gürsey [32], see also Takabayasi [65, 66].

Indeed, replacing in definition (3.126) the spinor components ψ by η and components ψ^+ by η^+, we obtain

$$\begin{aligned} j'^s = \mathrm{i}\eta^+\gamma^s\eta &= \mathrm{i}[\alpha^2\psi^+\gamma^s\psi + \mu^2\psi^+\gamma^5\gamma^s\gamma^5\psi + \alpha\mu\psi^+(\gamma^5\gamma^s + \gamma^s\gamma^5)\psi] \\ &= \mathrm{i}(\alpha^2 + \mu^2)\psi^+\gamma^s\psi = (\alpha^2 + \mu^2)j^s. \end{aligned}$$

Under the transformation of this equation it is necessary to take into account that the Dirac matrices satisfy the equations (see (3.11))

$$\gamma^5\gamma^i + \gamma^i\gamma^5 = 0, \quad \gamma^5\gamma^i\gamma^5 = \gamma^i.$$

In a similar way, the validity of the following equalities can be shown

$$\begin{aligned} S'^i &= \eta^+\overset{*}{\gamma}{}^i\eta = (\alpha^2 + \mu^2)S^i, \\ C'^i &= \eta^T E\gamma^i\eta = (\alpha^2 + \mu^2)C^i, \\ \rho' &= (S'_i S'^i)^{1/2} = (\alpha^2 + \mu^2)\rho, \end{aligned}$$

from which it follows:

$$\frac{1}{\rho'}p'^i = \frac{1}{\rho}p^i, \quad \frac{1}{\rho'}q'^i = \frac{1}{\rho}q^i, \quad \frac{1}{\rho'}S'^i = \frac{1}{\rho}S^i, \quad \frac{1}{\rho'}j'^i = \frac{1}{\rho}j^i$$

and consequently $\breve{\boldsymbol{e}}'_a = \breve{\boldsymbol{e}}_a$.

If $\rho \neq 0$, then the components of the complex tensors C^i, C^{ij} and the components of the real antisymmetric tensor M^{ij} can be expressed in terms of the vector components π^i, ξ^i, σ^i, u^i and the invariants Ω, N connected with the invariants ρ, η by the second relation in (3.66). Indeed, from definitions (3.126) and (3.129) it follows

$$C^j = q^j + \mathrm{i}p^j = \rho(\xi^j + \mathrm{i}\pi^j). \tag{3.136}$$

To obtain an expression of the tensor components C^{ij} in terms of π^i, ξ^i, σ^i, u^i we contract the equations (g) in (3.62) with the pseudotensor Levi-Civita components ε^{ijks} with respect to the indices i, j. We have

$$-NC^{ij} + \frac{1}{2}\varepsilon^{ijks}\Omega C_{ks} = C^i S^j - C^j S^i. \tag{3.137}$$

From the equations (g) in (3.62) and (3.137) it follows

$$\rho^2 C^{ij} = \Omega\varepsilon^{ijks}S_k C_s + N(S^i C^j - S^j C^i).$$

From this taking into account the equations (h) in (3.62), (3.136), and definitions (3.129), we get

$$C^{ij} = \Omega\left(\pi^i u^j - \pi^j u^i\right) - N\left(\xi^i \sigma^j - \xi^j \sigma^i\right)$$
$$+ \mathrm{i}\left[-\Omega\left(\xi^i u^j - \xi^j u^i\right) + N\left(\sigma^i \pi^j - \sigma^j \pi^i\right)\right]. \qquad (3.138)$$

Similar transformations of the equations (k) in (3.60) give the following expression for the real components M^{ij}:

$$M^{ij} = \Omega\left(\pi^i \xi^j - \pi^j \xi^i\right) + N\left(\sigma^i u^j - \sigma^j u^i\right). \qquad (3.139)$$

Since the tensors $\boldsymbol{C}$ are completely determined by the spinor $\boldsymbol{\psi}$, then from equalities (3.136) and (3.138) it follows that specifying of the vector components π^i, ξ^i, σ^i, u^i and two invariants Ω, N, at least one of which is nonzero, also completely determines the spinor $\boldsymbol{\psi}$. However, the connection between a spinor $\boldsymbol{\psi}$ and quantities π^i, ξ^i, σ^i, u^i, Ω, N is not one-to-one, since the vector components π^i, ξ^i, σ^i, u^i are determined only if $\rho^2 = \Omega^2 + N^2 \neq 0$.[10]

Since the components of vectors $\breve{\pi}^a$, $\breve{\xi}^a$, $\breve{\sigma}^a$, and $\breve{u}^a$ in the proper basis $\breve{\boldsymbol{e}}_a$ have the form

$$\breve{\pi}^a = (1, 0, 0, 0), \quad \breve{\xi}^a = (0, 1, 0, 0),$$
$$\breve{\sigma}^a = (0, 0, 1, 0), \quad \breve{u}^a = (0, 0, 0, 1),$$

then from Eqs. (3.129) and (3.139) it follows that the real tensors $\boldsymbol{D}$ defined by spinor $\boldsymbol{\psi}$ in accordance with formulas (3.58), in the basis $\breve{\boldsymbol{e}}_a$ are defined by the components

$$\breve{M}^{ab} = \left\| \begin{matrix} 0 & \Omega & 0 & 0 \\ -\Omega & 0 & 0 & 0 \\ 0 & 0 & 0 & N \\ 0 & 0 & -N & 0 \end{matrix} \right\|, \qquad \begin{aligned} \breve{j}^a &= \left(0, 0, 0, \sqrt{\Omega^2 + N^2}\right), \\ \breve{S}^a &= \left(0, 0, \sqrt{\Omega^2 + N^2}, 0\right). \end{aligned} \qquad (3.140)$$

[10]The quantities $p^i = \rho\pi^i$, $q^i = \rho\xi^i$, $S^i = \rho\sigma^i$, $j^i = \rho u^i$ are defined by relations(3.126) and for $\rho = 0$. However, one-to-one connection between $\boldsymbol{\psi}$ and quantities Ω, N, p^i, q^i, S^i, j^i does not exist either. For example, for semispinors $\Omega = N = 0$, $p^i = q^i = 0$. One nonzero isotropic vector with components $S^i = j^i$ or $S^i = -j^i$ remaining in this case from tetrad $\rho\breve{\boldsymbol{e}}_a$, determines the semispinor components up to a factor $\exp(\mathrm{i}\varphi)$ only.

The components of complex tensors $\boldsymbol{C}$ determined by a spinor $\boldsymbol{\psi}$ in accordance with formulas (3.51), in basis $\breve{\boldsymbol{e}}_a$ are defined by the formulae

$$\breve{C}^{ab} = \begin{Vmatrix} 0 & 0 & -\mathrm{i}N & \Omega \\ 0 & 0 & -N & -\mathrm{i}\Omega \\ \mathrm{i}N & N & 0 & 0 \\ -\Omega & \mathrm{i}\Omega & 0 & 0 \end{Vmatrix},$$

$$\breve{C}^{a} = \left(\mathrm{i}\sqrt{\Omega^2 + N^2}, \sqrt{\Omega^2 + N^2}, 0, 0\right), \tag{3.141}$$

which can be easily obtained by means of Eqs. (3.136) and (3.138).

Let us calculate components $\breve{\psi}^A$ of a spinor $\boldsymbol{\psi}$ in the proper basis $\breve{\boldsymbol{e}}_a$. For this purpose it is possible to use formulae (3.54), which in the case under consideration can be written in the form (for $\eta_C = \delta_{BC}$)

$$\breve{\psi}^A = \frac{\breve{\psi}^{BA}}{\sqrt{\breve{\psi}^{BB}}}, \tag{3.142}$$

$$\|\breve{\psi}^{BA}\| = \frac{1}{4}\left[\rho\,(\mathrm{i}\gamma_1 + \gamma_2) + N\,(\gamma_{23} + \mathrm{i}\gamma_{13}) + \Omega\,(\mathrm{i}\gamma_{24} - \gamma_{14})\right] E^{-1}.$$

When writing the second formula in (3.142) it is taken into account that tensors $\boldsymbol{C}$ in the proper basis $\breve{\boldsymbol{e}}_a$ are defined by components (3.141).

If the invariant spintensors E and γ_i are determined by matrices (3.24) and (3.25), then for quantities $\breve{\psi}^{BA}$ in (3.142) we have

$$\breve{\psi}^{BA} = -\frac{1}{2}\rho \begin{Vmatrix} 0 & 0 & 0 & 0 \\ 0 & \exp \mathrm{i}\eta & 0 & 1 \\ 0 & 0 & 0 & 0 \\ 0 & 1 & 0 & \exp(-\mathrm{i}\eta) \end{Vmatrix}.$$

Here the invariants ρ, η are connected with the invariants Ω, N by the equality

$$\Omega + \mathrm{i}N = \rho \exp \mathrm{i}\eta. \tag{3.143}$$

Using the first formula in (3.142) we obtain the following expression for $\breve{\psi}^A$:

$$\breve{\psi}^A = \mathrm{i}\sqrt{\tfrac{1}{2}\rho} \begin{Vmatrix} 0 \\ \exp\left(\frac{\mathrm{i}}{2}\eta\right) \\ 0 \\ \exp\left(-\frac{\mathrm{i}}{2}\eta\right) \end{Vmatrix} \tag{3.144}$$

The validity of formulas (3.144) can also be established directly, substituting (3.144) into definitions (3.56) and (3.64), that gives expressions (3.140) and (3.141).

By means of the proper tetrad $\breve{\boldsymbol{e}}_a$ it is possible to introduce the proper complex null tetrad $\breve{\boldsymbol{e}}_a^\circ$, which is determined by vectors with components l^i, n^i, m^i, $\dot{m}^i$:

$$\begin{aligned} \sqrt{2}\, l^i &= u^i + \sigma^i, \quad \sqrt{2}\, m^i = \pi^i - \mathrm{i}\xi^i, \\ \sqrt{2}\, n^i &= u^i - \sigma^i, \quad \sqrt{2}\, \dot{m}^i = \pi^i + \mathrm{i}\xi^i. \end{aligned} \tag{3.145}$$

By virtue of definitions (3.145) and conditions (3.130), the following equations are satisfied

$$\begin{aligned} \dot{m}_i m^i &= -l_i n^i = 1, \\ l_i l^i = n_i n^i = m_i m^i &= l_i m^i = n_i m^i = 0, \end{aligned} \tag{3.146}$$

which are equivalent to the equation

$$g^{ij} = -l^i n^j - l^j n^i + m^i \dot{m}^j + m^j \dot{m}^i,$$

in which g^{ij} are the components of the metric tensor of the pseudo-Euclidean space E_4^1.

Equations (3.131) for the vectors of the orthonormal tetrad $\breve{\boldsymbol{e}}_a$ pass into the following equations for the vectors of the null tetrad $\breve{\boldsymbol{e}}_a^\circ$:

$$\begin{aligned} m^i \dot{m}^j - m^j \dot{m}^i &= \mathrm{i}\varepsilon^{ijks} l_k n_s, \\ l^i n^j - l^j n^i &= \mathrm{i}\varepsilon^{ijks} m_k \dot{m}_s, \\ m^i n^j - m^j n^i &= \mathrm{i}\varepsilon^{ijks} m_k n_s, \\ m^i l^j - m^j l^i &= -\mathrm{i}\varepsilon^{ijks} m_k l_s. \end{aligned}$$

3.4.2 Field of Proper Tetrads, Determined by Field of a First-Rank Spinor in the Minkowski Space

Let us consider the point four-dimensional pseudo-Euclidean space with the metric signature $(+, +, +, -)$ (the Minkowski space), referred to a Cartesian coordinate system with the variables x^i and a covariant orthonormal vector basis $\boldsymbol{\vartheta}_i$. Let $\boldsymbol{\psi} = \boldsymbol{\psi}(x^i)$ be the field of a four-components spinor in the Minkowski space, and this field corresponds to a field of the real tensors $\boldsymbol{D}(x^i)$ and a field of the complex tensors $\boldsymbol{C}(x^i)$ At all points of the Minkowski space, in which the invariant $\rho^2 = \Omega^2 + N^2$ of the spinor $\boldsymbol{\psi}$ is nonzero, it is possible to introduce a proper orthonormal basis with the vectors $\breve{\boldsymbol{e}}_a(x^i)$ determined by formulas (3.128), (3.129), (3.126). The Ricci rotation coefficients $\breve{\Delta}_{s,ij}$, corresponding to the orthonormal bases $\breve{\boldsymbol{e}}_a(x^i)$, in the Cartesian system coordinates x^i are defined by the relation

$$\breve{\Delta}_{s,ij} = \frac{1}{2} g_{ab} \big(\breve{h}_i{}^a \partial_s \breve{h}_j{}^b - \breve{h}_j{}^a \partial_s \breve{h}_i{}^b \big),$$

where coefficients $\breve{h}_i{}^a$ are determined by the matrix (3.133), $\partial_s = \partial/\partial x^s$ is the symbol of a partial derivative with respect to variable x^s. Replacing here $\breve{h}_i{}^a$ by formula (3.133), we obtain the following expression for the Ricci rotation coefficients $\breve{\Delta}_{s,ij}$

$$\breve{\Delta}_{s,ij} = \frac{1}{2} \big(\pi_i \partial_s \pi_j - \pi_j \partial_s \pi_i + \xi_i \partial_s \xi_j - \xi_j \partial_s \xi_i + \sigma_i \partial_s \sigma_j - \sigma_j \partial_s \sigma_i - u_i \partial_s u_j + u_j \partial_s u_i \big). \tag{3.147}$$

The Ricci rotation coefficients (3.147) corresponding to the proper bases $\breve{\boldsymbol{e}}_a(x^i)$, one can express directly in terms of the spinor field $\boldsymbol{\psi}(x^i)$ by the relation

$$\breve{\Delta}_{s,ij} = \frac{1}{\Omega^2 + N^2} \big[\Omega \big(\psi^+ \gamma_{ij} \partial_s \psi - \partial_s \psi^+ \cdot \gamma_{ij} \psi \big) + N \big(\psi^+ \gamma^5 \gamma_{ij} \partial_s \psi - \partial_s \psi^+ \cdot \gamma^5 \gamma_{ij} \psi \big) \big], \tag{3.148}$$

in which $\gamma_{ij} = \gamma_{[i} \gamma_{j]}$. Relation (3.148) will be proved in Sect. 3.6.

Thus, specifying in the Minkowski space the field of a four-component first-rank spinor $\boldsymbol{\psi}(x^i)$ defines the system of the proper orthonormal bases $\breve{\boldsymbol{e}}_a(x^i)$ by formulas (3.128), (3.129), and (3.126). The Ricci rotation coefficients for such system of bases are defined by relation (3.147) or relation (3.148).

From formula (3.147) we get the identities, which are checked directly taking into account equations (3.130):

$$\partial_s \pi_i = -\breve{\Delta}_{s,i}{}^j \pi_j, \quad \partial_s \xi_i = -\breve{\Delta}_{s,i}{}^j \xi_j,$$
$$\partial_s \sigma_i = -\breve{\Delta}_{s,i}{}^j \sigma_j, \quad \partial_s u_i = -\breve{\Delta}_{s,i}{}^j u_j. \tag{3.149}$$

Along with the Ricci rotation coefficients $\breve{\Delta}_{s,ij}$, defined by relation (3.147), further on the coefficients $\breve{\Delta}_{a,bc}$ will also be used:

$$\breve{\Delta}_{a,bc} = \breve{h}^s{}_a \breve{h}^i{}_b \breve{h}^j{}_c \breve{\Delta}_{s,ij},$$

where the scale factors $\breve{h}^i{}_b$ are determined by matrix (3.133). By means of Eqs. (3.147) and (3.130) for the symbols $\breve{\Delta}_{a,bc}$ one can find

$$\begin{aligned} \breve{\Delta}_{a,12} &= \xi^i \breve{\partial}_a \pi_i, & \breve{\Delta}_{a,31} &= \pi^i \breve{\partial}_a \sigma_i, \\ \breve{\Delta}_{a,23} &= \sigma^i \breve{\partial}_a \xi_i, & \breve{\Delta}_{a,14} &= u^i \breve{\partial}_a \pi_i, \\ \breve{\Delta}_{a,24} &= u^i \breve{\partial}_a \xi_i, & \breve{\Delta}_{a,34} &= u^i \breve{\partial}_a \sigma_i. \end{aligned} \tag{3.150}$$

Here $\breve{\partial}_a = \breve{h}^i{}_a \partial_i$ is the directional derivative

$$\breve{\partial}_1 = \pi^i \partial_i, \quad \breve{\partial}_2 = \xi^i \partial_i, \quad \breve{\partial}_3 = \sigma^i \partial_i \quad \breve{\partial}_4 = u^i \partial_i.$$

Equations (3.149) for vectors of the orthonormal tetrad $\breve{\boldsymbol{e}}_a(x^i)$ can be written in the components of vectors of the null tetrad $\breve{\boldsymbol{e}}^\circ_a(x^i)$ as follows:

$$\begin{aligned} Dl^i &= -(\varepsilon + \dot{\varepsilon})l^i + \dot{\kappa} m^i + \kappa \dot{m}^i, & Dn^i &= (\varepsilon + \dot{\varepsilon})n^i - \pi m^i - \dot{\pi}\dot{m}^i, \\ \Delta l^i &= -(\gamma + \dot{\gamma})l^i + \dot{\tau} m^i + \tau \dot{m}^i, & \Delta n^i &= (\gamma + \dot{\gamma})n^i - \nu m^i - \dot{\nu}\dot{m}^i, \\ \delta l^i &= -(\dot{\alpha} + \beta)l^i + \dot{\varrho} m^i + \sigma \dot{m}^i, & \delta n^i &= (\dot{\alpha} + \beta)n^i - \mu m^i - \dot{\lambda}\dot{m}^i, \\ \dot{\delta} l^i &= -(\alpha + \dot{\beta})l^i + \dot{\sigma} m^i + \varrho \dot{m}^i, & \dot{\delta} n^i &= (\alpha + \dot{\beta})n^i - \lambda m^i - \dot{\mu}\dot{m}^i, \end{aligned}$$

$$\begin{aligned} Dm^i &= -\dot{\pi} l^i + \kappa n^i + (\dot{\varepsilon} - \varepsilon)m^i, \\ \Delta m^i &= -\dot{\nu} l^i + \tau n^i + (\dot{\gamma} - \gamma)m^i, \\ \delta m^i &= -\dot{\lambda} l^i + \sigma n^i + (\dot{\alpha} - \beta)m^i, \\ \dot{\delta} m^i &= -\dot{\mu} l^i + \varrho n^i + (\dot{\beta} - \alpha)m^i. \end{aligned} \tag{3.151}$$

The differential operators D, Δ, δ, $\dot{\delta}$ in Eqs. (3.151) are defined by the relations

$$D = l^i \partial_i, \quad \Delta = n^i \partial_i, \quad \delta = m^i \partial_i, \quad \dot{\delta} = \dot{m}^i \partial_i,$$

while the coefficients α, β ... in the right-hand side of Eqs. (3.151), called the spin-coefficients, are defined as follows

$$\begin{aligned} 2\alpha &= n^i \dot{\delta} l_i - \dot{m}^i \dot{\delta} m_i, & \kappa &= m^i D l_i, & \pi &= -\dot{m}^i D n_i, \\ 2\beta &= n^i \delta l_i - \dot{m}^i \delta m_i, & \tau &= m^i \Delta l_i, & \nu &= -\dot{m}^i \Delta n_i, \\ 2\gamma &= n^i \Delta l_i - \dot{m}^i \Delta m_i, & \sigma &= m^i \delta l_i, & \mu &= -\dot{m}^i \delta n_i, \\ 2\varepsilon &= n^i D l_i - \dot{m}^i D m_i, & \varrho &= m^i \dot{\delta} l_i, & \lambda &= -\dot{m}^i \dot{\delta} n_i. \end{aligned} \tag{3.152}$$

These formulas for the spin-coefficients can easily be obtained by contracting equations (3.151) with components of the vectors l_i, n_i, m_i, $\dot{m}_i$, taking into account relations (3.146). It is necessary to distinguish the invariant ρ of the spinor field $\psi(x^i)$ and the spin-coefficient ϱ.

The operator of the derivative ∂_i can be expressed in terms of the differential operators D, Δ, δ, $\dot{\delta}$:

$$\partial_i = \left(- l_i n^j - l^j n_i + m_i \dot{m}^j + \dot{m}_i m^j \right) \partial_j = -l_i \Delta - n_i D + m_i \dot{\delta} + \dot{m}_i \delta.$$

Equation (3.151) and formula (3.152) for the spin–coefficients are valid for any null tetrads $\boldsymbol{e}^{\circ}_{a}(x^i)$ with components of the vectors of l_i, n_i, m_i, $\dot{m}_i$, satisfying Eqs. (3.146).

Direct calculation shows that the Ricci rotation coefficients $\breve{\Delta}_{a,bc}$ and the spin–coefficients are connected by the linear equations

$$\breve{\Delta}_{1,12} = \frac{\mathrm{i}}{\sqrt{2}}(-\dot{\alpha} + \alpha - \dot{\beta} + \beta), \quad \breve{\Delta}_{1,34} = \frac{1}{\sqrt{2}}(\dot{\alpha} + \alpha + \dot{\beta} + \beta),$$

$$\breve{\Delta}_{1,23} = \frac{\mathrm{i}}{2\sqrt{2}}(\dot{\sigma} - \sigma + \dot{\varrho} - \varrho - \dot{\mu} + \mu - \dot{\lambda} + \lambda),$$

$$\breve{\Delta}_{1,31} = \frac{1}{2\sqrt{2}}(\dot{\sigma} + \sigma + \dot{\varrho} + \varrho + \dot{\mu} + \mu + \dot{\lambda} + \lambda),$$

$$\breve{\Delta}_{1,14} = \frac{1}{2\sqrt{2}}(-\dot{\sigma} - \sigma - \dot{\varrho} - \varrho + \dot{\mu} + \mu + \dot{\lambda} + \lambda),$$

$$\breve{\Delta}_{1,24} = \frac{\mathrm{i}}{2\sqrt{2}}(\dot{\sigma} - \sigma + \dot{\varrho} - \varrho + \dot{\mu} - \mu + \dot{\lambda} - \lambda),$$

$$\breve{\Delta}_{2,12} = \frac{1}{\sqrt{2}}(\dot{\alpha} + \alpha - \dot{\beta} - \beta), \quad \breve{\Delta}_{2,34} = \frac{\mathrm{i}}{\sqrt{2}}(\dot{\alpha} - \alpha - \dot{\beta} + \beta),$$

$$\breve{\Delta}_{2,23} = \frac{1}{2\sqrt{2}}(\dot{\sigma} + \sigma - \dot{\varrho} - \varrho - \dot{\mu} - \mu + \dot{\lambda} + \lambda),$$

$$\breve{\Delta}_{2,31} = \frac{\mathrm{i}}{2\sqrt{2}}(-\dot{\sigma} + \sigma + \dot{\varrho} - \varrho - \dot{\mu} + \mu + \dot{\lambda} - \lambda),$$

$$\breve{\Delta}_{2,14} = \frac{\mathrm{i}}{2\sqrt{2}}(\dot{\sigma} - \sigma - \dot{\varrho} + \varrho - \dot{\mu} + \mu + \dot{\lambda} - \lambda),$$

$$\breve{\Delta}_{2,24} = \frac{1}{2\sqrt{2}}(\dot{\sigma} + \sigma - \dot{\varrho} - \varrho + \dot{\mu} + \mu - \dot{\lambda} - \lambda),$$

$$\breve{\Delta}_{3,12} = \frac{\mathrm{i}}{\sqrt{2}}(-\dot{\varepsilon} + \varepsilon + \dot{\gamma} - \gamma), \quad \breve{\Delta}_{3,34} = \frac{1}{\sqrt{2}}(\dot{\varepsilon} + \varepsilon - \dot{\gamma} - \gamma),$$

$$\breve{\Delta}_{3,23} = \frac{\mathrm{i}}{2\sqrt{2}}(\dot{\kappa} - \kappa - \dot{\tau} + \tau - \dot{\pi} + \pi + \dot{\nu} - \nu),$$

$$\Delta_{3,31} = \frac{1}{2\sqrt{2}}(\dot{\kappa} + \kappa - \dot{\tau} - \tau + \dot{\pi} + \pi - \dot{\nu} - \nu),$$

$$\breve{\Delta}_{3,14} = \frac{1}{2\sqrt{2}}(-\dot{\kappa} - \kappa + \dot{\tau} + \tau + \dot{\pi} + \pi - \dot{\nu} - \nu),$$

$$\breve{\Delta}_{3,24} = \frac{\mathrm{i}}{2\sqrt{2}}(\dot{\kappa} - \kappa - \dot{\tau} + \tau + \dot{\pi} - \pi - \dot{\nu} + \nu),$$

$$\breve{\Delta}_{4,12} = \frac{\mathrm{i}}{\sqrt{2}}(-\dot{\varepsilon} + \varepsilon - \dot{\gamma} + \gamma), \quad \breve{\Delta}_{4,34} = \frac{1}{\sqrt{2}}(\dot{\varepsilon} + \varepsilon + \dot{\gamma} + \gamma),$$

$$\breve{\Delta}_{4,23} = \frac{\mathrm{i}}{2\sqrt{2}}(\dot{\kappa} - \kappa + \dot{\tau} - \tau - \dot{\pi} + \pi - \dot{\nu} + \nu),$$

$$\breve{\Delta}_{4,31} = \frac{1}{2\sqrt{2}}(\dot{\kappa} + \kappa + \dot{\tau} + \tau + \dot{\pi} + \pi + \dot{\nu} + \nu),$$

$$\breve{\Delta}_{4,14} = \frac{1}{2\sqrt{2}}(-\dot{\kappa} - \kappa - \dot{\tau} - \tau + \dot{\pi} + \pi + \dot{\nu} + \nu),$$

$$\breve{\Delta}_{4,24} = \frac{\mathrm{i}}{2\sqrt{2}}(\dot{\kappa} - \kappa + \dot{\tau} - \tau + \dot{\pi} - \pi + \dot{\nu} - \nu). \tag{3.153}$$

Let us give also an expression of the coefficients $\breve{\Delta}_{s,ij} = \breve{h}_s{}^a \breve{h}_i{}^b \breve{h}_j{}^c \breve{\Delta}_{a,bc}$ in terms of the vectors of the proper null tetrad

$$\breve{\Delta}_{s,ij} = \frac{1}{2}\big(m_i \partial_s \dot{m}_j - m_j \partial_s \dot{m}_i + \dot{m}_i \partial_s m_j - \dot{m}_j \partial_s m_i$$
$$- l_i \partial_s n_j + l_j \partial_s n_i - n_i \partial_s l_j + n_j \partial_s l_i\big).$$

3.4.3 Proper Bases (Tetrads), Determined by a Semispinor in the Space E_4^1

Let us consider in the pseudo-Euclidean space E_4^1 a four-component spinor $\boldsymbol{\psi}$ in the spinbasis $\overset{*}{\boldsymbol{\varepsilon}}_A$, which is determined by the Dirac matrices (3.24) and metric spinor (3.25). The components of spinor ψ in the spinbasis $\overset{*}{\boldsymbol{\varepsilon}}_A$ can be represented in the form

$$\psi = \begin{Vmatrix} \xi^A \\ \eta_{\dot{A}} \end{Vmatrix}, \quad \dot{A}, A = 1, 2,$$

where ξ^A, $\eta_{\dot{A}}$ determine in the pseudo-Euclidean space E_4^1 the two-component spinors with the fixed relative sign.

Relations (3.126) and (3.129) determining the orthonormal tetrads $\breve{\boldsymbol{e}}_a$ in notations of two-component spinors have the form

$$\begin{aligned}
\rho\pi^j &= \sigma^j_{\dot{B}A}\left(-\eta^{\dot{B}}\xi^A - \dot{\eta}^{\dot{A}}\dot{\xi}^B\right),\\
\rho\xi^j &= \mathrm{i}\sigma^j_{\dot{B}A}\left(-\eta^{\dot{B}}\xi^A + \dot{\eta}^{\dot{A}}\dot{\xi}^B\right),\\
\rho\sigma^j &= \sigma^j_{\dot{B}A}\left(\dot{\xi}^B\xi^A - \eta^{\dot{B}}\dot{\eta}^{\dot{A}}\right),\\
\rho u^j &= \sigma^j_{\dot{B}A}\left(-\dot{\xi}^B\xi^A - \eta^{\dot{B}}\dot{\eta}^{\dot{A}}\right),
\end{aligned} \tag{3.154}$$

and for the invariants Ω, N of the spinor $\boldsymbol{\psi}$ we have

$$\Omega + \mathrm{i}N = \rho\exp\mathrm{i}\eta = 2\varepsilon_{BA}\xi^B\dot{\eta}^{\dot{A}} = 2\det\begin{Vmatrix}\xi^1 & \dot{\eta}^{\dot{1}}\\ \xi^2 & \dot{\eta}^{\dot{2}}\end{Vmatrix}. \tag{3.155}$$

The components of invariant spintensors $\sigma^j_{\dot{B}A}$ in definitions (3.154) are defined by the equalities (σ^α are the two-dimensional Pauli matrices, I is the unit two-dimensional matrix)

$$\|\sigma^4_{\dot{B}A}\| = -I, \quad \|\sigma^\alpha_{\dot{B}A}\| = -\sigma^\alpha, \quad \alpha = 1, 2, 3.$$

Components (3.145) of the vectors of the complex null tetrad $\breve{\boldsymbol{e}}^\circ_a$ in the notations of two-component spinors are written as follows[11]

$$l^i = -\frac{\sqrt{2}}{\rho}\sigma^i_{\dot{B}A}\eta^{\dot{B}}\dot{\eta}^{\dot{A}}, \quad n^i = -\frac{\sqrt{2}}{\rho}\sigma^i_{\dot{B}A}\dot{\xi}^B\xi^A,$$
$$m^i = -\frac{\sqrt{2}}{\rho}\sigma^i_{\dot{B}A}\eta^{\dot{B}}\xi^A. \tag{3.156}$$

Relations (3.154) and (3.156) put into correspondence to the two-component spinor of the first-rank with undotted contravariant components ξ^A some set of the orthonormal tetrads $\{\breve{\boldsymbol{e}}_a\}$ or null tetrads $\{\breve{\boldsymbol{e}}^\circ_a\}$, obtained by using in (3.154), (3.156) all $\eta_{\dot{A}}$ satisfying the condition $\rho \neq 0$. In the same way, these relations put into correspondence to the two-component spinor of the first-rank with dotted covariant components $\eta_{\dot{A}}$ the set tetrads $\{\breve{\boldsymbol{e}}_a\}$ or $\{\breve{\boldsymbol{e}}^\circ_a\}$ determined by formulas (3.154), (3.156) when using all ξ^A satisfying the condition $\rho \neq 0$.

The sets of tetrads defined above $\{\breve{\boldsymbol{e}}_a\}$, $\{\breve{\boldsymbol{e}}^\circ_a\}$, corresponding to the spinor $\boldsymbol{\psi}$ with the fixed components ξ^A (or to the spinor $\boldsymbol{\psi}$ with the fixed components $\eta_{\dot{A}}$), it is possible to narrow by specifying the invariants ρ, η of the spinor $\boldsymbol{\psi}$. Let us consider,

[11]The null tetrad $\breve{\boldsymbol{e}}^\circ_a$, determined by the equations analogous to Eqs. (3.156) and (3.154) for the spinor fields with given invariants $\rho = 2$, $\eta = 0$, was used in [31].

in particular, the set of four-component spinors $\boldsymbol{\psi}$ with fixed components ξ^A, for which invariants ρ, η are the same. From condition $\rho \neq 0$ and definition (3.155) it follows that $\det \begin{Vmatrix} \xi^1 & \dot{\eta}^{\dot{1}} \\ \xi^2 & \dot{\eta}^{\dot{2}} \end{Vmatrix} \neq 0$ and, therefore, spinors η and $\varepsilon\dot{\xi}$ are linearly independent. Therefore, an arbitrary linear transformation component $\eta_{\dot{A}}$ can be written in the form

$$\eta'_{\dot{A}} = \eta_{\dot{A}} + A\eta_{\dot{A}} + B\varepsilon_{\dot{A}\dot{B}}\dot{\xi}^B,$$

where A, B are arbitrary complex numbers. It is easy to show that from condition $\rho' = \rho$, $\eta' = \eta$ it follows $A = 0$ and, thus, the set $\{\psi\}$ with the fixed components ξ^A and with the given invariants ρ, η can be determined by the components

$$\{\psi\} = \begin{Vmatrix} \xi^A \\ \eta_{\dot{A}} + B\varepsilon_{\dot{A}\dot{B}}\dot{\xi}^B \end{Vmatrix}.$$

From definitions (3.155), (3.154) it follows that when transforming

$$\xi'^A = \xi^A, \quad \eta'_{\dot{A}} = \eta_{\dot{A}} + B\varepsilon_{\dot{A}\dot{B}}\dot{\xi}^B \tag{3.157}$$

the quantities ρ, η do not change, while the components of the vectors of tetrad $\breve{\boldsymbol{e}}_a$ are transformed as

$$\begin{aligned} \pi'_i &= \pi_i + (u_i - \sigma_i)\,\mathrm{Re}\,B, \\ \xi'_i &= \xi_i + (u_i - \sigma_i)\,\mathrm{Im}\,B, \\ \sigma'_i &= \sigma_i\left(1 - \frac{1}{2}\dot{B}B\right) + \frac{1}{2}\dot{B}Bu_i + \pi_i\,\mathrm{Re}\,B + \xi_i\,\mathrm{Im}\,B, \\ u'_i &= u_i\left(1 + \frac{1}{2}\dot{B}B\right) - \frac{1}{2}\dot{B}B\sigma_i + \pi_i\,\mathrm{Re}\,B + \xi_i\,\mathrm{Im}\,B. \end{aligned} \tag{3.158}$$

The vectors of the null tetrad $\breve{\boldsymbol{e}}^{\circ}_a$ under transformations (3.157) are transformed by formulas

$$\begin{aligned} n'_i &= n_i, \\ m'_i &= m_i + Bn_i, \\ l'_i &= l_i + \dot{B}Bn_i + \dot{B}m_i + B\dot{m}_i. \end{aligned} \tag{3.159}$$

Thus, the two-component spinor with the undotted covariant components ξ^A determines an orthonormal tetrad $\breve{\boldsymbol{e}}_a$ and an complex null tetrad $\breve{\boldsymbol{e}}^{\circ}_a$ up to transformations (3.158), (3.159), that depend on an arbitrary complex parameter B.

It is obvious that the similar statement is valid also for the components of a spinor with the dotted index $\eta_{\dot{A}}$. In this case from definitions (3.155), (3.154) it follows that when transforming

$$\eta'_{\dot{A}} = \eta_{\dot{A}}, \quad \xi'^{A} = \xi^{A} + A\varepsilon^{AB}\dot{\eta}_{\dot{B}}$$

the quantities ρ, η do not change, and the components of the vectors of tetrad $\breve{\boldsymbol{e}}_a$ are transformated as

$$\begin{aligned}
\pi'_i &= \pi_i + (u_i + \sigma_i) \operatorname{Re} A, \\
\xi'_i &= \xi_i - (u_i + \sigma_i) \operatorname{Im} A, \\
\sigma'_i &= \sigma_i \left(1 - \frac{1}{2}\dot{A}A\right) - \frac{1}{2}\dot{A}Au_i - \pi_i \operatorname{Re} A + \xi_i \operatorname{Im} A, \\
u'_i &= u_i \left(1 + \frac{1}{2}\dot{A}A\right) + \frac{1}{2}\dot{A}A\sigma_i + \pi_i \operatorname{Re} A - \xi_i \operatorname{Im} A.
\end{aligned} \tag{3.160}$$

The components of vectors of the tetrad $\breve{\boldsymbol{e}}^{\circ}_a$ in this case are transformed by formulas

$$\begin{aligned}
l'_i &= l_i, \\
m'_i &= m_i + Al_i, \\
n'_i &= n_i + \dot{A}Al_i + \dot{A}m_i + A\dot{m}_i.
\end{aligned} \tag{3.161}$$

Thus, the two-component spinor with the dotted covariant components $\eta_{\dot{A}}$ determines an orthonormal tetrad $\breve{\boldsymbol{e}}_a$ and an complex null tetrad $\breve{\boldsymbol{e}}^{\circ}_a$ up to transformations (3.160), (3.161), that depend on an arbitrary complex parameter A.

In accordance with formulas (3.144) and (3.89), the spinors ξ and η are defined in basis $\{\breve{\boldsymbol{e}}_a(x^i)\}$ by the components

$$\xi^A = \left\| \begin{matrix} 0 \\ \mathrm{i}\sqrt{\frac{1}{2}\rho}\, \exp\left(\frac{\mathrm{i}}{2}\eta\right) \end{matrix} \right\|, \quad \eta_{\dot{A}} = \left\| \begin{matrix} 0 \\ \mathrm{i}\sqrt{\frac{1}{2}\rho}\, \exp\left(-\frac{\mathrm{i}}{2}\eta\right) \end{matrix} \right\|.$$

The proper tetrads $\{\breve{\boldsymbol{e}}_a(x^i)\}$, $\{\breve{\boldsymbol{e}}^{\circ}_a(x^i)\}$, corresponding to the spinor fields $\xi^A(x^i)$ (or $\eta_{\dot{A}}(x^i)$), it is possible to specify also, using instead of conditions $\rho = \text{const}$, $\eta = \text{const}$ others additional invariant algebraic or differential conditions on the fields $\xi^A(x^i)$, $\eta_{\dot{A}}(x^i)$. Such additional conditions on field $\xi^A(x^i)$, $\eta_{\dot{A}}(x^i)$ can be formulated in the form of equations on the fields of scalars and vectors $\rho(x^i)$, $\eta(x^i)$, $\breve{\boldsymbol{e}}_a(x^i)$, $\breve{\boldsymbol{e}}^{\circ}_a(x^i)$.

3.4.4 Pseudo-Orthogonal Transformations of Proper Bases of a Spinor Field

Let $\boldsymbol{\psi}$ be a four-component spinor of the first-rank in the pseudo-Euclidean space E_4^1 referred to an orthonormal basis $\mathbf{\ni}_i$. In the spinbasis defined by spintensors (3.24) and (3.25), the spinor components ψ can be represented in the form

$$\psi = \begin{Vmatrix} \xi^A \\ \eta_{\dot{A}} \end{Vmatrix}, \qquad \dot{A}, A = 1, 2,$$

where ξ^A and $\eta_{\dot{A}}$ determine in the space E_4^1 two-component spinors with the fixed relative sign.

Let us consider the gauge transformation of the spinor $\boldsymbol{\psi}$, which is written with the aid of the components ξ^A and $\eta^{\dot{A}}$:

$$\begin{aligned} \xi'^A &= \quad \alpha\xi^A - \beta\dot{\eta}^{\dot{A}}, \\ \dot{\eta}'^{\dot{A}} &= -\gamma\xi^A + \delta\dot{\eta}^{\dot{A}}, \end{aligned} \tag{3.162}$$

where α, β, γ, and δ are arbitrary complex numbers satisfying the equation $\alpha\delta - \beta\gamma = 1$.

It is obvious that transformations (3.162) form a group.

Transformation (3.162) by means of the components ψ^A and ψ^{+A} is written as follows

$$\psi'^A = \frac{1}{2}(\alpha + \dot{\delta})\psi^A + \frac{\mathrm{i}}{2}(\alpha - \dot{\delta})\gamma^{5A}{}_B\psi^B - \frac{1}{2}(\beta + \dot{\gamma})\psi^{+A} + \frac{\mathrm{i}}{2}(-\beta + \dot{\gamma})\gamma^{5A}{}_B\psi^{+B}. \tag{3.163}$$

For the corresponding transformation of the conjugate components ψ^{+A} we have

$$\begin{aligned} \psi'^{+A} = -\frac{1}{2}(\dot{\beta} + \gamma)\psi^A + \frac{\mathrm{i}}{2}(\dot{\beta} - \gamma)\gamma^{5A}{}_B\psi^B \\ + \frac{1}{2}(\dot{\alpha} + \delta)\psi^{+A} + \frac{\mathrm{i}}{2}(-\dot{\alpha} + \delta)\gamma^{5A}{}_B\psi^{+B}. \end{aligned}$$

Thus, transformations (3.163) are not linear transformations in the space of four-component spinors, since in the right-hand side of Eq. (3.163) along with components ψ^A there are the components of conjugate spinor ψ^{+A}.

Let the spinor ψ corresponds to the invariants Ω, N and the proper orthonormal basis $\breve{\boldsymbol{e}}_a$ defined by relations (3.154) and (3.155), while the spinor ψ' corresponds to the invariants Ω', N' and the proper orthonormal basis $\breve{\boldsymbol{e}}'_a$. For the complex invariant $\Omega' + \mathrm{i}N'$ we have

$$\Omega' + \mathrm{i}N' = 2e_{BA}\xi'^B\dot{\eta}'^{\dot{A}}. \tag{3.164}$$

Replacing here the components of spinors ξ', η' by formula (3.162), we find that invariants Ω, N do not change under the gauge transformation (3.162):

$$\Omega' + \mathrm{i}N' = (\alpha\delta - \beta\gamma)2e_{BA}\xi^B\dot{\eta}^{\dot{A}} = \Omega + \mathrm{i}N. \tag{3.165}$$

From this it follows that the invariants ρ and η also do not change under these transformations

$$\rho' = \rho, \quad \eta' = \eta.$$

The calculations similar to (3.164)–(3.165) give the following formulas for the transformation of the components of the vectors π^i, ξ^i, σ^i, u^i, corresponding to transformation (3.162):

$$\begin{aligned}
\pi'^i &= l^1{}_1\pi^i + l^2{}_1\xi^i + l^3{}_1\sigma^i + l^4{}_1u^i,\\
\xi'^i &= l^1{}_2\pi^i + l^2{}_2\xi^i + l^3{}_2\sigma^i + l^4{}_2u^i,\\
\sigma'^i &= l^1{}_3\pi^i + l^2{}_3\xi^i + l^3{}_3\sigma^i + l^4{}_3u^i,\\
u'^i &= l^1{}_4\pi^i + l^2{}_4\xi^i + l^3{}_4\sigma^i + l^4{}_4u^i,
\end{aligned} \tag{3.166}$$

Here $l^b{}_a$ are defined by the matrix

$$l^b{}_a = \frac{1}{2}\left\|\begin{matrix}
\dot\alpha\delta + \dot\beta\gamma + \dot\gamma\beta + \dot\delta\alpha & \mathrm{i}(-\dot\alpha\delta - \dot\beta\gamma + \dot\gamma\beta + \dot\delta\alpha) & \dot\alpha\beta + \dot\beta\alpha - \dot\gamma\delta - \dot\delta\gamma & -\dot\alpha\beta - \dot\beta\alpha - \dot\gamma\delta - \dot\delta\gamma\\
\mathrm{i}(\dot\alpha\delta - \dot\beta\gamma + \dot\gamma\beta - \dot\delta\alpha) & \dot\alpha\delta - \dot\beta\gamma - \dot\gamma\beta + \dot\delta\alpha & \mathrm{i}(\dot\alpha\beta - \dot\beta\alpha - \dot\gamma\delta + \dot\delta\gamma) & \mathrm{i}(-\dot\alpha\beta + \dot\beta\alpha - \dot\gamma\delta + \dot\delta\gamma)\\
\dot\alpha\gamma - \dot\beta\delta + \dot\gamma\alpha - \dot\delta\beta & \mathrm{i}(-\dot\alpha\gamma + \dot\beta\delta + \dot\gamma\alpha - \dot\delta\beta) & \dot\alpha\alpha - \dot\beta\beta - \dot\gamma\gamma + \dot\delta\delta & -\dot\alpha\alpha + \dot\beta\beta - \dot\gamma\gamma + \dot\delta\delta)\\
-\dot\alpha\gamma - \dot\beta\delta - \dot\gamma\alpha - \dot\delta\beta & \mathrm{i}(\dot\alpha\gamma + \dot\beta\delta - \dot\gamma\alpha - \dot\delta\beta) & -\dot\alpha\alpha - \dot\beta\beta + \dot\gamma\gamma + \dot\delta\delta & \dot\alpha\alpha + \dot\beta\beta + \dot\gamma\gamma + \dot\delta\delta
\end{matrix}\right\|. \tag{3.167}$$

Transformation (3.166) can be written as the transformation of the scale factors determined by relations (3.133):

$$\breve{h}'^i{}_a = l^b{}_a\breve{h}^i{}_b. \tag{3.168}$$

Relation (3.168) means that the proper bases $\breve{\boldsymbol{e}}_a$ of the spinor field $\boldsymbol{\psi}$ under transformations (3.162) are subjected to the transformation

$$\breve{\boldsymbol{e}}'_a = l^b{}_a\breve{\boldsymbol{e}}_b. \tag{3.169}$$

The transformation from orthonormal bases $\breve{\boldsymbol{e}}_a$ and $\breve{\boldsymbol{e}}'_a$ to the orthonormal basis $\boldsymbol{\vartheta}_i$ of the pseudo-Euclidean space E^1_4 is the restricted pseudo-orthogonal Lorentz transformation. Therefore and transformation (3.169) also is the restricted pseudo-orthogonal transformation. The pseudo-orthogonality of the matrix $\|l^b{}_a\|$ also follows directly from definition (3.167).

It is easy to see that to the Pauli transformation

$$\begin{aligned} \xi'^A &= \alpha\xi^A - \beta\dot{\eta}^{\dot{A}}, \\ \dot{\eta}'^{\dot{A}} &= \dot{\beta}\xi^A + \dot{\alpha}\dot{\eta}^{\dot{A}}, \end{aligned} \quad \text{or} \quad \psi'^A = \alpha\psi^A - \mathrm{i}\beta\gamma^{5A}{}_B\psi^{+B}, \quad \dot{\alpha}\alpha + \dot{\beta}\beta = 1,$$

corresponds the three-dimensional orthogonal transformation of the vectors $\breve{\boldsymbol{e}}_1$, $\breve{\boldsymbol{e}}_2$, and $\breve{\boldsymbol{e}}_3$. In this case the transformation coefficients $l^b{}_a$ are defined by the matrix

$$l^b{}_a = \begin{Vmatrix} \frac{1}{2}(\dot{\alpha}^2 - \dot{\beta}^2 - \beta^2 + \alpha^2) & \frac{\mathrm{i}}{2}(-\dot{\alpha}^2 + \dot{\beta}^2 - \beta^2 + \alpha^2) & \dot{\alpha}\beta + \dot{\beta}\alpha & 0 \\ \frac{\mathrm{i}}{2}(\dot{\alpha}^2 + \dot{\beta}^2 - \beta^2 - \alpha^2) & \frac{1}{2}(\dot{\alpha}^2 + \dot{\beta}^2 + \beta^2 + \alpha^2) & \mathrm{i}(\dot{\alpha}\beta - \dot{\beta}\alpha) & 0 \\ -\dot{\alpha}\dot{\beta} - \alpha\beta & \mathrm{i}(\dot{\alpha}\dot{\beta} - \alpha\beta) & \dot{\alpha}\alpha - \dot{\beta}\beta & 0 \\ 0 & 0 & 0 & 1 \end{Vmatrix}.$$

To the transformation of spinors

$$\begin{aligned} \xi'^A &= \quad \alpha\xi^A - \beta\dot{\eta}^{\dot{A}}, \\ \dot{\eta}'^{\dot{A}} &= -\dot{\beta}\xi^A + \dot{\alpha}\dot{\eta}^{\dot{A}}, \end{aligned} \quad \text{or} \quad \psi'^A = \alpha\psi^A - \beta\psi^{+A}, \quad \dot{\alpha}\alpha - \dot{\beta}\beta = 1 \qquad (3.170)$$

corresponds the three-dimensional pseudo-orthogonal transformation of the vectors $\breve{\boldsymbol{e}}_1$, $\breve{\boldsymbol{e}}_2$, and $\breve{\boldsymbol{e}}_4$. In this case for the transformation coefficients $l^b{}_a$ we have

$$l^b{}_a = \begin{Vmatrix} \frac{1}{2}(\dot{\alpha}^2 + \dot{\beta}^2 + \beta^2 + \alpha^2) & \frac{\mathrm{i}}{2}(-\dot{\alpha}^2 - \dot{\beta}^2 + \beta^2 + \alpha^2) & 0 & -\dot{\alpha}\beta - \dot{\beta}\alpha \\ \frac{\mathrm{i}}{2}(\dot{\alpha}^2 - \dot{\beta}^2 + \beta^2 - \alpha^2) & \frac{1}{2}(\dot{\alpha}^2 - \dot{\beta}^2 - \beta^2 + \alpha^2) & 0 & \mathrm{i}(\dot{\alpha}\beta + \dot{\beta}\alpha) \\ 0 & 0 & 1 & 0 \\ -\dot{\alpha}\dot{\beta} - \alpha\beta & \mathrm{i}(\dot{\alpha}\dot{\beta} - \alpha\beta) & 0 & \dot{\alpha}\alpha + \dot{\beta}\beta \end{Vmatrix}. \qquad (3.171)$$

If coefficients α, β, γ, and δ in transformation (3.162) are real then the vector $\breve{\boldsymbol{e}}_2$ does not change $\breve{\boldsymbol{e}}'_2 = \breve{\boldsymbol{e}}_2$ while the vectors $\breve{\boldsymbol{e}}_1$, $\breve{\boldsymbol{e}}_3$, $\breve{\boldsymbol{e}}_4$ are subjected to the arbitrary three-dimensional pseudo-orthogonal transformation.

Let us now consider a restricted pseudo-orthogonal transformation of the basis $\boldsymbol{\vartheta}_i$ of the pseudo-Euclidean space E^1_4:

$$\boldsymbol{\vartheta}'_i = l^j{}_i\boldsymbol{\vartheta}_j, \quad \det\|l^j{}_i\| = 1, \quad l^4{}_4 \geqslant 1. \qquad (3.172)$$

Transformation (3.172) corresponds to the transformation S of the spinors ψ determined by Eqs. (3.43). If the spintensors E and γ_i are defined by formulas (3.24) and (3.25), then the matrix of the spinor transformation S is written in the form

$$S = \pm \begin{Vmatrix} A & \\ & (\dot{A}^T)^{-1} \end{Vmatrix}, \tag{3.173}$$

where A is an unimodular two-dimensional matrix

$$A = \begin{Vmatrix} \alpha & \beta \\ \gamma & \delta \end{Vmatrix}, \qquad \alpha\delta - \beta\gamma = 1. \tag{3.174}$$

Replacing in definitions (3.129) and (3.126) the components of the spinor ψ by the formula $\psi' = S\psi$, where S is defined by equalities (3.173) and (3.174), for transformation of the vectors of the proper tetrad we find

$$\begin{aligned} \pi'_i &= l^j{}_i \pi_j, & \xi'_i &= l^j{}_i \xi_j, \\ \sigma'_i &= l^j{}_i \sigma_j, & u'_i &= l^j{}_i u_j, \end{aligned} \tag{3.175}$$

where the transformation coefficients $l^j{}_i$ are defined by matrix (3.167). Relation (3.175) can also be written as the transformation of the scale factors $\breve{h}_i{}^a$ determined by matrix (3.133):

$$\breve{h}'^{a}_i = l^j{}_i \breve{h}_j{}^a. \tag{3.176}$$

Thus, the coefficients $l^j{}_i$ of the pseudo-orthogonal transformation (3.176) of the proper basis $\breve{\boldsymbol{e}}_a$ turns out to be the same, as in equality (3.168), corresponding to the gauge transformation (3.162) of the spinor ψ.

3.5 Complex Orthogonal Vector Triads, Defined by a Spinor Field

Let us define three antisymmetric complex tensors of the second rank by the contravariant components α^{ij}, β^{ij}, and λ^{ij} in the orthonormal basis $\boldsymbol{\mathit{Э}}_i$ by means of the vectors of the proper orthonormal tetrad $\breve{\boldsymbol{e}}_a$ determined by a first-rank spinor $\boldsymbol{\psi}$ by formulas (3.128), (3.129), and (3.126):

$$\begin{aligned} \alpha^{ij} &= -u^i \pi^j + u^j \pi^i - \mathrm{i}\varepsilon^{ijks} u_k \pi_s, \\ \beta^{ij} &= -u^i \xi^j + u^j \xi^i - \mathrm{i}\varepsilon^{ijks} u_k \xi_s, \\ \lambda^{ij} &= -u^i \sigma^j + u^j \sigma^i - \mathrm{i}\varepsilon^{ijks} u_k \sigma_s. \end{aligned} \tag{3.177}$$

Consider first some simple algebraic properties of the tensors determined by components (3.177). Direct calculation shows that from definitions (3.177) and conditions (3.130) it follows that the tensors with components α^{ij}, β^{ij}, and λ^{ij} are antiself-dual

$$\frac{1}{2}\varepsilon^{ijks}\alpha_{ks} = -\mathrm{i}\alpha^{ij}, \quad \frac{1}{2}\varepsilon^{ijks}\beta_{ks} = -\mathrm{i}\beta^{ij}, \quad \frac{1}{2}\varepsilon^{ijks}\lambda_{ks} = -\mathrm{i}\lambda^{ij}$$

and satisfy the invariant bilinear equations

$$\alpha^{i}{}_{j}\alpha^{js} = \beta^{i}{}_{j}\beta^{js} = \lambda^{i}{}_{j}\lambda^{js} = g^{is}, \tag{3.178}$$

and also the following equations

$$\alpha^{in}\beta_{n}{}^{j} = -\mathrm{i}\lambda^{ij}, \quad \beta^{in}\lambda_{n}{}^{j} = -\mathrm{i}\alpha^{ij}, \quad \lambda^{in}\alpha_{n}{}^{j} = -\mathrm{i}\beta^{ij}. \tag{3.179}$$

A calculation of the invariants of the tensors determined by components (3.177) gives

$$\frac{1}{4}\alpha_{ij}\alpha^{ij} = \frac{1}{4}\beta_{ij}\beta^{ij} = \frac{1}{4}\lambda_{ij}\lambda^{ij} = -1,$$
$$\frac{1}{8}\varepsilon_{ijks}\alpha^{ij}\alpha^{ks} = \frac{1}{8}\varepsilon_{ijks}\beta^{ij}\beta^{ks} = \frac{1}{8}\varepsilon_{ijks}\lambda^{ij}\lambda^{ks} = \mathrm{i}.$$

and

$$\det\|\alpha_{ij}\| = \det\|\beta_{ij}\| = \det\|\lambda_{ij}\| = -1. \tag{3.180}$$

By means of Eqs. (3.52), (3.60), and (3.62) the components of tensors (3.177) can be expressed in terms of the components of antisymmetric tensors C^{ij} and M^{ij} determined by relations (3.51) and (3.58):

$$\alpha^{ij} = \frac{1}{\Omega + \mathrm{i}N}\left(\operatorname{Re} C^{ij} + \frac{\mathrm{i}}{2}\varepsilon^{ijks}\operatorname{Re} C_{ks}\right),$$
$$\beta^{ij} = -\frac{1}{\Omega + \mathrm{i}N}\left(\operatorname{Im} C^{ij} + \frac{\mathrm{i}}{2}\varepsilon^{ijks}\operatorname{Im} C_{ks}\right),$$
$$\lambda^{ij} = \frac{\mathrm{i}}{\Omega + \mathrm{i}N}\left(M^{ij} + \frac{\mathrm{i}}{2}\varepsilon^{ijks}M_{ks}\right). \tag{3.181}$$

The inverse relations have the form

$$C^{ij} = \Omega\operatorname{Re}\alpha^{ij} - N\operatorname{Im}\alpha^{ij} + \mathrm{i}\big(-\Omega\operatorname{Re}\beta^{ij} + N\operatorname{Im}\beta^{ij}\big),$$
$$M^{ij} = N\operatorname{Re}\lambda^{ij} + \Omega\operatorname{Im}\lambda^{ij}.$$

Since a spinor ψ is completely defined by specifying the tensor components C^{ij} and M^{ij},[12] from these equalities it follows that specifying the invariants Ω, N and the antisymmetric tensors with components α^{ij}, β^{ij}, λ^{ij} also completely defines a spinor ψ.

The Ricci rotation coefficients $\breve{\Delta}_{k,ij}$ corresponding to the proper tetrads $\breve{\boldsymbol{e}}_a$ of a spinor field $\psi(x^i)$, can be expressed in terms of the components of tensors (3.177). We give an expression for the quantities $\breve{\Delta}_{k,ij} + \frac{\mathrm{i}}{2}\varepsilon_{ijmn}\breve{\Delta}_{k,}{}^{mn}$, used in the sequel, in terms of the tensor components α^{ij}, β^{ij}, and λ^{ij}:

$$\breve{\Delta}_{k,ij} + \frac{\mathrm{i}}{2}\varepsilon_{ijmn}\breve{\Delta}_{k,}{}^{mn} = \frac{1}{4}\big(- \alpha_{is}\partial_k\alpha_j{}^s + \alpha_{js}\partial_k\alpha_i{}^s \\ - \beta_{is}\partial_k\beta_j{}^s + \beta_{js}\partial_k\beta_i{}^s - \lambda_{is}\partial_k\lambda_j{}^s + \lambda_{js}\partial_k\lambda_i{}^s\big). \qquad (3.182)$$

The derivation of relation(3.182) is quite cumbersome, however the validity of this relation can be established directly substituting (3.177) in (3.182) and this gives (3.147).

It is obvious that the quantities $\breve{\Delta}_{k,ij} + \frac{\mathrm{i}}{2}\varepsilon_{ijmn}\breve{\Delta}_{k,}{}^{mn}$ are antiself-dual over the indices i, j:

$$\frac{1}{2}\varepsilon_{ijmn}\left(\breve{\Delta}_{k,}{}^{mn} + \frac{\mathrm{i}}{2}\varepsilon^{mnpq}\breve{\Delta}_{k,pq}\right) = -\mathrm{i}\left(\breve{\Delta}_{k,ij} + \frac{\mathrm{i}}{2}\varepsilon_{ijmn}\breve{\Delta}_{k,}{}^{mn}\right).$$

Let us introduce the notations for the tensor components α^{ij}, β^{ij}, and λ^{ij}:

$$\alpha^\alpha = \alpha^{\alpha 4}, \quad \beta^\alpha = \beta^{\alpha 4}, \quad \lambda^\alpha = \lambda^{\alpha 4}, \qquad \alpha = 1, 2, 3.$$

Since the tensors with components α^{ij}, β^{ij}, λ^{ij} are antiself-dual, we obtain that the matrices of the components α^{ij}, β^{ij}, λ^{ij} are defined only by the quantities α^α, β^α, and λ^α

$$\alpha^{ij} = \begin{Vmatrix} 0 & \mathrm{i}\alpha^3 & -\mathrm{i}\alpha^2 & \alpha^1 \\ -\mathrm{i}\alpha^3 & 0 & \mathrm{i}\alpha^1 & \alpha^2 \\ \mathrm{i}\alpha^2 & -\mathrm{i}\alpha^1 & 0 & \alpha^3 \\ -\alpha^1 & -\alpha^2 & -\alpha^3 & 0 \end{Vmatrix}, \quad \beta^{ij} = \begin{Vmatrix} 0 & \mathrm{i}\beta^3 & -\mathrm{i}\beta^2 & \beta^1 \\ -\mathrm{i}\beta^3 & 0 & \mathrm{i}\beta^1 & \beta^2 \\ \mathrm{i}\beta^2 & -\mathrm{i}\beta^1 & 0 & \beta^3 \\ -\beta^1 & -\beta^2 & -\beta^3 & 0 \end{Vmatrix},$$

$$\lambda^{ij} = \begin{Vmatrix} 0 & \mathrm{i}\lambda^3 & -\mathrm{i}\lambda^2 & \lambda^1 \\ -\mathrm{i}\lambda^3 & 0 & \mathrm{i}\lambda^1 & \lambda^2 \\ \mathrm{i}\lambda^2 & -\mathrm{i}\lambda^1 & 0 & \lambda^3 \\ -\lambda^1 & -\lambda^2 & -\lambda^3 & 0 \end{Vmatrix}. \qquad (3.183)$$

[12]As it was already noted (see p. 134), specifying only the tensor components C^{ij} determines two spinors with components ψ and $\mathrm{i}\gamma^5\psi$. Under transformation $\psi \rightarrow \mathrm{i}\gamma^5\psi$ the components M^{ij} pass into $-M^{ij}$. Therefore specifying C^{ij} and M^{ij} completely determines the spinor ψ.

If the spinbasis is defined by the matrices E and γ_a in the form (3.24) and (3.25), then using definitions (3.181) and formulas (3.56), (3.64), for the components α^α, β^α, λ^α one can find

$$\begin{aligned}
\alpha^1 &= \frac{1}{\Omega + \mathrm{i}N}(\psi^1\psi^1 - \psi^2\psi^2 + \dot{\psi}^3\dot{\psi}^3 - \dot{\psi}^4\dot{\psi}^4),\\
\alpha^2 &= \frac{\mathrm{i}}{\Omega + \mathrm{i}N}(\psi^1\psi^1 + \psi^2\psi^2 - \dot{\psi}^3\dot{\psi}^3 - \dot{\psi}^4\dot{\psi}^4),\\
\alpha^3 &= \frac{2}{\Omega + \mathrm{i}N}(-\psi^1\psi^2 - \dot{\psi}^3\dot{\psi}^4),\\
\beta^1 &= \frac{\mathrm{i}}{\Omega + \mathrm{i}N}(\psi^1\psi^1 - \psi^2\psi^2 - \dot{\psi}^3\dot{\psi}^3 + \dot{\psi}^4\dot{\psi}^4),\\
\beta^2 &= \frac{1}{\Omega + \mathrm{i}N}(-\psi^1\psi^1 - \psi^2\psi^2 - \dot{\psi}^3\dot{\psi}^3 - \dot{\psi}^4\dot{\psi}^4),\\
\beta^3 &= \frac{2\mathrm{i}}{\Omega + \mathrm{i}N}(-\psi^1\psi^2 + \dot{\psi}^3\dot{\psi}^4),\\
\lambda^1 &= \frac{2}{\Omega + \mathrm{i}N}(-\dot{\psi}^3\psi^2 - \dot{\psi}^4\psi^1),\\
\lambda^2 &= \frac{2\mathrm{i}}{\Omega + \mathrm{i}N}(\dot{\psi}^3\psi^2 - \dot{\psi}^4\psi^1),\\
\lambda^3 &= \frac{2}{\Omega + \mathrm{i}N}(-\dot{\psi}^3\psi^1 + \dot{\psi}^4\psi^2),
\end{aligned} \tag{3.184}$$

where $\Omega + \mathrm{i}N = 2(\dot{\psi}^3\psi^1 + \dot{\psi}^4\psi^2)$.

Let us establish the transformation law of the quantities α^α, β^α, and λ^α under the restricted Lorentz transformation (3.172) of the orthonormal basis $\mathcal{Э}_i$ of the pseudo-Euclidean space E_4^1. If the spinbasis is defined by the spintensors (3.24) and (3.25), then the transformation matrix S of the spinor components is defined by the relations (3.173) and (3.174). Replacing the components ψ in definitions (3.184) by the formula $\psi' = S\psi$ and taking into account that the invariants Ω and N do not change under such transformations, for the transformation of the quantities α_α, β_α, and λ_α we get

$$\alpha'_\alpha = \ell^\beta{}_\alpha \alpha_\beta, \quad \beta'_\alpha = \ell^\beta{}_\alpha \beta_\beta, \quad \lambda'_\alpha = \ell^\beta{}_\alpha \lambda_\beta, \tag{3.185}$$

where the complex matrix $\mathcal{L} = \|\ell^\beta{}_\alpha\|$ has the form

$$\mathcal{L} = \begin{Vmatrix} \frac{1}{2}(\alpha^2 - \beta^2 - \gamma^2 + \delta^2) & \frac{\mathrm{i}}{2}(\alpha^2 - \beta^2 + \gamma^2 - \delta^2) & -\alpha\gamma + \beta\delta \\ \frac{\mathrm{i}}{2}(-\alpha^2 - \beta^2 + \gamma^2 + \delta^2) & \frac{1}{2}(\alpha^2 + \beta^2 + \gamma^2 + \delta^2) & \mathrm{i}(\alpha\gamma + \beta\delta) \\ -\alpha\beta + \gamma\delta & -\mathrm{i}(\alpha\beta + \gamma\delta) & \alpha\delta + \beta\gamma \end{Vmatrix}. \tag{3.186}$$

From definition (3.186) it follows that the product $\mathcal{L}^T\mathcal{L}$ is the unit matrix $\mathcal{L}^T\mathcal{L} = I$. Consequently, $\mathcal{L}$ is the complex orthogonal matrix $\mathcal{L}^T = \mathcal{L}^{-1}$. Thus, under the restricted Lorentz transformation (3.172) of the bases $\mathcal{D}_i$ the quantities λ^α, α^α, and β^α are subjected to the complex orthogonal transformation (3.185).

In particular, to the Lorentz transformation

$$L = \begin{Vmatrix} \cos\varphi & -\sin\varphi & & \\ \sin\varphi & \cos\varphi & & \\ & & \cosh\xi & \sinh\xi \\ & & \sinh\xi & \cosh\xi \end{Vmatrix},$$

(rotation in the $\mathcal{D}_1$, $\mathcal{D}_2$ plane through an angle φ and boost along the $\mathcal{D}_3$ direction) there corresponds the spinor transformation A and the complex orthogonal transformation $\mathcal{L}$ [34]

$$A = \begin{Vmatrix} \exp[\frac{\mathrm{i}}{2}(\varphi + \mathrm{i}\xi)] & 0 \\ 0 & \exp[-\frac{\mathrm{i}}{2}(\varphi + \mathrm{i}\xi)] \end{Vmatrix},$$

$$\mathcal{L} = \begin{Vmatrix} \cos(\varphi + \mathrm{i}\xi) & -\sin(\varphi + \mathrm{i}\xi) & 0 \\ \sin(\varphi + \mathrm{i}\xi) & \cos(\varphi + \mathrm{i}\xi) & 0 \\ 0 & 0 & 1 \end{Vmatrix}.$$

It is easy to verify that the set of all matrices $\mathcal{L}$, corresponding to the restricted Lorentz group, forms a group, which realizes a representation of the restricted Lorentz group in the three-dimensional complex Euclidean space E_3^+.

Let $\mathcal{E}_\alpha$ ($\alpha = 1, 2, 3$) be vectors of an orthonormal basis in the space E_3^+. The metric tensor of the space E_3^+ in the basis $\mathcal{E}_\alpha$ is defined by Kronecker deltas $\delta^{\alpha\beta}$, $\delta_{\alpha\beta}$. Consider in the space E_3^+ three vectors

$$\breve{\mathcal{E}}_1 = \alpha^\alpha \mathcal{E}_\alpha, \quad \breve{\mathcal{E}}_2 = \beta^\alpha \mathcal{E}_\alpha, \quad \breve{\mathcal{E}}_3 = \lambda^\alpha \mathcal{E}_\alpha$$

with components α^α, β^α, and λ^α defined by equalities (3.184).

Taking into account notations (3.183), it is easy to find that Eqs. (3.178) and (3.179) for $i = j$ can be written in the form

$$\begin{aligned} \alpha_\mu \alpha^\mu &= \beta_\mu \beta^\mu = \lambda_\mu \lambda^\mu = 1, \\ \alpha_\mu \beta^\mu &= \alpha_\mu \lambda^\mu = \beta_\mu \lambda^\mu = 0, \end{aligned} \tag{3.187}$$

or

$$\alpha^\mu \alpha^\nu + \beta^\mu \beta^\nu + \lambda^\mu \lambda^\nu = \delta^{\mu\nu}.$$

Thus, the components α^{μ}, β^{μ}, and λ^{μ} define in the space E_3^+ three complex orthonormal vectors (complex triad). Equations (3.179) for $i \neq j$ pass into the equations

$$\alpha^{\alpha} = \varepsilon^{\alpha\beta\gamma}\beta_{\beta}\lambda_{\gamma}, \quad \beta^{\alpha} = \varepsilon^{\alpha\beta\gamma}\lambda_{\beta}\alpha_{\gamma}, \quad \lambda^{\alpha} = \varepsilon^{\alpha\beta\gamma}\alpha_{\beta}\beta_{\gamma}.$$

It is clear that specifying of vectors of the triad $\breve{\boldsymbol{\mathcal{E}}}_{\alpha}$ and invariants Ω, N completely defines the spinor $\boldsymbol{\psi}$, since the components of tensors α^{ij}, β^{ij}, and λ^{ij} are defined by equalities (3.183) in terms of the components of vectors of the triad $\breve{\boldsymbol{\mathcal{E}}}_{\alpha}$,

Let us now consider the field of a spinor the first-rank $\psi(x^i)$ in the Cartesian coordinate system of the Minkowski space to which there the field of the complex vector triads $\breve{\boldsymbol{\mathcal{E}}}_{\alpha}(x^i)$ corresponds. Let $\delta_{k,\alpha\beta}$ be the Ricci rotation coefficients, corresponding to the field $\breve{\boldsymbol{\mathcal{E}}}_{\alpha}(x^i)$. By definition we have

$$\delta_{k,\alpha\beta} = \frac{1}{2}(\alpha_{\alpha}\partial_k\alpha_{\beta} - \alpha_{\beta}\partial_k\alpha_{\alpha} + \beta_{\alpha}\partial_k\beta_{\beta} - \beta_{\beta}\partial_k\beta_{\alpha} + \lambda_{\alpha}\partial_k\lambda_{\beta} - \lambda_{\beta}\partial_k\lambda_{\alpha}). \tag{3.188}$$

From definition (3.188) and conditions (3.187) it follows

$$\partial_k\alpha_{\alpha} = -\delta_{k,\alpha}{}^{\beta}\alpha_{\beta}, \quad \partial_k\beta_{\alpha} = -\delta_{k,\alpha}{}^{\beta}\beta_{\beta}, \quad \partial_k\lambda_{\alpha} = -\delta_{k,\alpha}{}^{\beta}\lambda_{\beta}.$$

Using expression (3.182) for the quantities $\breve{\Delta}_{k,ij} + \frac{\mathrm{i}}{2}\varepsilon_{ijmn}\breve{\Delta}_{k,}{}^{mn}$ and expression (3.183) for the components of tensors α^{ij}, β^{ij}, and λ^{ij}, we find the connection between the Ricci rotation coefficients $\delta_{k,\alpha\beta}$ and $\breve{\Delta}_{k,ij}$:

$$\breve{\Delta}_{k,ij} + \frac{\mathrm{i}}{2}\varepsilon_{ijmn}\breve{\Delta}_{k,}{}^{mn} = \begin{Vmatrix} 0 & \delta_{k,12} & -\delta_{k,31} & -\mathrm{i}\delta_{k,23} \\ -\delta_{k,12} & 0 & \delta_{k,23} & -\mathrm{i}\delta_{k,31} \\ \delta_{k,13} & -\delta_{k,23} & 0 & -\mathrm{i}\delta_{k,12} \\ \mathrm{i}\delta_{k,23} & \mathrm{i}\delta_{k,31} & \mathrm{i}\delta_{k,12} & 0 \end{Vmatrix}$$

or

$$\delta_{k,\alpha\beta} = \breve{\Delta}_{k,\alpha\beta} + \frac{\mathrm{i}}{2}\varepsilon_{\alpha\beta mn}\breve{\Delta}_{k,}{}^{mn}. \tag{3.189}$$

Let us give also an expression of the Ricci rotation coefficients $\delta_{k,\alpha\beta}$ directly in terms of the spinor field $\boldsymbol{\psi}$:

$$\delta_{k,\alpha\beta} = \frac{1}{\Omega + \mathrm{i}N}\left[\psi^{+}\left(I + \mathrm{i}\gamma^5\right)\gamma_{\alpha\beta}\partial_k\psi - \partial_k\psi^{+}\cdot\left(I + \mathrm{i}\gamma^5\right)\gamma_{\alpha\beta}\psi\right]. \tag{3.190}$$

Expression (3.190) can be obtained, using expression (3.148) for $\breve{\Delta}_{k,ij}$ and definition (3.189).

3.5.1 The Group of Orthogonal Transformations of the Complex Vector Triad $\breve{\boldsymbol{\mathcal{E}}}_\alpha$

Let us represent the components of the first-rank spinor ψ in spinbasis (3.24), (3.25) in the form

$$\psi = \begin{Vmatrix} \xi^A \\ \eta_{\dot{A}} \end{Vmatrix}, \qquad \dot{A}, A = 1, 2, \tag{3.191}$$

where ξ^A and $\eta_{\dot{A}}$ define in the pseudo-Euclidean space E_4^1 two-component spinors with the fixed relative sign.

Let us consider a group of the gauge transformations of the four-component spinor $\boldsymbol{\psi}$

$$\begin{aligned} \xi'^A &= \alpha\xi^A - \beta\dot{\eta}^{\dot{A}}, \\ \dot{\eta}'^{\dot{A}} &= -\gamma\xi^A + \delta\dot{\eta}^{\dot{A}}, \end{aligned} \tag{3.192}$$

where α, β, γ, and δ are arbitrary complex numbers satisfying the equation $\alpha\delta - \beta\gamma = 1$.

Let a spinor $\boldsymbol{\psi}$ corresponds to an orthonormal triad $\breve{\boldsymbol{\mathcal{E}}}_\alpha$, and a spinor $\boldsymbol{\psi}'$ corresponds to an orthonormal triad $\breve{\boldsymbol{\mathcal{E}}}'_\alpha$. Let us calculate the orthogonal transformation from the basis $\breve{\boldsymbol{\mathcal{E}}}'_\alpha$ to the basis $\breve{\boldsymbol{\mathcal{E}}}_\alpha$. Replacing in definitions (3.184) the components of the spinor ψ by formulas (3.191) and (3.192) and taking into account the invariance of the quantity $\Omega + \mathrm{i}N$ under transformation (3.192), we obtain

$$\begin{aligned} \alpha'^\alpha &= \ell^1{}_1\alpha^\alpha + \ell^2{}_1\beta^\alpha + \ell^3{}_1\lambda^\alpha, \\ \beta'^\alpha &= \ell^1{}_2\alpha^\alpha + \ell^2{}_2\beta^\alpha + \ell^3{}_2\lambda^\alpha, \\ \lambda'^\alpha &= \ell^1{}_3\alpha^\alpha + \ell^2{}_3\beta^\alpha + \ell^3{}_3\lambda^\alpha. \end{aligned} \tag{3.193}$$

The complex coefficients $\ell^\alpha{}_\beta$ in these formulas are defined by the three-dimensional orthogonal matrix (3.186).

We introduce the matrices of the scale factors $H^\alpha{}_\beta$ and $H'^\alpha{}_\beta$:

$$H^\alpha{}_\beta = \begin{Vmatrix} \alpha^1 & \beta^1 & \lambda^1 \\ \alpha^2 & \beta^2 & \lambda^2 \\ \alpha^3 & \beta^3 & \lambda^3 \end{Vmatrix}, \qquad H'^\alpha{}_\beta = \begin{Vmatrix} \alpha'^1 & \beta'^1 & \lambda'^1 \\ \alpha'^2 & \beta'^2 & \lambda'^2 \\ \alpha'^3 & \beta'^3 & \lambda'^3 \end{Vmatrix},$$

connecting bases $\breve{\boldsymbol{\mathcal{E}}}_\alpha$,$\breve{\boldsymbol{\mathcal{E}}}'_\alpha$ and $\boldsymbol{\mathcal{E}}_\alpha$ by the equalities

$$\breve{\boldsymbol{\mathcal{E}}}_\alpha = H^\beta{}_\alpha \boldsymbol{\mathcal{E}}_\beta, \quad \breve{\boldsymbol{\mathcal{E}}}'_\alpha = H'^\beta{}_\alpha \boldsymbol{\mathcal{E}}_\beta.$$

Then transformation (3.193) can be written in the form

$$H'^{\beta}{}_{\alpha} = \ell^{\gamma}{}_{\alpha} H^{\beta}{}_{\gamma}.$$

Therefore the connection between bases $\breve{\mathcal{E}}_{\alpha}$ and $\breve{\mathcal{E}}'_{\alpha}$ alpha is determined by the complex coefficients $\ell^{\beta}{}_{\alpha}$:

$$\breve{\mathcal{E}}'_{\alpha} = \ell^{\beta}{}_{\alpha} \breve{\mathcal{E}}_{\beta}.$$

3.6 Expression for Derivatives of a Spinor Field in Terms of Derivatives of Tensor Fields

The relations obtained in the previous sections allow to express derivatives with respect to x^i of a field $\psi(x^i)$ in the Minkowski space in terms of derivatives of the different scalars, vectors and tensors determined by the field $\psi(x^i)$. Below we obtain expressions for derivatives of the first-rank spinor field in terms of derivatives of its invariants and the Ricci rotation coefficients of the proper tetrads, and also in terms of derivatives of the complex tensors $\boldsymbol{C}$. The formulas, obtained here, are very useful and important in various applications.

3.6.1 Expressions for Derivatives of the First-Rank Spinor Field in Terms of Derivatives of Its Invariants and the Ricci Rotation Coefficients of Proper Tetrads

Let $\psi(x^i)$ be the contravariant components of the first-rank spinor calculated in the Cartesian coordinate system with variables x^i and orthonormal vector bases $\boldsymbol{\ni}_i$ in Minkowski space. Let us introduce the proper bases $\breve{\boldsymbol{e}}_a(x^i)$ of the spinor field $\psi(x^i)$ connected with the bases $\boldsymbol{\ni}_i$ by the scale factors

$$\breve{\boldsymbol{e}}_a(x^i) = \breve{h}^i{}_a \boldsymbol{\ni}_i, \qquad \boldsymbol{\ni}_i = \breve{h}_i{}^a \breve{\boldsymbol{e}}_a(x^i),$$

which are defined by matrices (3.133). Transformation from basis $\boldsymbol{\ni}_i$ to basis $\breve{\boldsymbol{e}}_a$ is the restricted Lorentz transformation.

We denote the column from the contravariant components of the spinor, calculated in the proper basis $\breve{\boldsymbol{e}}_a$, by the symbol $\breve{\psi}$. The components ψ and $\breve{\psi}$ are connected by the spinor transformation

$$\breve{\psi} = S\psi, \tag{3.194}$$

where for the matrix S according to definitions (3.43) we have

$$\breve{h}^j{}_a S\gamma_j = \gamma_a S, \qquad S^T E S = E. \tag{3.195}$$

If the Dirac matrices γ_i and the metric spinor E are determined by equalities (3.24) and (3.25), then the components of the spinor $\breve{\psi}$ are defined by equalities (3.144). Differentiating equalities (3.144) with respect to x^α, we get

$$\partial_i \breve{\psi} = \frac{1}{2}\big(I\partial_i \ln\rho - \gamma^5\partial_i\eta\big)\breve{\psi}. \tag{3.196}$$

It is obvious that relation (3.196) is valid also for any choice of the matrices E and γ^i.

From (3.194), taking into account equality (3.196), we find

$$\partial_i\psi = \partial_i S^{-1}\cdot\breve{\psi} + S^{-1}\frac{1}{2}\big(I\partial_i \ln\rho - \gamma^5\partial_i\eta\big)\breve{\psi}. \tag{3.197}$$

From the invariance of spintensor γ^5 under restricted Lorentz transformations, the matrix S in (3.194) commutes with γ^5, i.e., $S\gamma^5 = \gamma^5 S$. Therefore equality (3.197) it is possible to rewrite in the form

$$\partial_i\psi = \partial_i S^{-1}\cdot S\psi + \frac{1}{2}\big(I\partial_i \ln\rho - \gamma^5\partial_i\eta\big)\psi. \tag{3.198}$$

To calculate the quantity $\partial_i S^{-1}$ in this equality we differentiate the first equation in (3.195) with respect to x^s:

$$\partial_s\breve{h}^j{}_a\cdot S\gamma_j + \breve{h}^j{}_a\partial_s S\gamma_j = \gamma_a\partial_s S.$$

From this, after a transformation taking into account the definition of the Ricci rotation coefficients (see (2.37))

$$\breve{\Delta}_{s,ij} = \frac{1}{2}\big(\breve{h}_i{}^a\partial_s\breve{h}_{ja} - \breve{h}_j{}^a\partial_s\breve{h}_{ia}\big) \equiv \breve{h}_i{}^a\partial_s\breve{h}_{ja}, \tag{3.199}$$

we find

$$\breve{\Delta}_{s,i}{}^j\gamma_j = \gamma_i\left(S^{-1}\partial_s S\right) - \left(S^{-1}\partial_s S\right)\gamma_i. \tag{3.200}$$

Differentiation of the second relation in (3.195) after simple transformations gives

$$\left(S^{-1}\partial_s S\right)^T E + E\left(S^{-1}\partial_s S\right) = 0. \tag{3.201}$$

From Eqs. (3.200) and (3.201) it follows

$$S^{-1}\partial_s S = \frac{1}{4}\breve{\Delta}_{s,ij}\gamma^{ij}.$$

Whence for derivatives $\partial_s S$ and $\partial_s S^{-1}$ we get

$$\partial_s S = \frac{1}{4}\breve{\Delta}_{s,ij} S\gamma^{ij}, \quad \partial_s S^{-1} = -\frac{1}{4}\breve{\Delta}_{s,ij}\gamma^{ij} S^{-1}. \tag{3.202}$$

Replacing in (3.198) derivatives $\partial_i S^{-1}$ according to equality (3.202), for derivatives of an arbitrary first-rank spinor field, specified by the contravariant components ψ in the basis $\boldsymbol{\ni}_i$, we obtain the following relation[88, 89]

$$\partial_s \psi = \left(\frac{1}{2} I \partial_s \ln \rho - \frac{1}{2}\gamma^5 \partial_s \eta - \frac{1}{4}\breve{\Delta}_{s,ij}\gamma^{ij}\right)\psi. \tag{3.203a}$$

For the conjugate spinor field with the covariant components $\psi^+(x^\alpha)$ we have

$$\partial_s \psi^+ = \psi^+\left(\frac{1}{2} I \partial_s \ln \rho - \frac{1}{2}\gamma^5 \partial_s \eta + \frac{1}{4}\breve{\Delta}_{s,ij}\gamma^{ij}\right). \tag{3.203b}$$

The Ricci rotation coefficients $\breve{\Delta}_{s,ij}$ in formulas (3.203) correspond to proper bases $\breve{\boldsymbol{e}}_a(x^i)$ of the spinor field $\boldsymbol{\psi}(x^i)$ and are defined by the relation

$$\breve{\Delta}_{s,ij} = \frac{1}{2}\big(\pi_i \partial_s \pi_j - \pi_j \partial_s \pi_i + \xi_i \partial_s \xi_j - \xi_j \partial_s \xi_i + \sigma_i \partial_s \sigma_j - \sigma_j \partial_s \sigma_i - u_i \partial_s u_j + u_j \partial_s u_i\big),$$

which can be obtained from definition (3.199) replacing in it the coefficients $\breve{h}_i{}^a$ by formulas (3.133).

To write these formulas in a curvilinear coordinate system of the pseudo-Euclidean space or in the Riemannian space it suffices to replace in them the symbol of partial derivative ∂_i by the symbol of the covariant derivative ∇_i (see (5.105) and (5.106) in Chap. 5).

Formulas (3.203) give the expression of derivatives of the spinor fields $\boldsymbol{\psi}(x^s)$ and $\boldsymbol{\psi}^+(x^s)$ in terms of derivatives of the invariants $\rho(x^s)$, $\eta(x^s)$, and the vector fields $\pi^i(x^s)$, $\xi^i(x^s)$, $\sigma^i(x^s)$, $u^i(x^s)$ determined by the fields $\boldsymbol{\psi}(x^s)$ and $\boldsymbol{\psi}^+(x^s)$. These formulas are the identities by virtue of definitions (3.143), (3.129), (3.126) of the quantities $\rho(x^s)$, $\eta(x^s)$, $\pi^i(x^s)$, $\xi^i(x^s)$, $\sigma^i(x^s)$, $u^i(x^s)$. These formulas are very convenient for tensor reformulations of differential spinor equations.

Using relation (3.203a), it is not difficult to obtain also expressions for derivatives of the fields of semispinors $\psi_{(I)} = \frac{1}{2}(I + \mathrm{i}\gamma^5)\psi$ and $\psi_{(II)} = \frac{1}{2}(I - \mathrm{i}\gamma^5)\psi$:

$$\partial_s \psi_{(I)} = \frac{1}{2}\left[I\partial_s(\ln\rho + \mathrm{i}\eta) - \frac{1}{4}\left(\breve{\Delta}_{s,ij} + \frac{\mathrm{i}}{2}\varepsilon_{ijmn}\breve{\Delta}_{s,}{}^{mn}\right)\gamma^{ij}\right]\psi_{(I)},$$

$$\partial_s \psi_{(II)} = \frac{1}{2}\left[I\partial_s(\ln\rho - \mathrm{i}\eta) - \frac{1}{4}\left(\breve{\Delta}_{s,ij} - \frac{\mathrm{i}}{2}\varepsilon_{ijmn}\breve{\Delta}_{s,}{}^{mn}\right)\gamma^{ij}\right]\psi_{(II)} \qquad (3.204)$$

or, in notations (3.88) and (3.89) of the two-component spinors

$$\partial_s \xi^A = \frac{1}{2}\xi^A\partial_s(\ln\rho + \mathrm{i}\eta) + \frac{\mathrm{i}}{8}\left(\breve{\Delta}_{s,ij} + \frac{\mathrm{i}}{2}\varepsilon_{ijmn}\breve{\Delta}_{s,}{}^{mn}\right)\sigma^A{}_B{}^{ij}\xi^B,$$

$$\partial_s \eta_{\dot{A}} = \frac{1}{2}\eta_{\dot{A}}\partial_s(\ln\rho - \mathrm{i}\eta) + \frac{\mathrm{i}}{8}\left(\breve{\Delta}_{s,ij} - \frac{\mathrm{i}}{2}\varepsilon_{ijmn}\breve{\Delta}_{s,}{}^{mn}\right)\dot{\sigma}^B{}_A{}^{ij}\eta_{\dot{B}}.$$

The components of spintensors $\sigma^A{}_B{}^{ij}$ are defined by equality (3.99).

Bearing in mind the symmetry properties (3.104) of spintensors σ^{ij}, formulas for $\partial_s\xi^A$ and $\partial_s\eta_{\dot{A}}$ can be written in a more compact form

$$\partial_s \xi^A = \frac{1}{2}\xi^A\partial_s(\ln\rho + \mathrm{i}\eta) + \frac{\mathrm{i}}{4}\breve{\Delta}_{s,ij}\sigma^A{}_B{}^{ij}\xi^B,$$

$$\partial_s \eta_{\dot{A}} = \frac{1}{2}\eta_{\dot{A}}\partial_s(\ln\rho - \mathrm{i}\eta) + \frac{\mathrm{i}}{4}\breve{\Delta}_{s,ij}\dot{\sigma}^B{}_A{}^{ij}\eta_{\dot{B}}. \qquad (3.205)$$

Using relations (3.203) it is easy to show the validity of expression (3.148) for the Ricci rotation coefficients, corresponding to the proper bases $\breve{\boldsymbol{e}}_a$. Indeed, identities (3.11) imply the relations

$$\gamma^{ij}\gamma^5 - \gamma^5\gamma^{ij} = 0,$$

$$\gamma^{ij}\gamma^{ks} + \gamma^{ks}\gamma^{ij} = 2\left[-I\varepsilon^{ijks} + \gamma^5\left(g^{is}g^{jk} - g^{ik}g^{js}\right)\right]. \qquad (3.206)$$

Replacing in the right-hand side of formula (3.148) derivatives $\partial_i\psi$ and $\partial_i\psi^+$ by formulas (3.203) and taking into account relations (3.206) we find

$$\frac{1}{\Omega^2 + N^2}\left[\Omega\left(\psi^+\gamma_{ij}\partial_s\psi - \partial_s\psi^+\cdot\gamma_{ij}\psi\right) + N\left(\psi^+\gamma^5\gamma_{ij}\partial_s\psi - \partial_s\psi^+\cdot\gamma^5\gamma_{ij}\psi\right)\right]$$

$$\equiv \frac{1}{2(\Omega^2 + N^2)}\breve{\Delta}_{s,mn}\left[\Omega\psi^+\left(I\delta^{mn}_{ij} - \gamma^5\varepsilon_{ij}{}^{mn}\right) + N\left(\gamma^5\delta^{mn}_{ij} + I\varepsilon_{ij}{}^{mn}\right)\right]. \qquad (3.207)$$

Here $\delta^{mn}_{ij} = \delta^m_i\delta^n_j - \delta^n_i\delta^m_j$; coefficients $\breve{\Delta}_{s,mn}$ are defined by equalities (3.147).

Taking into account definitions (3.58) and (3.59) of the invariants Ω and N, it is possible to write the right-hand side of formula (3.207) in the form

$$\frac{1}{2(\Omega^2+N^2)}\breve{\Delta}_{s,mn}\big[\Omega\big(\Omega\delta_{ij}^{mn}-N\varepsilon_{ij}{}^{mn}\big)+N\big(N\delta_{ij}^{mn}+\Omega\varepsilon_{ij}{}^{mn}\big)\big]=\breve{\Delta}_{s,ij}.$$

Thus, formula (3.148) is proved.

3.6.2 Expressions for Derivatives of Spinor Fields in Terms of Derivatives of Complex Tensor Fields

Since a first-rank spinor $\boldsymbol{\psi}$ is completely defined by complex tensors $\boldsymbol{C}$, it is obvious that derivatives $\partial_i\boldsymbol{\psi}$ can be expressed also in terms of derivatives $\partial_i\boldsymbol{C}$. In order to obtain such expression, we consider the obvious identity

$$\psi^A(\psi^B\partial_i\psi^E+\psi^E\partial_i\psi^B)=\psi^A\partial_i(\psi^B\psi^E). \tag{3.208}$$

Let us contract this identity with components of spintensor $\gamma^j_{BA}e_{DE}$ with respect to the indices A, B, E. Contraction of the left-hand side of identity (3.208) gives

$$\gamma^j_{BA}e_{DE}\psi^A\left(\psi^B\partial_i\psi^E+\psi^E\partial_i\psi^B\right)=C^j\partial_i\psi_D+\left(\gamma^j_{BA}\psi^A\partial_i\psi^B\right)\psi_D. \tag{3.209}$$

Using the symmetry property of the components of spintensors $\gamma^j_{BA}=\gamma^j_{AB}$, we find

$$\gamma^j_{BA}\psi^A\partial_i\psi^B=\frac{1}{2}\gamma^j_{BA}\left(\psi^A\partial_i\psi^B+\psi^B\partial_i\psi^A\right)=\frac{1}{2}\partial_iC^j$$

and equality (3.209) can be continued

$$\gamma^j_{BA}e_{DE}\psi^A\left(\psi^B\partial_i\psi^E+\psi^E\partial_i\psi^B\right)=C^j\partial_i\psi_D+\frac{1}{2}\psi_D\partial_iC^j. \tag{3.210}$$

Contraction of the right-hand side of identity (3.208) taking into account the second identity in (C.1) gives

$$\gamma^j_{BA}e_{DE}\psi^A\partial_i(\psi^B\psi^E)=\frac{1}{4}\bigg(e_{DA}\partial_iC^j+\gamma^{sj}_{DA}\partial_iC_s$$
$$-\gamma^s_{DA}\partial_iC_s{}^j+\frac{1}{2}\varepsilon^{jsmn}\overset{*}{\gamma}_{sDA}\partial_iC_{mn}\bigg)\psi^A. \tag{3.211}$$

Taking into account formulas (3.210) and (3.211), we obtain that the result of contraction of identity (3.208) with components of spintensor $\gamma^{j}_{DE}e_{BA}$ gives the following expression

$$C^j\partial_i\psi_D = \frac{1}{4}\Big(- e_{DA}\partial_i C^j + \gamma^{sj}_{DA}\partial_i C_s - \gamma^{s}_{DA}\partial_i C_s{}^j + \frac{1}{2}\varepsilon^{jsmn}\overset{*}{\gamma}_{sDA}\partial_i C_{mn}\Big)\psi^A \tag{3.212}$$

or, in the matrix notations

$$C^j\partial_i\psi = \frac{1}{4}\left(-I\partial_i C^j + \gamma^{sj}\partial_i C_s - \gamma_s\partial_i C^{sj} + \frac{1}{2}\varepsilon^{jsmn}\overset{*}{\gamma}_s\partial_i C_{mn}\right)\psi. \tag{3.213}$$

For the conjugate spinor field we have

$$\dot{C}^j\partial_i\psi^+ = \frac{1}{4}\psi^+\left(-I\partial_i\dot{C}^j - \gamma^{sj}\partial_i\dot{C}_s + \gamma_s\partial_i\dot{C}^{sj} + \frac{1}{2}\varepsilon^{jsmn}\overset{*}{\gamma}_s\partial_i\dot{C}_{mn}\right).$$

To obtain the invariant expression $\partial_i\psi$ in terms of $\partial_i C$ it is enough to contract equation (3.213) with components of an arbitrary vector η_j satisfying the condition $\eta_j C^j \neq 0$ (for example, it is possible to take $\eta_j = \dot{C}_j$ if the invariant ρ of the spinor $\boldsymbol{\psi}$ is not equal to zero $\rho \neq 0$)

$$\partial_i\psi = \frac{1}{4\eta_n C^n}\eta_j\left(-I\partial_i C^j + \gamma^{sj}\partial_i C_s - \gamma_s\partial_i C^{sj} + \frac{1}{2}\varepsilon^{jsmn}\overset{*}{\gamma}_s\partial_i C_{mn}\right)\psi.$$

Completely analogous to the derivation of relation (3.212) given here, one can show that the contraction of identity (3.208) with components of spintensor $\gamma^{js}_{DE}e_{BA}$ with respect to the indices E, B, A gives the following relation

$$C^{js}\partial_i\psi_D = \frac{1}{4}\Big(- e_{DA}\partial_i C^{js} + \gamma^{s}_{DA}\partial_i C^j - \gamma^{j}_{DA}\partial_i C^s + \gamma^{sn}_{DA}\partial_i C_n{}^j - \gamma^{jn}_{DA}\partial_i C_n{}^s - \varepsilon^{jsmn}\overset{*}{\gamma}_{nDA}\partial_i C_m + \frac{1}{2}\varepsilon^{jsmn}\gamma^{5}_{DA}\partial_i C_{mn}\Big)\psi^A. \tag{3.214}$$

In the matrix notations formula (3.214) has the form

$$C^{js}\partial_i\psi = \frac{1}{4}\Big(- I\partial_i C^{js} + \gamma^{s}\partial_i C^j - \gamma^{j}\partial_i C^s + \gamma^{sn}\partial_i C_n{}^j - \gamma^{jn}\partial_i C_n{}^s - \varepsilon^{jsmn}\overset{*}{\gamma}_{n}\partial_i C_m + \frac{1}{2}\varepsilon^{jsmn}\gamma^{5}\partial_i C_{mn}\Big)\psi. \tag{3.215}$$

For the conjugate spinor field we find

$$\dot{C}^{js}\partial_i\psi^+ = \frac{1}{4}\psi^+\Big(- I\partial_i\dot{C}^{js} - \gamma^s\partial_i\dot{C}^j + \gamma^j\partial_i\dot{C}^s - \gamma^{sn}\partial_i\dot{C}_n{}^j$$
$$+ \gamma^{jn}\partial_i\dot{C}_n{}^s - \varepsilon^{jsmn}\overset{*}{\gamma}_n\partial_i\dot{C}_m + \frac{1}{2}\varepsilon^{jsmn}\gamma^5\partial_i\dot{C}_{mn}\Big).$$

From (3.215) we get the expression for the derivative of the spinor components

$$\partial_i\psi = \frac{1}{4\eta_{pq}C^{pq}}\eta_{js}\Big(- I\partial_i C^{js} + \gamma^s\partial_i C^j - \gamma^j\partial_i C^s + \gamma^{sn}\partial_i C_n{}^j$$
$$- \gamma^{jn}\partial_i C_n{}^s - \varepsilon^{jsmn}\overset{*}{\gamma}_n\partial_i C_m + \frac{1}{2}\varepsilon^{jsmn}\gamma^5\partial_i C_{mn}\Big)\psi, \qquad (3.216)$$

in which the components of an arbitrary antisymmetric tensor η_{pq} satisfy the condition $\eta_{pq}C^{pq} \neq 0$.

3.7 Invariant Subspaces of Spinors

Let us consider in the pseudo-Euclidean space E_4^1 an equation that is invariant under the restricted Lorentz transformations of orthonormal bases Э_i and linear with respect to the components of the first-rank spinor ψ^A and the components of the conjugate spinor ψ^{+A}:

$$\psi^A = \mathrm{i}\eta\gamma^{5A}{}_B\psi^B + \mu\psi^{+A}\exp\mathrm{i}\theta, \qquad (3.217)$$

where η, μ, and θ are arbitrary real numbers, connected by the equation

$$\eta^2 + \mu^2 = 1. \qquad (3.218)$$

It is not difficult to verify that if χ are the components of an arbitrary spinor of the first-rank in the space E_4^1, then the spinor of the first-rank, determined by the contravariant components

$$\psi^A = \chi^A + \mathrm{i}\eta\gamma^{5A}{}_B\chi^B + \mu\chi^{+A}\exp\mathrm{i}\theta,$$

satisfies Eq. (3.217) if coefficients η and μ satisfy Eq. (3.218). Thus, Eq. (3.217) is solvable with respect to ψ.

Since Eq. (3.217) is invariant under transformations of the restricted Lorentz group, it defines in the space of spinors some subspace that is invariant with respect to this group.

In spinbasis $\overset{*}{\boldsymbol{\varepsilon}}_A$, in which the components of spintensors γ_i and β are defined by matrices (3.81) and (3.82), Eqs. (3.217) are written as follows

$$\begin{aligned}(1-\eta)\psi^1 &= -\mu\dot{\psi}^4 \exp \mathrm{i}\theta,\\ (1-\eta)\psi^2 &= \mu\dot{\psi}^3 \exp \mathrm{i}\theta,\\ (1+\eta)\psi^3 &= \mu\dot{\psi}^2 \exp \mathrm{i}\theta,\\ (1+\eta)\psi^4 &= -\mu\dot{\psi}^1 \exp \mathrm{i}\theta.\end{aligned}$$

From this it follows that the subspace under consideration is possible to determine in notations of two-component spinors by the following equation

$$\dot{\eta}^{\dot{A}} = H\xi^A. \tag{3.219}$$

Here ξ^A and $\eta^{\dot{A}}$ determine the two-component spinors with a fixed relative sign, corresponding to the spinor ψ by formulas (3.89); H is the coefficient, connected with numbers η, μ, and θ in formula (3.217) by the relation

$$H = \frac{1-\eta}{\mu}\exp(-\mathrm{i}\theta).$$

The complex tensors $\boldsymbol{C}$ and the real tensors $\boldsymbol{D}$, corresponding to the spinors from subspace (3.217), are connected by the following relations

$$\begin{aligned} S^i &= -\eta j^i, \quad C^s = \mathrm{i}\mu j^s \exp \mathrm{i}\theta, \quad \Omega = N = 0,\\ \mu M^{jm} &= -\mathrm{i}\left(C^{jm} - \frac{\mathrm{i}}{2}\eta\varepsilon^{jmks}C_{ks}\right)\exp(-\mathrm{i}\theta),\\ \mu C^{jm} &= \mathrm{i}\left(M^{jm} + \frac{\mathrm{i}}{2}\eta\varepsilon^{jmks}M_{ks}\right)\exp \mathrm{i}\theta. \end{aligned} \tag{3.220}$$

Let us get, for example, the first equation in (3.220). For this purpose we replace the spinor components ψ^A in definition (3.59) of the vector components S^i by formula (3.217):

$$S^i = -\overset{*}{\gamma}{}^i_{AB}\psi^{+A}\psi^B = -\mathrm{i}\eta\overset{*}{\gamma}{}^i_{AB}\gamma^{5B}{}_C\psi^{+A}\psi^C - \mu\overset{*}{\gamma}{}^i_{AB}\psi^{+A}\psi^{+B}\exp \mathrm{i}\theta. \tag{3.221}$$

Since the components of the spintensor $\overset{*}{\gamma}{}^i_{AB}$ are antisymmetric in the indices A, B, the second term in the right-hand side of formula (3.221) is equal to zero; replacing the product of matrices $\overset{*}{\gamma}{}^i\gamma^5$ in (3.221) according to relations (3.11) and taking into account definition (3.57) for the components of the vector j^i, we find

$$S^i = \mathrm{i}\eta\gamma^i_{AC}\psi^{+A}\psi^C = -\eta j^i.$$

In a similar way, one obtains the other equations in (3.220). It is not difficult to obtain all relations (3.220) using also relation (3.219) and definitions (3.111), (3.112) for the tensors $\boldsymbol{C}$ and $\boldsymbol{D}$.

In particular, it follows from Eqs. (3.220)

$$\left(C^j e^{-\mathrm{i}\theta}\right)^{\cdot} = -C^j e^{-\mathrm{i}\theta} \quad \text{i. e.} \quad \mathrm{Re}\left(C^j e^{-\mathrm{i}\theta}\right) = 0.$$

Subspace (3.217) for $\eta = 0$ and $\mu = 1$ is defined by the equation

$$\psi^A = \psi^{+A} \exp \mathrm{i}\theta. \tag{3.222}$$

In the spinbasis, in which matrices γ^i and β are defined by equalities (3.28) and (3.29), Eq. (3.222) is written in the form

$$\mathrm{Im}\left[\psi^A \exp\left(-\frac{\mathrm{i}}{2}\theta\right)\right] = 0.$$

In particular, for $\theta = 0$ the spinor components in this spinbasis are real. Therefore the spinor determined by a components ψ^A satisfying equations $\psi^{+A} = \psi^A$ in arbitrary spinbasis is called the real spinor (or the Majorana spinor).

Equations (3.220) for spinors from subspace (3.222) pass into the following equations:

$$\Omega = N = S^i = 0,$$
$$C^m = \mathrm{i} j^m \exp \mathrm{i}\theta, \quad C^{mn} = \mathrm{i} M^{mn} \exp \mathrm{i}\theta.$$

Thus, tensors $\boldsymbol{C}$ and $\boldsymbol{D}$ determined by the spinors from subspace (3.222), have the special form

$$\boldsymbol{C} = \{j^i, M^{ij}\}\mathrm{i} \exp \mathrm{i}\theta,$$
$$\boldsymbol{D} = \{0, j^i, M^{ij}, 0, 0\}.$$

The algebraic equations (3.52), (3.53), (3.60), and (3.62) connecting the tensor components $\boldsymbol{C}$ and $\boldsymbol{D}$, in this case pass into the equations

$$j_i j^i = 0, \quad M_{ij} M^{ij} = 0, \quad \varepsilon_{ijks} M^{ij} M^{ks} = 0,$$
$$j_i M^{ij} = 0, \quad \varepsilon^{ijks} j_i M_{jk} = 0, \quad j^i j^n = M^i{}_j M^{nj}. \tag{3.223}$$

The last equation in (3.223) and the condition $j^4 \geqslant 0$ completely determine the vector components j^i in terms of the vector components M^{ij}. Therefore spinors from subspace (3.222) are completely defined by the real invariant θ and the real antisymmetric tensor of the second rank with zero invariants

$$M_{ij} M^{ij} = 0, \quad \varepsilon_{ijks} M^{ij} M^{ks} = 0. \tag{3.224}$$

The real spinors in E_4^1 are completely defined only by the antisymmetric tensor of the second rank with components M^{ij}, which satisfy the invariant algebraic equations (3.224).

For $\mu = 0$, $\eta = \pm 1$ Eq. (3.217) are written in the form

$$\psi = \mathrm{i}\gamma^5\psi \quad \text{or} \quad \psi = -\mathrm{i}\gamma^5\psi$$

and define the subspaces of semispinors. The tensors $\boldsymbol{C}$ and $\boldsymbol{D}$ for semispinors have the form

$$\begin{aligned} \boldsymbol{C} = \{0, C^{ij}\}, \quad C_{ij} &= \pm\frac{\mathrm{i}}{2}\varepsilon_{ijks}C^{ks}, \\ \boldsymbol{D} &= \{0, j^i, 0, \mp j^i, 0\}, \end{aligned} \tag{3.225}$$

while the components of tensors j^i and C^{ij} satisfy the equations

$$\begin{aligned} j_i j^i = 0, \quad j_i C^{ij} = 0, \\ 2j^i j^s = C^i{}_m \dot{C}^{sm}. \end{aligned}$$

Subspaces of semispinors are considered in detail in Sect. 3.3 of this chapter.

Considering the compatibility conditions for equations of a general form that are linear with respect to ψ and ψ^+ and invariant under the restricted Lorentz group

$$\alpha\psi^A + \eta\gamma^{5A}{}_B\psi^B + \mu\psi^{+A} + \varkappa\gamma^{5A}{}_B\psi^{+B} = 0, \tag{3.226}$$

one can show that the subspace of spinors defined by Eqs. (3.217) is the most general subspace of form (3.226) invariant under the restricted Lorentz group. In particular, invariant subspaces determined by the equations

$$\begin{aligned} \psi^A &= a\psi^{+A} + \mathrm{i}b\gamma^{5A}{}_B\psi^{+B}, \\ \dot{a}a - \dot{b}b &= 1, \quad \dot{a}b + a\dot{b} = 0, \end{aligned}$$

or by the equations

$$\begin{aligned} \psi^A &= \mathrm{i}\gamma^{5A}{}_B\left(\eta\psi^B + b\psi^{+B}\right), \\ \dot{\eta}\eta - \dot{b}b &= 1, \quad \dot{\eta} = \eta, \end{aligned}$$

are identical to the invariant subspace determined by Eqs. (3.217) and (3.218).

3.8 Spinors in Pseudo-Euclidean Space E_4^3

In the pseudo-Euclidean space of the index three E_4^3 (with the metric signature $(-,-,-,+)$) the Dirac matrices γ_j satisfy the equation

$$\gamma_i\gamma_j + \gamma_j\gamma_i = 2g_{ij}I, \tag{3.227}$$

in which the components of the metric tensor g_{ij} calculated in an orthonormal basis $\mathit{Э}_i$ of the space E_4^3, are defined by the matrix

$$g_{ij} = \begin{Vmatrix} -1 & 0 & 0 & 0 \\ 0 & -1 & 0 & 0 \\ 0 & 0 & -1 & 0 \\ 0 & 0 & 0 & 1 \end{Vmatrix}. \tag{3.228}$$

Definition (3.12) of the metric spinor E does not depend on the metric signature of the pseudo-Euclidean space. The invariant spinor of the second rank $\beta = \|\beta_{\dot{B}A}\|$ in the space E_4^3 is defined by the equations

$$\dot{\gamma}_i^T = \beta\gamma_i\beta^{-1}, \quad \dot{\beta}^T = \beta, \tag{3.229}$$

which differ from Eqs. (3.17) determining the spinor β in the space E_4^1. Due to definitions (3.229) also the equations are fulfilled

$$(\beta\gamma_i)^{\cdot} = (\beta\gamma_i)^T, \qquad (\beta\gamma_{ij})^{\cdot} = -(\beta\gamma_{ij})^T,$$
$$(\beta\overset{*}{\gamma}_i)^{\cdot} = -(\beta\overset{*}{\gamma}_i)^T, \quad (\beta\gamma^5)^{\cdot} = (\beta\gamma^5)^T.$$

The invariant spinor of the second rank $\Pi = E^{-1}\beta^T = \|\Pi^B{}_{\dot{A}}\|$ in the space E_4^3 satisfies the equations

$$\begin{gathered} \Pi\dot{\Pi} = I, \\ \dot{\gamma}_i = -\Pi^{-1}\gamma_i\Pi, \qquad \dot{\gamma}_{ij} = -\Pi^{-1}\gamma_{ij}\Pi, \\ (\overset{*}{\gamma}_i)^{\cdot} = -\Pi^{-1}\overset{*}{\gamma}_i\Pi, \quad (\gamma^5)^{\cdot} = -\Pi^{-1}\gamma^5\Pi, \end{gathered} \tag{3.230}$$

which also differ from the corresponding Eqs. (3.20) in the space E_4^1.

It is easy to see that if γ_j' are the Dirac matrices in the space E_4^1, satisfying Eqs. (3.227) with components of the metric tensor $g_{ij} = \mathrm{diag}(1, 1, 1, -1)$, then the matrices $\gamma_j = -\mathrm{i}\gamma_j'$ satisfy Eq. (3.227) with components of the metric tensor $g_{ij} = \mathrm{diag}(-1, -1, -1, 1)$ of the space E_4^3. If to the Dirac matrices γ_j' in the space E_4^1 there correspond the invariant spinors β and Π satisfying Eqs. (3.17)–(3.20), then the same matrices β, Π satisfy Eqs. (3.229)–(3.230) in which $\gamma_j = -\mathrm{i}\gamma'_j$. Thus, if between the Dirac matrices in spaces E_4^1 and E_4^3 the correspondence $\gamma_j \to -\mathrm{i}\gamma_j$ is established, then the components of the invariant spinors β and Π in these spaces are defined by identical matrices.

Equations (3.10) and (3.11) and all identities with the γ-matrices in the appendix C in the spaces E_4^1 and E_4^3 are written identically (of course, with the corresponding metric tensor in these relations). The proof of these identities and all relations given below in the space E_4^3 does not differ from the corresponding proofs in the space E_4^1.

The complex tensors $\boldsymbol{C} = \{C^i \mathfrak{D}_i, C^{ij} \mathfrak{D}_i \mathfrak{D}_j\}$, defined by the spinor of the first-rank $\boldsymbol{\psi}$ in the space E_4^3 one can define by the components

$$C^i = -\mathrm{i}\gamma^i_{BA}\psi^B\psi^A = -\mathrm{i}\psi^T E\gamma^i\psi,$$

$$C^{ij} = -\mathrm{i}\gamma^{ij}_{BA}\psi^B\psi^A = -\mathrm{i}\psi^T E\gamma^{ij}\psi.$$

In this case an expression of the components of the second rank spinor $\psi^{BA} = \psi^B\psi^A$ in terms of the components of tensors C^i, C^{ij} has the form

$$\psi^{BA} = \frac{1}{4}\left(-\mathrm{i}C^i\gamma_i^{BA} + \frac{\mathrm{i}}{2}C^{ij}\gamma_{ij}^{BA}\right).$$

The algebraic equations (3.52) and (3.53) connecting tensors $\boldsymbol{C}$ in the space E_4^1, do not change in a transition to the space E_4^3.

Let us define in the pseudo-Euclidean space E_4^3 the real tensors components $\boldsymbol{D} = \{\Omega, j^i, M^{ij}, S^i, N\}$ by the relations

$$\Omega = -e_{AB}\psi^{+A}\psi^B = \psi^+\psi,$$

$$j^i = -\gamma^i_{AB}\psi^{+A}\psi^B = \psi^+\gamma^i\psi,$$

$$M^{ij} = \mathrm{i}\gamma^{ij}_{AB}\psi^{+A}\psi^B = -\mathrm{i}\psi^+\gamma^{ij}\psi,$$

$$S^i = \mathrm{i}\overset{*}{\gamma}{}^i_{AB}\psi^{+A}\psi^B = -\mathrm{i}\psi^+\overset{*}{\gamma}{}^i\psi,$$

$$N = -\gamma^5_{AB}\psi^{+A}\psi^B = \psi^+\gamma^5\psi.$$

Here $\psi^+ = \dot{\psi}^T\beta$ is a row of the components of the conjugate spinor. In this case for the components of the second rank spinor $\psi^{\dot{B}A} = \dot{\psi}^B\psi^A$ we have

$$\|\psi^{\dot{B}A}\|^T = \frac{1}{4}\left(\Omega I + j^s\gamma_s - \frac{\mathrm{i}}{2}M^{js}\gamma_{js} - \mathrm{i}S^i\overset{*}{\gamma}_i + N\gamma^5\right)\beta^{-1}.$$

The algebraic equations connecting the real tensors $\boldsymbol{D}$ in the space E_4^3 take the form

$$
\begin{aligned}
&a.\quad j_i j^i = \Omega^2 + N^2,\\
&b.\quad S_i S^i = -\Omega^2 - N^2,\\
&c.\quad S_i j^i = 0,\\
&d.\quad \frac{1}{2} M_{ij} M^{ij} = \Omega^2 - N^2,\\
&e.\quad \frac{1}{4} \varepsilon_{ijks} M^{ij} M^{ks} = 2\Omega N,\\
&f.\quad \Omega j^i = -\frac{1}{2} \varepsilon^{ijks} S_j M_{ks},\\
&g.\quad N j^i = S_j M^{ij},\\
&h.\quad \Omega S_i = -\frac{1}{2} \varepsilon_{ijks} j^j M^{ks},\\
&i.\quad N S_i = j^n M_{in},\\
&j.\quad j^i j^j = S^i S^j - M^i{}_s M^{js} + \Omega^2 g^{ij},\\
&k.\quad \Omega M_{ij} + \frac{1}{2} N \varepsilon_{ijks} M^{ks} = -\varepsilon_{ijks} j^k S^s,\\
&l.\quad M^{ij} M^{ks} = \left(\Omega^2 + N^2\right)\left(g^{ik} g^{js} - g^{is} g^{jk}\right) - \frac{1}{4} \varepsilon^{ijpq} \varepsilon^{ksmn} M_{pq} M_{mn}\\
&\qquad - g^{ik}\left(j^s j^j - S^s S^j\right) + g^{jk}\left(j^s j^i - S^s S^i\right) + g^{is}\left(j^j j^k - S^j S^k\right)\\
&\qquad\qquad - g^{js}\left(j^i j^k - S^i S^k\right),\\
&m.\quad M^i{}_j M^{sj} - \frac{1}{4} g^{is} M_{jq} M^{jq} = \frac{1}{2} g^{is}(\Omega^2 + N^2) - j^i j^s + S^i S^s. \qquad (3.231)
\end{aligned}
$$

The equations (c), (d), (e), (k) in (3.231) in the transition from the space E_4^1 to the space E_4^3 remain unchanged.

The orthonormal proper basis $\breve{\boldsymbol{e}}_a = \{\pi^i, \xi^i, \sigma^i, u^i\}$ of the spinor field $\boldsymbol{\psi}$ in the space E_4^3 is defined by the relations

$$
\begin{aligned}
\rho \pi^i &= \operatorname{Im} C^i = -\frac{1}{2} \gamma^i_{AB} (\psi^A \psi^B + \psi^{+A} \psi^{+B}),\\
\rho \xi^i &= \operatorname{Re} C^i = -\frac{\mathrm{i}}{2} \gamma^i_{AB} (\psi^A \psi^B - \psi^{+A} \psi^{+B}),\\
\rho \sigma^i &= \mathrm{i} \overset{*}{\gamma}{}^i_{AB} \psi^{+A} \psi^B, \qquad \rho u^i = -\gamma^i_{AB} \psi^{+A} \psi^B,
\end{aligned}
$$

in which $\rho = \sqrt{\Omega^2 + N^2}$.

If the matrices γ_j in the space E_4^3 differ from matrices (3.24) by a factor $-\mathrm{i}$ and the metric spinor E is determined by matrix (3.25), then the spinor components in the proper basis $\breve{\boldsymbol{e}}_a$ are defined by the invariants Ω, N by formulas (3.144).

The contravariant components of vectors of C^a, j^a, S^a and tensors M^{ab}, C^{ab} in the proper basis $\breve{\boldsymbol{e}}_a$ are defined by equalities (3.140) and (3.141).

The Ricci rotation coefficients $\breve{\Delta}_{s,ij}$ corresponding to proper bases $\breve{\boldsymbol{e}}_a$ in the space E_4^3 are defined by the relation

$$\breve{\Delta}_{s,ij} = \frac{1}{2}\big(- \pi_i \partial_s \pi_j + \pi_j \partial_s \pi_i - \xi_i \partial_s \xi_j + \xi_j \partial_s \xi_i \\ - \sigma_i \partial_s \sigma_j + \sigma_j \partial_s \sigma_i + u_i \partial_s u_j - u_j \partial_s u_i \big). \qquad (3.232)$$

For the partial derivatives $\partial_i \psi$ in the space E_4^3 are valid relation (3.203), in which the Ricci rotation coefficients are defined by formulas (3.232), and the matrices γ^i are defined by Eqs. (3.227) and (3.228).

Chapter 4
Spinors in Three-Dimensional Euclidean Spaces

4.1 Spinor Representation of the Orthogonal Transformation Group of the Three-Dimensional Complex Euclidean Space

4.1.1 *Algebra of γ-Matrices*

For the three-dimensional complex Euclidean vector space E_3^+ the dimension of the corresponding spinor space is equal to 2, while the components of invariant spin-tensors $\overset{\circ}{\gamma}_\alpha = \|\overset{\circ}{\gamma}{}_{\alpha A}^{B}\|$ (α = 1, 2, 3; A, B = 1, 2) are represented by two-dimensional matrices. By definition, the matrix $\overset{\circ}{\gamma}_3$ expresses in terms of the matrix product $\overset{\circ}{\gamma}_1$ and $\overset{\circ}{\gamma}_2$ as follows

$$\overset{\circ}{\gamma}_3 = -\mathrm{i}\overset{\circ}{\gamma}_1\overset{\circ}{\gamma}_2. \tag{4.1}$$

The traces of the matrices $\overset{\circ}{\boldsymbol{\gamma}}$ satisfy the following equalities

$$\mathrm{tr}\,\overset{\circ}{\gamma}_\alpha = 0, \quad \mathrm{tr}\left(\overset{\circ}{\gamma}_\alpha\overset{\circ}{\gamma}_\beta\right) = 2\delta_{\alpha\beta},$$

Relations (1.15) for the three-dimensional space E_3^+ are written as

$$\begin{aligned}
\overset{\circ}{\gamma}_\alpha\overset{\circ}{\gamma}_\beta &= \overset{\circ}{\gamma}_{\alpha\beta} + \delta_{\alpha\beta}I,\\
\overset{\circ}{\gamma}_\alpha\overset{\circ}{\gamma}_{\mu\nu} &= \overset{\circ}{\gamma}_{\alpha\mu\nu} + \delta_{\alpha\mu}\overset{\circ}{\gamma}_\nu - \delta_{\alpha\nu}\overset{\circ}{\gamma}_\mu,\\
\overset{\circ}{\gamma}_{\mu\nu}\overset{\circ}{\gamma}_\alpha &= \overset{\circ}{\gamma}_{\alpha\mu\nu} - \delta_{\alpha\mu}\overset{\circ}{\gamma}_\nu + \delta_{\alpha\nu}\overset{\circ}{\gamma}_\mu.
\end{aligned} \tag{4.2}$$

V. A. Zhelnorovich, *Theory of Spinors and Its Application in Physics and Mechanics*, https://doi.org/10.1007/978-3-030-27836-6_4

Here the matrices of spintensor components $\overset{\circ}{\gamma}_{\mu\nu}$, $\overset{\circ}{\gamma}_{\alpha\mu\nu}$ are defined by the relations

$$\overset{\circ}{\gamma}_{\mu\nu} = \overset{\circ}{\gamma}_{[\mu}\overset{\circ}{\gamma}_{\nu]}, \quad \overset{\circ}{\gamma}_{\alpha\mu\nu} = \overset{\circ}{\gamma}_{[\alpha}\overset{\circ}{\gamma}_{\mu}\overset{\circ}{\gamma}_{\nu]}. \tag{4.3}$$

In the three-dimensional space E_3^+ the matrices $\overset{\circ}{\gamma}_{\alpha}$, $\overset{\circ}{\gamma}_{\mu\nu}$, and $\overset{\circ}{\gamma}_{\alpha\mu\nu}$ directly by virtue of definitions (4.1), (4.3) satisfy the equalities

$$\overset{\circ}{\gamma}_{\mu\nu} = \mathrm{i}\varepsilon_{\mu\nu\alpha}\overset{\circ}{\gamma}{}^{\alpha}, \quad \overset{\circ}{\gamma}_{\alpha\mu\nu} = \mathrm{i}\varepsilon_{\alpha\mu\nu}I,$$
$$\overset{\circ}{\gamma}{}^{\alpha} = -\frac{\mathrm{i}}{2}\varepsilon^{\alpha\mu\nu}\overset{\circ}{\gamma}_{\mu\nu}, \tag{4.4}$$

where $\varepsilon^{\alpha\mu\nu}$ are the components of the three-dimensional antisymmetric Levi-Civita pseudotensor:

$$\varepsilon^{\alpha\mu\nu} = 1, \quad \text{if substitution} \begin{pmatrix} \alpha & \mu & \nu \\ 1 & 2 & 3 \end{pmatrix} \text{ is even,}$$

$$\varepsilon^{\alpha\mu\nu} = -1, \quad \text{if substitution} \begin{pmatrix} \alpha & \mu & \nu \\ 1 & 2 & 3 \end{pmatrix} \text{ is odd,}$$

$$\varepsilon^{\alpha\mu\nu} = 0, \quad \text{if among indices } \alpha,\ \mu,\ \nu \text{ at least two coincide}$$

The components $\varepsilon^{\alpha\beta\lambda}$ by virtue of this definition satisfy the identities

$$\varepsilon_{\alpha\beta\lambda}\varepsilon^{\mu\nu\rho} = \det \begin{Vmatrix} \delta^{\mu}_{\alpha} & \delta^{\mu}_{\beta} & \delta^{\mu}_{\lambda} \\ \delta^{\nu}_{\alpha} & \delta^{\nu}_{\beta} & \delta^{\nu}_{\lambda} \\ \delta^{\rho}_{\alpha} & \delta^{\rho}_{\beta} & \delta^{\rho}_{\lambda} \end{Vmatrix}, \quad \begin{array}{c} \varepsilon_{\alpha\beta\lambda}\varepsilon^{\mu\nu\lambda} = \delta^{\mu}_{\alpha}\delta^{\nu}_{\beta} - \delta^{\nu}_{\alpha}\delta^{\mu}_{\beta}, \\ \varepsilon_{\alpha\beta\lambda}\varepsilon^{\mu\beta\lambda} = 2\delta^{\mu}_{\alpha}, \quad \varepsilon_{\alpha\beta\lambda}\varepsilon^{\alpha\beta\lambda} = 6. \end{array}$$

Equations (4.2) by means of equalities (4.4) can be written as

$$\overset{\circ}{\gamma}_{\alpha}\overset{\circ}{\gamma}_{\beta} = \mathrm{i}\varepsilon_{\alpha\beta\mu}\overset{\circ}{\gamma}{}^{\mu} + \delta_{\alpha\beta}I,$$
$$\overset{\circ}{\gamma}_{\alpha}\overset{\circ}{\gamma}_{\mu\nu} = \mathrm{i}\varepsilon_{\alpha\mu\nu}I + \delta_{\alpha\mu}\overset{\circ}{\gamma}_{\nu} - \delta_{\alpha\nu}\overset{\circ}{\gamma}_{\mu},$$
$$\overset{\circ}{\gamma}_{\mu\nu}\overset{\circ}{\gamma}_{\alpha} = \mathrm{i}\varepsilon_{\alpha\mu\nu}I - \delta_{\alpha\mu}\overset{\circ}{\gamma}_{\nu} + \delta_{\alpha\nu}\overset{\circ}{\gamma}_{\mu}.$$

If matrix $\overset{\circ}{\gamma}_1$ is symmetric, while $\overset{\circ}{\gamma}_2$ is antisymmetric

$$\overset{\circ}{\gamma}{}_1^T = \overset{\circ}{\gamma}_1, \quad \overset{\circ}{\gamma}{}_2^T = -\overset{\circ}{\gamma}_2, \tag{4.5}$$

then the metric spinor E satisfying the equation

$$\overset{\circ}{\gamma}{}^{T}_{\alpha} = -E\overset{\circ}{\gamma}_{\alpha}E^{-1}, \tag{4.6}$$

can be determined by the equality

$$E = \|e_{BA}\| = \mathrm{i}\overset{\circ}{\gamma}_2. \tag{4.7}$$

It is seen from definitions (4.5) and (4.7) that the components e_{BA} of the metric spinor E are antisymmetric, while the components of the spintensor $\overset{\circ}{\gamma}_{\alpha AB}$ are symmetric with respect to the indices A, B:

$$e_{AB} = -e_{BA}, \quad \overset{\circ}{\gamma}_{\alpha AB} = \overset{\circ}{\gamma}_{\alpha BA}. \tag{4.8}$$

In the various calculations it is often necessary to use also the following relations with the matrices $\overset{\circ}{\gamma}_{\alpha}$ and E:

$$\begin{aligned}
2e_{DE}e_{BA} &= e_{DA}e_{BE} + \overset{\circ}{\gamma}{}^{\alpha}_{DA}\overset{\circ}{\gamma}_{\alpha BE},\\
2\overset{\circ}{\gamma}_{\alpha DE}e_{BA} &= e_{DA}\overset{\circ}{\gamma}_{\alpha BE} + \overset{\circ}{\gamma}_{\alpha DA}e_{BE} - \mathrm{i}\varepsilon_{\alpha\beta\eta}\overset{\circ}{\gamma}{}^{\beta}_{DA}\overset{\circ}{\gamma}{}^{\eta}_{BE},\\
2e_{DE}\overset{\circ}{\gamma}_{\alpha BA} &= e_{DA}\overset{\circ}{\gamma}_{\alpha BE} + \overset{\circ}{\gamma}_{\alpha DA}e_{BE} + \mathrm{i}\varepsilon_{\alpha\beta\eta}\overset{\circ}{\gamma}{}^{\beta}_{DA}\overset{\circ}{\gamma}{}^{\eta}_{BE},\\
2\overset{\circ}{\gamma}{}^{\alpha}_{DE}\overset{\circ}{\gamma}{}^{\beta}_{BA} &= \delta^{\alpha\beta}\big(e_{DA}e_{BE} - \overset{\circ}{\gamma}{}^{\eta}_{DA}\overset{\circ}{\gamma}_{\eta BE}\big) + \overset{\circ}{\gamma}{}^{\alpha}_{DA}\overset{\circ}{\gamma}{}^{\beta}_{BE} + \overset{\circ}{\gamma}{}^{\beta}_{DA}\overset{\circ}{\gamma}{}^{\alpha}_{BE}\\
&\quad + \mathrm{i}\varepsilon^{\alpha\beta\eta}\big(\overset{\circ}{\gamma}_{\eta DA}e_{BE} - e_{DA}\overset{\circ}{\gamma}_{\eta BE}\big).
\end{aligned} \tag{4.9}$$

The first relation in (4.9) is the Pauli identity (see (1.19)) written for the two-dimensional matrices $\overset{\circ}{\gamma}_{\alpha}$. The remaining relations in (4.9) are obtained from the first one by contracting with components of spintensors $\overset{\circ}{\boldsymbol{\gamma}}_{\alpha}$. We note also the following relations, which are a corollary of identities (4.9):

$$\begin{aligned}
&\overset{\circ}{\gamma}_{\alpha DE}\overset{\circ}{\gamma}{}^{\alpha}_{BA} - e_{DE}e_{BA} = -\big(\overset{\circ}{\gamma}_{\alpha DA}\overset{\circ}{\gamma}{}^{\alpha}_{BE} - e_{DA}e_{BE}\big),\\
&e_{DE}\overset{\circ}{\gamma}_{\alpha BA} - \overset{\circ}{\gamma}_{\alpha DE}e_{BA} - \mathrm{i}\varepsilon_{\alpha\beta\lambda}\overset{\circ}{\gamma}{}^{\beta}_{DE}\overset{\circ}{\gamma}{}^{\lambda}_{BA}\\
&\qquad = -\big(e_{DA}\overset{\circ}{\gamma}_{\alpha BE} - \overset{\circ}{\gamma}_{\alpha DA}e_{BE} - \mathrm{i}\varepsilon_{\alpha\beta\lambda}\overset{\circ}{\gamma}{}^{\beta}_{DA}\overset{\circ}{\gamma}{}^{\lambda}_{BE}\big).
\end{aligned}$$

The matrices $\overset{\circ}{\gamma}_{\alpha}$ can be determined by the equalities[1]

$$\overset{\circ}{\gamma}_1 = \sigma_1 = \begin{Vmatrix} 0 & 1\\ 1 & 0 \end{Vmatrix}, \quad \overset{\circ}{\gamma}_2 = \sigma_2 = \begin{Vmatrix} 0 & -\mathrm{i}\\ \mathrm{i} & 0 \end{Vmatrix}, \quad \overset{\circ}{\gamma}_3 = \sigma_3 = \begin{Vmatrix} 1 & 0\\ 0 & -1 \end{Vmatrix}. \tag{4.10}$$

[1]Matrices (4.10) are called the Pauli matrices and are usually designated by σ_{α}.

If $\overset{\circ}{\gamma}_\alpha = \sigma_\alpha$, then components of the metric spinor E can be determined by the matrix

$$E = \|e_{BA}\| = \begin{Vmatrix} 0 & 1 \\ -1 & 0 \end{Vmatrix}. \tag{4.11}$$

4.1.2 Spinors in the Three-Dimensional Complex Euclidean Space E_3^+

Let $\mathbf{\ni}_\alpha$ ($\alpha = 1, 2, 3$) be an orthonormal basis in the complex Euclidean space E_3^+. The spinor representation of the complex orthogonal group SO_3^+ of transformations

$$\mathbf{\ni}'_\alpha = l^\beta{}_\alpha \mathbf{\ni}_\beta \tag{4.12}$$

is determined in the space E_3^+ by the group of the pairs $\{\pm S\}$, (i.e., by the factor group $S/(\pm I)$), satisfying the equations

$$l^\beta{}_\alpha \overset{\circ}{\gamma}_\beta = S^{-1} \overset{\circ}{\gamma}_\alpha S, \quad S^T E S = E. \tag{4.13}$$

The Greek indices α, β in Eqs. (4.13) take the values 1, 2, 3.

The invariant geometric object of the form $\boldsymbol{\psi} = \pm\psi^A \boldsymbol{\varepsilon}_A$, where the pairs of the contravariant components $\pm\psi^A$ and spinbases $\pm\{\boldsymbol{\varepsilon}_A\}$ are referred to an orthonormal basis $\mathbf{\ni}_\alpha$ and are transformed according to the formula

$$\pm\,\psi'^B = \pm S^B{}_A \psi^A, \quad \pm\boldsymbol{\varepsilon}'_B = \pm Z^A{}_B \boldsymbol{\varepsilon}_A, \tag{4.14}$$

under orthogonal transformation (4.12) of the basis $\mathbf{\ni}_\alpha$, is called a spinor of the first rank in the three-dimensional complex Euclidean space E_3^+. The quantities $S^B{}_A$ in Eqs. (4.14) are the elements of the matrix S, determined by Eqs. (4.13), $Z^A{}_B$ are the elements of the inverse matrix S^{-1}.

The covariant components of a spinor ψ_B are defined with the aid of the metric spinor E that is determined by Eq. (4.6):

$$\psi_B = e_{BA}\psi^A.$$

If the components of the metric spinor E are determined by equality (4.11), then the covariant and contravariant components of a spinor are connected as follows $\psi_1 = \psi^2, \quad \psi_2 = -\psi^1$.

4.1.3 Spinor Representation of the Orthogonal Transformation Group of Bases of the Real Euclidean Space

The invariant spinor of the second rank $\beta = \|\beta_{\dot{B}A}\|$ in the three-dimensional real vector Euclidean space E_3^0 according to results of Sect. 1.8, Chap. 1, is determined by the equations

$$(\overset{\circ}{\gamma}{}_{\alpha}^{T})^{\cdot} = \beta \overset{\circ}{\gamma}_{\alpha} \beta^{-1}, \quad \beta^{T} = \dot{\beta}.$$

The matrices of the spintensor components $\beta \overset{\circ}{\gamma}_{\alpha}$ in the space E_3^0 are Hermitian $(\beta \overset{\circ}{\gamma}_{\alpha})^{\cdot} = (\beta \overset{\circ}{\gamma}_{\alpha})^{T}$.

If the matrices $\overset{\circ}{\gamma}_{\alpha}$ are Hermitian, then the components of the invariant spinor $\beta = \|\beta_{\dot{A}B}\|$ can be determined by the unit matrix

$$\beta = I = \begin{Vmatrix} 1 & 0 \\ 0 & 1 \end{Vmatrix}.$$

In this case for the covariant and contravariant components of the conjugate spinor we have

$$\psi^{+} = \|\psi^{+}_{A}\| = (\dot{\psi}^{1}, \dot{\psi}^{2}), \quad \bar{\psi} = \|\psi^{+A}\| = \begin{Vmatrix} -\dot{\psi}^{2} \\ \dot{\psi}^{1} \end{Vmatrix}.$$

It is easy to see that for the components of the conjugate spinors, the equality is fulfilled $(\psi^{+A})^{+} = -\psi^{A}$.

It is not difficult to calculate in an explicit form the matrix of the spinor transformation S, satisfying Eqs. (4.13) in which $l^{\beta}{}_{\alpha}$ determine a real proper orthogonal transformation of bases $\mathbf{Э}_{\alpha}$ in the space E_3^0. As is known, an arbitrary three-dimensional proper orthogonal transformation can be defined by the coefficients $l^{\beta}{}_{\alpha}$:

$$l^{\beta}{}_{\alpha} = \begin{Vmatrix} \cos\varphi_2\cos\varphi_1 - \cos\theta\sin\varphi_2\sin\varphi_1 & -\sin\varphi_2\cos\varphi_1 - \cos\theta\cos\varphi_2\sin\varphi_1 & \sin\varphi_1\sin\theta \\ \cos\varphi_2\sin\varphi_1 + \cos\theta\sin\varphi_2\cos\varphi_1 & -\sin\varphi_2\sin\varphi_1 + \cos\theta\cos\varphi_2\cos\varphi_1 & -\cos\varphi_1\sin\theta \\ \sin\varphi_2\sin\theta & \cos\varphi_2\sin\theta & \cos\theta \end{Vmatrix}, \tag{4.15}$$

where φ_1, φ_2, and θ are the Euler angles determining rotation from the basis $\mathbf{Э}_{\alpha}$ to basis $\mathbf{Э}'_{\alpha} = l^{\beta}{}_{\alpha}\mathbf{Э}_{\beta}$ in the space E_3^0. If the matrices E and $\overset{\circ}{\gamma}_{\alpha}$ are determined by equalities (4.10) and (4.11), then substituting in Eqs. (4.13) coefficients $l^{\beta}{}_{\alpha}$, defined according to (4.15), we find an expression for the transformation S of the spinor components

$$S = \left\| \begin{matrix} \cos\frac{\theta}{2}\, e^{\frac{i}{2}(\varphi_1+\varphi_2)} & i\sin\frac{\theta}{2}\, e^{-\frac{i}{2}(\varphi_1-\varphi_2)} \\ i\sin\frac{\theta}{2}\, e^{\frac{i}{2}(\varphi_1-\varphi_2)} & \cos\frac{\theta}{2}\, e^{-\frac{i}{2}(\varphi_1+\varphi_2)} \end{matrix} \right\|. \tag{4.16}$$

The matrix S, determined by equality (4.16), is unitary and unimodular

$$\dot{S}^T S = I, \quad \det S = 1. \tag{4.17}$$

One can give another expression for the spinor transformation S. As is well known, the arbitrary proper orthogonal transformation (4.12) in the three-dimensional Euclidean space can be realized by rotation on some angle φ around of a fixed axis $\boldsymbol{n}$. Let n_α be components of a unit vector ($n_\alpha n^\alpha = 1$), directed along the axis $\boldsymbol{n}$. Then the coefficients $l^\beta{}_\alpha$ of the arbitrary proper orthogonal transformation can be represented as

$$l^\beta{}_\alpha = \delta^\beta_\alpha \cos\varphi + n^\beta n_\alpha (1-\cos\varphi) - \varepsilon^{\alpha\beta\lambda} n_\lambda \sin\varphi, \quad \det\|l^\beta{}_\alpha\| = +1. \tag{4.18}$$

The vector with components n_α is an eigenvector of an orthogonal matrix defined by the components of (4.18)

$$l^\beta{}_\alpha n_\beta = n_\alpha.$$

It is not difficult to obtain formula (4.18) using the Lagrange–Sylvester formula for the tensor function $L = \|l^\beta{}_\alpha\| = \exp(\varphi N)$, where N is the antisymmetric matrix $N = \|\varepsilon^{\alpha\beta\lambda} n_\lambda\|$.

It is obvious that a pair n_α, φ and a pair $-n_\alpha$, $-\varphi$ determine the same orthogonal transformation. If the orthogonal matrix $\|l^\beta{}_\alpha\|$ is given, then the angle of rotation φ and the components of the unit vector n_α are determined as follows

$$\cos\varphi = \frac{1}{2}(l^\alpha{}_\alpha - 1), \quad n_\alpha = \frac{1}{2\sin\varphi}\varepsilon_{\alpha\beta\lambda} l^{\beta\lambda}.$$

Using Eqs. (4.13) in which matrices E and $\overset{\circ}{\gamma}_\alpha$ are determined by equalities (4.10) and (4.11), it is possible to calculate the matrix of the spinor transformation S, which corresponds to the proper orthogonal transformation (4.18):

$$S = I\cos\frac{\varphi}{2} - i n^\alpha \sigma_\alpha \sin\frac{\varphi}{2} \equiv \exp\left(-\frac{i}{2}\varphi n^\alpha \sigma_\alpha\right).$$

4.2 Tensor Representation of Spinors in the Three-Dimensional Euclidean Spaces

4.2.1 Tensor Representation of Spinors in the Three-Dimensional Complex Euclidean Space

From the symmetry properties (4.8) it follows that complex tensors $\boldsymbol{C}$, determined by a spinor of the first rank $\boldsymbol{\psi}$ in the three-dimensional complex Euclidean space E_3^+, contain only a complex vector $\boldsymbol{C} = C^\alpha\, \boldsymbol{\vartheta}_\alpha$ with components C^α determined by the equality

$$C^\alpha = \overset{\circ}{\gamma}{}^\alpha_{AB}\psi^A\psi^B = \psi^T E\overset{\circ}{\gamma}{}^\alpha\psi. \tag{4.19}$$

Contracting the first relation in (4.9) with the product of spinor components $\psi^A\psi^B\psi^D\psi^E$ with respect to the indices A, B, D, E, we find that the components of the vector C^α by virtue of definition (4.19) satisfy the equation

$$C_\alpha C^\alpha = 0. \tag{4.20}$$

Thus, the complex vector determined by the components C^α is isotropic.

Components of a spinor ψ^A are expressed in terms of the components of the vector C^α according to the formula

$$\psi^A = \frac{\psi^{BA}\eta_B}{\pm\sqrt{\psi^{CD}\eta_C\eta_D}}, \qquad \psi^{BA} = -\frac{1}{2}C^\alpha\overset{\circ}{\gamma}{}^{BA}_\alpha. \tag{4.21}$$

Here η_C $(C = 1, 2)$ are arbitrary complex numbers, satisfying the condition $\psi^{CD}\eta_C\eta_D \neq 0$; the components of spintensors $\overset{\circ}{\gamma}{}^{BA}_\alpha$ are determined by the relation $\overset{\circ}{\gamma}{}^{BA}_\alpha = e^{AC}\overset{\circ}{\gamma}{}^B_{\alpha C}$.

The second formula in (4.21) in matrix notations has the form

$$\|\psi^{BA}\| = \frac{1}{2}C^\alpha\overset{\circ}{\gamma}_\alpha E^{-1}. \tag{4.22}$$

By virtue of identity (4.20), the quantities ψ^A, determined by the first formula in (4.21), are independent of the choice of numbers η_B.

Thus, the following theorem is valid.

Theorem *The first-rank spinor* $\boldsymbol{\psi}$ *in the three-dimensional complex Euclidean space* E_3^+ *with components* $\pm\psi^A$*, defined up to a common sign, is equivalent to an isotropic complex vector* $\boldsymbol{C}$*. The one-to-one relationship between components of the spinor* $\pm\psi^A$ *and those of the vector* C^α*, invariant under orthogonal transformations of the basis* $\boldsymbol{\vartheta}_\alpha$*, is performed by relations* (4.19) *and* (4.21).

If the components of invariant spintensors E and $\overset{\circ}{\gamma}_\alpha$ are determined by matrices (4.10) and (4.11), then formula (4.19) gives

$$C^1 = \psi^1\psi^1 - \psi^2\psi^2,$$
$$C^2 = \mathrm{i}(\psi^1\psi^1 + \psi^2\psi^2),$$
$$C^3 = -2\psi^1\psi^2.$$

In this case formula (4.22) takes the form

$$\psi^{11} = \frac{1}{2}(C^1 - \mathrm{i}C^2), \quad \psi^{22} = -\frac{1}{2}(C^1 + \mathrm{i}C^2),$$
$$\psi^{12} = \psi^{21} = -\frac{1}{2}C^3.$$

Tensors $\boldsymbol{K}$, determined by two spinors of the first rank $\boldsymbol{\psi}$ and $\boldsymbol{\chi}$, in the three-dimensional complex space E_3^+ contain only a complex scalar K and a complex vector with components K^α that are defined by the equalities

$$K = e_{AB}\chi^A\psi^B, \quad K^\alpha = \overset{\circ}{\gamma}{}^\alpha_{AB}\chi^A\psi^B, \tag{4.23}$$

or, in matrix notations

$$K = \chi^T E\psi, \quad K^\alpha = \chi^T E\overset{\circ}{\gamma}{}^\alpha\psi.$$

Contracting the first relation (4.9) with the product of spinor components $\psi^A\chi^B\chi^D\psi^E$, we find that by virtue of definitions (4.23) the components K and K^α satisfy the equation

$$K_\alpha K^\alpha = K^2. \tag{4.24}$$

For components of the second rank spinor $\chi^A\psi^B$ it is possible to write

$$\chi^A\psi^B = -\frac{1}{2}\left(Ke^{AB} + K^\alpha\overset{\circ}{\gamma}{}^{AB}_\alpha\right). \tag{4.25}$$

If the matrices E and $\overset{\circ}{\gamma}_\alpha$ are determined by equalities (4.10) and (4.11), then for the components K, K^α we have

$$K = \chi^1\psi^2 - \chi^2\psi^1, \quad K^1 = \chi^1\psi^1 - \chi^2\psi^2,$$
$$K^2 = \mathrm{i}(\chi^1\psi^1 + \chi^2\psi^2), \quad K^3 = -\chi^1\psi^2 - \chi^2\psi^1.$$

Relations (4.25) in this case take the form

$$\chi^1\psi^1 = \frac{1}{2}(K^1 - \mathrm{i}K^2), \quad \chi^2\psi^2 = -\frac{1}{2}(K^1 + \mathrm{i}K^2),$$
$$\chi^1\psi^2 = \frac{1}{2}(K - K^3), \quad \chi^2\psi^1 = -\frac{1}{2}(K + K^3).$$

Contracting identities (4.9) with components $\psi^B\chi^A\chi^D\psi^E$, we find that the components K and K^α are expressed in terms of the components of the vector C^α and $C'^\alpha = \mathring{\gamma}^\alpha_{BA}\chi^B\chi^A = \chi^T E\mathring{\gamma}^\alpha\chi$ as follows

$$2K^2 = -C'_\alpha C^\alpha, \quad 2KK^\alpha = \mathrm{i}\varepsilon^{\alpha\beta\eta}C'_\beta C_\eta,$$
$$2K^\alpha K^\beta = -\delta^{\alpha\beta}C'_\eta C^\eta + C'^\alpha C^\beta + C'^\beta C^\alpha.$$

The scalar K and the vectors with components K^α, C^α, C'^α are connected also by the equations

$$K_\alpha C^\alpha = 0, \quad KC^\alpha - \mathrm{i}\varepsilon^{\alpha\beta\eta}K_\beta C_\eta = 0,$$
$$K_\alpha C'^\alpha = 0, \quad KC'^\alpha + \mathrm{i}\varepsilon^{\alpha\beta\eta}K_\beta C'_\eta = 0,$$
$$C'^\alpha C^\beta = -K^2\delta^{\alpha\beta} + K^\alpha K^\beta - \mathrm{i}\varepsilon^{\alpha\beta\eta}KK_\eta,$$

which are also obtained by contracting identities (4.9) with components of spinors ψ and χ.

4.2.2 Tensor Representation of Spinors in the Three-Dimensional Real Euclidean Space

For spinors in the three-dimensional real Euclidean space E_3^0, as well as in the complex Euclidean space E_3^+, it is possible to determine the vector $\boldsymbol{C}$ by formula (4.19). All relations between the vector $\boldsymbol{C}$ and the spinor of the first rank $\boldsymbol{\psi}$, which are fulfilled in the space E_3^+, are fulfilled and in the space E_3^0. Together with the complex vector $\boldsymbol{C}$ for spinors in the real space E_3^0 it is possible to determine also the real tensors $\boldsymbol{D}$: the scalar ρ and the vector $\boldsymbol{j} = j^\alpha \boldsymbol{\mathsf{Э}}_\alpha$. The components ρ and j^α in an orthonormal basis $\boldsymbol{\mathsf{Э}}_\alpha$ of the space E_3^0 can be determined by the equalities

$$\rho = -e_{AB}\psi^{+A}\psi^B, \quad j^\alpha = -\mathring{\gamma}^\alpha_{AB}\psi^{+A}\psi^B. \tag{4.26}$$

or, in a matrix form:

$$\rho = \psi^+\psi, \quad j^\alpha = \psi^+\mathring{\gamma}^\alpha\psi. \tag{4.27}$$

Here ψ^+ is a row of the covariant components of the conjugate first rank spinor ψ^+_A. By virtue of definitions (4.26) the components ρ and j^α satisfy the equation

$$j_\alpha j^\alpha = \rho^2, \tag{4.28}$$

which is obtained by contracting the first identity in (4.9) with components of the spinor $\psi^A \psi^{+B} \psi^{+D} \psi^E$.

Further, we will also use the relations

$$\rho\psi = j^\alpha \sigma_\alpha \psi, \quad j_\alpha \psi = \rho\sigma_\alpha \psi - \mathrm{i}\varepsilon_{\alpha\beta\eta} j^\beta \sigma^\eta \psi, \tag{4.29}$$

which are obtained by contracting identities (4.9) with components of the spinor $\psi^{+D}\psi^E\psi^A$ with respect to the indices A, D, E.

The components of a spinor ψ^A are determined by the scalar ρ and the components of the vector j^α that satisfy Eq. (4.28), up to phase $\exp \mathrm{i}\varphi$, where φ is an arbitrary real number

$$\psi^A = \frac{\psi^{\dot{B}A}}{\sqrt{\psi^{\dot{B}B}}} \exp(\mathrm{i}\varphi), \quad \psi^{\dot{B}A} = \frac{1}{2}\left(\rho\beta^{A\dot{B}} + j^\alpha \overset{\circ}{\gamma}{}_\alpha^{A\dot{B}}\right). \tag{4.30}$$

The second formula in (4.30) in matrix notations can be written as follows

$$\|\psi^{\dot{B}A}\|^T = \frac{1}{2}\left(\rho I + j^\alpha \overset{\circ}{\gamma}_\alpha\right)\beta^{-1}. \tag{4.31}$$

If the components of invariant spintensors E, $\overset{\circ}{\gamma}_\alpha$ are determined by matrices (4.10), (4.11), and $\beta = I$, then the scalar ρ and the vector components j^α according to definitions (4.26) are expressed in terms of the spinor components ψ^1, ψ^2 and the complex conjugate components $\dot{\psi}^1$, $\dot{\psi}^2$ as follows

$$\begin{aligned}
\rho &= \dot{\psi}^1\psi^1 + \dot{\psi}^2\psi^2,\\
j^1 &= \dot{\psi}^1\psi^2 + \dot{\psi}^2\psi^1,\\
j^2 &= \mathrm{i}(\dot{\psi}^2\psi^1 - \dot{\psi}^1\psi^2),\\
j^3 &= \dot{\psi}^1\psi^1 - \dot{\psi}^2\psi^2.
\end{aligned} \tag{4.32}$$

Formula (4.31) determining the components of a spinor of the second rank $\psi^{\dot{B}A}$ in terms of components ρ, j^α, in this case gives

$$\psi^{\dot{1}1} = \frac{1}{2}(\rho + j^3), \quad \psi^{\dot{1}2} = \frac{1}{2}(j^1 + \mathrm{i}j^2),$$
$$\psi^{\dot{2}2} = \frac{1}{2}(\rho - j^3), \quad \psi^{\dot{2}1} = \frac{1}{2}(j^1 - \mathrm{i}j^2).$$

From definition (4.32) it follows that for any nonzero spinor the inequality is fulfilled

$$\rho = \dot{\psi}^1 \psi^1 + \dot{\psi}^2 \psi^2 > 0. \tag{4.33}$$

The scalar ρ and the vector components j^α are completely determined in terms of components of the complex vector C^α by the condition $\rho > 0$ and the equations

$$\begin{aligned} 2\rho^2 &= \dot{C}_\alpha C^\alpha, \quad 2\rho j^\alpha = -\mathrm{i}\varepsilon^{\alpha\beta\eta} \dot{C}_\beta C_\eta, \\ 2 j^\alpha j^\beta &= \delta^{\alpha\beta} \dot{C}_\eta C^\eta - \dot{C}^\alpha C^\beta - \dot{C}^\beta C^\alpha. \end{aligned} \tag{4.34}$$

Here $\dot{C}_\alpha$ is the complex conjugate components C_α defined in terms of the contravariant components of the conjugate spinor ψ^{+A} by the formula $\dot{C}_\alpha = -\overset{\circ}{\gamma}_{\alpha AB} \psi^{+A} \psi^{+B}$.

The vectors with components j^α and C^α defined by a spinor $\boldsymbol{\psi}$ are also connected by the following equations

$$\begin{aligned} C_\alpha j^\alpha = 0, \quad & j^\alpha C^\beta - j^\beta C^\alpha = -\mathrm{i}\varepsilon^{\alpha\beta\eta} \rho C_\eta, \\ & \rho C^\alpha = \mathrm{i}\varepsilon^{\alpha\beta\eta} j_\beta C_\eta. \end{aligned} \tag{4.35}$$

It is possible to obtain Eqs. (4.34) and (4.35) by contracting identities (4.9) with components of spinors $\psi^A \psi^{+B} \psi^{+D} \psi^E$, $\psi^{+A} \psi^{+B} \psi^D \psi^E$.

The real tensors $\boldsymbol{D} = \{\rho, j^\alpha \boldsymbol{\Im}_\alpha\}$, defined by a spinor $\boldsymbol{\psi}$ and the real tensors $\boldsymbol{D}' = \{\rho', j'^\alpha \boldsymbol{\Im}_\alpha\}$, defined by a spinor $\boldsymbol{\chi}$

$$\rho' = -e_{AB} \chi^{+A} \chi^B, \quad j'^\alpha = -\overset{\circ}{\gamma}{}^\alpha_{AB} \chi^{+A} \chi^B,$$

one can express in terms of the tensors $\boldsymbol{K}$:

$$\begin{aligned} 2\rho\rho' &= K\dot{K} + K_\alpha \dot{K}^\alpha, \\ 2\rho j'^\alpha &= -K\dot{K}^\alpha - K^\alpha \dot{K} + \mathrm{i}\varepsilon^{\alpha\beta\eta} K_\beta \dot{K}_\eta, \\ 2\rho' j^\alpha &= K\dot{K}^\alpha + K^\alpha \dot{K} + \mathrm{i}\varepsilon^{\alpha\beta\eta} K_\beta \dot{K}_\eta, \\ 2 j'^\beta j^\alpha &= \delta^{\alpha\beta}\big(-K\dot{K} + K_\lambda \dot{K}^\lambda\big) - K^\alpha \dot{K}^\beta - K^\beta \dot{K}^\alpha - \mathrm{i}\varepsilon^{\alpha\beta\eta}\big(K\dot{K}_\eta - K_\eta \dot{K}\big). \end{aligned} \tag{4.36}$$

The tensors $\boldsymbol{D}$, $\boldsymbol{D}'$, $\boldsymbol{K}$ are also connected by the equations

$$\begin{aligned} \rho K = j_\alpha K^\alpha, \quad & \rho K^\alpha - j^\alpha K = \mathrm{i}\varepsilon^{\alpha\beta\eta} j_\beta K_\eta, \\ -j^\alpha K^\beta + j^\beta K^\alpha &= \mathrm{i}\varepsilon^{\alpha\beta\eta}\big(\rho K_\eta - j_\eta K\big), \end{aligned}$$

$$\rho' K = -j'_\alpha K^\alpha, \quad \rho' K^\alpha + j'^\alpha K = \mathrm{i}\varepsilon^{\alpha\beta\eta} j'_\beta K_\eta,$$
$$-j'^\alpha K^\beta + j'^\beta K^\alpha = \mathrm{i}\varepsilon^{\alpha\beta\eta}(\rho' K_\eta + j'_\eta K). \tag{4.37}$$

Equations (4.36) and (4.37) are obtained by contracting identities (4.9) with components of spinors $\chi^A \chi^{+B} \psi^D \psi^{+E}$ and $\psi^A \chi^B \psi^{+D} \psi^E$.

The relations determining a spinor ψ in terms of tensors $\boldsymbol{K}$, $\boldsymbol{D}'$, $\boldsymbol{C}'$, in the space E_3^0 are obtained by contraction of identities (4.9) with components $\chi^D \chi^{+E} \psi^A$ and $\chi^D \chi^E \psi^A$:

$$\rho' \psi^B = -\frac{1}{2}(K e^{AB} + K^\alpha \overset{\circ}{\gamma}{}^{AB}_\alpha)\chi^+_A,$$
$$j'^\eta \psi^B = \frac{1}{2}(K^\eta e^{AB} + K \overset{\circ}{\gamma}{}^{\eta AB} - \mathrm{i} K_\alpha \varepsilon^{\eta\alpha\mu} \overset{\circ}{\gamma}{}^{AB}_\mu)\chi^+_A,$$
$$C'^\eta \psi^B = \frac{1}{2}(-K^\eta e^{AB} - K \overset{\circ}{\gamma}{}^{\eta AB} + \mathrm{i} K_\alpha \varepsilon^{\eta\alpha\mu} \overset{\circ}{\gamma}{}^{AB}_\mu)\chi_A.$$

4.3 Proper Orthonormal Bases for a Spinor Field in the Three-Dimensional Euclidean Spaces

4.3.1 The Proper Orthonormal Vector Basis Determined by a First-Rank Spinor

Let $\boldsymbol{\psi}$ be a first-rank spinor in the three-dimensional real Euclidean space E_3^0, referred to an orthonormal basis $\mathbf{Э}_\alpha$. Let us represent the complex components of the vector C^α, determined by the spinor $\boldsymbol{\psi}$ according to formula (4.19), in the form $C^\alpha = p^\alpha + \mathrm{i} q^\alpha$, where p^α and q^α determine two real vectors in the space E_3^0:

$$p^\alpha = \mathrm{Re}\, C^\alpha = \frac{1}{2}\overset{\circ}{\gamma}{}^\alpha_{AB}\left(\psi^A \psi^B - \psi^{+A}\psi^{+B}\right),$$
$$q^\alpha = \mathrm{Im}\, C^\alpha = \frac{\mathrm{i}}{2}\overset{\circ}{\gamma}{}^\alpha_{AB}\left(-\psi^A \psi^B - \psi^{+A}\psi^{+B}\right). \tag{4.38}$$

Equations (4.20), (4.28), (4.34), and (4.35) imply that the three vectors with components p^α, q^α and j^α are mutually orthogonal and have the equal moduli

$$j_\alpha p^\alpha = j_\alpha q^\alpha = p_\alpha q^\alpha = 0,$$
$$p_\alpha p^\alpha = q_\alpha q^\alpha = j_\alpha j^\alpha = \rho^2. \tag{4.39}$$

It is easy to obtain from Eqs. (4.34) and (4.35) that the components of the vectors p^α, q^α, and j^α are also connected by the relations

$$\begin{aligned}\rho p^\lambda &= \varepsilon^{\lambda\alpha\beta} q_\alpha j_\beta,\\ \rho q^\lambda &= \varepsilon^{\lambda\alpha\beta} j_\alpha p_\beta,\\ \rho j^\lambda &= \varepsilon^{\lambda\alpha\beta} p_\alpha q_\beta.\end{aligned} \tag{4.40}$$

From the orthogonality conditions (4.39) it follows that for any nonzero spinor of the first rank $\boldsymbol{\psi}$ the vectors with components

$$\pi^\alpha = \frac{1}{\rho} p^\alpha, \quad \xi^\alpha = \frac{1}{\rho} q^\alpha, \quad n^\alpha = \frac{1}{\rho} j^\alpha \tag{4.41}$$

form an orthonormal basis $\breve{\boldsymbol{e}}_a$ in the space E_3^0:

$$\breve{\boldsymbol{e}}_1 = \pi^\alpha \boldsymbol{\vartheta}_\alpha, \quad \breve{\boldsymbol{e}}_2 = \xi^\alpha \boldsymbol{\vartheta}_\alpha, \quad \breve{\boldsymbol{e}}_3 = n^\alpha \boldsymbol{\vartheta}_\alpha, \tag{4.42}$$

which we shall call the proper basis of the spinor $\boldsymbol{\psi}$.

The scale factors connecting basis $\boldsymbol{\vartheta}_\lambda$ and $\breve{\boldsymbol{e}}_a$:

$$\breve{\boldsymbol{e}}_a = \breve{h}^\lambda{}_a \boldsymbol{\vartheta}_\lambda, \quad \boldsymbol{\vartheta}_\lambda = \breve{h}_\lambda{}^a \breve{\boldsymbol{e}}_a \tag{4.43}$$

in accordance with (4.42) and (4.43), are determined as follows

$$\|\breve{h}^\lambda{}_a\| = \|\breve{h}_\lambda{}^a\| = \begin{Vmatrix} \pi^1 & \xi^1 & n^1 \\ \pi^2 & \xi^2 & n^2 \\ \pi^3 & \xi^3 & n^3 \end{Vmatrix} = \frac{1}{\rho} \begin{Vmatrix} p^1 & q^1 & j^1 \\ p^2 & q^2 & j^2 \\ p^3 & q^3 & j^3 \end{Vmatrix}. \tag{4.44}$$

Thus, $\breve{h}^\lambda{}_a = \delta^{\lambda\theta} \delta_{ab} \breve{h}_\theta{}^b$.

From Eqs. (4.40) it follows

$$\det \|\breve{h}_\lambda{}^a\| = \det \begin{Vmatrix} \pi^1 & \xi^1 & n^1 \\ \pi^2 & \xi^2 & n^2 \\ \pi^3 & \xi^3 & n^3 \end{Vmatrix} = 1.$$

Therefore a nonzero spinor of the first rank $\boldsymbol{\psi}$ in the three-dimensional real Euclidean space E_3^0 determines a rotation given by the proper orthogonal transformation from the basis $\boldsymbol{\vartheta}_\alpha$ to basis $\breve{\boldsymbol{e}}_a$.

If matrices $\overset{\circ}{\gamma}_\alpha$ and E are determined by equalities (4.10) and (4.11) then, using formula (4.21), it is possible to calculate the contravariant components $\breve{\psi}$ of a spinor $\boldsymbol{\psi}$ in the proper basis $\breve{\boldsymbol{e}}_a$:

$$\breve{\psi} = \pm \begin{Vmatrix} \sqrt{\rho} \\ 0 \end{Vmatrix}. \tag{4.45}$$

Using definitions (4.41) and (4.38), it is not difficult to verify that both spinor $\boldsymbol{\psi}$ and $\eta\boldsymbol{\psi}$, where η is an arbitrary real non-zero number, correspond to the same proper bases $\breve{\boldsymbol{e}}_a$ and, consequently, to the same scale factors $\breve{h}^{\lambda}{}_a$.

4.3.2 Orthogonal Transformations Group of the Proper Basis of a Spinor Field

Let ψ^A be the contravariant components of a spinor of the first rank $\boldsymbol{\psi}$ in the three-dimensional Euclidean space E_3^0, referred to an orthonormal vector basis $\boldsymbol{\ni}_\alpha$, $\alpha = 1, 2, 3$. Let us consider a group of the gauge transformations of the spinor components

$$\begin{aligned} \psi'^A &= \ \dot{\alpha}\psi^A + \dot{\beta}\bar{\psi}^A, \\ \bar{\psi}'^A &= -\beta\psi^A + \alpha\bar{\psi}^A, \end{aligned} \tag{4.46}$$

where the generally complex coefficients α and β are connected by the relation

$$\dot{\alpha}\alpha + \dot{\beta}\beta = 1.$$

The invariant ρ of the spinor $\boldsymbol{\psi}$, defined by equality (4.26), under transformation (4.46) does not vary:

$$\begin{aligned} \rho' = -e_{AB}\bar{\psi}'^A\psi'^B &= -e_{AB}\big(-\beta\psi^A + \alpha\bar{\psi}^A\big)\big(\dot{\alpha}\psi^B + \dot{\beta}\bar{\psi}^B\big) \\ &= -\big(\dot{\alpha}\alpha + \dot{\beta}\beta\big)e_{AB}\bar{\psi}^A\psi^B = \rho. \end{aligned}$$

The similar calculations give that the vector components π^α, ξ^α, and n^α, determined by equalities (4.41), under the gauge transformation (4.46) are transformed as follows

$$\begin{aligned} \pi'^\lambda &= l^1{}_1\pi^\lambda + l^2{}_1\xi^\lambda + l^3{}_1\sigma^\lambda, \\ \xi'^\lambda &= l^1{}_2\pi^\lambda + l^2{}_2\xi^\lambda + l^3{}_2\sigma^\lambda, \\ n'^\lambda &= l^1{}_3\pi^\lambda + l^2{}_3\xi^\lambda + l^3{}_3\sigma^\lambda. \end{aligned} \tag{4.47}$$

Coefficients $l^b{}_a$ in these formulas are determined by the matrix

$$L = \begin{Vmatrix} \frac{1}{2}(\alpha^2 + \dot{\alpha}^2 - \beta^2 - \dot{\beta}^2) & \frac{\mathrm{i}}{2}(\alpha^2 - \dot{\alpha}^2 + \beta^2 - \dot{\beta}^2) & -\alpha\beta - \dot{\alpha}\dot{\beta} \\ \frac{\mathrm{i}}{2}(\dot{\alpha}^2 - \alpha^2 + \beta^2 - \dot{\beta}^2) & \frac{1}{2}(\alpha^2 + \dot{\alpha}^2 + \beta^2 + \dot{\beta}^2) & \mathrm{i}(\alpha\beta - \dot{\alpha}\dot{\beta}) \\ \dot{\alpha}\beta + \dot{\beta}\alpha & \mathrm{i}(\alpha\dot{\beta} - \dot{\alpha}\beta) & \dot{\alpha}\alpha - \dot{\beta}\beta \end{Vmatrix}. \tag{4.48}$$

By means of the scale factors (4.44) formulas (4.47) can be written in the form

$$\breve{h}'^{\lambda}{}_{a} = l^{b}{}_{a}\breve{h}^{\lambda}{}_{b}. \tag{4.49}$$

From this it follows that under the gauge transformation (4.46) the vectors $\boldsymbol{e}_a$ of the proper basis of a spinor $\boldsymbol{\psi}$ are subjected to the transformation

$$\breve{\boldsymbol{e}}'_a = l^{b}{}_{a}\breve{\boldsymbol{e}}_b. \tag{4.50}$$

Since the bases $\breve{\boldsymbol{e}}'_a$ and $\breve{\boldsymbol{e}}_a$ are orthonormal, transformation (4.50) is orthogonal. The orthogonality of the matrix $\|l^{b}{}_{a}\|$ follows also directly from definition (4.48).

Let us now consider a proper orthogonal transformation of the basis $\boldsymbol{\ni}_\alpha$ of the Euclidean space E_3:

$$\boldsymbol{\ni}'_\alpha = l^{\beta}{}_{\alpha}\boldsymbol{\ni}_\beta, \tag{4.51}$$

With accordance to definitions (4.16) and (4.17), the matrix S of the spinor transformation $\psi' = S\psi$ corresponding to the transformation (4.51), in the general case can be represented as

$$S = \begin{Vmatrix} \alpha & \beta \\ -\dot{\beta} & \dot{\alpha} \end{Vmatrix}, \qquad \dot{\alpha}\alpha + \dot{\beta}\beta = 1. \tag{4.52}$$

Substituting in definitions (4.41), (4.38), and (4.26) the spinor components ψ by the formula $\psi' = S\psi$, where S is determined according to (4.52), for the transformation of the vector components of the proper basis $\breve{\boldsymbol{e}}_a$ we find

$$\pi'_\alpha = l^{\beta}{}_{\alpha}\pi_\beta, \quad \xi'_\alpha = l^{\beta}{}_{\alpha}\xi_\beta, \quad n'_\alpha = l^{\beta}{}_{\alpha}n_\beta,$$

where coefficients $l^{\beta}{}_{\alpha}$ are determined in terms of α and β by matrix (4.48). This transformation can be written by means of the scale factors (4.44) as follows

$$\breve{h}'_{\lambda}{}^{a} = l^{\beta}{}_{\lambda}\breve{h}_{\beta}{}^{a}. \tag{4.53}$$

Thus, the matrix $l^{\beta}{}_{\lambda}$ of the orthogonal transformation (4.53) of the proper basis $\breve{\boldsymbol{e}}_a$ turns out to be the same as in Eq. (4.49), corresponding to the gauge transformation (4.46) of the spinor ψ.

4.3.3 The Angular Velocity Vector of Rotation of the Proper Basis $\breve{e}_a$

Let us suppose that the components of the first-rank spinor ψ are given in an orthonormal basis $\mathcal{Э}_\alpha$ of the space E_3^0 as functions of some scalar parameter t. Consider a vector determined in the basis $\mathcal{Э}_\alpha$ by the components

$$\begin{aligned}\breve{\Omega}^\alpha &= \frac{\mathrm{i}}{\rho}\left[\psi^+ \overset{\circ}{\gamma}{}^\alpha \frac{d}{dt}\psi - \left(\frac{d}{dt}\psi^+\right)\overset{\circ}{\gamma}{}^\alpha \psi\right] \equiv \\ &\equiv \frac{\mathrm{i}}{\rho}\overset{\circ}{\gamma}{}^\alpha_{BA}\left(-\psi^{+B}\frac{d}{dt}\psi^A + \psi^A\frac{d}{dt}\psi^{+B}\right),\end{aligned} \tag{4.54}$$

where the invariant ρ is determined by equality (4.27).

It is not difficult to see that by virtue of definition (4.54) is carried out the equality

$$\psi^{+D}\psi^E\breve{\Omega}^\alpha = \frac{\mathrm{i}}{\rho}\overset{\circ}{\gamma}{}^\alpha_{BA}\left[-\psi^{+B}\psi^E\frac{d}{dt}\left(\psi^{+D}\psi^A\right) + \psi^E\psi^A\frac{d}{dt}\left(\psi^{+D}\psi^{+B}\right)\right]. \tag{4.55}$$

Contracting equality (4.55) with respect to indices D, E with components of the metric spinor e_{DE}, in view of the third identity in (4.9) we can express $\breve{\Omega}^\alpha$ in terms of the vector components p^α, q^α, and j^α:

$$\breve{\Omega}^\alpha = \frac{1}{2\rho^2}\varepsilon^{\alpha\beta\eta}\left(p_\beta\frac{d}{dt}p_\eta + q_\beta\frac{d}{dt}q_\eta + j_\beta\frac{d}{dt}j_\eta\right)$$

or, in terms of the components of unit vectors (4.41):

$$\breve{\Omega}^\alpha = \frac{1}{2}\varepsilon^{\alpha\beta\eta}\left(\pi_\beta\frac{d}{dt}\pi_\eta + \xi_\beta\frac{d}{dt}\xi_\eta + n_\beta\frac{d}{dt}n_\eta\right). \tag{4.56}$$

From (4.56) it follows

$$\begin{gathered}\frac{d}{dt}\pi^\alpha = \varepsilon^{\alpha\beta\lambda}\breve{\Omega}_\beta\pi_\lambda, \quad \frac{d}{dt}\xi^\alpha = \varepsilon^{\alpha\beta\lambda}\breve{\Omega}_\beta\xi_\lambda, \\ \frac{d}{dt}n^\alpha = \varepsilon^{\alpha\beta\lambda}\breve{\Omega}_\beta n_\lambda.\end{gathered} \tag{4.57}$$

Equations (4.57) are fulfilled identically by virtue of the definition of the vector components π^α, ξ^α, n^α, and $\breve{\Omega}^\alpha$.

If E_3^0 is the physical three-dimensional space and t is absolute time, Eqs. (4.57) imply that the vector with components $\breve{\Omega}^\alpha$ determined by equality (4.54), is the angular velocity vector of the rotation of the proper basis $\breve{e}_a$ relative to the basis $\mathcal{Э}_\alpha$.

4.3.4 Derivatives of Spinors with Respect to Time in the Rotating Orthonormal Basis

Let $\mathbf{Э}_\lambda$ be some fixed orthonormal basis in the physical three-dimensional Euclidean space, while an orthonormal basis $\overset{\circ}{\mathbf{Э}}_a$ rotates concerning the basis $\mathbf{Э}_\lambda$ with an angular velocity determined by the angular velocity vector $\overset{\circ}{\boldsymbol{\Omega}} = \overset{\circ}{\Omega}{}^\lambda \mathbf{Э}_\lambda = \overset{\circ}{\Omega}{}^a \overset{\circ}{\mathbf{Э}}_a$.

Assuming that bases $\overset{\circ}{\mathbf{Э}}_a$ and $\mathbf{Э}_\lambda$ are connected by the equality

$$\overset{\circ}{\mathbf{Э}}_a = \overset{\circ}{h}{}^\lambda{}_a \mathbf{Э}_\lambda, \tag{4.58}$$

where scale factors $\overset{\circ}{h}{}^\lambda{}_a$ determine an orthogonal matrix, we express components of the angular velocity vector $\overset{\circ}{\boldsymbol{\Omega}}$ in terms of the scale factors $\overset{\circ}{h}{}^\lambda{}_a$. By virtue of the definition of the vector $\overset{\circ}{\boldsymbol{\Omega}}$ we have

$$\frac{d}{dt}\overset{\circ}{\mathbf{Э}}_a = \overset{\circ}{\boldsymbol{\Omega}} \times \overset{\circ}{\mathbf{Э}}_a = \varepsilon_{\lambda\mu\nu}\overset{\circ}{\Omega}{}^\lambda \overset{\circ}{h}{}^\mu{}_a \mathbf{Э}^\nu = -\varepsilon_{abc}\overset{\circ}{\Omega}{}^b \overset{\circ}{\mathbf{Э}}{}^c. \tag{4.59}$$

Differentiating Eq. (4.58), we find

$$\frac{d}{dt}\overset{\circ}{\mathbf{Э}}_a = \mathbf{Э}_\lambda \frac{d}{dt}\overset{\circ}{h}{}^\lambda{}_a = \left(\delta_{\beta\lambda}\overset{\circ}{h}{}^\beta{}_b \frac{d}{dt}\overset{\circ}{h}{}^\lambda{}_a\right)\overset{\circ}{\mathbf{Э}}{}^b.$$

Using the orthogonality conditions of the scale factors $\overset{\circ}{h}{}^\lambda{}_a$, the last equation can be written also as

$$\frac{d}{dt}\overset{\circ}{\mathbf{Э}}_a = \frac{1}{2}\delta_{\beta\lambda}\left(\overset{\circ}{h}{}^\beta{}_b \frac{d}{dt}\overset{\circ}{h}{}^\lambda{}_a - \overset{\circ}{h}{}^\beta{}_a \frac{d}{dt}\overset{\circ}{h}{}^\lambda{}_b\right)\overset{\circ}{\mathbf{Э}}{}^b.$$

Comparing this equation with Eq. (4.59), for the components $\overset{\circ}{\Omega}{}^a$ of the angular velocity vector in the rotating basis $\overset{\circ}{\mathbf{Э}}_a$ we obtain

$$\overset{\circ}{\Omega}{}^a = -\frac{1}{2}\varepsilon^{abc}\delta_{\beta\lambda}\overset{\circ}{h}{}^\beta{}_b \frac{d}{dt}\overset{\circ}{h}{}^\lambda{}_c.$$

It is not difficult to find the following expression for the components $\overset{\circ}{\Omega}{}^\lambda$ of the angular velocity vector in the fixed basis $\mathbf{Э}_\lambda$ in terms of the scale factors

$$\overset{\circ}{\Omega}{}^\lambda = \frac{1}{2}\varepsilon^{\lambda\mu\nu}\delta_{ab}\overset{\circ}{h}{}_\mu{}^a \frac{d}{dt}\overset{\circ}{h}{}_\nu{}^b.$$

Let us denote the spinbasis corresponding to the vector basis $\mathbf{Э}_\lambda$ by the symbol $\boldsymbol{\varepsilon}_A$; the spinbasis, corresponding to the vector basis $\overset{\circ}{\mathbf{Э}}_a$, we denote by the symbol $\overset{\circ}{\boldsymbol{\varepsilon}}_A$. A matrix $S^{-1} = \|Z^B{}_A\|$ connecting spinbases $\boldsymbol{\varepsilon}_A$ and $\overset{\circ}{\boldsymbol{\varepsilon}}_A$:

$$\overset{\circ}{\boldsymbol{\varepsilon}}_A = Z^B{}_A \boldsymbol{\varepsilon}_B, \tag{4.60}$$

according to (4.13) is determined by the equations

$$\overset{\circ}{h}{}^\lambda{}_a \overset{\circ}{\gamma}_\lambda = S^{-1} \overset{\circ}{\gamma}_a S, \quad S^T E S = E. \tag{4.61}$$

Let $\boldsymbol{\psi}$ be an arbitrary first-rank spinor determined in the spinbasis $\boldsymbol{\varepsilon}_A$ by contravariant components ψ^A, while in the spinbasis $\overset{\circ}{\boldsymbol{\varepsilon}}_A$ by contravariant components $\overset{\circ}{\psi}{}^A$:

$$\boldsymbol{\psi} = \pm\psi^A \boldsymbol{\varepsilon}_A = \pm\overset{\circ}{\psi}{}^A \overset{\circ}{\boldsymbol{\varepsilon}}_A.$$

We assume that the components ψ^A and $\overset{\circ}{\psi}{}^A$ are given as functions of time t.

Let us find a connection between the derivative of a spinor $\boldsymbol{\psi}$ with respect to time $d\boldsymbol{\psi}/dt$, calculated in the fixed basis $\mathbf{Э}_\lambda$ and the derivative $d'\boldsymbol{\psi}/dt$, calculated in the rotating basis $\overset{\circ}{\mathbf{Э}}_a$. By definition, the derivative of a spinor $\boldsymbol{\psi}$ with respect to time in the rotating basis $\overset{\circ}{\mathbf{Э}}_a$ is a spinor of the same rank with components $d\overset{\circ}{\psi}{}^A/dt$ in the spinbasis $\overset{\circ}{\boldsymbol{\varepsilon}}_A$:

$$\frac{d'\boldsymbol{\psi}}{dt} = \pm\overset{\circ}{\boldsymbol{\varepsilon}}_A \frac{d\overset{\circ}{\psi}{}^A}{dt}. \tag{4.62}$$

According to the conditions we have

$$\frac{d}{dt}\mathbf{Э}_\lambda = 0, \quad \frac{d}{dt}\boldsymbol{\varepsilon}_A = 0,$$
$$\frac{d'}{dt}\overset{\circ}{\mathbf{Э}}_a = 0, \quad \frac{d'}{dt}\overset{\circ}{\boldsymbol{\varepsilon}}_A = 0.$$

Differentiating relations (4.61) and performing identical transformations, we obtain the equations

$$\left(S\frac{d}{dt}S^{-1}\right)^T E + E\left(S\frac{d}{dt}S^{-1}\right) = 0,$$

$$\left(S\frac{d}{dt}S^{-1}\right)\overset{\circ}{\gamma}_a - \overset{\circ}{\gamma}_a\left(S\frac{d}{dt}S^{-1}\right) = \left(\delta_{\beta\lambda}\overset{\circ}{h}{}^\beta{}_b \frac{d}{dt}\overset{\circ}{h}{}^\lambda{}_a\right)\overset{\circ}{\gamma}{}^b = \varepsilon_{abc}\overset{\circ}{\Omega}{}^c\overset{\circ}{\gamma}{}^b,$$

from which it follows

$$S\frac{d}{dt}S^{-1} = -\frac{\mathrm{i}}{2}\overset{\circ}{\Omega}{}^{a}\overset{\circ}{\gamma}_{a}. \tag{4.63}$$

From Eq. (4.63) we obtain expressions for the derivatives of the matrices S and S^{-1} with respect to time:

$$\frac{d}{dt}S = \frac{\mathrm{i}}{2}\overset{\circ}{\Omega}{}^{a}\overset{\circ}{\gamma}_{a}S, \quad \frac{d}{dt}S^{-1} = -\frac{\mathrm{i}}{2}\overset{\circ}{\Omega}{}^{a}S^{-1}\overset{\circ}{\gamma}_{a}. \tag{4.64}$$

Differentiating Eq. (4.60), in view of equalities (4.64) we find

$$\frac{d}{dt}\overset{\circ}{\boldsymbol{\varepsilon}}_{A} = \boldsymbol{\varepsilon}_{B}\frac{d}{dt}Z^{B}{}_{A} = \overset{\circ}{\boldsymbol{\varepsilon}}_{C}S^{C}{}_{B}\frac{d}{dt}Z^{B}{}_{A} = -\frac{\mathrm{i}}{2}\overset{\circ}{\Omega}{}^{a}\overset{\circ}{\gamma}{}^{C}_{aA}\overset{\circ}{\boldsymbol{\varepsilon}}_{C}. \tag{4.65}$$

Taking into account expression (4.65) for the derivative of $\overset{\circ}{\boldsymbol{\varepsilon}}_{A}$ and definition (4.62) for the derivative $d'\boldsymbol{\psi}/dt$, we obtain

$$\begin{aligned}\frac{d}{dt}\psi^{A}\boldsymbol{\varepsilon}_{A} = \frac{d}{dt}\overset{\circ}{\psi}{}^{A}\overset{\circ}{\boldsymbol{\varepsilon}}_{A} &= \overset{\circ}{\boldsymbol{\varepsilon}}_{A}\frac{d}{dt}\overset{\circ}{\psi}{}^{A} - \frac{\mathrm{i}}{2}\overset{\circ}{\Omega}{}^{a}\overset{\circ}{\gamma}{}^{B}_{aA}\overset{\circ}{\boldsymbol{\varepsilon}}_{B}\overset{\circ}{\psi}{}^{A}\\ &= \frac{d'}{dt}\overset{\circ}{\psi}{}^{A}\overset{\circ}{\boldsymbol{\varepsilon}}_{A} - \frac{\mathrm{i}}{2}\overset{\circ}{\Omega}{}^{a}\overset{\circ}{\gamma}{}^{B}_{aA}\overset{\circ}{\boldsymbol{\varepsilon}}_{B}\overset{\circ}{\psi}{}^{A}.\end{aligned}$$

Thus, derivatives $d'\boldsymbol{\psi}/dt$ and $d\boldsymbol{\psi}/dt$ are connected by the relation

$$\frac{d}{dt}\psi^{A}\boldsymbol{\varepsilon}_{A} = \left(\delta^{B}_{A}\frac{d'}{dt} - \frac{\mathrm{i}}{2}\overset{\circ}{\Omega}{}^{a}\overset{\circ}{\gamma}{}^{B}_{aA}\right)\overset{\circ}{\psi}{}^{A}\overset{\circ}{\boldsymbol{\varepsilon}}_{B}. \tag{4.66}$$

Direct verification shows that for the vectors $p^{\lambda}\boldsymbol{\vartheta}_{\lambda}$, $q^{\lambda}\boldsymbol{\vartheta}_{\lambda}$, and $j^{\lambda}\boldsymbol{\vartheta}_{\lambda}$ determined by a spinor $\boldsymbol{\psi}$, Eqs. (4.66) implies the usual relations for derivatives of vectors with respect to time [62]

$$\begin{aligned}\frac{d}{dt}p^{\lambda}\boldsymbol{\vartheta}_{\lambda} &= \left(\frac{d'}{dt}\overset{\circ}{p}{}^{a} + \varepsilon^{abc}\overset{\circ}{\Omega}_{b}\overset{\circ}{p}_{c}\right)\overset{\circ}{\boldsymbol{\vartheta}}_{a},\\ \frac{d}{dt}q^{\lambda}\boldsymbol{\vartheta}_{\lambda} &= \left(\frac{d'}{dt}\overset{\circ}{q}{}^{a} + \varepsilon^{abc}\overset{\circ}{\Omega}_{b}\overset{\circ}{q}_{c}\right)\overset{\circ}{\boldsymbol{\vartheta}}_{a},\\ \frac{d}{dt}j^{\lambda}\boldsymbol{\vartheta}_{\lambda} &= \left(\frac{d'}{dt}\overset{\circ}{j}{}^{a} + \varepsilon^{abc}\overset{\circ}{\Omega}_{b}\overset{\circ}{j}_{c}\right)\overset{\circ}{\boldsymbol{\vartheta}}_{a}.\end{aligned}$$

4.3.5 The Ricci Rotation Coefficients for the Proper Bases

Let us consider in the three-dimensional real Euclidean point space G_3 a field of the first rank spinor $\boldsymbol{\psi}(x^\alpha)$, given by components $\psi^A(x^\alpha)$ in a cartesian coordinate system with the variables x^α. The field of a spinor $\boldsymbol{\psi}(x^\alpha)$ determines in each point of the space G_3 the proper orthonormal bases $\breve{\boldsymbol{e}}_a(x^\alpha)$, defined in terms of $\boldsymbol{\psi}$ by formulas (4.42), (4.41), (4.38), and (4.19).

Consider the quantities $\breve{\Delta}_{\lambda,\alpha\beta}$, determined by the field of a spinor $\boldsymbol{\psi}(x^\alpha)$ and conjugate spinor $\boldsymbol{\psi}^+(x^\alpha)$:

$$\breve{\Delta}_{\lambda,\alpha\beta} = \frac{\mathrm{i}}{\rho}\varepsilon_{\alpha\beta\eta}\overset{\circ}{\gamma}{}^{\eta}{}_{BA}\left(-\psi^{+B}\partial_\lambda\psi^A + \psi^A\partial_\lambda\psi^{+B}\right). \tag{4.67}$$

From definition (4.67) it follows the obvious identity

$$\psi^{+D}\psi^E\breve{\Delta}_{\lambda,\alpha\beta} = \frac{\mathrm{i}}{\rho}\varepsilon_{\alpha\beta\eta}\overset{\circ}{\gamma}{}^{\eta}{}_{BA}\left[-\psi^{+B}\psi^E\partial_\lambda(\psi^{+D}\psi^A) + \psi^E\psi^A\partial_\lambda(\psi^{+D}\psi^{+B})\right], \tag{4.68}$$

Contracting this equality with components of the metric spinor e_{DE} with respect to the indices D, E, taking into account the third identity in (4.9), we find

$$\breve{\Delta}_{\lambda,\alpha\beta} = \frac{1}{2\rho^2}\left(p_\alpha\partial_\lambda p_\beta - p_\beta\partial_\lambda p_\alpha + q_\alpha\partial_\lambda q_\beta - q_\beta\partial_\lambda q_\alpha + j_\alpha\partial_\lambda j_\beta - j_\beta\partial_\lambda j_\alpha\right)$$

or

$$\breve{\Delta}_{\lambda,\alpha\beta} = \frac{1}{2}\left(\pi_\alpha\partial_\lambda\pi_\beta - \pi_\beta\partial_\lambda\pi_\alpha + \xi_\alpha\partial_\lambda\xi_\beta - \xi_\beta\partial_\lambda\xi_\alpha + n_\alpha\partial_\lambda n_\beta - n_\beta\partial_\lambda n_\alpha\right). \tag{4.69}$$

Using notations (4.44), one can write formula (4.69) as follows

$$\breve{\Delta}_{\lambda,\alpha\beta} = \frac{1}{2}\delta_{bc}\left(\breve{h}_\alpha{}^b\partial_\lambda\breve{h}_\beta{}^c - \breve{h}_\beta{}^b\partial_\lambda\breve{h}_\alpha{}^c\right). \tag{4.70}$$

Equations (4.70) are the definition of the Ricci rotation coefficients corresponding to the system of the proper bases $\breve{\boldsymbol{e}}_a(x^\lambda)$. By means of the Ricci rotation coefficients one can calculate the derivatives of the vectors of the proper basis

$$\partial_\lambda\breve{\boldsymbol{e}}_a = \left(\breve{h}^\mu{}_a\breve{h}^\nu{}_b\breve{\Delta}_{\lambda,\mu\nu}\right)\breve{\boldsymbol{e}}^b = \breve{\Delta}_{\lambda,ab}\breve{\boldsymbol{e}}^b. \tag{4.71}$$

If we replace the vectors $\breve{\boldsymbol{e}}_a$ in Eqs. (4.71) in terms of vectors $\boldsymbol{\ni}_\alpha$ by formulas (4.42), then we obtain the equations connecting the derivatives $\partial_\lambda\pi_\alpha$, $\partial_\lambda\xi_\alpha$, $\partial_\lambda n_\alpha$ and the vector components π_α, ξ_α, n_α of the proper basis of the spinor field

$$\partial_\lambda\pi_\alpha = -\breve{\Delta}_{\lambda,\alpha}{}^\beta\pi_\beta, \quad \partial_\lambda\xi_\alpha = -\breve{\Delta}_{\lambda,\alpha}{}^\beta\xi_\beta,$$
$$\partial_\lambda n_\alpha = -\breve{\Delta}_{\lambda,\alpha}{}^\beta n_\beta. \tag{4.72}$$

Equations (4.72) are obtained also directly by contracting definition (4.69) with components of the vectors π^α, ξ^α, n^α.

Instead of the Ricci rotation coefficients $\breve{\Delta}_{\lambda,\alpha\beta}$ it is convenient to use the dual quantities $\breve{\Delta}_{\lambda\alpha}$:

$$\breve{\Delta}_{\lambda\alpha} = \frac{1}{2}\varepsilon_\alpha{}^{\mu\nu}\breve{\Delta}_{\lambda,\mu\nu} = \frac{\mathrm{i}}{\rho}\Big[\psi^+\overset{\circ}{\gamma}_\alpha\partial_\lambda\psi - (\partial_\lambda\psi^+)\overset{\circ}{\gamma}_\alpha\psi\Big]. \tag{4.73}$$

It is not difficult to see, that $\breve{\Delta}_{\lambda,\mu\nu} = \varepsilon_{\mu\nu\alpha}\breve{\Delta}_\lambda{}^\alpha$.

4.4 The Expression of Derivatives of the Spinor Field in Terms of Derivatives of Vector Fields

Let us obtain an expression for derivatives of a spinor field $\boldsymbol{\psi}(x^\alpha, t)$ in terms of derivatives of the vectors of the proper basis $\breve{\boldsymbol{e}}_a$ and the spinor invariant ρ. Let the first-rank spinor $\boldsymbol{\psi}$ in an orthonormal basis $\boldsymbol{\mathfrak{Э}}_\alpha$ of the real point Euclidean space has the contravariant components ψ^A, while in the proper basis $\breve{\boldsymbol{e}}_a$ the same spinor has the contravariant components $\breve{\psi}^A$, defined by relation (4.45). Differentiating relation (4.45) with respect to variables x^λ and t, we find

$$\partial_\lambda\breve{\psi} = \frac{1}{2}\breve{\psi}\,\partial_\lambda\ln\rho, \qquad \frac{\partial}{\partial t}\breve{\psi} = \frac{1}{2}\breve{\psi}\,\frac{\partial}{\partial t}\ln\rho. \tag{4.74}$$

The components of spinors $\breve{\psi}$ and ψ are connected by the transformation

$$\breve{\psi} = S\psi. \tag{4.75}$$

The transformation matrix S is determined by the equations (see (4.13))

$$\breve{h}^\lambda{}_a S\overset{\circ}{\gamma}_\lambda = \overset{\circ}{\gamma}_a S, \qquad S^T E S = E, \tag{4.76}$$

where the coefficients $\breve{h}^\lambda{}_a$ are determined by matrix (4.44). Similarly to the derivation of the corresponding formula in the space E_4^1 for derivatives $\partial S/\partial t$, $\partial_\lambda S$ using Eqs. (4.76) one can find

$$\frac{\partial}{\partial t}S = \frac{\mathrm{i}}{2}\breve{\Omega}^\alpha S\overset{\circ}{\gamma}_\alpha, \qquad \partial_\lambda S = \frac{1}{4}\breve{\Delta}_{\lambda,\alpha\beta}S\overset{\circ}{\gamma}^\alpha\overset{\circ}{\gamma}^\beta. \tag{4.77}$$

Here the quantities $\breve{\Delta}_{\lambda,\alpha\beta}$ and $\breve{\Omega}^{\alpha}$ are determined in terms of derivatives of the proper basis $\breve{e}_a$ by relations (4.56) and (4.69). Bearing in mind formulas (4.74), (4.75), and (4.77) for derivatives $\partial\psi/\partial t$ and $\partial_\lambda\psi$ we find

$$\frac{\partial}{\partial t}\psi = \frac{1}{2}\left(I\frac{\partial}{\partial t}\ln\rho - \mathrm{i}\breve{\Omega}^{\alpha}\overset{\circ}{\gamma}_{\alpha}\right)\psi,$$
$$\partial_\lambda\psi = \frac{1}{2}\left(I\partial_\lambda\ln\rho - \frac{1}{2}\breve{\Delta}_{\lambda,\alpha\beta}\overset{\circ}{\gamma}{}^{\alpha}\overset{\circ}{\gamma}{}^{\beta}\right)\psi, \tag{4.78}$$

where I is the unit two-dimensional matrix.

For the conjugate field ψ^+ the equalities are valid

$$\frac{\partial}{\partial t}\psi^+ = \frac{1}{2}\psi^+\left(I\frac{\partial}{\partial t}\ln\rho + \mathrm{i}\breve{\Omega}^{\alpha}\overset{\circ}{\gamma}_{\alpha}\right),$$
$$\partial_\lambda\psi^+ = \frac{1}{2}\psi^+\left(I\partial_\lambda\ln\rho + \frac{1}{2}\breve{\Delta}_{\lambda,\alpha\beta}\overset{\circ}{\gamma}{}^{\alpha}\overset{\circ}{\gamma}{}^{\beta}\right). \tag{4.79}$$

The second equations in (4.78) and (4.79), taking into account definition (4.73) and the first equality in (4.4), can be written in the form

$$\partial_\lambda\psi = \frac{1}{2}\left(I\partial_\lambda\ln\rho - \frac{\mathrm{i}}{2}\breve{\Delta}_{\lambda\alpha}\overset{\circ}{\gamma}{}^{\alpha}\right)\psi,$$
$$\partial_\lambda\psi^+ = \frac{1}{2}\psi^+\left(I\partial_\lambda\ln\rho + \frac{\mathrm{i}}{2}\breve{\Delta}_{\lambda\alpha}\overset{\circ}{\gamma}{}^{\alpha}\right).$$

We can give a simpler (though more formal) derivation of Eqs. (4.78). Let us consider the identity obtained by contracting the fourth identity in (4.9) with components of the metric tensor $g_{\alpha\beta}$ with respect to the indices α and β

$$2\overset{\circ}{\gamma}{}^{\alpha}_{DE}\overset{\circ}{\gamma}_{\alpha BA} = 3e_{DA}e_{BE} - \overset{\circ}{\gamma}{}^{\alpha}_{DA}\overset{\circ}{\gamma}_{\alpha BE}. \tag{4.80}$$

The contraction of identity (4.80) with quantities $\psi^{+D}\psi^{A}\partial\psi^{E}/\partial t$ with respect to the indices D, A, E gives

$$2\left(\overset{\circ}{\gamma}_{\alpha DE}\psi^{+D}\frac{\partial}{\partial t}\psi^{E}\right)\overset{\circ}{\gamma}{}^{\alpha}_{BA}\psi^{A} = -3\rho\frac{\partial}{\partial t}\psi_B + j_\alpha\overset{\circ}{\gamma}{}^{\alpha}_{BE}\frac{\partial}{\partial t}\psi^{E}. \tag{4.81}$$

Let us transform the left-hand side of Eq. (4.81). In view of definitions (4.26) and (4.54) we have

$$2\left(\overset{\circ}{\gamma}_{\alpha DE}\psi^{+D}\frac{\partial}{\partial t}\psi^{E}\right)\overset{\circ}{\gamma}{}^{\alpha}_{BA}\psi^{A}=$$
$$=\overset{\circ}{\gamma}_{\alpha DE}\left[\psi^{+D}\frac{\partial}{\partial t}\psi^{E}-\psi^{E}\frac{\partial}{\partial t}\psi^{+D}+\frac{\partial}{\partial t}(\psi^{+D}\psi^{E})\right]\overset{\circ}{\gamma}{}^{\alpha}_{BA}\psi^{A}=$$
$$=\left(\mathrm{i}\rho\breve{\Omega}_{\alpha}-\frac{\partial}{\partial t}j_{\alpha}\right)\overset{\circ}{\gamma}{}^{\alpha}_{BA}\psi^{A}. \tag{4.82}$$

Using relation (4.82), one can write Eq. (4.81) as follows (in a matrix form)

$$\mathrm{i}\rho\breve{\Omega}^{\alpha}\overset{\circ}{\gamma}_{\alpha}\psi=-3\rho\frac{\partial}{\partial t}\psi+\frac{\partial}{\partial t}\left(j^{\alpha}\overset{\circ}{\gamma}_{\alpha}\psi\right). \tag{4.83}$$

Replacing the last term in Eq. (4.83) by formula (4.29), we find

$$\mathrm{i}\rho\breve{\Omega}^{\alpha}\overset{\circ}{\gamma}_{\alpha}\psi=-2\rho\frac{\partial}{\partial t}\psi+\psi\frac{\partial}{\partial t}\rho. \tag{4.84}$$

Dividing identity (4.84) by 2ρ, we obtain the first relation in (4.78). The second relation in (4.78) is obtained in a similar way.

The derivative of the spinor components can be expressed also in terms of the complex vector components $C^{\alpha}=\rho(\pi^{\alpha}+\mathrm{i}\xi^{\alpha})$. For this purpose we contract identity

$$\psi^{A}\left(\psi^{B}d\psi^{E}+\psi^{E}d\psi^{B}\right)=\psi^{A}d\left(\psi^{B}\psi^{E}\right)$$

with components $\sigma^{\alpha}_{BA}e_{DE}$ with respect to the indices A, B, E. As a result, after simple transformations taking into account the third identity in (4.9) we obtain

$$C_{\alpha}\frac{\partial}{\partial t}\psi=\frac{\mathrm{i}}{2}\varepsilon_{\alpha\beta\lambda}\sigma^{\beta}\psi\frac{\partial}{\partial t}C^{\lambda}. \tag{4.85}$$

The contraction of relation (4.85) with complex conjugate components $\dot{C}^{\alpha}$ gives

$$\frac{\partial}{\partial t}\psi=-\frac{\mathrm{i}}{4\rho^{2}}\varepsilon_{\alpha\beta\lambda}\sigma^{\alpha}\psi\dot{C}^{\beta}\frac{\partial}{\partial t}C^{\lambda}. \tag{4.86}$$

Here it is taken into account that $C_{\alpha}\dot{C}^{\alpha}=2\rho^{2}$.

Formulas (4.78), (4.86) reduce the problem on the tensor formulation of differential spinor equations in three-dimensional space to simple algebraic operations.

Chapter 5
Tensor Forms of Differential Spinor Equations

5.1 Some Relativistically Invariant Equations

Let us consider the four-dimensional pseudo-Euclidean Minkowski space with the metric signature $(+, +, +, -)$ referred to an Cartesian coordinate system with the variables x^i and holonomic vector basis $\mathfrak{Э}_i$ $(i = 1, 2, 3, 4)$. The contravariant and covariant components of the metric tensor of the Minkowski space calculated in the coordinate system x^i are defined by the matrix

$$g_{ij} = g^{ij} = \begin{Vmatrix} 1 & 0 & 0 & 0 \\ 0 & 1 & 0 & 0 \\ 0 & 0 & 1 & 0 \\ 0 & 0 & 0 & -1 \end{Vmatrix}.$$

Consider the following differential equation which is written in the matrix form

$$G^j \partial_j \psi + F(\varkappa^{\mathcal{A}})\psi = 0. \tag{5.1}$$

Here ψ is the column from n unknown functions; $\partial_j = \partial/\partial x^j$. $F(\varkappa^{\mathcal{A}})$ and G^j are the given quadratic generally complex matrices of the order n; matrix $F(\varkappa^{\mathcal{A}})$ can be a function of the components of various tensors or spinors $\varkappa^{\mathcal{A}}$, which must be set or be defined with the aid of the additional equations.

Consider an arbitrary Lorentz transformation of the variables of the Cartesian coordinate system x^i to variables y^i also of the Cartesian coordinate system

$$y^j = b^j{}_i x^i, \quad x^j = l^j{}_i y^i. \tag{5.2}$$

Matrices $\|l^j{}_i\|$ and $\|b^j{}_i\|$ are mutual inverse $\|l^j{}_i\| = \|b^j{}_i\|^{-1}$.

V. A. Zhelnorovich, *Theory of Spinors and Its Application in Physics and Mechanics*, https://doi.org/10.1007/978-3-030-27836-6_5

Suppose that under the Lorentz transformation (5.2) the functions $\psi = \psi(x^i)$ appearing in Eq. (5.1) are transformed as follows

$$\psi'(y^i) = S\psi(x^i), \tag{5.3}$$

where the set of transformations S corresponding to all Lorentz transformations (5.2), forms a group realizing some representation of the transformation group (5.2), and transformations S do not depend on the variables of the coordinate system x^i.

Equation (5.1) is invariant under the Lorentz transformations (5.2), if Eq. (5.1) retains its form in the coordinate system with the variables y^i

$$G^j \frac{\partial}{\partial y^j}\psi' + F(\varkappa'^A)\psi' = 0. \tag{5.4}$$

Let us find conditions under which Eq. (5.1) is invariant under the Lorentz transformations (5.2). Replacing in Eqs. (5.1) the functions ψ according to equality (5.3) and the derivative ∂_i by the formula

$$\frac{\partial}{\partial x^j} = b^i{}_j \frac{\partial}{\partial y^i},$$

we write down Eq. (5.1) in the form

$$G^j b^i{}_j \frac{\partial}{\partial y^i}(S^{-1}\psi') + F(\varkappa^A)S^{-1}\psi' = 0.$$

Multiplying this equation from the left by some nondegenerate matrix V, in general case depending on transformation (5.3), we find

$$b^i{}_j V G^j S^{-1} \frac{\partial}{\partial y^i}\psi' + V F(\varkappa^A)S^{-1}\psi' = 0. \tag{5.5}$$

Comparing Eq. (5.5) with Eq. (5.4), we find that if the conditions are satisfied

$$V F(\varkappa^A)S^{-1} = F(\varkappa'^A), \quad b^i{}_j V G^j S^{-1} = G^i, \tag{5.6}$$

then Eq. (5.5) coincide with Eq. (5.4). Thus, if there is the transformation V such that equalities (5.6) are satisfied, then Eq. (5.1) are invariant under the Lorentz transformations (5.2).

It is easy to show that the set of transformations $\{V\}$ corresponding to all Lorentz transformations (5.2), is a group realizing some representation of the Lorentz group.

If Eq. (5.1) is invariant under transformations of the subgroup $\boldsymbol{L}^{\uparrow}$ of the Lorentz group, then are said that Eq. (5.1) is relativistically invariant.

For simplicity and having in mind the physical applications, we consider here a case, when the matrix F in Eqs. (5.1) has the form

$$F = \varkappa I + \mathrm{i}\varkappa_j G^j + \frac{\mathrm{i}}{2}\varkappa_{sj} G^{js} + \overset{*}{\varkappa}_j \overset{*}{G}{}^j + \overset{*}{\varkappa} G^5, \tag{5.7}$$

where generally complex coefficients $\varkappa$, $\varkappa_j$, $\varkappa_{ij}$, $\overset{*}{\varkappa}_j$, and $\overset{*}{\varkappa}$ form, respectively, the field of the scalar, vector, antisymmetric tensor of the second rank, pseudo-vector and pseudo-scalar in the Minkowski space. Matrices G^{js}, $\overset{*}{G}{}^j$, G^5 are defined by the equalities

$$G^5 = -\frac{1}{24}\varepsilon_{ijks} G^i G^j G^k G^s,$$

$$G^{js} = \frac{1}{2}(G^j G^s - G^s G^j), \quad \overset{*}{G}{}^j = -\frac{1}{6}\varepsilon^{jsmn} G_s G_m G_n.$$

If the matrix F is determined by Eq. (5.7), then the first equation in (5.6) takes the form

$$V\left(\varkappa I + \mathrm{i}\varkappa_j G^j + \frac{\mathrm{i}}{2}\varkappa_{sj} G^{js} + \overset{*}{\varkappa}_j \overset{*}{G}{}^j + \overset{*}{\varkappa} G^5\right) S^{-1}$$

$$= \varkappa' I + \mathrm{i}\varkappa'_j G^j + \frac{\mathrm{i}}{2}\varkappa'_{sj} G^{js} + \overset{*}{\varkappa}{}'_j \overset{*}{G}{}^j + \overset{*}{\varkappa}{}' G^5.$$

Replacing here the components $\varkappa$ by formulas of a tensor transformation

$$\varkappa = \varkappa', \quad \varkappa_j = b^i{}_j \varkappa'_i, \quad \varkappa_{js} = b^j{}_i b^s{}_m \varkappa'_{im},$$

$$\overset{*}{\varkappa}_j = \Delta\, b^i{}_j \overset{*}{\varkappa}{}'_i, \quad \overset{*}{\varkappa} = \Delta\, \overset{*}{\varkappa}{}',$$

($\Delta = \det \|b^j{}_i\|$), we obtain the system of equations

$$V S^{-1} = I, \quad b^j{}_i V G^i S^{-1} = G^j, \tag{5.8}$$

$$b^j{}_i b^s{}_m V G^{im} S^{-1} = G^{js}, \quad \Delta\, b^j{}_i V \overset{*}{G}{}^i S^{-1} = \overset{*}{G}{}^j, \quad \Delta V G^5 S^{-1} = G^5.$$

The first equation (5.8) holds if in matrix (5.7) coefficient $\varkappa$ is not equal to zero $\varkappa \neq 0$; in this case $V = S$. The second equation in (5.8) coincides with the second equation in (5.6), and the last three equations in (5.8) as it is easy to see, are a corollary of the second equation in (5.6).

Thus, the system of equations (5.8) at $\varkappa \neq 0$ is reduced to the equations

$$V = S, \quad b^j{}_i S G^i S^{-1} = G^j. \tag{5.9}$$

If $\varkappa = 0$, then the equation $V = S$ may not hold (for example, $V \neq S$ for the Weyl equations for the two-component spinor).

Let us write Eqs. (5.9) for the small transformations (5.2). Up to first-order small quantities we have

$$b^j{}_i = \delta_i^j - \delta\varepsilon_i{}^j, \quad S = I + \frac{1}{2}A^{ij}\delta\varepsilon_{ij}. \tag{5.10}$$

Here I is the unit matrix of the order n, $\delta\varepsilon_{ij} = -\delta\varepsilon_{ji}$ are small parameters, $\delta\varepsilon^j{}_i = g^{js}\delta\varepsilon_{si}$.

Inserting expressions (5.10) for $b^j{}_i$ and S into the second equation in (5.9), we rewrite the conditions of invariancy of Eqs. (5.1) in the form (up to first-order small quantities)

$$G^i A^{js} - A^{js}G^i - g^{ij}G^s + g^{si}G^j = 0. \tag{5.11}$$

Further we shall consider only the case when the square matrices G^i in Eqs. (5.1) are given and satisfy the equation

$$G^i G^j + G^j G^i = 2g^{ij} I. \tag{5.12}$$

As it was already noted (Chap. 1, Sect. 1.1), the order of the square matrices G^i satisfying Eq. (5.12) is equal to $4q$, $q \geqslant 1$ is a positive integer. Consider the case when the order of the matrices G^i is minimum and equal to 4. In this case equations (5.11) determine the infinitesimal operators A^{ij} up to an arbitrary unit matrix

$$A^{ij} = \frac{1}{4}(G^i G^j - G^j G^i) + a^{ij} I. \tag{5.13}$$

Here a^{ij} are arbitrary numbers, I is the unit four-dimensional matrix. The matrices A^{ij} (for $a^{ij} = 0$), defined by formulas (5.13), are infinitesimal operators of the spinor representation of the Lorentz group (see Eq. (3.44). Therefore, if the order of matrices G^i is equal to 4, then Eq. (5.1) is relativistically invariant, if ψ is the four-component spinor. This case in more details is considered in following sections.

If the order of the matrices G^i is greater than four, then a solution of Eqs. (5.11) for infinitesimal operators A^{ij} is essentially no unique, and there exist solutions of these equations that define both the spinor representations and the tensor ones. For example, if the order of matrices of G^i is equal to eight, then Eq. (5.1) is relativistically invariant, if ψ is an object consisting of two four-component spinors. An example of the relativistically invariant tensor equations which are written in the form of the matrix equations (5.1) with the eight-dimensional matrices G^i satisfying

Eqs. (5.12) are the equations

$$\partial_j\left(F^{ij}+Fg^{ij}\right)+\varkappa_j\left(F^{ij}+Fg^{ij}\right)-\overset{*}{\varkappa}_j\left(\frac{1}{2}\varepsilon^{ijks}F_{ks}+\overset{*}{F}g^{ij}\right)=0,$$

$$\partial_j\left(\frac{1}{2}\varepsilon^{ijks}F_{ks}+\overset{*}{F}g^{ij}\right)+\varkappa_j\left(\frac{1}{2}\varepsilon^{ijks}F_{ks}+\overset{*}{F}g^{ij}\right)+\overset{*}{\varkappa}_j\left(F^{ij}+Fg^{ij}\right)=0, \tag{5.14}$$

where F is a scalar, F^{ij} are the components of a second rank antisymmetric tensor, $\overset{*}{F}$ is a pseudo-scalar, ε^{ijks} are the components of the Levi-Civita pseudotensor.

Indeed, Eqs. (5.14) are written in the form of the equations

$$G^j\partial_j\psi+\left(\varkappa_jG^j+\overset{*}{\varkappa}_j\overset{*}{G}{}^j\right)\psi=0, \tag{5.15}$$

if we take as ψ the column of the components F, F^{12}, F^{31}, F^{23}, F^{14}, F^{24}, F^{34}, $\overset{*}{F}$ and as G^i the real matrices

$$G^1=\left\|\begin{array}{cccccccc} 0&-1&0&0&&&& \\ -1&0&0&0&&&& \\ 0&0&0&1&&&& \\ 0&0&1&0&&&& \\ &&&&0&-1&0&0 \\ &&&&-1&0&0&0 \\ &&&&0&0&0&1 \\ &&&&0&0&1&0 \end{array}\right\|,\quad G^2=\left\|\begin{array}{cccccccc} 1&0&0&0&&&& \\ 0&-1&0&0&&&& \\ 0&0&1&0&&&& \\ 0&0&0&-1&&&& \\ &&&&1&0&0&0 \\ &&&&0&-1&0&0 \\ &&&&0&0&1&0 \\ &&&&0&0&0&-1 \end{array}\right\|,$$

$$G^3=\left\|\begin{array}{cccccccc} 0&0&0&1&&&& \\ 0&0&1&0&&&& \\ 0&1&0&0&&&& \\ 1&0&0&0&&&& \\ &&&&0&0&0&-1 \\ &&&&0&0&-1&0 \\ &&&&0&-1&0&0 \\ &&&&-1&0&0&0 \end{array}\right\|,\quad G^4=\left\|\begin{array}{cccccccc} &&&&0&1&0&0 \\ &&&&-1&0&0&0 \\ &&&&0&0&0&-1 \\ &&&&0&0&1&0 \\ 0&1&0&0&&&& \\ -1&0&0&0&&&& \\ 0&0&0&-1&&&& \\ 0&0&1&0&&&& \end{array}\right\|. \tag{5.16}$$

Matrices (5.16) satisfy Eq. (5.12).

We note that if to put in Eqs. (5.14) $\varkappa_j=\overset{*}{\varkappa}_j=F=\overset{*}{F}=0$, then Eqs. (5.14) are written in the form of the Maxwell equations for a free electromagnetic field

$$\partial_jF^{ij}=0,\quad \varepsilon^{ijks}\partial_jF_{ks}=0.$$

Thus, the Maxwell equations may be written as the matrix equations

$$G^j \partial_j \psi = 0,$$

in which matrices G^i satisfying Eq. (5.12) are defined by equalities (5.16), while ψ is the column of components 0, F^{12}, F^{31}, F^{23}, F^{14}, F^{24}, F^{34}, 0.

It is possible that Eqs. (5.14) are of physical interest and with $F \neq 0$ and $\overset{*}{F} \neq 0$.

Another example of the relativistically invariant tensor equations written in the form of Eqs. (5.15) with the eight-dimensional matrices G^i satisfying Eq. (5.12), are the equations

$$\partial_k \big(- \varepsilon^{ijks} \overset{*}{F}_s + g^{ik} F^j - g^{jk} F^i \big) + \varkappa_k \big(- \varepsilon^{ijks} \overset{*}{F}_s + g^{ik} F^j - g^{jk} F^i \big) + \overset{*}{\varkappa}_k \big(g^{ik} \overset{*}{F}{}^j - g^{jk} \overset{*}{F}{}^i + \varepsilon^{ijks} F_s \big) = 0,$$

$$\partial_i F^i + \varkappa_i F^i + \overset{*}{\varkappa}_i \overset{*}{F}{}^i = 0, \quad \partial_i \overset{*}{F}{}^i + \varkappa_i \overset{*}{F}{}^i - \overset{*}{\varkappa}_i F^i = 0,$$

in which F^i are the components of a four-dimensional vector, $\overset{*}{F}{}^i$ are the components of a four-dimensional pseudo-vector. These equations can be written in the form of the matrix equations (5.15), if we take as ψ the column of the components F^2, $-F^1$, $\overset{*}{F}{}^4$, $\overset{*}{F}{}^3$, $-\overset{*}{F}{}^3$, F^4, $\overset{*}{F}{}^1$, $-\overset{*}{F}{}^2$ and as G^i the real matrices (5.16).

System of the relativistically invariant tensor equations which are written in the form (5.1) with the sixteen-dimensional matrices G^i satisfying Eqs. (5.12), is considered in [82].

5.2 Spinor Differential Equations in the Minkowski Space

Let us consider in the Minkowski space the following differential equations

$$\gamma^{Bj}{}_A \partial_j \psi^A + \left(\varkappa \delta^B_A + \mathrm{i} \varkappa_j \gamma^{Bj}{}_A + \frac{\mathrm{i}}{2} \varkappa_{js} \gamma^{Bjs}{}_A + \overset{*}{\varkappa}_j \overset{*}{\gamma}{}^{Bj}{}_A + \overset{*}{\varkappa} \gamma^{5B}{}_A \right) \psi^A = 0 \tag{5.17}$$

or, in the matrix form

$$\gamma^j \partial_j \psi + \left(\varkappa I + \mathrm{i} \varkappa_j \gamma^j + \frac{\mathrm{i}}{2} \varkappa_{js} \gamma^{js} + \overset{*}{\varkappa}_j \overset{*}{\gamma}{}^j + \overset{*}{\varkappa} \gamma^5 \right) \psi = 0. \tag{5.18}$$

Here $\psi = \psi(x^i)$ is the field of a four-component spinor in the Minkowski space, defined in the Cartesian coordinate system by four complex components $\psi^A =$

$\psi^A(x^i)$; $\gamma^j = \|\gamma^{Bj}_{\ \ A}\|$ are the four-dimensional Dirac matrices; I is the unit four-dimensional matrix; the matrices γ^{sj}, $\overset{*}{\gamma}{}^j$, γ^5 are defined by the equalities

$$\gamma^{js} = \gamma^{[j}\gamma^{s]}, \quad \overset{*}{\gamma}{}^j = -\frac{1}{6}\varepsilon^{jksm}\gamma_k\gamma_s\gamma_m,$$
$$\gamma^5 = -\frac{1}{24}\varepsilon^{ijks}\gamma_i\gamma_j\gamma_k\gamma_s.$$

The coefficients $\varkappa$, $\varkappa_j$, $\varkappa_{js}$, $\overset{*}{\varkappa}_j$, and $\overset{*}{\varkappa}$ in Eqs. (5.17) may be arbitrary real differentiable functions forming the fields of a scalar, a vector, an antisymmetric second rank tensor, a pseudo-vector and a pseudo-scalar, respectively. The quantities $\varkappa, \varkappa_j, \varkappa_{js}, \overset{*}{\varkappa}_j$, and $\overset{*}{\varkappa}$ may be the given functions or must be defined by the additional equations.

If $\varkappa$ = const, $\varkappa_j = \varkappa_{js} = \overset{*}{\varkappa}_j = \overset{*}{\varkappa} = 0$, then Eq. (5.17) is the Dirac equation used in the relativistic theory of electron

$$\gamma^j\partial_j\psi + m\psi = 0, \quad m = \text{const}\,. \tag{5.19}$$

If $\overset{*}{\varkappa}_j = \lambda S_j$, $\varkappa = \varkappa_j = \varkappa_{js} = \overset{*}{\varkappa} = 0$, where the components of the pseudo-vector S_j are determined by equality (3.59), λ is a constant, then Eq. (5.17) is the Heisenberg equation:

$$\gamma^j\partial_j\psi + \lambda S^j\overset{*}{\gamma}_j\psi = 0, \quad \lambda = \text{const.} \tag{5.20}$$

This equation was used in the nonlinear theory of elementary particles [17, 39].

Equations (5.17) are used further in mechanics of the magnetizable spin fluids (see Chap. 6).

Taking the complex conjugate of Eq. (5.18) and replacing matrices $\dot{\gamma}^j, \dot{\gamma}^{js}$, $(\overset{*}{\gamma}_i)^{\dot{}}$, and $\dot{\gamma}^5$ by formulas (3.20), we get

$$\Pi^{-1}\gamma^j\Pi\partial_j\dot{\psi} + \Pi^{-1}\left(\varkappa I - \mathrm{i}\varkappa_j\gamma^j - \frac{\mathrm{i}}{2}\varkappa_{js}\gamma^{js} + \overset{*}{\varkappa}_j\overset{*}{\gamma}{}^j + \overset{*}{\varkappa}\gamma^5\right)\Pi\dot{\psi} = 0. \tag{5.21}$$

Multiplying Eq. (5.21) from the left by Π and taking into account definition (3.48) of the contravariant components $\bar{\psi} = \Pi\dot{\psi} = \|\psi^{+A}\|$, we write down Eq. (5.21) in the form

$$\gamma^j\partial_j\bar{\psi} + \left(\varkappa I - \mathrm{i}\varkappa_j\gamma^j - \frac{\mathrm{i}}{2}\varkappa_{js}\gamma^{js} + \overset{*}{\varkappa}_j\overset{*}{\gamma}{}^j + \overset{*}{\varkappa}\gamma^5\right)\bar{\psi} = 0.$$

Taking the Hermitian conjugate of Eq. (5.18) and replacing in result the matrices $\dot{\gamma}_i^T$, $\dot{\gamma}_{ij}^T$, ...by formulas (3.17), (3.18), we get the equation for the covariant components of the conjugate spinor

$$-\partial_j\psi^+\gamma^j+\psi^+\left(\varkappa I+\mathrm{i}\varkappa_j\gamma^j+\frac{\mathrm{i}}{2}\varkappa_{js}\gamma^{js}+\overset{*}{\varkappa}_j\overset{*}{\gamma}^j+\overset{*}{\varkappa}\gamma^5\right)=0. \tag{5.22}$$

Here $\psi^+=\dot{\psi}^T\beta$ is the row of the covariant components of conjugate spinor ψ_A^+.

Equation (5.18) can be written in the two-dimensional matrix notations with the aid of the components of semispinors. For this purpose we multiply Eq. (5.18) by the matrix $i\gamma^5$. Taking into account relation (3.11) we get

$$-\mathrm{i}\gamma^j\partial_j(\gamma^5\psi)+\left(-\mathrm{i}\overset{*}{\varkappa}I+\mathrm{i}\overset{*}{\varkappa}_j\gamma^j-\frac{1}{2}\varkappa_{js}\gamma^{js}\gamma^5+\varkappa_j\gamma^j\gamma^5+\mathrm{i}\varkappa\gamma^5\right)\psi=0. \tag{5.23}$$

Adding and subtracting Eqs. (5.18) and (5.23), we find

$$\begin{aligned}\gamma^j\partial_j\big[(I-\mathrm{i}\gamma^5)\psi\big]+\Big[(\varkappa-\mathrm{i}\overset{*}{\varkappa})(I+\mathrm{i}\gamma^5)+\frac{\mathrm{i}}{2}\varkappa_{js}\gamma^{js}(I+\mathrm{i}\gamma^5)\\+\mathrm{i}(\varkappa_j+\overset{*}{\varkappa}_j)\gamma^j(I-\mathrm{i}\gamma^5)\Big]\psi=0,\\ \gamma^j\partial_j\big[(I+\mathrm{i}\gamma^5)\psi\big]+\Big[(\varkappa+\mathrm{i}\overset{*}{\varkappa})(I-\mathrm{i}\gamma^5)+\frac{\mathrm{i}}{2}\varkappa_{js}\gamma^{js}(I-\mathrm{i}\gamma^5)\\+\mathrm{i}(\varkappa_j-\overset{*}{\varkappa}_j)\gamma^j(I+\mathrm{i}\gamma^5)\Big]\psi=0.\end{aligned} \tag{5.24}$$

Determining the components of semispinors $\psi_{(I)}$ and $\psi_{(II)}$ by the formulas

$$\psi_{(I)}=\frac{1}{2}(I+\mathrm{i}\gamma^5)\psi,\quad \psi_{(II)}=\frac{1}{2}(I-\mathrm{i}\gamma^5)\psi,$$

we can write down Eq. (5.24) in the form

$$\begin{aligned}\gamma^j\partial_j\psi_{(I)}+\mathrm{i}(\varkappa_j-\overset{*}{\varkappa}_j)\gamma^j\psi_{(I)}+(\varkappa+\mathrm{i}\overset{*}{\varkappa})\psi_{(II)}+\frac{\mathrm{i}}{2}\varkappa_{js}\gamma^{js}\psi_{(II)}=0,\\ \gamma^j\partial_j\psi_{(II)}+\mathrm{i}(\varkappa_j+\overset{*}{\varkappa}_j)\gamma^j\psi_{(II)}+(\varkappa-\mathrm{i}\overset{*}{\varkappa})\psi_{(I)}+\frac{\mathrm{i}}{2}\varkappa_{js}\gamma^{js}\psi_{(I)}=0.\end{aligned} \tag{5.25}$$

If the components of spintensors γ^j are determined by matrices (3.24), then Eq. (5.25) can be written as follows

$$\begin{aligned}\frac{1}{c}\frac{\partial}{\partial t}\xi+\sigma^\alpha\partial_\alpha\xi+\mathrm{i}(\varkappa+\mathrm{i}\overset{*}{\varkappa})\eta+\mathrm{i}\big[(\varkappa_\alpha-\overset{*}{\varkappa}_\alpha)\sigma^\alpha+(\varkappa_4-\overset{*}{\varkappa}_4)I\big]\xi\\+\left(\varkappa^{4\alpha}-\frac{\mathrm{i}}{2}\varepsilon^{\alpha\lambda\theta}\varkappa_{\lambda\theta}\right)\sigma_\alpha\eta=0,\end{aligned}$$

$$\frac{1}{c}\frac{\partial}{\partial t}\eta - \sigma^{\alpha}\partial_{\alpha}\eta + \mathrm{i}(\varkappa - \mathrm{i}\overset{*}{\varkappa})\xi + \mathrm{i}\big[-(\varkappa_{\alpha} + \overset{*}{\varkappa}_{\alpha})\sigma^{\alpha} + (\varkappa_{4} + \overset{*}{\varkappa}_{4})I\big]\eta$$
$$- \left(\varkappa^{4\alpha} + \frac{\mathrm{i}}{2}\varepsilon^{\alpha\lambda\theta}\varkappa_{\lambda\theta}\right)\sigma_{\alpha}\xi = 0. \tag{5.26}$$

Here $\varepsilon^{\alpha\lambda\theta}$ are the components of the Levi-Civita three-dimensional pseudotensor; $\xi = \begin{Vmatrix}\xi^{1}\\ \xi^{2}\end{Vmatrix}$, $\eta = \begin{Vmatrix}\eta_{\dot{1}}\\ \eta_{\dot{2}}\end{Vmatrix}$ are the two-component spinors corresponding to semispinors $\psi_{(I)}$, $\psi_{(II)}$:

$$\psi_{(I)} = \begin{Vmatrix}\xi^{1}\\ \xi^{2}\\ 0\\ 0\end{Vmatrix}, \quad \psi_{(II)} = \begin{Vmatrix}0\\ 0\\ \eta_{\dot{1}}\\ \eta_{\dot{2}}\end{Vmatrix}.$$

It is useful to write down Eq. (5.26) also in the explicit invariant form

$$\mathrm{i}\sigma^{j}_{\dot{B}A}\partial_{j}\xi^{A} + (\varkappa + \mathrm{i}\overset{*}{\varkappa})\eta_{\dot{B}} - (\varkappa_{j} - \overset{*}{\varkappa}_{j})\sigma^{j}_{\dot{B}A}\xi^{A} + \frac{1}{2}\varkappa^{sj}\sigma^{\dot{A}}{}_{\dot{B}sj}\eta_{\dot{A}} = 0,$$
$$\mathrm{i}\sigma^{B\dot{A}j}\partial_{j}\eta_{\dot{A}} + (\varkappa - \mathrm{i}\overset{*}{\varkappa})\xi^{B} - (\varkappa_{j} + \overset{*}{\varkappa}_{j})\sigma^{B\dot{A}j}\eta_{\dot{A}} + \frac{1}{2}\varkappa^{sj}\sigma^{B}{}_{Asj}\xi^{A} = 0.$$

Here the components of spintensors $\sigma^{j}_{\dot{B}A}$, $\sigma^{B\dot{A}j}$, $\sigma^{\dot{A}}{}_{\dot{B}sj}$, $\sigma^{B}{}_{Asj}$ are defined by equalities (3.96), (3.97), (3.99).

In the sequel, in connection with Eqs. (5.18) we shall consider the quantities $\mathcal{P}_{i}{}^{j}$ determined by the functions ψ and ψ^{+}:

$$\mathcal{P}_{i}{}^{j} = \alpha(\psi^{+}\gamma^{j}\partial_{i}\psi - \partial_{i}\psi^{+}\cdot\gamma^{j}\psi) = -\alpha\gamma^{j}_{AB}(\psi^{+A}\partial_{i}\psi^{B} - \psi^{B}\partial_{i}\psi^{+A}). \tag{5.27}$$

Here α is some constant. It is obvious that the quantities $\mathcal{P}_{i}{}^{j}$ form the components of the second rank tensor in the Minkowski space.

Let us calculate the divergence of the components $\mathcal{P}_{i}{}^{j}$. We have

$$\partial_{j}\mathcal{P}_{i}{}^{j} = \alpha\big[\psi^{+}\partial_{i}(\gamma^{j}\partial_{j}\psi) - \partial_{i}(\partial_{j}\psi^{+}\gamma^{j})\psi + (\partial_{j}\psi^{+}\gamma^{j})\partial_{i}\psi - \partial_{i}\psi^{+}(\gamma^{j}\partial_{j}\psi)\big]. \tag{5.28}$$

The replacement of the terms $\gamma^{j}\partial_{j}\psi$ and $\partial_{j}\psi^{+}\gamma^{j}$ in (5.28) in accordance with Eqs. (5.18) and (5.22) gives the following equation

$$\partial_{j}\mathcal{P}_{i}{}^{j} = -2\alpha\psi^{+}\partial_{i}\left(\varkappa I + \mathrm{i}\varkappa^{j}\gamma_{j} + \frac{\mathrm{i}}{2}\varkappa^{js}\gamma_{js} + \overset{*}{\varkappa}{}^{j}\overset{*}{\gamma}_{j} + \overset{*}{\varkappa}\gamma^{5}\right)\cdot\psi. \tag{5.29}$$

Taking into account definitions (3.58) Eq. (5.29) it is possible to write in the form

$$\partial_j \mathcal{P}_i{}^j + 2\alpha\left(\Omega\,\partial_i \varkappa + j^j \partial_i \varkappa_j + \frac{1}{2} M^{sj} \partial_i \varkappa_{sj} + S^j \partial_i \overset{*}{\varkappa}_j + N \partial_i \overset{*}{\varkappa}\right) = 0. \tag{5.30}$$

If functions $\varkappa$, $\varkappa_j$, $\varkappa_{sj}$, $\overset{*}{\varkappa}_j$, $\overset{*}{\varkappa}$ are connected by the relation

$$2\alpha\left(\Omega\,\partial_i \varkappa + j^j \partial_i \varkappa_j + \frac{1}{2} M^{sj} \partial_i \varkappa_{sj} + S^j \partial_i \overset{*}{\varkappa}_j + N \partial_i \overset{*}{\varkappa}\right) = \partial_j N_i{}^j,$$

then Eq. (5.30) determines the conservation law

$$\partial_j \left(\mathcal{P}_i{}^j + N_i{}^j\right) = 0. \tag{5.31}$$

Further (in Sect. 5.6) it will be shown that due to Eqs. (5.18) is also carried out the equation

$$\begin{aligned} \mathcal{P}^{ij} - \mathcal{P}^{ji} + \alpha\varepsilon^{ijks} \partial_k S_s + 2\alpha\big(\varkappa^i j^j - \varkappa^j j^i \\ + \varkappa^i{}_s M^{js} - \varkappa^j{}_s M^{is} + \overset{*}{\varkappa}{}^i S^j - \overset{*}{\varkappa}{}^j S^i\big) = 0, \end{aligned} \tag{5.32}$$

in which $\mathcal{P}^{ij} = g^{ik} \mathcal{P}_k{}^j$. For the physical spin 1/2 fields described by the Dirac equations or the Heisenberg equations, the components

$$P_i{}^j = \mathcal{P}_i{}^j + N_i{}^j = \alpha\left(\psi^+ \gamma^j \partial_i \psi - \partial_i \psi^+ \cdot \gamma^j \psi\right) + N_i{}^j \tag{5.33}$$

define the four-dimensional energy-momentum tensor, Eq. (5.31) is the law of conservation of energy-momentum.

5.3 Representation of Spinor Equations as Tensor Equations for the Components of Vectors of the Proper Basis

In Sect. 3.4 of Chap. 3 it was shown that the spinor of the first rank $\boldsymbol{\psi}$ with a nonzero invariant ρ is completely determined by specifying four orthonormal vectors of the proper basis $\breve{\boldsymbol{e}}_a$ with components π^i, ξ^i, σ^i, u^i and by two invariants ρ, η. In this connection differential equations for the spinor field $\boldsymbol{\psi}$ with nonzero invariant ρ can be presented as tensor equations for the components of vectors of the proper tetrad $\breve{\boldsymbol{e}}_a$ and two scalars ρ, η.

Below we give two various methods of obtaining of tensor equations for the components of vectors of the proper basis $\breve{\boldsymbol{e}}_a$ and invariants ρ, η. The first method is somewhat more complicated, but it shows that the tensor equations corresponding to the spinor equations, represent simply a writing of the spinor equations in the proper orthonormal bases $\breve{\boldsymbol{e}}_a$, determined by the field $\boldsymbol{\psi}$. This circumstance is important in the methodological relation and, in the same time, saves from the need of the direct proof of equivalence the spinor and tensor equations. The second method brings quicker to result and it is associated with the use of ready-made formulas (3.203) expressing derivatives of spinor fields in terms of derivatives of the tensor fields.

Further in this section we assume that the spinor field in Eqs. (5.18) possesses a nonzero invariant $\rho \neq 0$.

1. The First Method To obtain tensor equations, corresponding to the spinor equations (5.18), we write down the spinor equations (5.18) in the proper bases $\breve{\boldsymbol{e}}_a$ determined by the spinor field $\boldsymbol{\psi}$ in accordance with equalities (3.128), (3.129), (3.126)

$$\gamma^a \breve{\nabla}_a \breve{\psi} + \left(\varkappa I + \mathrm{i}\breve{\varkappa}_a \gamma^a + \frac{\mathrm{i}}{2} \breve{\varkappa}_{ab} \gamma^{ab} + \breve{\varkappa}^*_a \overset{*}{\gamma}{}^a + \overset{*}{\varkappa} \gamma^5 \right) \breve{\psi} = 0. \tag{5.34}$$

Here γ^a, $\breve{\nabla}_a$, $\breve{\psi}$, and coefficients $\breve{\varkappa}$ are calculated in bases $\breve{\boldsymbol{e}}_a$. In particular, we have

$$\breve{\varkappa}_a = \breve{h}^i{}_a \varkappa_i, \quad \breve{\varkappa}_{ab} = \breve{h}^i{}_a \breve{h}^j{}_b \varkappa_{ij}, \quad \breve{\varkappa}^*_a = \breve{h}^i{}_a \overset{*}{\varkappa}_i.$$

By virtue of equalities (3.133) we find for coefficients $\breve{\varkappa}_a, \breve{\varkappa}_{ab}, \overset{*}{\varkappa}_a$

$$\begin{gathered}
\breve{\varkappa}_1 = \pi^i \varkappa_i, \quad \breve{\varkappa}_2 = \xi^i \varkappa_i, \quad \breve{\varkappa}_3 = \sigma^i \varkappa_i, \quad \breve{\varkappa}_4 = u^i \varkappa_i, \\
\breve{\varkappa}^*_1 = \pi^i \overset{*}{\varkappa}_i, \quad \breve{\varkappa}^*_2 = \xi^i \overset{*}{\varkappa}_i, \quad \breve{\varkappa}^*_3 = \sigma^i \overset{*}{\varkappa}_i, \quad \breve{\varkappa}^*_4 = u^i \overset{*}{\varkappa}_i, \\
\breve{\varkappa}_{12} = \pi^i \xi^j \varkappa_{ij}, \quad \breve{\varkappa}_{23} = \xi^i \sigma^j \varkappa_{ij}, \quad \breve{\varkappa}_{31} = \sigma^i \pi^j \varkappa_{ij}, \\
\breve{\varkappa}_{14} = \pi^i u^j \varkappa_{ij}, \quad \breve{\varkappa}_{24} = \xi^i u^j \varkappa_{ij}, \quad \breve{\varkappa}_{34} = \sigma^i u^j \varkappa_{ij}.
\end{gathered} \tag{5.35}$$

Since the spintensors $\boldsymbol{\gamma}$ are invariant under the restricted Lorentz transformations, the numerical values γ^a, γ^{ab}, $\overset{*}{\gamma}{}^a$ in the basis $\breve{\boldsymbol{e}}_a$ are the same as γ^i, γ^{ij}, $\overset{*}{\gamma}{}^i$ in the basis $\boldsymbol{\vartheta}_i$.

From definition of the covariant derivative of the spinor field (see Chap. 2) it follows

$$\breve{\nabla}_a = \breve{h}^i{}_a (\partial_i - \Gamma_i) = \breve{\partial}_a - \frac{1}{4} \breve{\Delta}_{a,bc} \gamma^{bc}. \tag{5.36}$$

The Ricci rotation coefficients $\breve{\Delta}_{a,bc}$ in this equality are expressed in terms of the components of vectors π^i, ξ, σ^i, u^i by the equalities (3.150). Γ_i are the spinor connection coefficients. $\breve{\partial}_a = \breve{h}^i{}_a \partial_i$ is the symbol of the derivative in the direction of vectors of the tetrad $\breve{\boldsymbol{e}}_a$. Due to definition (3.133) we have

$$\breve{\partial}_1 = \pi^i \partial_i, \quad \breve{\partial}_2 = \xi^i \partial_i, \quad \breve{\partial}_3 = \sigma^i \partial_i, \quad \breve{\partial}_4 = u^i \partial_i.$$

Let us transform the first term $\gamma^a \breve{\nabla}_a \breve{\psi}$ in Eq. (5.34). Using definition (5.36) and formulas (3.11) for the product $\gamma^a \gamma^{bc}$ it is easy to find

$$\gamma^a \breve{\nabla}_a \breve{\psi} = \left(\gamma^a \breve{\partial}_a - \frac{1}{4} \breve{\Delta}_{a,bc} \gamma^a \gamma^{bc}\right) \breve{\psi} = \left(\gamma^a \breve{\partial}_a + \frac{1}{2} \breve{\Delta}_{b,a}{}^b \gamma^a - \frac{1}{4} \varepsilon^{abcd} \breve{\Delta}_{b,cd} \overset{*}{\gamma}_a\right) \breve{\psi}.$$

We assume further that E and γ^a in Eqs. (5.34) are determined by matrices (3.24) and (3.25). In this case the components of spinor $\breve{\psi}$ are defined by the equality (3.144). Differentiating expression (3.144) we obtain the expression for derivatives $\breve{\partial}_a \breve{\psi}$

$$\breve{\partial}_a \breve{\psi} = \frac{1}{2} \left(I \breve{\partial}_a \ln \rho - \gamma^5 \breve{\partial}_a \eta\right) \breve{\psi}. \tag{5.37}$$

Substituting in the expression $\gamma^a \breve{\nabla}_a \breve{\psi}$ the derivative $\breve{\partial}_a \breve{\psi}$ according to formula (5.37), we obtain

$$\gamma^a \breve{\nabla}_a \breve{\psi} = \frac{1}{2} \left[\gamma^a \left(\breve{\partial}_a \ln \rho + \breve{\Delta}_{b,a}{}^b\right) - \overset{*}{\gamma}_a \left(\breve{\partial}^a \eta + \frac{1}{2} \varepsilon^{abcd} \breve{\Delta}_{b,cd}\right)\right] \breve{\psi}. \tag{5.38}$$

Using relation (5.38), we write Eq. (5.34) in the form

$$\gamma^a \left(\breve{\partial}_a \ln \rho + \breve{\Delta}_{b,a}{}^b\right) \breve{\psi} - \overset{*}{\gamma}_a \left(\breve{\partial}^a \eta + \frac{1}{2} \varepsilon^{abcd} \breve{\Delta}_{b,cd}\right) \breve{\psi}$$
$$+ 2 \left(\varkappa I + \mathrm{i} \breve{\varkappa}_a \gamma^a + \frac{\mathrm{i}}{2} \breve{\varkappa}_{ab} \gamma^{ab} + \breve{\varkappa}^*_a \overset{*}{\gamma}{}^a + \overset{*}{\varkappa} \gamma^5\right) \breve{\psi} = 0. \tag{5.39}$$

After equating the real and imaginary parts of Eq. (5.39) to zero and taking into account definitions (3.24), (3.144), and (5.35) we obtain a system of eight real equations

$$\breve{\partial}_4 \ln \rho - \breve{\Delta}_{1,14} - \breve{\Delta}_{2,24} - \breve{\Delta}_{3,34} = 0,$$
$$\breve{\partial}_1 \ln \rho + \breve{\Delta}_{2,12} - \breve{\Delta}_{3,31} - \breve{\Delta}_{4,14} = 2\big[-\breve{\xi}^a \breve{\varkappa}_a - \breve{\pi}^a \breve{u}^b \big(\breve{\varkappa}_{ab} \cos \eta - \breve{\varkappa}^*_{ab} \sin \eta\big)\big],$$
$$\breve{\partial}_2 \ln \rho - \breve{\Delta}_{1,12} + \breve{\Delta}_{3,23} - \breve{\Delta}_{4,24} = 2\big[\breve{\pi}^a \breve{\varkappa}_a - \breve{\xi}^a \breve{u}^b \big(\breve{\varkappa}_{ab} \cos \eta - \breve{\varkappa}^*_{ab} \sin \eta\big)\big],$$
$$\breve{\partial}_3 \ln \rho + \breve{\Delta}_{1,31} - \breve{\Delta}_{2,23} - \breve{\Delta}_{4,34} = 2\big[\varkappa \sin \eta - \overset{*}{\varkappa} \cos \eta$$
$$- \breve{\sigma}^a \breve{u}^b \big(\breve{\varkappa}_{ab} \cos \eta - \breve{\varkappa}^*_{ab} \sin \eta\big)\big],$$
$$\breve{\partial}^4 \eta + \breve{\Delta}_{1,23} + \breve{\Delta}_{2,31} + \breve{\Delta}_{3,12} = -2\big(\breve{u}^a \breve{\varkappa}^*_a + \breve{\sigma}^a \breve{\varkappa}_a\big),$$
$$\breve{\partial}^1 \eta - \breve{\Delta}_{2,34} + \breve{\Delta}_{3,24} - \breve{\Delta}_{4,23} = 2\big[\breve{\pi}^a \breve{\varkappa}^*_a + \breve{\pi}^a \breve{u}^b \big(\breve{\varkappa}_{ab} \sin \eta + \breve{\varkappa}^*_{ab} \cos \eta\big)\big],$$
$$\breve{\partial}^2 \eta - \breve{\Delta}_{3,14} + \breve{\Delta}_{1,34} - \breve{\Delta}_{4,31} = 2\big[\breve{\xi}^a \breve{\varkappa}^*_a + \breve{\xi}^a \breve{u}^b \big(\breve{\varkappa}_{ab} \sin \eta + \breve{\varkappa}^*_{ab} \cos \eta\big)\big],$$

$$\breve{\partial}^3\eta - \breve{\Delta}_{1,24} + \breve{\Delta}_{2,14} - \breve{\Delta}_{4,12} = 2\big[\varkappa \cos\eta - \overset{*}{\varkappa}\sin\eta + \breve{u}^a\breve{\varkappa}_a + \breve{\sigma}^a\breve{\varkappa}^*_a + \breve{\sigma}^a\breve{u}^b\big(\breve{\varkappa}_{ab}\sin\eta + \breve{\varkappa}^*_{ab}\cos\eta\big)\big],$$

which can be written in the compact form

$$\breve{\partial}_a \ln\rho + \breve{\Delta}_{b,a}{}^{b} + \breve{M}_a = 0,$$
$$\breve{\partial}^a\eta + \frac{1}{2}\varepsilon^{abcd}\breve{\Delta}_{b,cd} + \breve{N}^a = 0. \tag{5.40}$$

The components of vectors $\breve{M}_a$, $\breve{N}^a$ in Eqs. (5.40) are defined by the relations

$$\breve{M}_a = 2\big[\breve{\sigma}_a(\overset{*}{\varkappa}\cos\eta - \varkappa\sin\eta) + \big(\breve{\pi}_a\breve{\xi}^b - \breve{\pi}^b\breve{\xi}_a\big)\breve{\varkappa}_b + \breve{u}^b\big(\breve{\varkappa}_{ab}\cos\eta - \breve{\varkappa}^*_{ab}\sin\eta\big)\big],$$
$$\breve{N}^a = 2\big[-\breve{\varkappa}^{*a} - \breve{\sigma}^a(\varkappa\cos\eta + \overset{*}{\varkappa}\sin\eta) + \big(\breve{u}^a\breve{\sigma}^b - \breve{u}^b\breve{\sigma}^a\big)\breve{\varkappa}_b - \breve{u}_b\big(\breve{\varkappa}^{ab}\sin\eta + \breve{\varkappa}^{*ab}\cos\eta\big)\big], \tag{5.41}$$

where $\breve{\varkappa}^{*ab} = \frac{1}{2}\varepsilon^{abcd}\,\breve{\varkappa}_{cd}$. Equations (5.40) are invariant tensor equations.

From the adduced derivation it follows that Eqs. (5.40) represent the writing of Eqs. (5.18) in basis $\breve{\boldsymbol{e}}_a$ and therefore are equivalent to them.

2. *The Second Method* Replacing in Eqs. (5.18) the derivatives $\partial_i\psi$ by formula (3.203a), we obtain

$$\frac{1}{2}\gamma^i\left(I\partial_i\ln\rho - \gamma^5\partial_i\eta - \frac{1}{2}\breve{\Delta}_{i,sj}\gamma^{sj}\right)\psi + \left(\varkappa I + \mathrm{i}\varkappa_j\gamma^j + \frac{\mathrm{i}}{2}\varkappa_{sj}\gamma^{sj} + \overset{*}{\varkappa}_j\overset{*}{\gamma}{}^j + \overset{*}{\varkappa}\gamma^5\right)\psi = 0. \tag{5.42}$$

By means of identities (see (3.11))

$$\gamma^i\gamma^{sj} = -\varepsilon^{isjm}\overset{*}{\gamma}_m + g^{is}\gamma^j - g^{ij}\gamma^s, \quad \gamma^i\gamma^5 = \overset{*}{\gamma}{}^i$$

Eq. (5.42) can be transformed to the form

$$\left(\partial_i\ln\rho + \breve{\Delta}_{j,i}{}^{j}\right)\gamma^i\psi - \left(\partial^i\eta + \frac{1}{2}\varepsilon^{ijmn}\breve{\Delta}_{j,mn}\right)\overset{*}{\gamma}_i\psi + 2\left(\varkappa I + \mathrm{i}\varkappa_j\gamma^j + \frac{\mathrm{i}}{2}\varkappa_{sj}\gamma^{sj} + \overset{*}{\varkappa}_j\overset{*}{\gamma}{}^j + \overset{*}{\varkappa}\gamma^5\right)\psi = 0. \tag{5.43}$$

Let us multiply Eq. (5.43) from the left by the spinor $\psi^+\left(\Omega\gamma^j - N\overset{*}{\gamma}{}^j\right)$ and from the right by the spinor $\psi^+\left(N\gamma^j + \Omega\overset{*}{\gamma}{}^j\right)$. As a result after equating the real and imaginary parts of Eq. (5.43) to zero we obtain the equations for the Ricci rotation coefficients $\breve{\Delta}_{i,j}{}^s$ of the proper tetrads $\breve{\boldsymbol{e}}_a$ and the invariants ρ, η:

$$\begin{aligned} \partial_i \ln\rho + \breve{\Delta}_{j,i}{}^j + M_i &= 0, \\ \partial^i\eta + \frac{1}{2}\varepsilon^{ijmn}\breve{\Delta}_{j,mn} + N^i &= 0. \end{aligned} \tag{5.44}$$

Here $M_i = \breve{h}_i{}^a\breve{M}_a$, $N^i = \breve{h}^i{}_a\breve{N}^a$; the quantities $\breve{M}_a$ and $\breve{N}^a$ are determined by equalities (5.41); $\breve{h}_i{}^a$ are the scale factors (3.133), determining proper tetrads $\breve{\boldsymbol{e}}_a$ of the spinor field $\psi(x^i)$.

By replacing Ricci rotation coefficients $\breve{\Delta}_{i,js}$ in Eqs. (5.44) by formula (3.147), after identical transformations we obtain the following system of equations

$$\begin{aligned} \partial_i \ln\rho + \pi_i\partial_j\pi^j + \xi_i\partial_j\xi^j + \sigma_i\partial_j\sigma^j - u_i\partial_j u^j + M_i &= 0, \\ \partial^i\eta - \frac{1}{2}\varepsilon^{ijms}\left(\pi_j\partial_m\pi_s + \xi_j\partial_m\xi_s + \sigma_j\partial_m\sigma_s - u_j\partial_m u_s\right) + N^i &= 0. \end{aligned} \tag{5.45}$$

Contracting the first equation in (5.45) with components of vectors π^i, ξ^i, σ^i, u^i, we get the four scalar equations that are linear in derivatives

$$\begin{aligned} \partial_i\rho\pi^i + \rho\pi^i M_i = 0, \quad \partial_i\rho\xi^i + \rho\xi^i M_i = 0, \\ \partial_i\rho\sigma^i + \rho\sigma^i M_i = 0, \quad \partial_i\rho u^i = 0. \end{aligned} \tag{5.46}$$

Here we take into account that due to definition (5.41) the equality $u^i M_i = 0$ is fulfilled. It is obvious that the first equation in (5.45) is equivalent to Eqs. (5.46).

Let us write out the obtained tensor equations (5.44) and Eqs. (5.45), (5.46) corresponding to the Dirac equation (5.19) and the Heisenberg equation (5.20). Equations (5.44) corresponding to the Dirac equation, have the form

$$\begin{aligned} \partial_i \ln\rho + \breve{\Delta}_{j,i}{}^j &= 2m\sigma_i \sin\eta, \\ \partial^i\eta + \frac{1}{2}\varepsilon^{ijmn}\breve{\Delta}_{j,mn} &= 2m\sigma^i\cos\eta. \end{aligned}$$

The complete system of equations (5.45), (5.46) corresponding to the Dirac equation, are written as follows

$$\begin{aligned} \partial_i\rho\pi^i = 0, \quad \partial_i\rho\xi^i = 0, \quad \partial_i\rho\sigma^i = 2m\rho\sin\eta, \quad \partial_i\rho u^i = 0, \\ \partial^i\eta - \frac{1}{2}\varepsilon^{ijms}\left(\pi_j\partial_m\pi_s + \xi_j\partial_m\xi_s + \sigma_j\partial_m\sigma_s - u_j\partial_m u_s\right) = 2m\sigma^i\cos\eta. \end{aligned}$$

To the spinor Heisenberg equations (5.20) there corresponds the following tensor equations in components ρ, η, and $\breve{\Delta}_{i,jk}$:

$$\partial_i \ln \rho + \breve{\Delta}_{j,i}{}^{j} = 0,$$

$$\partial^i \eta + \frac{1}{2}\varepsilon^{ijmn}\breve{\Delta}_{j,mn} = 2\lambda\rho\sigma^i$$

and the equations in the vectors components π^i , ξ^i , σ^i, u^i of the proper tetrad of the spinor field:

$$\partial_i \rho\pi^i = 0, \quad \partial_i \rho\xi^i = 0, \quad \partial_i \rho\sigma^i = 0, \quad \partial_i \rho u^i = 0,$$

$$\partial^i \eta - \frac{1}{2}\varepsilon^{ijms}\left(\pi_j \partial_m \pi_s + \xi_j \partial_m \xi_s + \sigma_j \partial_m \sigma_s - u_j \partial_m u_s\right) = 2\lambda\rho\sigma^i.$$

5.4 Representation of Spinor Equations as Tensor Equations for the Components of Vectors of the Complex Triad

In Sect. 3.5 of Chap. 3 it was shown that the spinor of the first rank in the pseudo-Euclidean space E_4^1 is completely defined by two invariants Ω, N and three complex three-dimensional orthonormal vectors (complex triad) with components α^a, β^a, λ^a. Therefore the spinor differential equations in the Minkowski space can be written as tensor equations on the invariants Ω, N and the complex components of vectors α^a, β^a, λ^a [88]. Here the complete system relativistically invariant tensor equations in components of vectors of complex triad and invariants Ω, N corresponding to the spinor equations (5.17) is established.

To obtain such equations we consider the system of the tensor equations (5.44), corresponding to the spinor equations (5.17). It is easy to see that Eqs. (5.44) can be written in the form of the single complex vector equation

$$\partial_i \ln(\Omega + \mathrm{i}N) + \breve{\Delta}_{j,i}{}^{j} + \frac{\mathrm{i}}{2}\varepsilon_{ijmn}\breve{\Delta}^{j,mn} + G_i = 0, \tag{5.47}$$

where $\Omega + \mathrm{i}N = \rho \exp \mathrm{i}\eta$ and

$$G_i = M_i + \mathrm{i}N_i = 2\big\{ - \mathrm{i}\overset{*}{\varkappa}_i + (\overset{*}{\varkappa} - \mathrm{i}\varkappa)\sigma_i e^{-\mathrm{i}\eta}$$
$$+ \left[\pi_i\xi^j - \pi^j\xi_i + \mathrm{i}\left(u_i\sigma^j - u^j\sigma_i\right)\right]\varkappa_j + u^j\left(\varkappa_{ij} - \mathrm{i}\overset{*}{\varkappa}_{ij}\right)e^{-\mathrm{i}\eta}\big\}.$$

Replacing in Eq. (5.47) the Ricci rotation coefficients by formula (3.182), we write Eq. (5.47) in the form of the following invariant tensor equation which is equivalent to Eq. (5.47):

$$\partial_i \ln(\Omega + \mathrm{i}N) + \frac{1}{2}\left(\alpha_{in}\partial_j\alpha^{nj} + \beta_{in}\partial_j\beta^{nj} + \lambda_{in}\partial_j\lambda^{nj}\right) + G_i = 0. \tag{5.48}$$

Let us contract equation (5.48) with components of tensor α^{si} with respect to the index i. Taking into account relations (3.178) and (3.179), we get

$$\alpha^{si}\partial_i \ln(\Omega + \mathrm{i}N) + \frac{1}{2}\left(\partial_j\alpha^{sj} - \mathrm{i}\lambda^s{}_n\partial_j\beta^{nj} + \mathrm{i}\beta^s{}_n\partial_j\lambda^{nj}\right) + \alpha^{si}G_i = 0. \tag{5.49}$$

Since due to identities (3.178), (3.179) the equality is fulfilled

$$-\mathrm{i}\lambda^s{}_n\partial_j\beta^{nj} = -\mathrm{i}\partial_j\left(\lambda^s{}_n\beta^{nj}\right) + \mathrm{i}\beta^{nj}\partial_j\lambda^s{}_n = \partial_j\alpha^{sj} + \mathrm{i}\beta^{nj}\partial_j\lambda^s{}_n,$$

Eq. (5.49) can be transformed to the form

$$\alpha^{si}\partial_i \ln(\Omega + \mathrm{i}N) + \partial_j\alpha^{sj} + \frac{\mathrm{i}}{2}\left(\beta^s{}_n\partial_j\lambda^{nj} + \beta^{nj}\partial_j\lambda^s{}_n\right) + \alpha^{si}G_i = 0. \tag{5.50}$$

Bearing in mind that by virtue of definitions (3.177) the identity is carried out

$$\beta^s{}_n\partial_j\lambda^{nj} + \beta^{nj}\partial_j\lambda^s{}_n = -\frac{1}{2}\beta_{ij}\partial^s\lambda^{ij},$$

which is checked directly, we finally write down Eq. (5.50) in the form

$$\frac{1}{\Omega + \mathrm{i}N}\partial_j[(\Omega + \mathrm{i}N)\alpha^{sj}] - \frac{\mathrm{i}}{4}\beta_{ij}\partial^s\lambda^{ij} + \alpha^{si}G_i = 0. \tag{5.51}$$

Contracting equation (5.48) with components β^{si} and components λ^{si} with respect to the index i, similar to the derivation (5.49)–(5.51) we can get the following equations

$$\begin{aligned}
&\frac{1}{\Omega + \mathrm{i}N}\partial_j[(\Omega + \mathrm{i}N)\beta^{sj}] - \frac{\mathrm{i}}{4}\lambda_{ij}\partial^s\alpha^{ij} + \beta^{si}G_i = 0,\\
&\frac{1}{\Omega + \mathrm{i}N}\partial_j[(\Omega + \mathrm{i}N)\lambda^{sj}] - \frac{\mathrm{i}}{4}\alpha_{ij}\partial^s\beta^{ij} + \lambda^{si}G_i = 0.
\end{aligned} \tag{5.52}$$

The determinants of the matrices of the component α^{si}, β^{si}, λ^{si} are equal to -1 (see equalities (3.180)), therefore the transition from Eqs. (5.48) to (5.51) and (5.52) is nondegenerate.

Let us replace in Eqs. (5.51) and (5.52) the components of the four-dimensional tensors α^{si}, β^{si}, λ^{si} in terms of the components of the three-dimensional vectors α^{μ}, β^{μ}, λ^{μ} in accordance with definitions (3.183). As a result Eq. (5.51) take the form

$$-\frac{1}{\Omega+\mathrm{i}N}\partial_\mu\big[(\Omega+\mathrm{i}N)\alpha^\mu\big]+\mathrm{i}\beta_\mu\partial^4\lambda^\mu-\alpha^\mu G_\mu=0, \tag{5.53}$$

$$\frac{1}{\Omega+\mathrm{i}N}\big[\partial_4(\Omega+\mathrm{i}N)\alpha^\mu+\mathrm{i}\varepsilon^{\mu\nu\theta}\partial_\nu(\Omega+\mathrm{i}N)\alpha_\theta\big]+\mathrm{i}\beta_\nu\partial^\mu\lambda^\nu+\alpha^{\mu j}G_j=0.$$

The first equation in (5.52) is written as follows

$$-\frac{1}{\Omega+\mathrm{i}N}\partial_\mu\big[(\Omega+\mathrm{i}N)\beta^\mu\big]+\mathrm{i}\lambda_\mu\partial^4\alpha^\mu-\beta^\mu G_\mu=0, \tag{5.54}$$

$$\frac{1}{\Omega+\mathrm{i}N}\big[\partial_4(\Omega+\mathrm{i}N)\beta^\mu+\mathrm{i}\varepsilon^{\mu\nu\theta}\partial_\nu(\Omega+\mathrm{i}N)\beta_\theta\big]+\mathrm{i}\lambda_\nu\partial^\mu\alpha^\nu+\beta^{\mu j}G_j=0.$$

For the second equation in (5.52) we have

$$-\frac{1}{\Omega+\mathrm{i}N}\partial_\mu\big[(\Omega+\mathrm{i}N)\lambda^\mu\big]+\mathrm{i}\alpha_\mu\partial^4\beta^\mu-\lambda^\mu G_\mu=0, \tag{5.55}$$

$$\frac{1}{\Omega+\mathrm{i}N}\big[\partial_4(\Omega+\mathrm{i}N)\lambda^\mu+\mathrm{i}\varepsilon^{\mu\nu\theta}\partial_\nu(\Omega+\mathrm{i}N)\lambda_\theta\big]+\mathrm{i}\alpha_\nu\partial^\mu\beta^\nu+\lambda^{\mu j}G_j=0.$$

In Eqs. (5.53)–(5.55) $\varepsilon^{\mu\nu\theta}$ are the components of the pseudotensor Levi-Civita. The first equations in (5.53)–(5.55) correspond to Eqs. (5.48), (5.51), (5.52) for s =4; the second equations in (5.53)–(5.55) correspond to Eqs. (5.48), (5.51), (5.52) for $s=1,2,3$.[1]

From the adduced derivation it is clear that Eqs. (5.48) or the each equation in (5.51), (5.52) form the complete system of the relativistically invariant differential equations.

5.5 Expression of the Tensor Components $\mathcal{P}_i{}^j$ in Terms of Components of the Real and Complex Tensors

In this section we get an expression of the components of tensor $\mathcal{P}_i{}^j$, determined by equalities (5.27), in terms of the components of the various real and complex tensors defined by spinor $\boldsymbol{\psi}$. For writing of the quantities $\mathcal{P}_i{}^j$ in the components of tensors $\boldsymbol{C}$, $\boldsymbol{D}$ we multiply equality (5.27) by $\psi^{+D}\psi^E$ and replace the terms $\psi^{+D}\psi^E(\psi^{+A}\partial_i\psi^B-\psi^B\partial_i\psi^{+A})$ in the right-hand side of the obtained equality

[1] The second equations in (5.53)–(5.55) with some special coefficients $\varkappa$ are obtained in a different way in [57]. Thus, the equations in [57] are the spatial part of the four-dimensional relativistically invariant vector equations (5.51), (5.52).

according to the obvious identity

$$\psi^{+D}\psi^{E}(\psi^{+A}\partial_i\psi^{B}-\psi^{B}\partial_i\psi^{+A})\equiv\psi^{+D}\psi^{+A}\partial_i(\psi^{B}\psi^{E})-\psi^{+D}\psi^{B}\partial_i(\psi^{+A}\psi^{E}).$$

As result we get

$$\psi^{+D}\psi^{E}\mathcal{P}_i{}^{j}=-\alpha\gamma^{j}_{AB}\big[\psi^{+D}\psi^{+A}\partial_i(\psi^{B}\psi^{E})-\psi^{+D}\psi^{B}\partial_i(\psi^{+A}\psi^{E})\big]. \qquad (5.56)$$

Contracting equality (5.56) with components of the invariant spintensors e_{DE}, $\overset{5}{\gamma}_{DE}$, and $\overset{s}{\gamma}_{DE}$ with respect to the indices D, E, after transformations by means of the identities (C.1) we get the following expressions for the components $\mathcal{P}_i{}^{j}$:

$$\begin{aligned}
\Omega\mathcal{P}_i{}^{j}&=\frac{\alpha}{2}\Big[-\partial_i(NS^{j})+M^{sj}\partial_i j_s-S^{j}\partial_i N-\frac{1}{2}(\dot{C}^{sj}\partial_i C_s+C^{sj}\partial_i\dot{C}^{s})\Big],\\
N\mathcal{P}_i{}^{j}&=\frac{\alpha}{2}\Big[\partial_i(\Omega S^{j})+S^{j}\partial_i\Omega\\
&\qquad+\frac{1}{2}\varepsilon^{jksm}\Big(-M_{ks}\partial_i j_m+\frac{1}{2}C_{ks}\partial_i\dot{C}_m+\frac{1}{2}\dot{C}_{ks}\partial_i C_m\Big)\Big],\\
j^{s}\mathcal{P}_i{}^{j}&=\frac{\alpha}{2}\Big\{-M^{sj}\partial_i\Omega-\frac{1}{2}\varepsilon^{sjkl}M_{kl}\partial_i N+\varepsilon^{sjkl}S_k\partial_i j_l+\frac{\mathrm{i}}{4}\Big[g^{sj}\Big(\dot{C}_k\partial_i C^{k}\\
&\qquad-C_k\partial_i\dot{C}^{k}+\frac{1}{2}\dot{C}_{kl}\partial_i C^{kl}-\frac{1}{2}C_{kl}\partial_i\dot{C}^{kl}\Big)+C^{s}\partial_i\dot{C}^{j}-\dot{C}^{s}\partial_i C^{j}\\
&\qquad+C^{j}\partial_i\dot{C}^{s}-\dot{C}^{j}\partial_i C^{s}+\dot{C}^{sk}\partial_i C_k{}^{j}-C^{sk}\partial_i\dot{C}_k{}^{j}+\dot{C}_k{}^{j}\partial_i C^{sk}\\
&\qquad-C_k{}^{j}\partial_i\dot{C}^{sk}\Big]\Big\}.
\end{aligned} \qquad (5.57)$$

We note the identity following from the last equality in (5.57):

$$j^{s}\mathcal{P}_i{}^{j}-j^{j}\mathcal{P}_i{}^{s}\equiv\alpha\Big(-M^{sj}\partial_i\Omega-\frac{1}{2}\varepsilon^{sjkm}M_{km}\partial_i N+\varepsilon^{sjkm}S_k\partial_i j_m\Big). \qquad (5.58)$$

If $\rho^2=\Omega^2+N^2\neq 0$, then adding the first equation in (5.57), multiplied by Ω/ρ, with the second equation in (5.57) multiplied by N/ρ, we get for components $\mathcal{P}_i{}^{j}$

$$\mathcal{P}_i{}^{j}=\alpha\Big[-S^{j}\partial_i\eta-\frac{1}{2}\mu^{js}\partial_i u_s+\frac{1}{4}(\dot{Z}^{js}\partial_i Z_s+Z^{js}\partial_i\dot{Z}_s)\Big]. \qquad (5.59)$$

The components u_s, μ^{js}, η, Z_s, Z^{js} in (5.59) are determined by equalities (3.66), (3.67). Using relation (5.59), for the function $\overset{\circ}{\Lambda}$ determined by the equality

$$\overset{\circ}{\Lambda} = \mathcal{P}_i{}^i \equiv \alpha\big(\psi^+\gamma^i\partial_i\psi - \partial_i\psi^+ \cdot \gamma^i\psi\big), \tag{5.60}$$

we obtain[2]

$$\overset{\circ}{\Lambda} = \alpha\left[-S^i\partial_i\eta - \frac{1}{2}\mu^{ij}\partial_i u_j + \frac{1}{4}\big(\dot{Z}^{ij}\partial_i Z_j + Z^{ij}\partial_i\dot{Z}_j\big)\right]. \tag{5.61}$$

Let us give now an expression for the components of tensor $\mathcal{P}_i{}^j$ only in terms of the components of the real tensors Ω, j^i, M^{ij}, S^i, N.

It is easy to see that the following identity is valid

$$\begin{aligned}\psi^{+D}\psi^E(\psi^{+B}d\psi^A - \psi^A d\psi^{+B}) - \psi^{+B}\psi^A(\psi^{+D}d\psi^E - \psi^E d\psi^{+D})\\ \equiv \psi^E\psi^{+B}d(\psi^{+D}\psi^A) - \psi^{+D}\psi^A d(\psi^{+B}\psi^E),\end{aligned} \tag{5.62}$$

which is carried out for any differentiable functions $\psi^A(x^i)$ and $\psi^{+A}(x^i)$.

Using identity (5.62), one can write for the components $\mathcal{P}_i{}^j$

$$\begin{aligned}\psi^{+D}\psi^E\mathcal{P}_i{}^j = -\alpha\gamma^j_{AB}\big[\psi^E\psi^{+A}\partial_i\big(\psi^{+D}\psi^B\big) - \psi^{+D}\psi^B\partial_i\big(\psi^E\psi^{+A}\big)\big]\\ - \mathrm{i}\alpha j^j\big(\psi^{+D}\partial_i\psi^E - \psi^E\partial_i\psi^{+D}\big).\end{aligned} \tag{5.63}$$

Let us now contract equation (5.63) with components of invariant spintensors e_{DE} and γ^5_{DE} with respect to the indices D, E. As a result of transformations with

---

[2]For the spinor equations of type (5.18), describing fields with half-integral spin (the Dirac equations, the Heisenberg equation, etc.), function $\Lambda = \overset{\circ}{\Lambda} + f(\psi, \psi^+)$ with the corresponding choice of an algebraic function f, is the Lagrangian. Formula (5.61) shows that the Lagrangian, describing fields of the half-integer spin, within the framework of known classical theories is represented in the form of sum $\Lambda = \Lambda_1 + \Lambda_2 + \Lambda_3 + \Lambda_{\text{int}}$, where

$$\Lambda_1 = -\alpha S^i\partial_i\eta + \frac{1}{2}\alpha\big(-S_iS^i + m_1^2\eta^2\big),$$

$$\Lambda_2 = -\frac{1}{2}\alpha\mu^{ij}\partial_i u_j + \frac{1}{4}\alpha\left(-\frac{1}{2}\mu_{ij}\mu^{ij} + m_2^2 u_i u^i\right),$$

$$\Lambda_3 = -\frac{1}{4}\alpha\big(\dot{Z}^{ij}\partial_i Z_j + Z^{ij}\partial_i\dot{Z}_j\big) + \frac{1}{4}\alpha\left(-\frac{1}{2}\dot{Z}_{ij}Z^{ij} + m_3^2\dot{Z}_iZ^i\right).$$

Here m_1, m_2, m_3 are constants. Functions Λ_1, Λ_2, Λ_3 with arbitrary quantities S_i, u_s, μ^{js}, η, Z_s, Z^{js} are the Lagrangians for the Proca equations describing, respectively, a neutral field of the spin 0, a neutral field of the spin 1 and a charged field of the spin 1. Function Λ_{int} does not depend on derivatives of tensor fields. A possible physical interpretation of this equality is considered in [76, 80].

the aid identities (C.1) we obtain

$$\varOmega \mathcal{P}_i{}^j = \alpha\Big[\mathrm{i} j^j e_{DE}\left(\psi^{+D}\partial_i\psi^E - \psi^E\partial_i\psi^{+D}\right) - S^j\partial_i N - M^{js}\partial_i j_s\Big], \qquad (5.64)$$

$$N\mathcal{P}_i{}^j = \alpha\Big[\mathrm{i} j^j \gamma^5_{DE}\left(\psi^{+D}\partial_i\psi^E - \psi^E\partial_i\psi^{+D}\right) + S^j\partial_i\varOmega - \frac{1}{2}\varepsilon^{jksq}M_{ks}\partial_i j_q\Big].$$

For a further transformation of expressions (5.64) we notice that from Eqs. (5.18) it follows[3]

$$\frac{\mathrm{i}}{2}e_{DE}(\psi^{+D}\partial_i\psi^E - \psi^E\partial_i\psi^{+D}) =$$

$$= \frac{1}{2}\partial_s M_i{}^s + \varkappa j_i - \varkappa_i\varOmega + \frac{1}{2}\varepsilon_{imks}\big(\overset{*}{\varkappa}{}^m M^{ks} + \varkappa^{mk}S^s\big),$$

$$\frac{\mathrm{i}}{2}\gamma^5_{DE}(\psi^{+D}\partial_i\psi^E - \psi^E\partial_i\psi^{+D}) =$$

$$= \frac{1}{4}\varepsilon_{imks}\partial^m M^{ks} - \varkappa_i N + \varkappa_{im}S^m - \overset{*}{\varkappa}{}^m M_{im} + \overset{*}{\varkappa} j_i. \qquad (5.65)$$

Replacing in (5.64) the terms with components of spinor ψ according to Eqs. (5.65), we find

$$\varOmega\mathcal{P}_i{}^j = \alpha\big(j^j\partial_s M_i{}^s - M^{js}\partial_i j_s - S^j\partial_i N\big) +$$

$$+ 2\alpha j^j\Big[\varkappa j_i - \varOmega\varkappa_i + \frac{1}{2}\varepsilon_{imks}\big(\overset{*}{\varkappa}{}^m M^{ks} + \varkappa^{mk}S^s\big)\Big],$$

$$N\mathcal{P}_i{}^j = \alpha\Big(\frac{1}{2}j^j\varepsilon_{imks}\partial^m M^{ks} - \frac{1}{2}\varepsilon^{jksq}M_{ks}\partial_i j_q + S^j\partial_i\varOmega\Big) +$$

$$+ 2\alpha j^j\big(-N\varkappa_i + \varkappa_{is}S^s - \overset{*}{\varkappa}{}^s M_{is} + \overset{*}{\varkappa} j_i\big). \qquad (5.66)$$

If $\rho^2 = \varOmega^2 + N^2 \neq 0$, then adding the first equation (5.66), multiplied by $\varOmega/\rho$, with the second equation (5.66) multiplied by N/ρ, we finally get

$$\mathcal{P}_i{}^j = \alpha\big[u_s\partial_i\mu^{js} + u^j\partial_s\mu_i{}^s - S^j\partial_i\eta + u^j(S_i u^k - S^k u_i)\partial_k\eta\big]$$

$$+ 2\alpha\Big[u_i u^j(\varOmega\varkappa + N\overset{*}{\varkappa}) - u^j\Big(\rho\varkappa_i - \frac{1}{2}\varepsilon_{inks}\mu^{nk}\overset{*}{\varkappa}{}^s\Big)$$

$$+ \frac{1}{\rho}u^j\Big(N\varkappa_{ik} + \frac{1}{2}\varOmega\varepsilon_{iksm}\varkappa^{sm}\Big)S^k\Big]. \qquad (5.67)$$

[3] To obtain Eqs. (5.65) it suffices to contract equation (5.17) with components of spintensors γ_i, $\overset{*}{\gamma}_i$ and use equality (3.11) (see also following section).

The components u_s, μ^{js}, η in Eq. (5.67) are determined by the equality (3.66).

Let us now get an expression of the tensor components $\breve{\mathcal{P}}_a{}^b = \breve{h}^i{}_a \breve{h}_j{}^b \mathcal{P}_i{}^j$ in the proper basis $\breve{\boldsymbol{e}}_a$ in terms of the components of vectors π^i, ξ^i, σ^i, u^i and scalars ρ, η. We have

$$\breve{\mathcal{P}}_a{}^b = \alpha(\breve{\psi}^+\gamma^b\breve{\nabla}_a\breve{\psi} - \breve{\nabla}_a\breve{\psi}^+ \cdot \gamma^b\breve{\psi}),$$

or, taking into account expression (5.36) for the covariant derivative

$$\breve{\mathcal{P}}_a{}^b = \alpha\left[\breve{\psi}^+\gamma^b\breve{\partial}_a\breve{\psi} - \breve{\partial}_a\breve{\psi}^+ \cdot \gamma^b\breve{\psi} - \frac{1}{4}\breve{\Delta}_{a,cd}\breve{\psi}^+(\gamma^b\gamma^{cd} + \gamma^{cd}\gamma^b)\breve{\psi}\right].$$

From identities (3.11) it follows

$$\gamma^b\gamma^{cd} + \gamma^{cd}\gamma^b = -2\varepsilon^{bcde}\overset{*}{\gamma}_e.$$

Using this equality and the definition $\breve{S}_e = \breve{\psi}^+\overset{*}{\gamma}_e\breve{\psi}$, we rewrite the equation for $\breve{\mathcal{P}}_a{}^b$ in the form

$$\breve{\mathcal{P}}_a{}^b = \alpha\left(\breve{\psi}^+\gamma^b\breve{\partial}_a\breve{\psi} - \breve{\partial}_a\breve{\psi}^+ \cdot \gamma^b\breve{\psi} + \frac{1}{2}\varepsilon^{bcde}\breve{\Delta}_{a,cd}\breve{S}_e\right).$$

Replacing here the derivatives $\breve{\partial}_a\breve{\psi}$ by formula (5.37), we find

$$\breve{\mathcal{P}}_a{}^b = \alpha\left(-\breve{S}^b\breve{\partial}_a\eta + \frac{1}{2}\varepsilon^{bcde}\breve{\Delta}_{a,cd}\breve{S}_e\right).$$

Using this relation, for the function $\overset{\circ}{\Lambda}$ determined by equality (5.60), we obtain

$$\overset{\circ}{\Lambda} = -\alpha S_i\left(\partial^i\eta + \frac{1}{2}\varepsilon^{ijmn}\breve{\Delta}_{j,mn}\right). \tag{5.68}$$

Let us notice now that by virtue of definition (3.66) of the tensor components μ^{ij} it is carried out the equality

$$\varepsilon^{ijks}S_s = u^i\mu^{jk} + u^j\mu^{ki} + u^k\mu^{ij}, \tag{5.69}$$

which can be obtained by contracting the last identity in (3.70) with components of the Levi-Civita pseudotensor ε^{ijks}. Replacing in (5.68) the contraction $\varepsilon^{ijks}S_s$ by formula (5.69) and using relation (3.149) for the components $\breve{\Delta}_{k,ij}u^j$, expression (5.68) for $\overset{\circ}{\Lambda}$ can be transformed to the form

$$\overset{\circ}{\Lambda} = \alpha(\psi^+\gamma^i\partial_i\psi - \partial_i\psi^+ \cdot \gamma^i\psi) \equiv \alpha\Big(-S^i\partial_i\eta - \mu^{ij}\partial_i u_j + \frac{1}{c}S^i\Omega_i\Big), \tag{5.70}$$

where the vector components Ω_i are defined by the equality

$$\Omega^i = \frac{1}{2}\varepsilon^{ijks} u_j \Omega_{ks}, \qquad \Omega_{ks} = cu^i \breve{\Delta}_{i,ks}.$$

The constant coefficient c is introduced into formula (5.70) for the convenience in connection with the further use of this formula (in the sequel c is the velocity of light).

In the theory of spin liquids the components Ω^i determine so called vector of the internal rotation.

5.6 Representation of Spinor Equations as Tensor Equations in the Components of the Real Tensors

As it was already noted, from the spinor equations it can be obtained the closed system of tensor equations in the components of the real tensors $\boldsymbol{D} = \{\Omega, j^i, M^{ij}, S^i, N\}$. Real tensors $\boldsymbol{D}$ in the known physical theories have a simple physical sense therefore such system of equations also is of interest.

To obtain such equations that are a consequence of Eqs. (5.17) we contract equation (5.17) with components of spintensors ψ^+_B, $\psi^+_C \gamma^{Ci}{}_B$, $\psi^+_C \gamma^{Cij}{}_B$, $\psi^+_C \overset{*}{\gamma}{}^{Ci}{}_B$, $\psi^+_C \gamma^{5C}{}_B$ with respect to the index B. As a result, after separating the real and imaginary parts and transformations by means of equality (3.11), we obtain

$$a.\quad \partial_i j^i = 0,$$

$$b.\quad \frac{1}{2}\gamma^i_{AB}(\psi^{+A}\partial_i\psi^B - \psi^B\partial_i\psi^{+A})$$
$$= \varkappa\,\Omega + \varkappa_i j^i + \frac{1}{2}\varkappa_{ij}M^{ij} + \overset{*}{\varkappa}_i S^i + \overset{*}{\varkappa} N,$$

$$c.\quad \frac{\mathrm{i}}{2}e_{AB}(\psi^{+A}\partial^i\psi^B - \psi^B\partial^i\psi^{+A})$$
$$= \frac{1}{2}\partial_j M^{ij} + \varkappa j^i - \varkappa^i\Omega + \frac{1}{2}\varepsilon^{ijks}(\overset{*}{\varkappa}_j M_{ks} + \varkappa_{jk} S_s),$$

$$d.\quad \frac{1}{2}\gamma^{ij}_{AB}(\psi^{+A}\partial_j\psi^B - \psi^B\partial_j\psi^{+A})$$
$$= \frac{1}{2}\partial^i\Omega + \varkappa_j M^{ij} + \varkappa^{ij} j_j + \overset{*}{\varkappa}{}^i N + \overset{*}{\varkappa} S^i,$$

$$e.\quad \frac{1}{2}(g^{ki}\gamma^j_{AB} - g^{kj}\gamma^i_{AB})(\psi^{+A}\partial_k\psi^B - \psi^B\partial_k\psi^{+A})$$
$$= \frac{1}{2}\varepsilon^{ijks}\partial_k S_s + \varkappa^i j^j - \varkappa^j j^i + \varkappa^i{}_s M^{js} - \varkappa^j{}_s M^{is} + \overset{*}{\varkappa}{}^i S^j - \overset{*}{\varkappa}{}^j S^i,$$

$$f.\quad \frac{\mathrm{i}}{2}\varepsilon^{ijks}\overset{*}{\gamma}_{ABk}\big(\psi^{+A}\partial_s\psi^B - \psi^B\partial_s\psi^{+A}\big)$$
$$= -\frac{1}{2}(\partial^i j^j - \partial^j j^i) + \varkappa M^{ij} + \varepsilon^{ijks}(\varkappa_k S_s + \overset{*}{\varkappa}_k j_s) + \Omega\varkappa^{ij}$$
$$- \frac{1}{2}\varepsilon^{ijks}\big(N\varkappa_{ks} - \overset{*}{\varkappa}M_{ks}\big),$$

$$g.\quad \frac{1}{4}\varepsilon^{ijks}\gamma_{ABks}\big(\psi^{+A}\partial_j\psi^B - \psi^B\partial_j\psi^{+A}\big)$$
$$= \frac{1}{2}\partial^i N - \varkappa S^i + \frac{1}{2}\varepsilon^{ijks}(\varkappa_j M_{ks} - \varkappa_{ks} j_j) - \overset{*}{\varkappa}{}^i\Omega,$$

$$h.\quad \frac{\mathrm{i}}{2}\gamma^5_{AB}\big(\psi^{+A}\partial^i\psi^B - \psi^B\partial^i\psi^{+A}\big)$$
$$= \frac{1}{4}\varepsilon^{ijks}\partial_j M_{ks} - \varkappa^i N + \varkappa^{ij} S_j - \overset{*}{\varkappa}_j M^{ij} + \overset{*}{\varkappa} j^i,$$

$$i.\quad \frac{\mathrm{i}}{2}\overset{*}{\gamma}{}^j_{AB}\big(\psi^{+A}\partial_j\psi^B - \psi^B\partial_j\psi^{+A}\big) = -\varkappa_i S^i - \overset{*}{\varkappa}_i j^i,$$

$$j.\quad \frac{1}{2}\partial_i S^i - \varkappa N - \frac{1}{4}\varepsilon^{ijks}\varkappa_{ij} M_{ks} + \overset{*}{\varkappa}\,\Omega = 0. \tag{5.71}$$

We give a more detailed derivation of one of Eqs. (5.71), for example, equation (e) in (5.71). It is convenient to carry out calculations using the matrix notations.

Multiplying Eq. (5.18) by $\psi^+\gamma^{kn}$ from the left and subtracting from the obtained result Eq. (5.22), multiplied from the right by $\gamma^{kn}\psi$, we obtain

$$\psi^+\gamma^{kn}\gamma^i\partial_i\psi + (\partial_i\psi^+)\gamma^i\gamma^{kn}\psi + \psi^+\big[\mathrm{i}\varkappa_j(\gamma^{kn}\gamma^j - \gamma^j\gamma^{kn})$$
$$+ \frac{\mathrm{i}}{2}\varkappa_{sj}(\gamma^{kn}\gamma^{sj} - \gamma^{sj}\gamma^{kn}) + \overset{*}{\varkappa}_j(\gamma^{kn}\overset{*}{\gamma}{}^j - \overset{*}{\gamma}{}^j\gamma^{kn}) + \overset{*}{\varkappa}(\gamma^{kn}\gamma^5 - \gamma^5\gamma^{kn})\big]\psi = 0.$$

Let us replace in this equation the products of the matrices $\boldsymbol{\gamma}$ by formulas (3.11):

$$\psi^+\big(-\varepsilon^{knim}\overset{*}{\gamma}_m + \gamma^k g^{ni} - \gamma^n g^{ki}\big)\partial_i\psi$$
$$+ \partial_i\psi^+\big(-\varepsilon^{knim}\overset{*}{\gamma}_m - \gamma^k g^{ni} + \gamma^n g^{ki}\big)\psi + 2\psi^+\big[\mathrm{i}\varkappa_j(\gamma^k g^{nj} - \gamma^n g^{kj})$$
$$+ \mathrm{i}\varkappa_{sj}(\gamma^{sk} g^{jn} - \gamma^{sn} g^{jk}) + \overset{*}{\varkappa}_j(\overset{*}{\gamma}{}^k g^{nj} - \overset{*}{\gamma}{}^n g^{kj})\big]\psi = 0.$$

Bearing in mind definitions (3.58) and (3.59) of the components of the real tensors Ω, j^i, μ^{ij}, S^i, N, we write the last equation in the form

$$\frac{1}{2}[\psi^+(\gamma^k g^{ni} - \gamma^n g^{ki})\partial_i\psi - \partial_i\psi^+ \cdot (\gamma^k g^{ni} - \gamma^n g^{ki})\psi]$$
$$= \frac{1}{2}\varepsilon^{knij}\partial_i S_j + \varkappa^k j^n - \varkappa^n j^k + \varkappa^k{}_s M^{ns} - \varkappa^n{}_s M^{ks} + \overset{*}{\varkappa}{}^k S^n - \overset{*}{\varkappa}{}^n S^k,$$

that coincides with the equation (e) in (5.71). The other equations in (5.71) are obtained similarly.

Calculating the determinant of the matrix of the transformation of Eqs. (5.17)–(5.71), it is possible to show that the system of equations (5.71) is equivalent to the spinor equations (5.17). As the complete system of the equations one can take, for example, the equations (a), (b), (e), (j) in (5.71).

The equations (a) and (j) in (5.71) are, respectively, the scalar and pseudo-scalar equations in the invariant tensor form. Bearing in mind definition (5.27) of the tensor component $\mathcal{P}_i{}^j$, the equation (e) in (5.71) can be written in the form

$$\mathcal{P}^{ij} - \mathcal{P}^{ji} + \alpha\varepsilon^{ijks}\partial_k S_s + 2\alpha\big(\varkappa^i j^j - \varkappa^j j^i$$
$$+ \varkappa^{is} M^j{}_s - \varkappa^{js} M^i{}_s + \overset{*}{\varkappa}{}^i S^j - \overset{*}{\varkappa}{}^j S^i\big) = 0. \qquad (5.72)$$

Equation (5.72), in which the components of the tensor $\mathcal{P}^{ij}$ are defined by equality (5.66), is the tensor equation in the components of the real tensors Ω, j^i, M^{ij}, S^i, N.

The equation (b) in (5.71) with the aid of the components $\mathcal{P}_i{}^j$ can be written as follows

$$\mathcal{P}_i{}^i = -2\alpha\left(\varkappa\Omega + \varkappa_i j^i + \frac{1}{2}\varkappa_{ij}M^{ij} + \overset{*}{\varkappa}_i S^i + \overset{*}{\varkappa} N\right). \qquad (5.73)$$

If the components of the tensor $\mathcal{P}_i{}^j$ in Eq. (5.73) are defined according to (5.67), then Eq. (5.73) is fulfilled identically. This is related to the fact that when we transformed expression (5.27) to the form (5.67) were used Eqs. (5.65) which are fulfilled by virtue of Eqs. (5.17). Therefore expressions for $\mathcal{P}_i{}^j$, determined by equalities (5.27) and (5.67), coincide only due to Eqs. (5.17). For the same reason the system of the equations (a), (j) in (5.71) and (5.72), in which the components $\mathcal{P}_i{}^j$ are determined by (5.67) is, in general, not closed.

To close the system of equations (a), (j) in (5.71) and (5.72) it is possible to take Eqs. (5.30) in which the components of the tensors $\mathcal{P}_i{}^j$ are expressed in terms of the components of the real tensors Ω, j^i, M^{ij}, S^i, N by formulas (5.66).

Using formula (5.67) for $\mathcal{P}_i{}^j$, it is possible to write Eq. (5.72) in the real components ρ, u^i, S^i, η. For this purpose we replace the components of tensor $\mathcal{P}^{ij}$ in Eq. (5.72) by formula (5.67), and the components $\varepsilon^{ijks}S_s$ by formula (5.69).

As a result after transformations we get

$$\frac{\rho}{g}\frac{d}{d\tau}\left(\frac{1}{\rho}\mu^{ij}\right) + \mu^{i}{}_{n}\overset{*}{F}{}^{nj} - \mu^{j}{}_{n}\overset{*}{F}{}^{ni} = 0. \tag{5.74}$$

Here $d/d\tau = cu^i\partial_i$; c and g are arbitrary nonzero constants; the components of the tensor $\overset{*}{F}{}^{ij}$ are defined by the equality

$$\overset{*}{F}{}^{ij} = \frac{c}{g}\left\{(\delta^i_m + u_m u^i)(\delta^j_n + u_n u^j)\left[\varepsilon^{mnks}u_k(-2\overset{*}{\varkappa}_s + \partial_s\eta) + \frac{2}{\rho}\left(\Omega\varkappa^{mn} - \frac{1}{2}N\varepsilon^{mnks}\varkappa_{ks}\right)\right] - \partial^i u^j + \partial^j u^i\right\}. \tag{5.75}$$

The constant coefficients c, g are introduced in Eqs. (5.74) and (5.75) for the sake of convenience in connection with the further applications.

Components $\overset{*}{F}{}^{ij}$ are antisymmetric in the indices i, j: $\overset{*}{F}{}^{ij} = -\overset{*}{F}{}^{ji}$.

Contracting equations (5.74) with components of the tensor $\varepsilon_{mijk}u^k$ with respect to the indices i, j, Eqs. (5.74) one can rewrite in the form

$$\rho\frac{d}{d\tau}\left(\frac{1}{\rho}S_m\right) - g\overset{*}{F}_{mj}S^j = 0. \tag{5.76}$$

Equations (5.74) also follows from Eqs. (5.76), therefore Eqs. (5.74) are equivalent to (5.76).

It is easy to see that due to definition (5.75) of the components $\overset{*}{F}{}^{ij}$, the contraction of Eqs. (5.76) with components of the vector u^k is fulfilled identically

$$u^m\left[\rho\frac{d}{d\tau}\left(\frac{1}{\rho}S_m\right) - g\overset{*}{F}_{mj}S^j\right] \equiv 0.$$

By virtue of the equation (b) in (3.60) and the anti-symmetry of the components $\overset{*}{F}_{mj}$, the contraction of Eqs. (5.76) with components of the vector S^m also is fulfilled identically

$$S^m\left[\rho\frac{d}{d\tau}\left(\frac{1}{\rho}S_m\right) - g\overset{*}{F}_{mj}S^j\right] \equiv 0.$$

Thus, Eqs. (5.76) contain in general case no more than two independent equations.

Equations (5.76) can be obtained also directly from Eqs. (5.44). Indeed, contracting the second equation in (5.44) with components of the tensor $\varepsilon^{mnsi}u_s$ with respect to the index i and taking into account definition (5.41) for N^a, we find

$$(-\breve{\Delta}^{mn,s} + \breve{\Delta}^{n,ms} - \breve{\Delta}^{s,mn})u_s - \varepsilon^{mnsi} u_s \partial_i \eta$$
$$= \frac{2}{\rho^2}\{(\varkappa\Omega + \overset{*}{\varkappa}N + \varkappa_i j^i)\mu^{mn} - \varepsilon^{mnsi} u_s[\rho^2 \overset{*}{\varkappa}_i + j^j(\Omega\overset{*}{\varkappa}_{ij} + N\varkappa_{ij})]\}.$$

Bearing in mind that $\breve{\Delta}^{m,ns} u_s = -\partial^m u^n$, this equation can be represented in the form

$$u_s \breve{\Delta}^{s,ij} = -\mathcal{F}^{ij}, \tag{5.77}$$

where the components of the antisymmetric tensor $\mathcal{F}^{ij}$ are defined by the following relation

$$\mathcal{F}^{ij} = (\delta^i_m + u^i u_m)(\delta^j_n + u^j u_n)\Big[\varepsilon^{mnks} u_k(-2\overset{*}{\varkappa}_s + \partial_s \eta)$$
$$+ \frac{2}{\rho}(\Omega\varkappa^{mn} - N\overset{*}{\varkappa}{}^{mn}) + \frac{2}{\rho^2}\mu^{mn}(\Omega\varkappa + N\overset{*}{\varkappa} + \varkappa_i j^i)\Big] - \partial^i u^j + \partial^j u^i.$$

From this one can obtain the equations for derivatives of tetrad vectors π^i, ξ^i, σ^i. For example, by contracting equation (5.77) with components σ^j with respect to the index j, we obtain Eq. (5.76), since due to definition μ^{mn} an identity is fulfilled

$$(\delta^i_m + u^i u_m)(\delta^j_n + u^j u_n)\mu^{mn} S_j \equiv \mu^{ij} S_j = 0.$$

For the convenience of references we write out separately a closed system of equations for the components of real tensors. This system is a corollary of Eqs. (5.17):

$$\partial_i \rho u^i = 0,$$
$$\partial_j \mathcal{P}_i{}^j + 2\alpha\big(\Omega\,\partial_i\varkappa + j^j\partial_i\varkappa_j + \frac{1}{2}M^{sj}\partial_i\varkappa_{sj} + S^j\partial_i\overset{*}{\varkappa}_j + N\partial_i\overset{*}{\varkappa}\big) = 0,$$
$$\rho\frac{d}{d\tau}\left(\frac{1}{\rho}S_m\right) - g\overset{*}{F}_{mj}S^j = 0,$$
$$\frac{1}{2}\partial_i S^i - \varkappa N - \frac{1}{4}\varepsilon^{ijks}\varkappa_{ij}M_{ks} + \overset{*}{\varkappa}\Omega = 0. \tag{5.78}$$

The components of the tensors $\mathcal{P}_i{}^j$ and $\overset{*}{F}_{mj}$ in Eqs. (5.78) are determined by equalities (5.67) and (5.75).

5.7 The Tensor Representation of the Spinor Weyl Equations

Let us consider in the Minkowski space referred to a Cartesian coordinate system the differential equations

$$\gamma^i \partial_i \psi = 0, \tag{5.79}$$

in which the components of the first rank spinor ψ satisfy the algebraic equations

$$\psi = \mathrm{i}\gamma^5\psi \quad \text{or} \quad \psi = -\mathrm{i}\gamma^5\psi.$$

We shall assume further that the metric spinor E and the spintensors γ^i are defined by matrices (3.81) and (3.82). In this case the components of an arbitrary spinor can be represented in the form

$$\psi = \begin{Vmatrix} \xi^A \\ \eta_{\dot{A}} \end{Vmatrix}, \qquad \dot{A}, A = 1, 2,$$

where the components ξ^A and $\eta_{\dot{A}}$ define the two-component spinors with the fixed relative sign.

Taking into account that spintensor γ^5 is defined by the diagonal matrix (3.82), we find that from condition $\psi = -\mathrm{i}\gamma^5\psi$ it follows $\xi^A = 0$, while from condition $\psi = \mathrm{i}\gamma^5\psi$ it follows $\eta_{\dot{A}} = 0$. Respectively, taking into account expression (3.94) of the Dirac matrices γ^i in terms of the Pauli matrices σ^α, we find that Eq. (5.79) with the additional condition $\psi = -\mathrm{i}\gamma^5\psi$ is written in the form

$$\sigma^{B\dot{A}i}\partial_i \eta_{\dot{A}} = 0, \tag{5.80}$$

while for $\psi = \mathrm{i}\gamma^5\psi$ in the form

$$\sigma^i_{\dot{B}A}\partial_i \xi^A = 0, \tag{5.81}$$

where $\sigma^i_{\dot{B}A}$ and $\sigma^{B\dot{A}i}$ are the components of the invariant spintensors, defined as

$$\sigma^i_{\dot{B}A} = \{-\sigma^\alpha, -I\}, \quad \sigma^{B\dot{A}i} = \{\sigma^\alpha, -I\}.$$

Here σ^α are the Pauli matrices, I is the two-dimensional unit matrix.

Equations (5.80) and (5.81) for the two-component spinors ξ and η are called the Weyl equations. Equations (5.80) and (5.81) are a particular case of Eqs. (5.18) for $\psi = \pm\mathrm{i}\gamma^5\psi$, $\varkappa = \varkappa_j = \varkappa_{ij} = \overset{*}{\varkappa}_j = \overset{*}{\varkappa} = 0$.

The current vector for the fields described by the Weyl equations is defined by the components j^i:

$$j^i = -\sigma^i_{\dot{B}A}\dot{\xi}^B\xi^A \quad \text{or} \quad j^i = -\sigma^{B\dot{A}i}\dot{\eta}_{\dot{B}}\eta_{\dot{A}}, \tag{5.82}$$

which by virtue of the Weyl equations satisfy the conservation law

$$\partial_i j^i = 0.$$

The components of the energy-momentum tensor, corresponding to the Weyl equation (5.80), have the form

$$P_i{}^j = \frac{\mathrm{i}}{2}\sigma^{B\dot{A}j}\left(\dot{\eta}_{\dot{B}}\partial_i\eta_{\dot{A}} - \eta_{\dot{A}}\partial_i\dot{\eta}_{\dot{B}}\right). \tag{5.83}$$

The components of the energy-momentum tensor, corresponding to the Weyl equation (5.81), have the form

$$P_i{}^j = \frac{\mathrm{i}}{2}\sigma^j_{\dot{B}A}\left(\dot{\xi}^B\partial_i\xi^A - \xi^A\partial_i\dot{\xi}^B\right). \tag{5.84}$$

It is easy to be convinced that by virtue of the Weyl equations the components of the energy-momentum tensors $P_i{}^j$ satisfy the conservation law

$$\partial_j P_i{}^j = 0 \tag{5.85}$$

and

$$\partial_k\left(x^j P^{ik} - x^i P^{jk} \mp \frac{1}{2}\varepsilon^{ijks} j_s\right) = P^{ij} - P^{ji} \mp \frac{1}{2}\varepsilon^{ijks}\partial_k j_s = 0. \tag{5.86}$$

Here and in the subsequent equations the upper sign corresponds to the Weyl equation (5.81), the lower sign corresponds to Eq. (5.80).

Due to the Weyl equations the trace of the energy-momentum tensors is equal to zero

$$P_i{}^i = 0. \tag{5.87}$$

Definitions (5.83), (5.84) and Eqs. (5.85)–(5.87) are the particular case of definitions (5.33), (5.27) and Eqs. (5.31), (5.32), (5.73) for $\psi = \pm\mathrm{i}\gamma^5\psi$ and $\varkappa = \varkappa_j = \varkappa_{ij} = \overset{*}{\varkappa}_j = \overset{*}{\varkappa} = 0, \alpha = 1/2$. In this case in formula (5.33) the quantities $N_i{}^j$ can be put equal to zero and, consequently, $P_i{}^j = \mathcal{P}_i{}^j$ for $\psi = \pm\mathrm{i}\gamma^5\psi, \alpha = 1/2$.

Let us now derive the tensor equations in the components of the complex tensor C^{ij} equivalent to the Weyl equations.[4] For this purpose we contract the Weyl equations (5.81) with components of spintensor $\sigma^{C\dot{B}i}\xi_C$ with respect to the index $\dot{B}$. Replacing the contraction $\sigma^{C\dot{B}i}\sigma^{\;j}_{\dot{B}A}$ by formula (3.105), we obtain

$$\xi_C\big(\mathrm{i}\sigma^C{}_A{}^{ij} - g^{ij}\delta^C_A\big)\partial_j\xi^A = 0. \tag{5.88}$$

Bearing in mind the definition of the tensor component $C^{ij} = -\mathrm{i}\sigma^{ij}_{AC}\xi^A\xi^C$, due to symmetry properties $\sigma^{BA}_{ij} = \sigma^{AB}_{ij}$, $\varepsilon_{BA} = -\varepsilon_{AB}$ we have

$$\mathrm{i}\sigma^C{}_A{}^{ij}\xi_C\partial_j\xi^A = -\frac{\mathrm{i}}{2}\sigma^{ij}_{AC}\partial_j(\xi^C\xi^A) = \frac{1}{2}\partial_j C^{ij},$$

$$\xi_C\partial_j\xi^C = \frac{1}{2}\varepsilon_{BC}\big(-\xi^B\partial_j\xi^C + \xi^C\partial_j\xi^B\big).$$

Therefore Eq. (5.88) can be rewritten in the form

$$\partial_j C^{ij} + \varepsilon_{BC}(\xi^B\partial^i\xi^C - \xi^C\partial^i\xi^B) = 0. \tag{5.89}$$

Let us multiply Eq. (5.89) by $-\mathrm{i}\sigma^{mn}_{DA}\xi^D\xi^A = C^{mn}$ and, using the obvious identity

$$\xi^A(\xi^B d\xi^C - \xi^C d\xi^B) \equiv \xi^B d(\xi^A\xi^C) - \xi^C d(\xi^B\xi^A),$$

we transform Eq. (5.89) to the form

$$C^{mn}\partial_j C^{ij} + \mathrm{i}\sigma^{mn}_{DA}\varepsilon_{BC}\big[\xi^D\xi^B\partial^i\big(\xi^A\xi^C\big) - \xi^C\xi^D\partial^i\big(\xi^B\xi^A\big)\big] = 0$$

From this, taking into account the symmetry property $\varepsilon_{BC} = -\varepsilon_{CB}$, we find

$$C^{mn}\partial_j C^{ij} - 2\mathrm{i}\sigma^{mn}_{DA}\varepsilon_{BC}\xi^C\xi^D\partial^i\big(\xi^B\xi^A\big) = 0.$$

Replacing in this equation the product of spintensors $\sigma^{mn}_{DA}\varepsilon_{BC}$ by formula (3.108), we obtain

$$C^{mn}\partial_j C_i{}^j + C^m{}_j\partial_i C^{nj} = 0. \tag{5.90}$$

[4] Tensor equations in the components of the complex tensor C^{ij}, equivalent to the Weyl equations, are obtained in [74, 82], see also [50, p. 221]. In order to avoid misunderstanding we recall that the spinor in the Weyl and Dirac equations is considered as the geometric object in the Minkowski space and its components are defined up to a common sign.

To close equations (5.90) it is necessary to add to them the algebraic equations (see Sect. 3.3, Chap. 3)

$$C^{ij} = \frac{\mathrm{i}}{2}\varepsilon^{ijks}C_{ks}, \quad C_{ij}C^{ij} = 0.$$

Multiplying Eq. (5.81) by $\sigma^{s}_{\dot{B}A}\dot{\xi}^{B}\xi^{A} = -j^{s}$ and carrying out the similar transformations, it is possible to get the invariant tensor equation [74]

$$j^{s}\partial_{j}C^{qj} = j_{j}\partial^{q}C^{js}, \tag{5.91}$$

The components C^{ij} and j^{i} are related by the algebraic equations

$$2j^{i}j^{s} = C^{i}{}_{m}\dot{C}^{sm}. \tag{5.92}$$

The tensor equations for the components of the tensor $C^{ij} = -\mathrm{i}\sigma^{\dot{B}\dot{A}ij}\eta_{\dot{B}}\eta_{\dot{A}}$ and the vector $j^{s} = -\sigma^{B\dot{A}s}\dot{\eta}_{\dot{B}}\eta_{\dot{A}}$, corresponding to the Weyl spinor equations (5.80), are obtained by a similar way and can also be written in the form (5.90), (5.91), but in this case they must be supplemented by the algebraic equations

$$\begin{aligned} C^{ij} &= -\frac{\mathrm{i}}{2}\varepsilon^{ijks}C_{ks}, \quad C_{ij}C^{ij} = 0, \\ 2j^{i}j^{s} &= C^{i}{}_{m}\dot{C}^{sm}, \end{aligned} \tag{5.93}$$

Taking into account definition (3.116) Eq. (5.91) for $s = 4$ is written as follows[5]

$$\begin{aligned} j^{4}\partial_{\alpha}p^{\alpha} &= j_{\alpha}\partial_{4}p^{\alpha}, \\ j^{4}\left(\partial_{4}p^{\alpha} \pm \mathrm{i}\varepsilon^{\alpha\beta\lambda}\partial_{\beta}p_{\lambda}\right) &= j_{\lambda}\partial^{\alpha}p^{\lambda}. \end{aligned} \tag{5.94}$$

We note one more tensor equation

$$j^{s}\partial_{q}j^{k} - j^{k}\partial_{q}j^{s} = \frac{1}{2}\left(\dot{C}^{sk}\partial_{j}C_{q}{}^{j} + C^{sk}\partial_{j}\dot{C}_{q}{}^{j}\right), \tag{5.95}$$

which is obtained by transformation of Eq. (5.89) for the spinor field ξ multiplied by $\dot{C}^{ij} = \mathrm{i}\dot{\sigma}^{ij}_{AC}\dot{\xi}^{A}\dot{\xi}^{C}$, or by an analogous transformation of Eq. (5.90) for the spinor field η.

As already noted, definitions (5.83) and (5.84) of the components of the energy-momentum tensor are obtained from formula (5.27) if to put in it $\alpha = 1/2$ and $\psi = \pm\mathrm{i}\gamma^{5}\psi$. Therefore, using relations (3.113), (3.118) and $\alpha = 1/2$, from

[5] The second equation in (5.94) for the neutrino is obtained a different way in [56] (where the notation $v^{i} = j^{i}/j^{4}$ is used). See also [44, 45, 58].

definition (5.57) of the components of the tensor $\mathcal{P}_i{}^j$ we get the formula for the components of the energy-momentum tensor for the Weyl equations

$$j^s P_i{}^j \equiv \frac{1}{4}\Big[\mp \varepsilon^{sjkl} j_k \partial_i j_l + \frac{\mathrm{i}}{2}\big(\dot{C}^{sk}\partial_i C_k{}^j - C^{sk}\partial_i \dot{C}_k{}^j\big)\Big]. \tag{5.96}$$

From this, in particular, it follows the algebraic identity connecting the components of the energy-momentum tensor

$$j_j P_i{}^j = 0, \tag{5.97}$$

Equation (5.97) is obtained directly by the contraction of Eq. (5.96) with respect to the indices j, s taking into account identity

$$\dot{C}^{jk}\partial_i C_{jk} \equiv 0$$

which is fulfilled due to the first equations in (5.92), (5.93).

Let us give also identity (5.58) for the case under consideration

$$j^s P_i{}^j - j^j P_i{}^s \equiv \mp\frac{1}{2}\varepsilon^{sjkm} j_k \partial_i j_m.$$

5.8 Spinor Differential Equations in the Four-Dimensional Riemannian Space

5.8.1 The Tensor Formalism

Consider the Riemannian space V_4 with the metric signature $(+, +, +, -)$, referred to a coordinate system with the variables x^i and with the covariant holonomic vector basis $\boldsymbol{\ni}_i(x^i)$. Let us introduce at each point x^i of the space V_4 the tangent pseudo-Euclidean space with an orthonormal basis $\boldsymbol{e}_a(x^i)$, $a = 1, 2, 3, 4$.

We will denote the indices of the tensor components, specified in the holonomic bases $\boldsymbol{\ni}_i$ by the Latin letters $i, j, k, \ldots$. The indices of tensor components specified in the local orthonormal bases $\boldsymbol{e}_a$ will be denoted by the first letters of the Latin alphabet a, b, c, d, e, f.

Bases $\boldsymbol{\ni}_i$ and $\boldsymbol{e}_a$ are connected by the scale factors

$$\boldsymbol{e}_a = h^i{}_a \boldsymbol{\ni}_i, \qquad \boldsymbol{\ni}_i = h_i{}^a \boldsymbol{e}_a.$$

Let g_{ij} be the covariant components of the metric tensor of the Riemannian space, calculated in the basis $\boldsymbol{\ni}_i$; the covariant components of the metric tensor calculated in orthonormal bases $\boldsymbol{e}_a$, are defined by the diagonal matrix $g_{ab} = \mathrm{diag}(1, 1, 1, -1)$.

The components of the metric tensor g_{ij} and g_{ab} are connected by the equalities

$$g_{ij} = h_i{}^a h_j{}^b g_{ab}, \quad g_{ab} = h^i{}_a h^j{}_b g_{ij}. \tag{5.98}$$

Since $\det \|g_{ab}\| = -1$, equality (2.23) in the Riemannian space V_4 is written as follows

$$-\det \|g_{ij}\| = \left(\det \|h_i{}^a\|\right)^2. \tag{5.99}$$

The parallel transport of the tensors specified by components in an orthonormal basis $\boldsymbol{e}_a$, is defined by means of the Ricci rotation coefficients $\Delta_{i,bc}$ which can be defined in terms of the scale factors by the equality (see Chap. 2, Sect. 2.2)

$$\Delta_{i,bc} = \frac{1}{2}\big[h^j{}_c\big(\partial_i h_{jb} - \partial_j h_{ib}\big) - h^j{}_b\big(\partial_i h_{jc} - \partial_j h_{ic}\big) + h_i{}^a h^j{}_b h^s{}_c\big(\partial_j h_{sa} - \partial_s h_{ja}\big)\big]. \tag{5.100}$$

Parallel transport of the spinors specified by the components ψ in orthonormal basis $\boldsymbol{e}_a$, in the Riemannian space is defined by the spinor connection coefficients γ_i (the Fock–Ivanenko coefficients)

$$\Gamma_i = \frac{1}{4}\Delta_{i,bc}\gamma^{bc}, \quad \gamma^{bc} = \frac{1}{2}(\gamma^b\gamma^c - \gamma^c\gamma^b), \tag{5.101}$$

where the Ricci rotation coefficients $\Delta_{i,bc}$ are defined by equality (5.100); γ^a are the Dirac matrices satisfying equation

$$\gamma^a\gamma^b + \gamma^b\gamma^a = 2g^{ab}I, \tag{5.102}$$

do not depend on variables x^i.

Formulas for the covariant derivatives of the first rank spinor fields $\boldsymbol{\psi}(x^i)$ and $\boldsymbol{\psi}^+(x^i)$, specified in an arbitrary orthonormal basis $\boldsymbol{e}_a(x^i)$, have the form

$$\nabla_s\psi = \partial_s\psi - \frac{1}{4}\Delta_{s,bc}\gamma^{bc}\psi,$$
$$\nabla_s\psi^+ = \partial_s\psi^+ + \frac{1}{4}\Delta_{s,bc}\psi^+\gamma^{bc}. \tag{5.103}$$

The vectors $\breve{\boldsymbol{e}}_a = \{\pi^i\,\boldsymbol{\ni}_i, \xi^i\,\boldsymbol{\ni}_i, \sigma^i\,\boldsymbol{\ni}_i, u^i\,\boldsymbol{\ni}_i\}$ of the proper basis of the spinor field $\psi(x^i)$ are defined by the components

$$\rho\pi^i = \mathrm{Im}(\psi^T E\gamma^i\psi),$$
$$\rho\xi^i = \mathrm{Re}(\psi^T E\gamma^i\psi),$$

$$\rho\sigma^i = \psi^+\gamma^i\gamma^5\psi,$$
$$\rho u^i = \mathrm{i}\psi^+\gamma^i\psi,$$
$$\rho\exp\mathrm{i}\eta = \psi^+\psi + \mathrm{i}\psi^+\gamma^5\psi. \tag{5.104}$$

Here the matrices $\gamma^i = h^i{}_a\gamma^a$ depend on variables x^i and by virtue of Eqs. (5.98), (5.102) satisfy the equation

$$\gamma^i\gamma^j + \gamma^j\gamma^i = 2g^{ij}I.$$

If the Dirac matrices γ_a and the metric spinor E are defined by equalities (3.24), (3.25), then components of the spinor $\breve{\psi}$, calculated in the proper basis $\breve{\boldsymbol{e}}_a$, are determined by the relations

$$\breve{\psi}^1 = 0, \quad \breve{\psi}^2 = \mathrm{i}\sqrt{\tfrac{1}{2}\rho}\,\exp\left(\tfrac{\mathrm{i}}{2}\eta\right),$$
$$\breve{\psi}^3 = 0, \quad \breve{\psi}^4 = \mathrm{i}\sqrt{\tfrac{1}{2}\rho}\,\exp\left(-\tfrac{\mathrm{i}}{2}\eta\right).$$

Formulas (3.203) for derivatives of components of the spinor field in the Riemannian space are written in the form

$$\nabla_s\psi = \left(\frac{1}{2}I\partial_s\ln\rho - \frac{1}{2}\gamma^5\partial_s\eta - \frac{1}{4}\breve{\Delta}_{s.ij}\gamma^{ij}\right)\psi,$$
$$\nabla_s\psi^+ = \psi^+\left(\frac{1}{2}I\partial_s\ln\rho - \frac{1}{2}\gamma^5\partial_s\eta + \frac{1}{4}\breve{\Delta}_{s.ij}\gamma^{ij}\right), \tag{5.105}$$

where the covariant derivatives $\nabla_s\psi$, $\nabla_s\psi^+$ are defined by equalities (5.103), while the Ricci rotation coefficients $\breve{\Delta}_{s.ij}$ correspond to proper orthonormal bases $\breve{\boldsymbol{e}}_a(x^i)$ of the spinor field $\boldsymbol{\psi}(x^i)$

$$\breve{\Delta}_{s.ij} = \frac{1}{2}\big(\pi_i\nabla_s\pi_j - \pi_j\nabla_s\pi_i + \xi_i\nabla_s\xi_j - \xi_j\nabla_s\xi_i$$
$$+ \sigma_i\nabla_s\sigma_j - \sigma_j\nabla_s\sigma_i - u_i\nabla_s u_j + u_j\nabla_s u_i\big). \tag{5.106}$$

The derivation of these formulas in the Riemannian space insignificantly differs from them derivation in pseudo-Euclidean space.

To write of the first order spinor differential equations (5.18) in the Riemannian space it suffices to replace the symbol of the partial derivative ∂_i in these equations by the symbol of the covariant derivative ∇_i. Thus, Eq. (5.18) in an arbitrary coordinate system with variables x^i in the Riemannian space is written as follows

$$\gamma^i \nabla_i \psi + \left(\varkappa I + \mathrm{i}\varkappa_j \gamma^j + \frac{\mathrm{i}}{2}\varkappa_{js}\gamma^{js} + \overset{*}{\varkappa}_j \overset{*}{\gamma}{}^j + \overset{*}{\varkappa}\gamma^5 \right) \psi = 0. \tag{5.107}$$

Here the components of a spinor ψ are calculated in the orthonormal bases $\boldsymbol{e}_a$. Let us transform operator $\gamma^i \nabla_i$ in Eqs. (5.107). According to definition (5.103) we have

$$\gamma^i \nabla_i = \gamma^i \left(\partial_i - \frac{1}{4}\Delta_{i,bc}\gamma^{bc} \right) = \gamma^i \partial_i - \frac{1}{4}\Delta_{i,js}\gamma^i \gamma^{js}. \tag{5.108}$$

Replacing the product of the matrices $\gamma^i \gamma^{js}$ in formula (5.108) in accordance with Eqs. (3.11), we get

$$\gamma^i \nabla_i = \gamma^i \left[\partial_i + \frac{1}{2}\left(\Delta_{j,i}{}^j I - \frac{1}{2}\varepsilon_{ijmn}\Delta^{j,mn}\gamma^5 \right) \right]. \tag{5.109}$$

Therefore Eq. (5.107) can be written down in the form

$$\gamma^i \left[\partial_i + \frac{1}{2}\left(\Delta_{j,i}{}^j I - \frac{1}{2}\varepsilon_{ijmn}\Delta^{j,mn}\gamma^5 \right) \right] \psi + \left(\varkappa I + \mathrm{i}\varkappa_j \gamma^j + \frac{\mathrm{i}}{2}\varkappa_{js}\gamma^{js} + \overset{*}{\varkappa}_j \overset{*}{\gamma}{}^j + \overset{*}{\varkappa}\gamma^5 \right) \psi = 0$$

or, using orthonormal bases $\boldsymbol{e}_a$:

$$\gamma^a \left[\nabla'_a + \frac{1}{2}\left(\Delta_{b,a}{}^b I - \frac{1}{2}\varepsilon_{abcd}\Delta^{b,cd}\gamma^5 \right) \right] \psi + \left(\varkappa I + \mathrm{i}\varkappa_a \gamma^a + \frac{\mathrm{i}}{2}\varkappa_{ab}\gamma^{ab} + \overset{*}{\varkappa}_a \overset{*}{\gamma}{}^a + \overset{*}{\varkappa}\gamma^5 \right) \psi = 0. \tag{5.110}$$

Here $\nabla'_a = h^i{}_a \nabla'_i$. For the derivative of the spinor field ψ we have $\nabla'_a \psi = \partial_a \psi = h^i{}_a \partial_i$.

The writing of the spinor equations in the Riemannian space in the form (5.107) and (5.110) is significantly founded on introduction to the Riemannian space the system of orthonormal tetrads $\boldsymbol{e}_a(x^i)$ in which are considered the components of the spinor ψ [6, 30, 36, 41, 42, 47]. For specifying of the system of orthonormal tetrads $\boldsymbol{e}_a(x^i)$ it is necessary to introduce sixteen functions—scale factors $h_i{}^a(x^i)$. Due to Eqs. (5.98) these coefficients with the given metric tensor of the Riemannian space contain six arbitrary functions (related to possibility of arbitrary Lorentz transformation of orthonormal bases $\boldsymbol{e}_a(x^i)$ with the fixed holonomic basis $\boldsymbol{\mathfrak{Э}}_i$ in the Riemannian space).

Therefore for the mathematical formulation of problems related to Eqs. (5.107), (5.110) it is necessary to specify the tetrads $\boldsymbol{e}_a(x^i)$ (or a tetrad gauge is said to be needed). A large number of such gauges are known. All these gauges are divided into two groups. Some gauges are defined by algebraic equations for scale factors. They allow us to exclude from the equations six additional arbitrary functions in scale factors, but all of them are not invariant under transformation of the variables x^i of the coordinate system. The other gauges are invariant with respect to the choice of the coordinate system x^i, however they are written in the form of differential equations for the scale factors, which complicates the initial system of equations.

As the tetrad gauge condition in the spinor equations one may accept [86] that tetrads $\boldsymbol{e}_a(x^i)$ coincide with the proper tetrads $\breve{\boldsymbol{e}}_a(x^i)$, which are determined by the spinor field $\psi(x^i)$ by formulas (5.104). Thus, this gauge is written in the form

$$\boldsymbol{e}_a(x^i) = \breve{\boldsymbol{e}}_a(x^i)$$

or in the form of the conditions directly on the scale factors

$$h^i{}_a = \breve{h}^i{}_a = \begin{Vmatrix} \pi^1 & \xi^1 & \sigma^1 & u^1 \\ \pi^2 & \xi^2 & \sigma^2 & u^2 \\ \pi^3 & \xi^3 & \sigma^3 & u^3 \\ \pi^4 & \xi^4 & \sigma^4 & u^4 \end{Vmatrix}, \tag{5.111}$$

where the components $\pi^i, \xi^i, \sigma^i, u^i$ determine the vectors of the proper orthonormal tetrad $\breve{\boldsymbol{e}}_a(x^i)$ and are calculated in terms of the spinor field $\psi(x^i)$ by formulas (5.104).

It is obvious that such tetrad gauge is purely algebraic and, at the same time, is invariant under transformations of the coordinate system x^i.

The writing of the spinor equations (5.107) in the pseudo-Euclidean space when using such tetrad gauge have been obtained in Sect. 5.3 and it has the form

$$\begin{aligned} &\breve{\partial}_a \ln \rho + \breve{\Delta}_{b,a}{}^b + \breve{M}_a = 0, \\ &\breve{\partial}^a \eta + \frac{1}{2}\varepsilon^{abcd}\breve{\Delta}_{b,cd} + \breve{N}^a = 0, \end{aligned} \tag{5.112}$$

where the components of the vectors $\breve{M}_a$, $\breve{N}^a$ are defined by equality (5.41). The Ricci rotation coefficients $\breve{\Delta}_{a,bc}$ for the system of proper tetrads $\breve{\boldsymbol{e}}_a$ of the spinor field are defined by Eqs. (3.150).

In the Riemannian space the corresponding equations may be also written in the form of Eqs. (5.112), however the Ricci rotation coefficients $\breve{\Delta}_{a,bc}$ in this case are defined as follows (see Chap. 2)

$$\breve{\Delta}_{a,bc} = \frac{1}{2}\big[\breve{h}^j{}_a(\breve{\partial}_b\breve{h}_{jc} - \breve{\partial}_c\breve{h}_{jb}) + \breve{h}^j{}_c(\breve{\partial}_a\breve{h}_{jb} + \breve{\partial}_b\breve{h}_{ja}) - \breve{h}^j{}_b(\breve{\partial}_a\breve{h}_{jc} + \breve{\partial}_c\breve{h}_{ja})\big], \quad (5.113)$$

Here the scale factors $\breve{h}^i{}_a$ are determined by matrix (5.111), $\breve{\partial}_a$ is the directional derivative along the vectors of the proper orthonormal tetrad $\breve{\boldsymbol{e}}_a$:

$$\breve{\partial}_1 = \pi^i\partial_i, \quad \breve{\partial}_2 = \xi^i\partial_i, \quad \breve{\partial}_3 = \sigma^i\partial_i, \quad \breve{\partial}_4 = u^i\partial_i.$$

On integration of the spinor equations in the Riemannian space it is useful to bear in mind the expression of the Ricci rotation coefficients (5.113) for system of proper tetrads $\breve{\boldsymbol{e}}_a$ which are written directly in components of the vectors π^i, ξ^i, σ^i, u^i:

$$\breve{\Delta}_{1,12} = \xi^i\breve{\partial}_1\pi_i - \pi^i\breve{\partial}_2\pi_i, \qquad \breve{\Delta}_{2,12} = -\pi^i\breve{\partial}_2\xi_i + \xi^i\breve{\partial}_1\xi_i,$$

$$\breve{\Delta}_{3,12} = \frac{1}{2}\big(\xi^i\breve{\partial}_3\pi_i - \pi^i\breve{\partial}_3\xi_i\big) + \frac{1}{2}\big(\sigma^i\breve{\partial}_1\xi_i + \xi^i\breve{\partial}_1\sigma_i - \sigma^i\breve{\partial}_2\pi_i - \pi^i\breve{\partial}_2\sigma_i\big),$$

$$\breve{\Delta}_{4,12} = \frac{1}{2}\big(\xi^i\breve{\partial}_4\pi_i - \pi^i\breve{\partial}_4\xi_i\big) + \frac{1}{2}\big(u^i\breve{\partial}_1\xi_i + \xi^i\breve{\partial}_1 u_i - u^i\breve{\partial}_2\pi_i - \pi^i\breve{\partial}_2 u_i\big),$$

$$\breve{\Delta}_{1,23} = \frac{1}{2}\big(\sigma^i\breve{\partial}_1\xi_i - \xi^i\breve{\partial}_1\sigma_i\big) + \frac{1}{2}\big(\pi^i\breve{\partial}_2\sigma_i + \sigma^i\breve{\partial}_2\pi_i - \pi^i\breve{\partial}_3\xi_i - \xi^i\breve{\partial}_3\pi_i\big),$$

$$\breve{\Delta}_{2,23} = \sigma^i\breve{\partial}_2\xi_i - \xi^i\breve{\partial}_3\xi_i, \qquad \breve{\Delta}_{3,23} = -\xi^i\breve{\partial}_3\sigma_i + \sigma^i\breve{\partial}_2\sigma_i,$$

$$\breve{\Delta}_{4,23} = \frac{1}{2}\big(\sigma^i\breve{\partial}_4\xi_i - \xi^i\breve{\partial}_4\sigma_i\big) + \frac{1}{2}\big(\sigma^i\breve{\partial}_2 u_i + u^i\breve{\partial}_2\sigma_i - \xi^i\breve{\partial}_3 u_i - u^i\breve{\partial}_3\xi_i\big),$$

$$\breve{\Delta}_{1,31} = -\sigma^i\breve{\partial}_1\pi_i + \pi^i\breve{\partial}_3\pi_i, \qquad \breve{\Delta}_{3,31} = \pi^i\breve{\partial}_3\sigma_i - \sigma^i\breve{\partial}_1\sigma_i,$$

$$\breve{\Delta}_{2,31} = \frac{1}{2}\big(\pi^i\breve{\partial}_2\sigma_i - \sigma^i\breve{\partial}_2\pi_i\big) + \frac{1}{2}\big(\pi^i\breve{\partial}_3\xi_i + \xi^i\breve{\partial}_3\pi_i - \xi^i\breve{\partial}_1\sigma_i - \sigma^i\breve{\partial}_1\xi_i\big),$$

$$\breve{\Delta}_{4,31} = \frac{1}{2}\big(\pi^i\breve{\partial}_4\sigma_i - \sigma^i\breve{\partial}_4\pi_i\big) + \frac{1}{2}\big(\pi^i\breve{\partial}_3 u_i + u^i\breve{\partial}_3\pi_i - \sigma^i\breve{\partial}_1 u_i - u^i\breve{\partial}_1\sigma_i\big),$$

$$\breve{\Delta}_{1,14} = u^i\breve{\partial}_1\pi_i - \pi^i\breve{\partial}_4\pi_i, \qquad \breve{\Delta}_{4,14} = -\pi^i\breve{\partial}_4 u_i + u^i\breve{\partial}_1 u_i,$$

$$\breve{\Delta}_{2,14} = \frac{1}{2}\big(u^i\breve{\partial}_2\pi_i - \pi^i\breve{\partial}_2 u_i\big) + \frac{1}{2}\big(\xi^i\breve{\partial}_1 u_i + u^i\breve{\partial}_1\xi_i - \pi^i\breve{\partial}_4\xi_i - \xi^i\breve{\partial}_4\pi_i\big),$$

$$\breve{\Delta}_{3,14} = \frac{1}{2}\big(u^i\breve{\partial}_3\pi_i - \pi^i\breve{\partial}_3 u_i\big) + \frac{1}{2}\big(\sigma^i\breve{\partial}_1 u_i + u^i\breve{\partial}_1\sigma_i - \pi^i\breve{\partial}_4\sigma_i - \sigma^i\breve{\partial}_4\pi_i\big),$$

$$\breve{\Delta}_{1,24} = \frac{1}{2}\big(u^i\breve{\partial}_1\xi_i - \xi^i\breve{\partial}_1 u_i\big) + \frac{1}{2}\big(\pi^i\breve{\partial}_2 u_i + u^i\breve{\partial}_2\pi_i - \pi^i\breve{\partial}_4\xi_i - \xi^i\breve{\partial}_4\pi_i\big),$$

$$\breve{\Delta}_{2,24} = u^i\breve{\partial}_2\xi_i - \xi^i\breve{\partial}_4\xi_i, \qquad \breve{\Delta}_{4,24} = -\xi^i\breve{\partial}_4 u_i + u^i\breve{\partial}_2 u_i,$$

$$\breve{\Delta}_{3,24} = \frac{1}{2}\left(u^i\breve{\partial}_3\xi_i - \xi^i\breve{\partial}_3 u_i\right) + \frac{1}{2}\left(\sigma^i\breve{\partial}_2 u_i + u^i\breve{\partial}_2\sigma_i - \xi^i\breve{\partial}_4\sigma_i - \sigma^i\breve{\partial}_4\xi_i\right),$$

$$\breve{\Delta}_{1,34} = \frac{1}{2}\left(u^i\breve{\partial}_1\sigma_i - \sigma^i\breve{\partial}_1 u_i\right) + \frac{1}{2}\left(\pi^i\breve{\partial}_3 u_i + u^i\breve{\partial}_3\pi_i - \pi^i\breve{\partial}_4\sigma_i - \sigma^i\breve{\partial}_4\pi_i\right),$$

$$\breve{\Delta}_{2,34} = \frac{1}{2}\left(u^i\breve{\partial}_2\sigma_i - \sigma^i\breve{\partial}_2 u_i\right) + \frac{1}{2}\left(\xi^i\breve{\partial}_3 u_i + u^i\breve{\partial}_3\xi_i - \xi^i\breve{\partial}_4\sigma_i - \sigma^i\breve{\partial}_4\xi_i\right),$$

$$\breve{\Delta}_{3,34} = u^i\breve{\partial}_3\sigma_i - \sigma^i\breve{\partial}_4\sigma_i, \qquad \breve{\Delta}_{4,34} = -\sigma^i\breve{\partial}_4 u_i + u^i\breve{\partial}_3 u_i. \tag{5.114}$$

In passing to pseudo-Euclidean space, Eqs. (5.114) pass into Eq. (3.150) by virtue of the orthonormality conditions (3.130) of the proper tetrad π^i, ξ^i, σ^i, u^i.

The tensor differential equations (5.45) and (5.46) corresponding to Eqs. (5.112), in the Riemannian space are written as follows

$$\nabla_i\rho\pi^i + \rho\pi^i M_i = 0, \quad \nabla_i\rho\xi^i + \rho\xi^i M_i = 0,$$
$$\nabla_i\rho\sigma^i + \rho\sigma^i M_i = 0, \quad \nabla_i\rho u^i = 0, \tag{5.115}$$
$$\nabla^i\eta - \frac{1}{2}\varepsilon^{ijms}\left(\pi_j\nabla_m\pi_s + \xi_j\nabla_m\xi_s + \sigma_j\nabla_m\sigma_s - u_j\nabla_m u_s\right) + N^i = 0.$$

Equations (5.115) in the Riemannian space are obtained from Eqs. (5.45), (5.46) in pseudo-Euclidean space by replacement of the partial derivative ∂_i on the covariant derivative ∇_i and they do not require using of orthonormal bases $\boldsymbol{e}_a$.

It is easy to see that the Christoffel symbols Γ^s_{ij} do not appear in the last equation in (5.115), and in other equations in (5.115) the Christoffel symbols appear only in the form of the contraction $g^{ij}\Gamma^s_{ij}$ (if to consider the covariant components of the vectors π^i, ξ^i, σ^i, u^i as unknown functions). Therefore in the harmonic coordinate systems in which by definition the relation $g^{ij}\Gamma^s_{ij} = 0$ is fulfilled, the Christoffel symbols Γ^s_{ij} do not enter into Eqs. (5.115).

Let us consider a symmetric tensor with components T_{ij}, defined in the holonomic basis $\boldsymbol{\mathcal{Э}}_i$ by the relation[6]

$$T_{ij} = \frac{1}{4}\left(\psi^+\gamma_i\nabla_j\psi + \psi^+\gamma_j\nabla_i\psi - \nabla_i\psi^+\cdot\gamma_j\psi - \nabla_j\psi^+\cdot\gamma_i\psi\right). \tag{5.116}$$

Replacing here the covariant derivative ∇_i according to formulas (5.103), we get

$$T_{ij} = \frac{1}{4}\Big[\psi^+\gamma_i\partial_j\psi + \psi^+\gamma_j\partial_i\psi - \partial_i\psi^+\cdot\gamma_j\psi - \partial_j\psi^+\cdot\gamma_i\psi$$
$$+ \frac{1}{2}S_m\left(\Delta_{i,ks}\varepsilon_j{}^{ksm} + \Delta_{j,ks}\varepsilon_i{}^{ksm}\right)\Big]. \tag{5.117}$$

[6]For the Dirac equations the components T_{ij} define the Einstein energy-momentum tensor.

The Ricci rotation coefficients $\Delta_{i,ks}$ in this expression are defined by scale factors $h_i{}^a$ of arbitrary orthonormal tetrads $\boldsymbol{e}_a$ by the formula (see (2.36)):

$$\Delta_{i,ks} = h_k{}^a h_s{}^c \Delta_{i,ac} = \frac{1}{2}\big[h_i{}^a\big(\partial_k h_{sa} - \partial_s h_{ka}\big) + h_k{}^a\big(\partial_i h_{sa} - \partial_s h_{ia}\big) - h_s{}^a\big(\partial_i h_{ka} - \partial_k h_{ia}\big)\big]$$

Replacing in definition (5.116) the covariant derivatives according to formulas (5.105), we obtain an expression of the components T_{ij} in terms of the invariants of the spinor field ρ, η and the Ricci rotation coefficients $\breve{\Delta}_{s,ij}$, determined by the proper orthonormal tetrads $\breve{\boldsymbol{e}}_a$ of the spinor field $\boldsymbol{\psi}$ by relation (5.106):

$$T_{ij} = \frac{1}{4}\rho\left[-\sigma_j\partial_i\eta - \sigma_i\partial_j\eta + \frac{1}{2}\sigma_m\left(\breve{\Delta}_{i,ks}\varepsilon_j{}^{ksm} + \breve{\Delta}_{j,ks}\varepsilon_i{}^{ksm}\right)\right]. \tag{5.118}$$

5.8.2 Formalism of the Spin-Coefficients

Consider in the Riemannian space V_4 an arbitrary smooth field of the orthonormal tetrads $\boldsymbol{e}_a(x^i)$ connected with the golonomic vector basis $\boldsymbol{\partial}_i$ by scale factors $\boldsymbol{e}_a = h^i{}_a \boldsymbol{\partial}_i$. Let us define the complex null tetrads $\boldsymbol{e}^{\circ}_a = \{l^i\boldsymbol{\partial}_i, n^i\boldsymbol{\partial}_i, m^i\boldsymbol{\partial}_i, \dot{m}^i\boldsymbol{\partial}_i\}$ whose components of vectors are defined in terms of scale factors of tetrad $\boldsymbol{e}_a$:

$$\sqrt{2}\,l^i = h^i{}_4 + h^i{}_3, \quad \sqrt{2}\,m^i = h^i{}_1 - \mathrm{i}h^i{}_2,$$
$$\sqrt{2}\,n^i = h^i{}_4 - h^i{}_3, \quad \sqrt{2}\,\dot{m}^i = h^i{}_1 + \mathrm{i}h^i{}_2.$$

From these definitions and orthonormality properties of the vectors of the tetrad $\boldsymbol{e}_a$ it follows that the components of the vectors l^i, n^i, m^i, $\dot{m}^i$ satisfy the equations

$$\dot{m}_i m^i = -l_i n^i = 1,$$
$$l_i l^i = n_i n^i = m_i m^i = l_i m^i = n_i m^i = 0. \tag{5.119}$$

Let us define in the Riemannian space invariant differential operators D, Δ, δ, $\dot{\delta}$ by the equalities

$$D = l^i\nabla'_i, \quad \Delta = n^i\nabla'_i, \quad \delta = m^i\nabla'_i, \quad \dot{\delta} = \dot{m}^i\nabla'_i \tag{5.120}$$

and a system of spin–coefficients, relating to an arbitrary complex null basis $\boldsymbol{e}^{\circ}_a$. These spin–coefficients are calculated by formulas (3.152), in which the operators D, Δ, δ, $\dot{\delta}$ are determined by equalities (5.120). The operator ∇'_i in the definitions act only upon the indices relating to the holonomic basis $\mathbf{\ni}_i$ (so that $\nabla'_i\psi = \partial_i\psi$, $\nabla'_i h_{ja} = \partial_i h_{ja} - \Gamma^s_{ij} h_{sa}$).

It is easy to show that in the Riemannian space the differential operators $\nabla'_a = h^i{}_a \nabla'_i$ are connected with operators D, Δ, δ, $\dot{\delta}$ by the relations

$$\nabla'_1 = \frac{1}{\sqrt{2}}(\delta + \dot{\delta}), \quad \nabla'_3 = \frac{1}{\sqrt{2}}(D - \Delta),$$

$$\nabla'_2 = \frac{\mathrm{i}}{\sqrt{2}}(\delta - \dot{\delta}), \quad \nabla'_4 = \frac{1}{\sqrt{2}}(D + \Delta).$$

Equations (3.151) expressing the derivatives of the vectors of an arbitrary null basis $\boldsymbol{e}^\circ_a$ in terms of the spin–coefficients do not change in passing to the Riemannian space, if the differential operators D, Δ, δ, $\dot{\delta}$ in these equations are determined by equalities (5.120); in the same way the Ricci rotation coefficients (5.114) and the spin–coefficients in the Riemannian space are connected by the same Eqs. (3.153), as in pseudo-Euclidean space.

For the writing of Eqs. (5.110) in the formalism of the spin–coefficients we replace in them the Ricci rotation coefficients $\Delta_{a,bc}$ in terms of the spin–coefficients by formulas (3.153), and operators ∇'_a in terms of operators D, Δ, δ, $\dot{\delta}$. Assuming that the Dirac matrices γ^a are defined by equalities (3.24) after simple transformations we get the system of equations

$$(\delta - \beta + \tau)\psi^4 - (\Delta - \mu + \gamma)\psi^3 - \frac{\mathrm{i}}{\sqrt{2}}\kappa^1 = 0,$$

$$(\dot{\delta} - \pi + \alpha)\psi^3 - (D - \varepsilon + \varrho)\psi^4 - \frac{\mathrm{i}}{\sqrt{2}}\kappa^2 = 0,$$

$$(\delta - \dot{\pi} + \dot{\alpha})\psi^2 + (D - \dot{\varepsilon} + \dot{\varrho})\psi^1 + \frac{\mathrm{i}}{\sqrt{2}}\kappa^3 = 0,$$

$$(\dot{\delta} - \dot{\beta} + \dot{\tau})\psi^1 + (\Delta - \dot{\mu} + \dot{\gamma})\psi^2 + \frac{\mathrm{i}}{\sqrt{2}}\kappa^4 = 0, \tag{5.121}$$

in which for the spinor components $\kappa = \{\kappa^A\}$ is introduced the notation

$$\kappa = \left(\varkappa I + \mathrm{i}\varkappa_a\gamma^a + \frac{\mathrm{i}}{2}\varkappa_{ab}\gamma^{ab} + \overset{*}{\varkappa}_a\overset{*}{\gamma}{}^a + \overset{*}{\varkappa}\gamma^5\right)\psi.$$

Equations (5.121) are the spinor equations (5.110) in formalism of the spin–coefficients. The spin–coefficients and operators D, Δ, δ, $\dot{\delta}$ in Eqs. (5.121) are defined by the arbitrary null tetrads $\boldsymbol{e}^\circ_a$.

Let us get now an expression for the tetrad components of the tensor, defined in the basis $\mathbf{\ni}_i$ by (5.117), in terms of spin–coefficients. The components T_{ij} of any real symmetric tensor of the second rank can be represented as an expansion in vectors of an arbitrary null tetrad $\boldsymbol{e}^{\circ}_{a}$ as follows

$$T_{ij} = T^{\circ}_{00} n_i n_j + \left(T^{\circ}_{11} - \frac{1}{4} T^{\circ}\right)(n_i l_j + n_j l_i) + \left(T^{\circ}_{11} + \frac{1}{4} T^{\circ}\right)(m_i \dot{m}_j + m_j \dot{m}_i)$$
$$+ T^{\circ}_{22} l_i l_j - T^{\circ}_{01}(\dot{m}_i n_j + \dot{m}_j n_i) - T^{\circ}_{10}(m_i n_j + m_j n_i) + T^{\circ}_{20} m_i m_j$$
$$+ T^{\circ}_{02} \dot{m}_i \dot{m}_j - T^{\circ}_{12}(l_i \dot{m}_j + l_j \dot{m}_i) - T^{\circ}_{21}(l_i m_j + l_j m_i). \qquad (5.122)$$

Using Eqs. (5.119), for the coefficients in expansion (5.122) it is easy to find

$$T^{\circ}_{11} - \frac{1}{4} T^{\circ} = l^i n^j T_{ij}, \qquad T^{\circ}_{00} = l^i l^j T_{ij},$$
$$T^{\circ}_{11} + \frac{1}{4} T^{\circ} = m^i \dot{m}^j T_{ij}, \quad T^{\circ}_{02} = \dot{T}^{\circ}_{20} = m^i m^j T_{ij},$$
$$T^{\circ}_{01} = \dot{T}^{\circ}_{10} = l^i m^j T_{ij}, \qquad T^{\circ}_{22} = n^i n^j T_{ij},$$
$$T^{\circ}_{12} = \dot{T}^{\circ}_{21} = n^i m^j T_{ij}, \qquad T^{\circ} = T_i{}^i.$$

The calculation of these quantities for the tensor with components (5.117) in the assumption that the invariant spintensors β and γ^i are determined by matrices (3.24) and (3.25), gives the following relations[7]

$$T^{\circ}_{00} = \frac{\mathrm{i}}{\sqrt{2}}[\dot{\psi}^2 D\psi^2 - \psi^2 D\dot{\psi}^2 + \dot{\psi}^3 D\psi^3 - \psi^3 D\dot{\psi}^3 + \dot{\kappa}(\dot{\psi}^2\psi^1 + \dot{\psi}^4\psi^3)$$
$$- \kappa(\dot{\psi}^1\psi^2 + \dot{\psi}^3\psi^4) + (\dot{\varepsilon} - \varepsilon)(\dot{\psi}^2\psi^2 - \dot{\psi}^3\psi^3)],$$
$$T^{\circ}_{10} = \frac{\mathrm{i}}{2\sqrt{2}}[\dot{\psi}^2\delta\psi^2 - \psi^2\delta\dot{\psi}^2 + \dot{\psi}^3\delta\psi^3 - \psi^3\delta\dot{\psi}^3 - \dot{\psi}^1 D\psi^2 + \psi^2 D\dot{\psi}^1$$
$$+ \dot{\psi}^3 D\psi^4 - \psi^4 D\dot{\psi}^3 - (\varrho + \dot{\varepsilon} + \varepsilon)(\dot{\psi}^1\psi^2 + \dot{\psi}^3\psi^4) + \dot{\sigma}(\dot{\psi}^2\psi^1 + \dot{\psi}^4\psi^3)$$
$$+ \dot{\kappa}(\dot{\psi}^4\psi^4 - \dot{\psi}^1\psi^1) + (\pi + \alpha - \dot{\beta})(\dot{\psi}^3\psi^3 - \dot{\psi}^2\psi^2)],$$
$$T^{\circ}_{22} = \frac{\mathrm{i}}{\sqrt{2}}[\dot{\psi}^1\Delta\psi^1 - \psi^1\Delta\dot{\psi}^1 + \dot{\psi}^4\Delta\psi^4 - \psi^4\Delta\dot{\psi}^4$$
$$+ (\dot{\gamma} - \gamma)(\dot{\psi}^4\psi^4 - \dot{\psi}^1\psi^1) + \nu(\dot{\psi}^2\psi^1 + \dot{\psi}^4\psi^3) - \dot{\nu}(\dot{\psi}^1\psi^2 + \dot{\psi}^3\psi^4)],$$
$$T^{\circ}_{20} = \frac{\mathrm{i}}{\sqrt{2}}[-\dot{\psi}^1\delta\psi^2 + \psi^2\delta\dot{\psi}^1 + \dot{\psi}^3\delta\psi^4 - \psi^4\delta\dot{\psi}^3$$
$$- (\alpha + \dot{\beta})(\dot{\psi}^1\psi^2 + \dot{\psi}^3\psi^4) + \dot{\sigma}(\dot{\psi}^4\psi^4 - \dot{\psi}^1\psi^1) + \lambda(\dot{\psi}^3\psi^3 - \dot{\psi}^2\psi^2)],$$

[7] In the Riemannian space with the metric signature $(-,-,-,+)$ the spin coefficients enter into Eq. (5.123), (5.121) with the opposite sign.

$$\begin{aligned}
T^{\circ}_{12} = \frac{\mathrm{i}}{2\sqrt{2}}\big[&\dot{\psi}^1\delta\psi^1 - \psi^1\delta\dot{\psi}^1 - \dot{\psi}^2\Delta\psi^1 + \psi^1\Delta\dot{\psi}^2 + \dot{\psi}^4\delta\psi^4 - \psi^4\delta\dot{\psi}^4 \\
&- \psi^3\Delta\dot{\psi}^4 + \dot{\psi}^4\Delta\psi^3 + \dot{\nu}(\dot{\psi}^2\psi^2 - \dot{\psi}^3\psi^3) - \dot{\lambda}(\dot{\psi}^1\psi^2 + \dot{\psi}^3\psi^4) \\
&+ (\dot{\alpha} - \beta - \tau)(\dot{\psi}^4\psi^4 - \dot{\psi}^1\psi^1) + (\mu + \dot{\gamma} + \gamma)(\dot{\psi}^2\psi^1 + \dot{\psi}^4\psi^3)\big], \\
T^{\circ}_{11} + \frac{1}{4}T_i{}^i = \frac{\mathrm{i}}{2\sqrt{2}}\big[&- \dot{\psi}^2\dot{\delta}\psi^1 + \psi^1\dot{\delta}\dot{\psi}^2 - \dot{\psi}^1\delta\psi^2 + \psi^2\delta\dot{\psi}^1 + \dot{\psi}^4\dot{\delta}\psi^3 \\
&- \psi^3\dot{\delta}\dot{\psi}^4 + \dot{\psi}^3\delta\psi^4 - \psi^4\delta\dot{\psi}^3 - (\dot{\alpha} + \beta)(\dot{\psi}^1\psi^2 + \dot{\psi}^3\psi^4) \\
&+ (\alpha + \dot{\beta})(\dot{\psi}^2\psi^1 + \dot{\psi}^4\psi^3) + (\dot{\varrho} - \varrho)(\dot{\psi}^4\psi^4 - \dot{\psi}^1\psi^1) \\
&+ (\dot{\mu} - \mu)(\dot{\psi}^2\psi^2 - \dot{\psi}^3\psi^3)\big], \\
T^{\circ}_{11} - \frac{1}{4}T_i{}^i = \frac{\mathrm{i}}{2\sqrt{2}}\big[&\dot{\psi}^1 D\psi^1 - \psi^1 D\dot{\psi}^1 + \dot{\psi}^2\Delta\psi^2 - \psi^2\Delta\dot{\psi}^2 + \dot{\psi}^3\Delta\psi^3 \\
&- \psi^3\Delta\dot{\psi}^3 + \dot{\psi}^4 D\psi^4 - \psi^4 D\dot{\psi}^4 + (\dot{\tau} + \pi)(\dot{\psi}^2\psi^1 + \dot{\psi}^4\psi^3) \\
&- (\tau + \dot{\pi})(\dot{\psi}^1\psi^2 + \dot{\psi}^3\psi^4) + (\dot{\varepsilon} - \varepsilon)(\dot{\psi}^4\psi^4 - \dot{\psi}^1\psi^1) \\
&+ (\dot{\gamma} - \gamma)(\dot{\psi}^2\psi^2 - \dot{\psi}^3\psi^3)\big].
\end{aligned} \tag{5.123}$$

The spin–coefficients and operators D, Δ, δ, $\dot{\delta}$ in equalities (5.123) are defined by arbitrary null tetrads $\overset{\circ}{\boldsymbol{e}}_a$.

When using Eqs. (5.121), (5.123) in the formalism of the spin-coefficients it is also necessary to specify a gauge of the null tetrads $\overset{\circ}{\boldsymbol{e}}_a(x^i)$. As the gauge of tetrads $\overset{\circ}{\boldsymbol{e}}_a(x^i)$ one can accept that tetrads $\overset{\circ}{\boldsymbol{e}}_a(x^i)$ coincide with the proper null tetrads $\breve{\overset{\circ}{\boldsymbol{e}}}_a(x^i)$, which are connected with the proper orthonormal tetrads of the spinor field $\boldsymbol{\psi}(x^i)$ by the algebraic equalities

$$\sqrt{2}\,l^i = u^i + \sigma^i, \quad \sqrt{2}\,m^i = \pi^i - \mathrm{i}\xi^i,$$
$$\sqrt{2}\,n^i = u^i - \sigma^i, \quad \sqrt{2}\,\dot{m}^i = \pi^i + \mathrm{i}\xi^i.$$

If we take as tetrads $\overset{\circ}{\boldsymbol{e}}_a(x^i)$ in Eqs. (5.121) the proper null tetrads $\breve{\overset{\circ}{\boldsymbol{e}}}_a(x^i)$, then Eqs. (5.121) take the following form (for simplicity we write out here the equations at $\kappa^A = m\psi^A$, $m = $ const, i.e., for the Dirac equation) [93]

$$\begin{aligned}
\delta G &= \beta - \tau, \\
\dot{\delta} G &= \pi - \alpha, \\
DG &= \varepsilon - \varrho - \frac{\mathrm{i}m}{\sqrt{2}}e^{\mathrm{i}\eta}, \\
\Delta G &= \mu - \gamma + \frac{\mathrm{i}m}{\sqrt{2}}e^{\mathrm{i}\eta},
\end{aligned} \tag{5.124}$$

where G is the complex invariant of the spinor field

$$G = \frac{1}{2}(\ln \rho - \mathrm{i}\eta),$$

and the spin–coefficients are defined by the proper null basis $\breve{\boldsymbol{e}}^{\circ}_{a}(x^i)$.

When using the gauge $\boldsymbol{e}^{\circ}_{a}(x^i) = \breve{\boldsymbol{e}}^{\circ}_{a}(x^i)$, definitions (5.123) are also considerable simplified and take the form

$$\begin{aligned}
T^{\circ}_{00} &= \frac{\rho}{2\sqrt{2}}[-D\eta + \mathrm{i}(\dot{\varepsilon} - \varepsilon)],\\
T^{\circ}_{11} &= \frac{\rho}{4\sqrt{2}}[D\eta - \Delta\eta + \mathrm{i}(\dot{\varepsilon} - \varepsilon + \dot{\gamma} - \gamma)],\\
T^{\circ}_{22} &= \frac{\rho}{2\sqrt{2}}[\Delta\eta + \mathrm{i}(\dot{\gamma} - \gamma)],\\
T^{\circ}_{01} &= \dot{T}^{\circ}_{10} = \frac{\rho}{4\sqrt{2}}[-\delta\eta + \mathrm{i}(-\kappa - \beta + \dot{\alpha} + \dot{\pi})],\\
T^{\circ}_{02} &= \dot{T}^{\circ}_{20} = \frac{\mathrm{i}\rho}{2\sqrt{2}}(-\sigma + \dot{\lambda}),\\
T^{\circ}_{12} &= T^{\circ}_{21} = \frac{\rho}{4\sqrt{2}}[\delta\eta + \mathrm{i}(-\beta - \tau + \dot{\alpha} + \dot{\nu})],\\
T^{\circ} &= T_i{}^{i}.
\end{aligned} \tag{5.125}$$

Equations (5.124) and (5.125) can be obtained from Eqs. (5.112) and (5.118), replacing in them the Ricci rotation coefficients of proper orthonormal bases in terms of the spin-coefficients of the proper null basis by formulas (3.153).

The spin-coefficients and the Ricci rotation coefficients are connected by linear equalities (3.153), therefore Eqs. (5.112), (5.118) and (5.124), (5.125) in the complexity are identical. However, the use of Eqs. (5.112), (5.118) seems to be more preferable, since these equations have compact invariant and comprehensible writing, unlike corresponding to them Eqs. (5.124), (5.125) in the formalism of the spin-coefficients.

In general relativity the functions $g_{ij}(x^s)$ are the desired unknown functions. Therefore the components of the tetrad vectors π_i, ξ_i, σ_i, u_i (and l_i, n_i, m_i, $\dot{m}_i$) must be considered as arbitrary functions of the variables x^i, since in this case the orthonormality conditions are fulfilled due to definition of the components of the metric tensor

$$g_{ij} = \pi_i\pi_j + \xi_i\xi_j + \sigma_i\sigma_j - u_iu_j = -l_in_j - l_jn_i + m_i\dot{m}_j + m_j\dot{m}_i.$$

5.8.3 Weyl Equations in the Riemannian Space

Let ψ be the four-component spinor field in the Riemannian space V_4, specified in spinbasis $\overset{*}{\boldsymbol{\varepsilon}}_A$ by the components

$$\psi = \begin{Vmatrix} \xi \\ \eta \end{Vmatrix}, \tag{5.126}$$

where ξ, η are the two-component spinor fields in the space V_4. Using expressions (3.94) for the matrices γ^{ab}, formula (5.103) for the covariant derivative of the spinor field ψ can be represented in the form

$$\nabla_i \psi = \partial_i \psi - \frac{1}{4} \Delta_{i,bc} \gamma^{bc} \psi = \begin{Vmatrix} \partial_i \xi \\ \partial_i \eta \end{Vmatrix} - \frac{1}{4} \Delta_{i,\alpha\beta} \begin{Vmatrix} \sigma^{[\alpha} \sigma^{\beta]} \xi \\ \sigma^{[\alpha} \sigma^{\beta]} \eta \end{Vmatrix} - \frac{1}{2} \Delta_{i,4\alpha} \begin{Vmatrix} -\sigma_\alpha \xi \\ \sigma_\alpha \eta \end{Vmatrix}.$$

From this it follows

$$\nabla_i \xi = \partial_i \xi - \frac{1}{4} \Delta_{i,\alpha\beta} \sigma^{[\alpha} \sigma^{\beta]} \xi + \frac{1}{2} \Delta_{i,4\alpha} \sigma^\alpha \xi,$$

$$\nabla_i \eta = \partial_i \eta - \frac{1}{4} \Delta_{i,\alpha\beta} \sigma^{[\alpha} \sigma^{\beta]} \eta - \frac{1}{2} \Delta_{i,4\alpha} \sigma^\alpha \eta.$$

Taking into account definition (3.100) of the spintensors $\sigma^{ij} = \|\sigma^B{}_A{}^{ij}\|$, formulas for $\nabla_i \xi$ and $\nabla_i \eta$ can be written down in the matrix form

$$\nabla_i \xi = \partial_i \xi + \frac{\mathrm{i}}{4} \Delta_{i,jk} \sigma^{jk} \xi,$$
$$\nabla_i \eta = \partial_i \eta + \frac{\mathrm{i}}{4} \Delta_{i,jk} (\dot{\sigma}^{jk})^T \eta, \tag{5.127}$$

or, in the component form

$$\nabla_i \xi^B = \partial_i \xi^B + \frac{\mathrm{i}}{4} \Delta_{i,jk} \sigma^B{}_A{}^{jk} \xi^A,$$

$$\nabla_i \eta_{\dot{A}} = \partial_i \eta_{\dot{A}} + \frac{\mathrm{i}}{4} \Delta_{i,jk} \sigma^{\dot{B}}{}_{\dot{A}}{}^{jk} \eta_{\dot{B}}.$$

The quantities $\nabla_i \xi^B$, $\nabla_i \eta_{\dot{A}}$ by definition are called the covariant derivatives of the two-component spinor fields ξ, η. The Ricci rotation coefficients $\Delta_{i,jk}$ in these equations are defined by the arbitrary orthonormal bases $\boldsymbol{e}_a(x^i)$.

Formulas (3.205) expressing derivatives of the components of semispinors in terms of the Ricci rotation coefficients of the proper tetrad and invariants ρ, η in the

Riemannian space are written as follows

$$\nabla_s \xi^B = \frac{1}{2}\xi^B \nabla_s(\ln \rho + \mathrm{i}\eta) + \frac{\mathrm{i}}{4}\breve{\Delta}_{s,ij}\sigma^B{}_A{}^{ij}\xi^A,$$

$$\nabla_s \eta_{\dot{A}} = \frac{1}{2}\eta_{\dot{A}} \nabla_s(\ln \rho - \mathrm{i}\eta) + \frac{\mathrm{i}}{4}\breve{\Delta}_{s,ij}\dot{\sigma}^B{}_A{}^{ij}\eta_{\dot{B}}. \tag{5.128}$$

The Ricci rotation coefficients $\breve{\Delta}_{s,ij}$ in Eqs. (5.128) are defined by the proper tetrad of the spinor field (5.126) by formula (5.106).

The Weyl equations (5.81) and the symmetric components of the Einstein energy-momentum tensor of the spinor field ξ in the Riemannian space are written in the following way

$$\sigma^i_{\dot{B}A}\nabla_i \xi^A = 0, \tag{5.129}$$

$$T_{ij} = \frac{\mathrm{i}}{4}\sigma_{\dot{B}Aj}\left(\dot{\xi}^B \nabla_i \xi^A - \xi^A \nabla_i \dot{\xi}^B\right) + \frac{\mathrm{i}}{4}\sigma_{\dot{B}Ai}\left(\dot{\xi}^B \nabla_j \xi^A - \xi^A \nabla_j \dot{\xi}^B\right).$$

The Weyl equations (5.80) and the symmetric components of the Einstein energy-momentum tensor of the spinor field η in the Riemannian space V_4 have the form

$$\sigma^{B\dot{A}i}\nabla_i \eta_{\dot{A}} = 0, \tag{5.130}$$

$$T_{ij} = \frac{\mathrm{i}}{4}\sigma_j^{B\dot{A}}\left(\dot{\eta}_{\dot{B}}\nabla_i \eta_{\dot{A}} - \eta_{\dot{A}}\nabla_i \dot{\eta}_{\dot{B}}\right) + \frac{\mathrm{i}}{4}\sigma_i^{B\dot{A}}\left(\dot{\eta}_{\dot{B}}\nabla_j \eta_{\dot{A}} - \eta_{\dot{A}}\nabla_j \dot{\eta}_{\dot{B}}\right).$$

Replacing in the first equation in (5.129) the covariant derivative $\nabla_i \xi^B$ by formula (5.127) and using formulas (3.105) for the contraction of spintensors σ^i and σ^{jk}, the Weyl equations for the spinor field ξ can be transformed to the form

$$\sigma^i_{\dot{B}A}\left[\partial_i \xi^A + \frac{1}{2}\left(\Delta_{j,i}{}^j + \frac{\mathrm{i}}{2}\varepsilon_{ijmn}\Delta^{j,mn}\right)\xi^A\right] = 0. \tag{5.131}$$

In the same way the Weyl equations for the dotted components of the spinor $\eta_{\dot{A}}$ are written down in the form:

$$\sigma^{B\dot{A}i}\left[\partial_i \eta_{\dot{A}} + \frac{1}{2}\left(\Delta_{j,i}{}^j - \frac{\mathrm{i}}{2}\varepsilon_{ijmn}\Delta^{j,mn}\right)\eta_{\dot{A}}\right] = 0. \tag{5.132}$$

The tensor equations, corresponding to the Weyl equations in the Riemannian space, are got by the replacement in Eqs. (5.90), (5.91), and (5.95) of the partial derivatives by covariant ones:

$$C^{mn}\nabla_j C_i{}^j + C^m{}_j \nabla_i C^{nj} = 0,$$

$$j^s \nabla_j C^{qj} = j_j \nabla^q C^{js},$$

$$j^s \nabla_q j^k - j^k \nabla_q j^s = \frac{1}{2}\left(\dot{C}^{sk}\nabla_j C_q{}^j + C^{sk}\nabla_j \dot{C}_q{}^j\right).$$

The Weyl equations and the components of the Einstein energy-momentum tensor (5.129), (5.130) in the Riemannian space in the formalism of the spin–coefficients in an arbitrary null basis $\boldsymbol{e}_a^\circ$ are obtained from Eqs. (5.121) if to put in them, respectively, $\psi^1 = \psi^2 = 0$, $\psi^3 = \eta_{\dot{1}}$, $\psi^4 = \eta_{\dot{2}}$, $\kappa^A = 0$ and $\psi^1 = \xi^1$, $\psi^2 = \xi^2$, $\psi^3 = \psi^4 = 0$, $\kappa^A = 0$. Thus, it is possible to get that the following equations correspond to the Weyl equations for the field ξ:

$$(\delta - \dot{\pi} + \dot{\alpha})\xi^2 + (D - \dot{\varepsilon} + \dot{\varrho})\xi^1 = 0,$$

$$(\dot{\delta} - \dot{\beta} + \dot{\tau})\xi^1 + (\Delta - \dot{\mu} + \dot{\gamma})\xi^2 = 0,$$

The components of the Einstein energy-momentum tensor in an arbitrary null basis $\boldsymbol{e}_a^\circ$ have the form

$$T^\circ_{00} = \frac{\mathrm{i}}{\sqrt{2}}\left[\dot{\xi}^2 D\xi^2 - \xi^2 D\dot{\xi}^2 + \dot{\kappa}\dot{\xi}^2\xi^1 - \kappa\dot{\xi}^1\xi^2 + (\dot{\varepsilon} - \varepsilon)\dot{\xi}^2\xi^2\right],$$

$$T^\circ_{10} = \frac{\mathrm{i}}{2\sqrt{2}}\left[\dot{\xi}^2\delta\xi^2 - \xi^2\delta\dot{\xi}^2 - \dot{\xi}^1 D\xi^2 + \xi^2 D\dot{\xi}^1 - (\varrho + \dot{\varepsilon} + \varepsilon)\dot{\xi}^1\xi^2 + \dot{\sigma}\dot{\xi}^2\xi^1 - \dot{\kappa}\dot{\xi}^1\xi^1 - (\pi + \alpha - \dot{\beta})\dot{\xi}^2\xi^2\right],$$

$$T^\circ_{22} = \frac{\mathrm{i}}{\sqrt{2}}\left[\dot{\xi}^1\Delta\xi^1 - \xi^1\Delta\dot{\xi}^1 - (\dot{\gamma} - \gamma)\dot{\xi}^1\xi^1 + \nu\dot{\xi}^2\xi^1 - \dot{\nu}\dot{\xi}^1\xi^2\right],$$

$$T^\circ_{20} = \frac{\mathrm{i}}{\sqrt{2}}\left[- \dot{\xi}^1\delta\xi^2 + \xi^2\delta\dot{\xi}^1 - (\alpha + \dot{\beta})\dot{\xi}^1\xi^2 - \dot{\sigma}\dot{\xi}^1\xi^1 - \lambda\dot{\xi}^2\xi^2\right],$$

$$T^\circ_{12} = \frac{\mathrm{i}}{2\sqrt{2}}\left[\dot{\xi}^1\delta\xi^1 - \xi^1\delta\dot{\xi}^1 - \dot{\xi}^2\Delta\xi^1 + \xi^1\Delta\dot{\xi}^2 + (\mu + \dot{\gamma} + \gamma)\dot{\xi}^2\xi^1 - \dot{\lambda}\dot{\xi}^1\xi^2 - (\dot{\alpha} - \beta - \tau)\dot{\xi}^1\xi^1 + \dot{\nu}\dot{\xi}^2\xi^2\right],$$

$$T^\circ_{11} + \frac{1}{4}T_i^{\ i} = \frac{\mathrm{i}}{2\sqrt{2}}\left[- \dot{\xi}^2\delta\xi^1 + \xi^1\dot{\delta}\dot{\xi}^2 - \dot{\xi}^1\delta\xi^2 + \xi^2\delta\dot{\xi}^1 - (\dot{\alpha} + \beta)\dot{\xi}^1\xi^2 + (\alpha + \dot{\beta})\dot{\xi}^2\xi^1 - (\dot{\varrho} - \varrho)\dot{\xi}^1\xi^1 + (\dot{\mu} - \mu)\dot{\xi}^2\xi^2\right],$$

$$T^\circ_{11} - \frac{1}{4}T_i^{\ i} = \frac{\mathrm{i}}{2\sqrt{2}}\left[\dot{\xi}^1 D\xi^1 - \xi^1 D\dot{\xi}^1 + \dot{\xi}^2\Delta\xi^2 - \xi^2\Delta\dot{\xi}^2 + (\dot{\tau} + \pi)\dot{\xi}^2\xi^1 - (\tau + \dot{\pi})\dot{\xi}^1\xi^2 - (\dot{\varepsilon} - \varepsilon)\dot{\xi}^1\xi^1 + (\dot{\gamma} - \gamma)\dot{\xi}^2\xi^2\right].$$

Here the differentiation operators D, Δ, δ, $\dot{\delta}$ are defined according to (5.120).

Let us write out also the Weyl equations for the spinor field η in an arbitrary null tetrad $\boldsymbol{e}_a^\circ$:

$$(\delta - \beta + \tau)\eta_{\dot{2}} - (\Delta - \mu + \gamma)\eta_{\dot{1}} = 0,$$

$$(\dot{\delta} - \pi + \alpha)\eta_{\dot{1}} - (D - \varepsilon + \varrho)\eta_{\dot{2}} = 0$$

and the components of the Einstein energy-momentum tensor (5.130) corresponding to them in the null tetrad $\boldsymbol{e}^{\circ}_{a}$:

$$
\begin{aligned}
T^{\circ}_{00} &= \frac{\mathrm{i}}{\sqrt{2}}\big[\dot\eta_{\dot 1} D\eta_{\dot 1} - \eta_{\dot 1} D\dot\eta_{\dot 1} + \dot\kappa\dot\eta_{\dot 2}\eta_{\dot 1} - \kappa\dot\eta_{\dot 1}\eta_{\dot 2} - (\dot\varepsilon - \varepsilon)\dot\eta_{\dot 1}\eta_{\dot 1}\big],\\
T^{\circ}_{10} &= \frac{\mathrm{i}}{2\sqrt{2}}\big[\dot\eta_{\dot 1}\dot\delta\eta_{\dot 1} - \eta_{\dot 1}\dot\delta\dot\eta_{\dot 1} + \dot\eta_{\dot 1} D\eta_{\dot 2} - \eta_{\dot 2} D\dot\eta_{\dot 1} - (\varrho + \dot\varepsilon + \varepsilon)\dot\eta_{\dot 1}\eta_{\dot 2} + \dot\sigma\dot\eta_{\dot 2}\eta_{\dot 1}\\
&\quad + \dot\kappa\dot\eta_{\dot 2}\eta_{\dot 2} + (\pi + \alpha - \dot\beta)\dot\eta_{\dot 1}\eta_{\dot 1}\big],\\
T^{\circ}_{22} &= \frac{\mathrm{i}}{\sqrt{2}}\big[\dot\eta_{\dot 2}\Delta\eta_{\dot 2} - \eta_{\dot 2}\Delta\dot\eta_{\dot 2} + (\dot\gamma - \gamma)\dot\eta_{\dot 2}\eta_{\dot 2} + \nu\dot\eta_{\dot 2}\eta_{\dot 1} - \dot\nu\dot\eta_{\dot 1}\eta_{\dot 2}\big],\\
T^{\circ}_{20} &= \frac{\mathrm{i}}{\sqrt{2}}\big[\dot\eta_{\dot 1}\dot\delta\eta_{\dot 2} - \eta_{\dot 2}\dot\delta\dot\eta_{\dot 1} - (\alpha + \dot\beta)\dot\eta_{\dot 1}\eta_{\dot 2} + \dot\sigma\dot\eta_{\dot 2}\eta_{\dot 2} + \lambda\dot\eta_{\dot 1}\eta_{\dot 1}\big],\\
T^{\circ}_{12} &= \frac{\mathrm{i}}{2\sqrt{2}}\big[\dot\eta_{\dot 2}\delta\eta_{\dot 2} - \eta_{\dot 2}\delta\dot\eta_{\dot 2} - \eta_{\dot 1}\Delta\dot\eta_{\dot 2} + \dot\eta_{\dot 2}\Delta\eta_{\dot 1} + (\mu + \dot\gamma + \gamma)\dot\eta_{\dot 2}\eta_{\dot 1}\\
&\quad - \dot\lambda\dot\eta_{\dot 1}\eta_{\dot 2} + (\dot\alpha - \beta - \tau)\dot\eta_{\dot 2}\eta_{\dot 2} - \dot\nu\dot\eta_{\dot 1}\eta_{\dot 1}\big],\\
T^{\circ}_{11} + \frac{1}{4}T_i{}^i &= \frac{\mathrm{i}}{2\sqrt{2}}\big[\dot\eta_{\dot 2}\dot\delta\eta_{\dot 1} - \eta_{\dot 1}\dot\delta\dot\eta_{\dot 2} + \dot\eta_{\dot 1}\delta\eta_{\dot 2} - \eta_{\dot 2}\delta\dot\eta_{\dot 1} - (\dot\alpha + \beta)\dot\eta_{\dot 1}\eta_{\dot 2}\\
&\quad + (\alpha + \dot\beta)\dot\eta_{\dot 2}\eta_{\dot 1} + (\dot\varrho - \varrho)\dot\eta_{\dot 2}\eta_{\dot 2} - (\dot\mu - \mu)\dot\eta_{\dot 1}\eta_{\dot 1}\big],\\
T^{\circ}_{11} - \frac{1}{4}T_i{}^i &= \frac{\mathrm{i}}{2\sqrt{2}}\big[\dot\eta_{\dot 1}\Delta\eta_{\dot 1} - \eta_{\dot 1}\Delta\dot\eta_{\dot 1} + \dot\eta_{\dot 2} D\eta_{\dot 2} - \eta_{\dot 2} D\dot\eta_{\dot 2} + (\dot\tau + \pi)\dot\eta_{\dot 2}\eta_{\dot 1}\\
&\quad - (\tau + \dot\pi)\dot\eta_{\dot 1}\eta_{\dot 2} + (\dot\varepsilon - \varepsilon)\dot\eta_{\dot 2}\eta_{\dot 2} - (\dot\gamma - \gamma)\dot\eta_{\dot 1}\eta_{\dot 1}\big].
\end{aligned}
$$

We obtain now the tensor formulation of the Weyl equations and an expression for the Einstein energy-momentum tensor (5.129), (5.130) in the components of the vectors of the proper tetrad $\breve{\boldsymbol{e}}_a$ determined by the two-component spinor field (see Sect. 3.4, Chap. 3). We consider at first equation (5.130) for the spinor field η. Let us replace the derivative $\nabla_i\eta_{\dot A}$ in the Weyl equations (5.130) by formula (5.128). For simplicity and for definiteness we accept that the invariants ρ, η in formulas (3.154), (3.156) and, therefore, in formula (5.128), are defined as follows

$$
\rho = \sqrt{2}, \qquad \eta = \text{const.} \tag{5.133}
$$

In this case formula (3.156) for the components of the vectors of the proper null tetrad $\breve{\boldsymbol{e}}^{\circ}_a$ pass into the following equalities

$$
\begin{aligned}
l^i = -\sigma^i_{\dot B A}\eta^{\dot B}\dot\eta^{A}, \quad n^i &= -\sigma^i_{\dot B A}\dot\xi^{\dot B}\xi^{A},\\
m^i &= -\sigma^i_{\dot B A}\eta^{\dot B}\xi^{A},
\end{aligned} \tag{5.134}
$$

while formula (5.128) for the derivative $\nabla_s \eta_{\dot{A}}$ takes the form

$$\nabla_s \eta_{\dot{A}} = \frac{\mathrm{i}}{4} \breve{\Delta}_{s,ij} \dot{\sigma}^{B}{}_{A}{}^{ij} \eta_{\dot{B}}. \tag{5.135}$$

Replacing the derivative $\nabla_s \eta_{\dot{A}}$ in Eq. (5.130) by formula (5.135) and making transformations with the aid of equality (3.105), we get (in the orthonormal tetrad $\boldsymbol{e}_a$)

$$\sigma^{A\dot{D}a}\left(\breve{\Delta}_{b,a}{}^{b} - \frac{\mathrm{i}}{2}\varepsilon_{abcd}\breve{\Delta}^{b,cd}\right)\eta_{\dot{D}} = 0.$$

Contraction of this equation with components of spinors $\dot{\eta}_{\dot{A}}$ and ξ_A gives the equations[8]

$$\begin{gathered} l^{a}\breve{\Delta}_{b,a}{}^{b} = 0, \quad \varepsilon^{abcd} l_a \breve{\Delta}_{bcd} = 0, \\ m^{a}\left(\breve{\Delta}_{b,a}{}^{b} - \frac{\mathrm{i}}{2}\varepsilon_{abcd}\breve{\Delta}^{b,cd}\right) = 0, \end{gathered} \tag{5.136}$$

while for the component of the Einstein energy-momentum tensor T_{ab} we get the following expression

$$T_{ab} = \frac{1}{8} l^{e}\left(\varepsilon_{acde}\breve{\Delta}_{b,}{}^{cd} + \varepsilon_{bcde}\breve{\Delta}_{a,}{}^{cd}\right). \tag{5.137}$$

From definitions (5.82) and (5.134) it follows that in this case the components of the vector l^i coincide with the components j^i of the current vector of the field $\eta_{\dot{A}}$. From Eqs. (3.149) and (3.145) it follows that the first equation in (5.136) is the continuity equation for current vector

$$\nabla_i l^i = 0.$$

In the same way, it is obtained that from the Weyl equations (5.129) for the spinor field ξ provided (5.133) it follows

$$\sigma^{a}_{\dot{B}A}\left(\breve{\Delta}_{b,a}{}^{b} + \frac{\mathrm{i}}{2}\varepsilon_{abcd}\breve{\Delta}^{b,cd}\right)\xi^{A} = 0$$

[8]The field of the spinor ξ is used here only to determine the tetrad $\breve{e}^{\circ}_{a}$ and is not related to the Weyl equation (5.129).

and

$$n^a \breve{\Delta}_{b,a}{}^b = 0, \quad \varepsilon^{abcd} n_a \breve{\Delta}_{b,cd} = 0,$$

$$m^a \left(\breve{\Delta}_{b,a}{}^b + \frac{\mathrm{i}}{2} \varepsilon_{abcd} \breve{\Delta}^{b,cd} \right) = 0, \tag{5.138}$$

The components of the Einstein energy-momentum tensor in this case are defined by the relation

$$T_{ab} = -\frac{1}{8} n^e \left(\varepsilon_{acde} \breve{\Delta}_{b,}{}^{cd} + \varepsilon_{bcde} \breve{\Delta}_{a,}{}^{cd} \right). \tag{5.139}$$

For the Weyl equations (5.129) the components n^i coincide with components j^i of the current vector of the field ξ^A, and the first equation in (5.138) is the continuity equation

$$\nabla_i n^i = 0$$

To write the Weyl equations and the Einstein energy-momentum tensor in the proper null tetrads $\breve{e}^{\circ}_a$ provided (5.133), we must replace in Eqs. (5.136), (5.137) and (5.138), (5.139) the Ricci rotation coefficients in terms of the spin–coefficients by formulas (3.153). As a result, the Weyl equations for the spinor field ξ in the proper null tetrad are written in the form

$$\pi = \alpha, \qquad \mu = \gamma, \tag{5.140}$$

while the components of the Einstein energy-momentum tensor in the form

$$\begin{aligned}
\breve{T}^{\circ}_{00} &= \frac{\mathrm{i}}{2}(\dot{\varepsilon} - \varepsilon), & \breve{T}^{\circ}_{11} &= \frac{\mathrm{i}}{4}(\dot{\gamma} - \gamma),\\
\breve{T}^{\circ}_{01} &= \frac{\mathrm{i}}{4}(\dot{\alpha} - \beta + \dot{\pi}), & \breve{T}^{\circ}_{10} &= \frac{\mathrm{i}}{4}(\dot{\beta} - \alpha - \pi),\\
\breve{T}^{\circ}_{12} &= \frac{\mathrm{i}}{4}\dot{\nu}, & \breve{T}^{\circ}_{21} &= -\frac{\mathrm{i}}{4}\nu,\\
\breve{T}^{\circ}_{22} &= 0, & \breve{T}^{\circ}_{02} &= \frac{\mathrm{i}}{2}\dot{\lambda},\\
\breve{T}^{\circ}_{20} &= -\frac{\mathrm{i}}{2}\lambda, & T_i{}^i &= \frac{\mathrm{i}}{2}(\dot{\mu} - \mu + \gamma - \dot{\gamma}) = 0.
\end{aligned} \tag{5.141}$$

The Weyl equations for the spinor field η in the proper null bases $\breve{e}^{\circ}_a$ provided (5.133) are obtained from Eqs. (5.136) and have the form

$$\varepsilon = \varrho, \qquad \beta = \tau. \tag{5.142}$$

The components of the Einstein energy-momentum tensor in the proper null tetrads $\breve{\boldsymbol{e}}^{\circ}_{a}$ for the spinor field ξ are determined by the relations:

$$\begin{aligned}
\breve{T}^{\circ}_{00} &= 0, & \breve{T}^{\circ}_{11} &= \frac{\mathrm{i}}{4}(\dot{\varepsilon} - \varepsilon),\\
\breve{T}^{\circ}_{01} &= -\frac{\mathrm{i}}{4}\kappa, & \breve{T}^{\circ}_{10} &= \frac{\mathrm{i}}{4}\dot{\kappa},\\
\breve{T}^{\circ}_{12} &= \frac{\mathrm{i}}{4}(\dot{\alpha} - \beta - \tau), & \breve{T}^{\circ}_{21} &= \frac{\mathrm{i}}{4}(\dot{\beta} - \alpha + \dot{\tau}),\\
\breve{T}^{\circ}_{22} &= \frac{\mathrm{i}}{2}(\dot{\gamma} - \gamma), & \breve{T}^{\circ}_{20} &= \frac{\mathrm{i}}{2}\dot{\sigma},\\
\breve{T}^{\circ}_{02} &= -\frac{\mathrm{i}}{2}\sigma, & T_{i}{}^{i} &= \frac{\mathrm{i}}{2}(\dot{\rho} - \rho + \varepsilon - \dot{\varepsilon}) = 0.
\end{aligned} \tag{5.143}$$

Within the framework of general relativity (GR) the Weyl equations in the tensor formulation contain at four unknown functions less, than the Weyl equations in their classical spinor formulation. This is related to the fact that for definition of the spinor fields in the Riemannian space a system of tetrads $\boldsymbol{e}_a$ (or $\boldsymbol{e}^{\circ}_{a}$) is introduced, which is defined up to the arbitrary Lorentz transformations depending on six arbitrary functions. At the tensor formulation of the Weyl equations a special system of tetrads $\breve{\boldsymbol{e}}_a$ (or $\breve{\boldsymbol{e}}^{\circ}_{a}$) is used as $\boldsymbol{e}_a$ ($\boldsymbol{e}^{\circ}_{a}$), which is defined by the spinor field only to within two real functions. Besides, within the formalism of the spin–coefficients, when spin–coefficients are considered as the arbitrary functions, the Weyl equations (5.131) or (5.132) (nonlinear differential equations in partial derivatives) are replaced in the theory proposed here by the linear algebraic equations (5.140) or (5.142). With these two circumstances is connected essential simplification of the theory of the massless spin 1/2 field.

It is obvious that all tensor systems of the equations in the components $\breve{\boldsymbol{e}}_a$, $\boldsymbol{e}^{\circ}_{a}$, C^{ij}, j^i corresponding to the Weyl equations, of course, are equivalent; however, use of various systems of unknown functions leads to tensor equations of varying complexity. Equations (5.140), (5.141) or (5.138), (5.139) and (5.142), (5.143) or (5.136), (5.137), seemingly, are most convenient for use in GR.

5.9 The Spinor Differential Equations in Three-Dimensional Euclidean Space

Let $\boldsymbol{\psi}$ be the first rank spinor in the three-dimensional Euclidean space E^0_3 referred to an orthonormal basis $\boldsymbol{\ni}_\alpha$. Let us assume that the spinor components $\boldsymbol{\psi}$ are determined as function of some scalar parameter s. Consider a differential spinor equation of the form

$$\frac{\mathrm{d}}{\mathrm{d}s}\psi = \frac{1}{2}(R_\alpha + \mathrm{i}H_\alpha)\sigma^\alpha\psi + \frac{1}{2}(Q_\alpha + \mathrm{i}N_\alpha)\sigma^\alpha\bar{\psi}, \tag{5.144}$$

in which the real components Q_α, N_α, R_α, and H_α, calculated in the basis $\mathcal{Э}_\alpha$, determine three-dimensional vectors in the space E_3^0; σ^α are the two-dimensional Pauli matrices; $\bar{\psi}$ is the column of the contravariant conjugate spinor components ψ^{+A}, $A = 1, 2$.

Let us obtain a writing of the spinor equations (5.144) in the form of equivalent tensor equations. For this purpose we replace the derivative $d\psi/ds$ in Eqs. (5.144) by formula (4.78). We obtain

$$\frac{1}{\rho}\frac{\mathrm{d}\rho}{\mathrm{d}s}\psi = \left[R_\alpha + \mathrm{i}(H_\alpha + \Omega_\alpha)\right]\sigma^\alpha\psi + (Q_\alpha + \mathrm{i}N_\alpha)\sigma^\alpha\bar{\psi}. \tag{5.145}$$

Here the vector components Ω^α are determined by the relation

$$\Omega^\alpha = \frac{1}{2}\varepsilon^{\alpha\beta\eta}\left(\pi_\beta\frac{\mathrm{d}}{\mathrm{d}s}\pi_\eta + \xi_\beta\frac{\mathrm{d}}{\mathrm{d}s}\xi_\eta + n_\beta\frac{\mathrm{d}}{\mathrm{d}s}n_\eta\right).$$

Since the spintensors σ^α are invariant under the orthogonal transformations of the basis of the space E_3^0, Eq. (5.145) in the proper basis $\breve{\boldsymbol{e}}_a$ can be written as follows

$$\frac{1}{\rho}\frac{\mathrm{d}\rho}{\mathrm{d}s}\breve{\psi} = \left[\breve{R}_a + \mathrm{i}(\breve{H}_a + \breve{\Omega}_a)\right]\sigma^a\breve{\psi} + (\breve{Q}_a + \mathrm{i}\breve{N}_a)\sigma^a\breve{\bar{\psi}}, \tag{5.146}$$

where the symbol $\breve{}$ means that the components noted by it are calculated in the proper basis $\breve{\boldsymbol{e}}_a$ determined by the equality (4.42). Taking into account definitions (4.45) of the components $\breve{\psi}$, Eq. (5.146) can be rewritten in the form of two complex equations

$$\frac{1}{\sqrt{\rho}}\frac{\mathrm{d}\rho}{\mathrm{d}s} = (\breve{Q}_1 + \mathrm{i}\breve{N}_1)\sqrt{\rho} - \mathrm{i}(\breve{Q}_2 + \mathrm{i}\breve{N}_2)\sqrt{\rho} + [\breve{R}_3 + \mathrm{i}(\breve{\Omega}_3 + \breve{H}_3)]\sqrt{\rho}, \tag{5.147}$$

$$\frac{1}{\sqrt{\rho}}\frac{\mathrm{d}\rho}{\mathrm{d}s} = [\breve{R}_1 + \mathrm{i}(\breve{\Omega}_1 + \breve{H}_1)]\sqrt{\rho} + \mathrm{i}[\breve{R}_2 + \mathrm{i}(\breve{\Omega}_2 + \breve{H}_2)]\sqrt{\rho} - (\breve{Q}_3 + \mathrm{i}\breve{N}_3)\sqrt{\rho}.$$

Separating in (5.147) the real and imaginary parts, we find

$$\begin{aligned} &\frac{1}{\rho}\frac{\mathrm{d}\rho}{\mathrm{d}s} = \breve{R}_3 + \breve{Q}_1 + \breve{N}_2, \\ &\breve{\Omega}_3 + \breve{H}_3 + \breve{N}_1 - \breve{Q}_2 = 0, \\ &\breve{\Omega}_2 + \breve{H}_2 + \breve{Q}_3 - \breve{R}_1 = 0, \\ &\breve{\Omega}_1 + \breve{H}_1 - \breve{N}_3 + \breve{R}_2 = 0. \end{aligned} \tag{5.148}$$

The first equation in (5.148) can be written in the form of a scalar equation

$$\frac{1}{\rho}\frac{\mathrm{d}\rho}{\mathrm{d}s} = \pi^\lambda Q_\lambda + \xi^\lambda N_\lambda + n^\lambda R_\lambda. \tag{5.149}$$

When transforming the first equation in (5.148) to the form (5.149) we must take into account the equalities

$$\breve{Q}_a = \breve{h}^\lambda{}_a Q_\lambda, \quad \breve{N}_a = \breve{h}^\lambda{}_a N_\lambda, \quad \breve{R}_a = \breve{h}^\lambda{}_a R_\lambda,$$

in which the scale factors $\breve{h}^\lambda{}_a$ are determined according to (4.44). Thus, for example, for $\breve{Q}_a$ we have

$$\breve{Q}_1 = \pi^\lambda Q_\lambda, \quad \breve{Q}_2 = \xi^\lambda Q_\lambda, \quad \breve{Q}_3 = n^\lambda Q_\lambda.$$

Bearing in mind that the equalities are fulfilled

$$\breve{\pi}_a = (1, 0, 0), \quad \breve{\xi}_a = (0, 1, 0), \quad \breve{n}_a = (0, 0, 1),$$

three last equations in (5.148) can be written as a single vector equation

$$\breve{\Omega}^a = -\breve{H}^a + \varepsilon^{abc}\big(\breve{\pi}_b \breve{Q}_c + \breve{\xi}_b \breve{N}_c + \breve{n}_b \breve{R}_c\big).$$

In an arbitrary orthonormal basis $\mathcal{Э}_\alpha$ of the space E_3^0 this equation is written as follows

$$\Omega^\alpha = -H^\alpha + \varepsilon^{\alpha\beta\lambda}\big(\pi_\beta Q_\lambda + \xi_\beta N_\lambda + n_\beta R_\lambda\big). \tag{5.150}$$

Thus, the spinor equations (5.144) can be represented in the form of system of the invariant tensor equations (5.149) and (5.150).

Equations (5.149) and (5.150) can be obtained as well without using bases $\breve{\boldsymbol{e}}_a$, by contracting equation (5.145) with components of spinors ψ, $\bar{\psi}$ and performing simple algebraic transformations. Such way, perhaps, is simpler, but the derivation used here has the great generality since it can be applied to the more complicated spinor equations.

Contracting equation (5.150) with respect to the index α with components of tensors $\varepsilon_{\alpha\beta\lambda}\pi^\lambda$, $\varepsilon_{\alpha\beta\lambda}\xi^\lambda$ and $\varepsilon_{\alpha\beta\lambda}n^\lambda$, taking into account the orthonormality conditions of the vectors with components π^α, ξ^α, n^α, we find equations for the derivatives of the components π^α, ξ^α, n^α [90]:

$$\begin{aligned}
\frac{\mathrm{d}\rho\pi^\alpha}{\mathrm{d}s} &= \rho Q^\alpha + \rho\varepsilon^{\alpha\beta\lambda}\big(\pi_\beta H_\lambda + \xi_\beta R_\lambda - n_\beta N_\lambda\big),\\
\frac{\mathrm{d}\rho\xi^\alpha}{\mathrm{d}s} &= \rho N^\alpha + \rho\varepsilon^{\alpha\beta\lambda}\big(-\pi_\beta R_\lambda + \xi_\beta H_\lambda + n_\beta Q_\lambda\big),\\
\frac{\mathrm{d}\rho n^\alpha}{\mathrm{d}s} &= \rho R^\alpha + \rho\varepsilon^{\alpha\beta\lambda}\big(\pi_\beta N_\lambda - \xi_\beta Q_\lambda + n_\beta H_\lambda\big).
\end{aligned} \tag{5.151}$$

Let us introduce the notations

$$T_\alpha{}^1 = Q_\alpha, \quad T_\alpha{}^2 = N_\alpha, \quad T_\alpha{}^3 = R_\alpha. \tag{5.152}$$

Then Eqs. (5.149) and (5.150) can be rewritten in the form

$$\frac{\mathrm{d}\ln\rho}{\mathrm{d}s} = T_\alpha{}^b \breve{h}^\alpha{}_b,$$

$$\Omega^\alpha = -H^\alpha + \varepsilon^{\alpha\beta\lambda} g_{ab} \breve{h}_\beta{}^a T_\lambda{}^b, \tag{5.153}$$

where $g_{ab} = \mathrm{diag}\,(1,1,1)$ are the components of the metric tensor of the Euclidean space E_3^0. The scale factors $\breve{h}_\beta{}^a$ are determined by matrix (4.44).

Equation (5.151) in notations (5.152) take the form

$$\frac{1}{\rho}\frac{\mathrm{d}}{\mathrm{d}s}\left(\rho \breve{h}_\theta{}^a\right) = T_\theta{}^a + \varepsilon_{\theta\beta\lambda}\left(\varepsilon^{abc}\breve{h}^\beta{}_b T^\lambda{}_c + \breve{h}^{\beta a} H^\lambda\right). \tag{5.154}$$

Equations (5.153) and (5.154) are invariant under the rotations of the basis $\mathcal{Э}_\alpha$ and rotations of the basis $\breve{\boldsymbol{e}}_a$ (which do not depend on parameter s).

5.10 Spinor Form of Some Equations of Mechanics

5.10.1 The Frenet-Serret Equations

We consider in the three-dimensional point Euclidean space a continuous differentiable curve ℓ with the arch length s. We introduce at each point of ℓ a trihedral consisting of the unit tangent vector with components ξ^α, the unit normal vector with components n^α and the unit binormal vector with components π^α. It is well known that the components of vectors of this trihedral satisfy the Frenet-Serret equations

$$\frac{\mathrm{d}\xi^\alpha}{\mathrm{d}s} = kn^\alpha, \qquad \frac{\mathrm{d}n^\alpha}{\mathrm{d}s} = -k\xi^\alpha + \varkappa\pi^\alpha, \qquad \frac{\mathrm{d}\pi^\alpha}{\mathrm{d}s} = -\varkappa\xi^\alpha, \tag{5.155}$$

where k is the curvature and $\varkappa$ is the torsion of the curve ℓ.

Let us consider the particular case of the spinor equations (5.144) under the conditions

$$Q_\alpha = N_\alpha = R_\alpha = 0, \quad H_\alpha = -k\pi_\alpha - \varkappa\xi_\alpha.$$

The spinor equations (5.144) corresponding to the case under consideration can be written as follows

$$\frac{\mathrm{d}}{\mathrm{d}s}\psi = \frac{1}{2}(\varkappa - \mathrm{i}k)\bar{\psi}. \tag{5.156}$$

When transforming equations (5.144) to the form (5.156) we must take into account the identities

$$\pi_\alpha \sigma^\alpha \psi = \bar{\psi}, \quad \xi_\alpha \sigma^\alpha \psi = \mathrm{i}\bar{\psi},$$

which follow from the definition of the vector components π_α, ξ_α. Equation (5.149), corresponding to the spinor equations (5.156), is written in the form

$$\frac{\mathrm{d}\rho}{\mathrm{d}s} = 0.$$

Thus, it follows from Eqs. (5.156)

$$\rho = \dot{\psi}^1 \psi^1 + \dot{\psi}^2 \psi^2 = \text{const}. \tag{5.157}$$

It is easy to see that Eq. (5.151) corresponding to the spinor equations (5.156), coincide with the Frenet-Serret equations (5.155) due to the equation $\rho = \text{const}$. Thus, the Frenet-Serret equations are a consequence of the spinor equations (5.156).[9]

Let us put now

$$\varkappa - \mathrm{i}k = 2R \exp \mathrm{i}F, \quad \mathrm{d}\tau = R\,\mathrm{d}s.$$

Then from Eqs. (5.156) it follows the linear second order equation for functions ψ:

$$\psi'' - \mathrm{i}F'\psi' + \psi = 0, \tag{5.158}$$

where the prime denotes the derivative with respect to τ. The replacement of unknown functions

$$\psi = \eta \exp(\mathrm{i}F/2)$$

transforms Eq. (5.158) to the standard form

$$\eta'' + A\eta = 0,$$

where

$$A = \frac{\mathrm{i}}{2}F'' + \frac{1}{4}(F')^2 + 1.$$

Thus, the integration of the Frenet-Serret equations (5.155) for nine functions π^α, ξ^α, n^α connected by the orthonormality conditions are reduced to the integration of

[9]The spinor equations of another form from which the Frenet-Serret equations also follow, were considered in [68].

the spinor equations (5.156) or (5.158) with two complex functions $\boldsymbol{\psi}$, connected by the single condition (5.157).

5.10.2 *Equations for the Rotating Rigid Body with the Fixed Point*

The kinematic equations describing the rotating rigid body with the fixed point, have the form

$$\frac{\mathrm{d}\boldsymbol{\pi}}{\mathrm{d}t} = \boldsymbol{\Omega} \times \boldsymbol{\pi}, \quad \frac{\mathrm{d}\boldsymbol{\xi}}{\mathrm{d}t} = \boldsymbol{\Omega} \times \boldsymbol{\xi}, \quad \frac{\mathrm{d}\boldsymbol{n}}{\mathrm{d}t} = \boldsymbol{\Omega} \times \boldsymbol{n}, \tag{5.159}$$

where t is time, the vector $\boldsymbol{\Omega}$ is the angular velocity of the rotation of the rigid body; the vectors $\boldsymbol{\pi}, \boldsymbol{\xi}, \boldsymbol{n}$ form the orthonormal basis fixed in the body. It is easy to see that Eqs. (5.159) are obtained from Eqs. (5.151), (5.149) under the conditions

$$s = t, \quad H_\alpha = -\Omega_\alpha, \quad Q_\alpha = N_\alpha = R_\alpha = 0.$$

Therefore tensor equations (5.159) are the consequence of the spinor equation

$$\frac{\mathrm{d}}{\mathrm{d}t}\psi = -\frac{\mathrm{i}}{2}\Omega^\alpha \sigma_\alpha \psi. \tag{5.160}$$

Let us put

$$\mathrm{d}\tau = \frac{1}{2}R\,\mathrm{d}t, \quad \Omega_1 + \mathrm{i}\Omega_2 = R\exp \mathrm{i}\omega, \quad G = \omega - \int \Omega_3\, dt.$$

Then replacement of unknown functions

$$\psi^1 = \eta_1 \exp\left(-\frac{\mathrm{i}}{2}\int \Omega_3\, dt\right), \quad \psi^2 = \eta_2 \exp\left(\frac{\mathrm{i}}{2}\int \Omega_3\, dt\right)$$

transforms Eqs. (5.160) to the form

$$\frac{\mathrm{d}\eta_1}{\mathrm{d}\tau} = -\mathrm{i}\eta_2 \exp(-\mathrm{i}G), \quad \frac{\mathrm{d}\eta_2}{\mathrm{d}\tau} = -\mathrm{i}\eta_1 \exp(\mathrm{i}G), \tag{5.161}$$

whence the equations follow (the prime $'$ denotes the derivative with respect to τ).

$$\eta_1'' + \mathrm{i}G'\eta_1' + \eta_1 = 0, \qquad \eta_2'' - \mathrm{i}G'\eta_2' + \eta_2 = 0. \tag{5.162}$$

Equation (5.149), corresponding to the spinor equations (5.160), is written in the form $\mathrm{d}\rho/\mathrm{d}t = 0$ and therefore

$$\rho = \dot{\psi}^1\psi^1 + \dot{\psi}^2\psi^2 \equiv \dot{\eta}_1\eta_1 + \dot{\eta}_2\eta_2 = \text{const.} \tag{5.163}$$

Thus, and in this case nine equations in (5.159) for the nine unknown functions π^α, ξ^α, n^α with the given functions Ω^α are reduced to Eqs. (5.161) or (5.162) for two complex functions η_1 and η_2 with constraint (5.163).

5.10.3 Landau-Lifshitz Equations with the Relaxation Term

The Landau-Lifshitz equations with the relaxation term have the form

$$\frac{\mathrm{d}}{\mathrm{d}t}M^\alpha = g\varepsilon^{\alpha\beta\lambda}M_\beta \overset{*}{H}_\lambda + R^\alpha. \tag{5.164}$$

Here M^α are the components of the volume density of magnetization vector of the continuous medium, $\overset{*}{H}_\lambda$ are the components of the effective strength vector of the magnetic field. g is constant, t is time. Equation (5.164) is used for a description of the magnetization of a continuous medium in the magnetic field.

Let us consider the spinor equation [82]

$$\frac{\mathrm{d}}{\mathrm{d}t}\psi = \frac{\mathrm{i}}{2}\left(g\overset{*}{H}{}^\alpha - \frac{\mathrm{i}}{M}R^\alpha\right)\sigma_\alpha\psi, \tag{5.165}$$

where $M = (M_\alpha M^\alpha)^{1/2}$. We parametrize the vector components M^α by the relation $M^\alpha = -\sigma^\alpha_{BA}\psi^{+B}\psi^A$. Then from (5.151) we get that Eq. (5.164) is the consequence of Eqs. (5.165). In some cases the spinor equation (5.165) is integrated more simply than the tensor equation (5.164).

5.11 Spinor Equations in the Orthogonal Coordinate Systems

In problems possessing the certain spatial symmetry it is convenient to use special curvilinear coordinate systems, in which due to the symmetry of the problem the required functions do not depend on one or several coordinates. In applications, orthogonal coordinate systems are especially often used, in which all coordinate lines are orthogonal among themselves. In an arbitrary curvilinear coordinate system of the Minkowski space and in the Riemannian space the spinor equations (5.18) are written in the form

$$\gamma^i \nabla_i \psi + \left(\varkappa I + \mathrm{i}\varkappa_j \gamma^j + \frac{\mathrm{i}}{2}\varkappa_{js}\gamma^{js} + \overset{*}{\varkappa}_j \overset{*}{\gamma}^j + \overset{*}{\varkappa}\gamma^5 \right) \psi = 0, \tag{5.166}$$

where ∇_i is the symbol of the covariant derivative determined by formula (5.103). In this section we give the writing of Eqs. (5.166) in an arbitrary orthogonal coordinate system and, in particular, in the cylindrical and spherical coordinate system.

Let y^i be the variables of the orthogonal coordinate system in the Minkowski space with the covariant vector basis $\mathcal{Э}_i$. By definition, all vectors $\mathcal{Э}_i$ are orthogonal to each other. It is obvious that in the orthogonal coordinate system the covariant components g_{ij} and contravariant components g^{ij} of the metric tensor of the Minkowski space are determined by diagonal matrices

$$g_{ij} = \begin{Vmatrix} g_{11} & 0 & 0 & 0 \\ 0 & g_{22} & 0 & 0 \\ 0 & 0 & g_{33} & 0 \\ 0 & 0 & 0 & g_{44} \end{Vmatrix}, \qquad g^{ij} = \begin{Vmatrix} g^{11} & 0 & 0 & 0 \\ 0 & g^{22} & 0 & 0 \\ 0 & 0 & g^{33} & 0 \\ 0 & 0 & 0 & g^{44} \end{Vmatrix}.$$

Since the matrix $\|g^{ij}\|$ of the contravariant components of the metric tensor is inverse to the matrix $\|g_{ij}\|$ of the covariant components, we have

$$g^{11} = \frac{1}{g_{11}}, \quad g^{22} = \frac{1}{g_{22}}, \quad g^{33} = \frac{1}{g_{33}}, \quad g^{44} = \frac{1}{g_{44}}.$$

The Christoffel symbols in the considered orthogonal coordinate system with variables y^i according to (2.2) are determined by the equalities

$$\Gamma^j_{jk} = \frac{1}{2} g^{jj} \partial_k g_{jj}, \quad \Gamma^j_{kk} = -\frac{1}{2} g^{jj} \partial_j g_{kk},$$
$$\Gamma^j_{jj} = \frac{1}{2} g^{jj} \partial_j g_{jj}, \quad \Gamma^j_{ik} = 0, \quad \text{for} \quad i \neq j \neq k. \tag{5.167}$$

In formulas (5.167) there is no summation over the indices j and k.

At each point of the Minkowski space, in addition to the orthogonal vector basis $\mathcal{Э}_i$, we introduce an orthonormal basis with vectors $\boldsymbol{e}_a$ directed along the corresponding vectors $\mathcal{Э}_i$. Since the moduli of the basis vectors $\mathcal{Э}_i$ are equal to $|g_{ii}|^{1/2}$, it is obvious that the introduced vectors $\boldsymbol{e}_a$ are connected with the vectors $\mathcal{Э}_i$ in the following way

$$\mathcal{Э}_1 = \sqrt{g_{11}}\, \boldsymbol{e}_1, \quad \mathcal{Э}_2 = \sqrt{g_{22}}\, \boldsymbol{e}_2,$$
$$\mathcal{Э}_3 = \sqrt{g_{33}}\, \boldsymbol{e}_3, \quad \mathcal{Э}_4 = \sqrt{-g_{44}}\, \boldsymbol{e}_4.$$

Therefore the scale factors $h_i{}^a$ for the introduced orthonormal basis $\boldsymbol{e}_a$ have the form

$$h_i{}^a = \left(\boldsymbol{\ni}_i, \boldsymbol{e}^a\right) = \begin{Vmatrix} \sqrt{g_{11}} & 0 & 0 & 0 \\ 0 & \sqrt{g_{22}} & 0 & 0 \\ 0 & 0 & \sqrt{g_{33}} & 0 \\ 0 & 0 & 0 & \sqrt{-g_{44}} \end{Vmatrix}. \tag{5.168}$$

The matrix of the scale factors $\|h^j{}_a\|$ is inverse of the matrix $\|h_j{}^a\|$

$$h^j{}_a = \left(\boldsymbol{\ni}^I, \boldsymbol{e}_a\right) = \begin{Vmatrix} (g_{11})^{-1/2} & 0 & 0 & 0 \\ 0 & (g_{22})^{-1/2} & 0 & 0 \\ 0 & 0 & (g_{33})^{-1/2} & 0 \\ 0 & 0 & 0 & (-g_{44})^{-1/2} \end{Vmatrix}. \tag{5.169}$$

The components of the vectors and tensors calculated in the orthonormal basis $\boldsymbol{e}_a$ are usually called the physical components. Note that the physical components of vectors or tensors have the same dimension, unlike their components, calculated in the orthogonal (or in others curvilinear) coordinate systems.

Using expressions (5.167) for the Christoffel symbols and expressions (5.168), (5.169) for the scale factors, it is possible to calculate the Ricci rotation coefficients $\Delta_{c,ab}$ by formulas (2.38):

$$\begin{aligned}
\Delta_{c,ab} &= 0 \quad \text{for } c \neq a, \;\; c \neq b, \\
\Delta_{4,a4} &= -\Delta_{4,4a} = -\frac{1}{\sqrt{-g_{44}g_{aa}}}\partial_a\sqrt{-g_{44}}, \\
\Delta_{a,4a} &= -\Delta_{a,a4} = \frac{1}{\sqrt{-g_{44}g_{aa}}}\partial_4\sqrt{g_{aa}}, \\
\Delta_{a,ba} &= -\Delta_{a,ab} = \frac{1}{\sqrt{g_{bb}g_{aa}}}\partial_b\sqrt{g_{aa}}.
\end{aligned} \tag{5.170}$$

In formulas (5.170) the indices a, b take values 1, 2, 3. In formulas (5.170) there is no summation over the indices a and b.

From formulas (5.170) it follows that the result of alternation over all indices of the Ricci rotation coefficients for the orthogonal coordinate systems is equal to zero

$$\Delta_{[c,ab]} \equiv \frac{1}{3}(\Delta_{c,ab} + \Delta_{a,bc} + \Delta_{b,ca}) = 0. \tag{5.171}$$

Let us give also expressions for the spinor connection coefficients Γ_i in the orthogonal coordinate systems (see (5.101))

$$\begin{aligned}
\Gamma_1 &= \frac{1}{2}\Big(-\frac{1}{\sqrt{g_{22}}}\partial_2\sqrt{g_{11}}\,\gamma^{12} + \frac{1}{\sqrt{g_{33}}}\partial_3\sqrt{g_{11}}\,\gamma^{31} - \frac{1}{\sqrt{-g_{44}}}\partial_4\sqrt{g_{11}}\,\gamma^{14}\Big),\\
\Gamma_2 &= \frac{1}{2}\Big(\frac{1}{\sqrt{g_{11}}}\partial_1\sqrt{g_{22}}\,\gamma^{12} - \frac{1}{\sqrt{g_{33}}}\partial_3\sqrt{g_{22}}\,\gamma^{23} - \frac{1}{\sqrt{-g_{44}}}\partial_4\sqrt{g_{22}}\,\gamma^{24}\Big),\\
\Gamma_3 &= \frac{1}{2}\Big(-\frac{1}{\sqrt{g_{11}}}\partial_1\sqrt{g_{33}}\,\gamma^{31} + \frac{1}{\sqrt{g_{22}}}\partial_2\sqrt{g_{33}}\,\gamma^{23} - \frac{1}{\sqrt{-g_{44}}}\partial_4\sqrt{g_{33}}\,\gamma^{34}\Big),\\
\Gamma_4 &= \frac{1}{2}\Big(-\frac{1}{\sqrt{g_{11}}}\partial_1\sqrt{-g_{44}}\,\gamma^{14} - \frac{1}{\sqrt{g_{22}}}\partial_2\sqrt{-g_{44}}\,\gamma^{24} - \frac{1}{\sqrt{g_{33}}}\partial_3\sqrt{-g_{44}}\,\gamma^{34}\Big).
\end{aligned} \tag{5.172}$$

The components of spintensors γ^{ab} in equalities (5.172) are calculated in the orthonormal bases $\boldsymbol{e}_a$ and do not depend on the variables y^i.

Equalities (5.172) for the spinor connection coefficients allow to calculate operator $\gamma^i\nabla_i$ in the spinor equations (5.166). A calculation gives

$$\gamma^i\nabla_i = \gamma^i\partial_i + \frac{1}{2\sqrt{-g}}\sum_{i=1}^{4}\sqrt{|g_{ii}|}\,\gamma^i\partial_i\left(\frac{\sqrt{-g}}{\sqrt{|g_{ii}|}}\right). \tag{5.173}$$

For the components of the spintensors γ^i in Eq. (5.173) we have $\gamma^i = h^i{}_a\gamma^a$.

Formula (5.173) for operator $\gamma^i\nabla_i$ in an arbitrary orthogonal coordinate system due to its special form can be obtained without using expression (5.172) for symbols Γ_i. Indeed, in an arbitrary curvilinear coordinate system for the operator $\gamma^i\nabla_i$ a relation is carried out (see (5.109))

$$\gamma^i\nabla_i = \gamma^i\partial_i - \frac{1}{4}\varepsilon^{abcd}\overset{*}{\gamma}_a\Delta_{b,cd} + \frac{1}{2}\Delta_{b,a}{}^b\gamma^a. \tag{5.174}$$

In an arbitrary orthogonal coordinate system in which Eq. (5.171) is fulfilled, formula (5.174) takes the form

$$\gamma^i\nabla_i = \gamma^i\partial_i + \frac{1}{2}\Delta_{b,a}{}^b\gamma^a. \tag{5.175}$$

Taking into account equality (2.39) for the Ricci rotation coefficients, we finally rewrite formula (5.175) in the form

$$\gamma^i\nabla_i = \gamma^i\partial_i + \frac{1}{2\sqrt{-g}}\partial_j(h^j{}_a\sqrt{-g}\,)\gamma^a \tag{5.176}$$

or

$$\gamma^i \nabla_i = \gamma^i \partial_i + \frac{1}{2\sqrt{-g}} \partial_j (\gamma^j \sqrt{-g}).$$

By virtue of definition (5.169) of the scale factors $h^j{}_a$ formulas (5.173) and (5.176) coincide.

Replacing the operator $\gamma^i \nabla_i$ in Eqs. (5.166) by formula (5.176), we find that in an arbitrary orthogonal coordinate system with variables y^i the spinor equations (5.166) are written in the following way

$$\gamma^i \partial_i \psi + \frac{1}{2\sqrt{-g}} \partial_j (h^j{}_a \sqrt{-g}) \gamma^a \psi + \left(\varkappa I + \mathrm{i}\varkappa_j \gamma^j + \frac{\mathrm{i}}{2} \varkappa_{js} \gamma^{js} + \overset{*}{\varkappa}_j \overset{*}{\gamma}{}^j + \overset{*}{\varkappa} \gamma^5 \right) \psi = 0.$$

Here coefficients $h^j{}_a$ are defined by matrix (5.169).

As an example, let us give the writing of Eqs. (5.166) in the cylindrical and spherical coordinate system.

5.11.1 *Cylindrical Coordinate System in the Pseudo-Euclidean Space*

In the Minkowski space the variables x^i of the Cartesian coordinate system are related to variables $y^1 = \rho$, $y^2 = \varphi$, y^3, y^4 of the cylindrical coordinate system by the equalities

$$x^1 = \rho \cos \varphi, \quad x^2 = \rho \sin \varphi, \quad x^3 = y^3, \quad x^4 = y^4.$$

The covariant and contravariant components of the metric tensor in the cylindrical coordinate system are determined by matrices

$$g_{ij} = \begin{Vmatrix} 1 & 0 & 0 & 0 \\ 0 & \rho^2 & 0 & 0 \\ 0 & 0 & 1 & 0 \\ 0 & 0 & 0 & -1 \end{Vmatrix}, \quad g^{ij} = \begin{Vmatrix} 1 & 0 & 0 & 0 \\ 0 & \rho^{-2} & 0 & 0 \\ 0 & 0 & 1 & 0 \\ 0 & 0 & 0 & -1 \end{Vmatrix}.$$

Assuming that vectors $\boldsymbol{e}_a$ of the nonholonomic orthonormal coordinate system are directed along the corresponding vectors $\boldsymbol{\partial}_i$ of the cylindrical coordinate system, for the scale factors $h^j{}_a$, $h_i{}^a$ we find

$$h_i{}^a = \begin{Vmatrix} 1 & 0 & 0 & 0 \\ 0 & \rho & 0 & 0 \\ 0 & 0 & 1 & 0 \\ 0 & 0 & 0 & 1 \end{Vmatrix}, \quad h^j{}_a = \begin{Vmatrix} 1 & 0 & 0 & 0 \\ 0 & \rho^{-1} & 0 & 0 \\ 0 & 0 & 1 & 0 \\ 0 & 0 & 0 & 1 \end{Vmatrix}.$$

For the Christoffel symbols of the cylindrical coordinate system we have

$$\Gamma^1_{22} = -\rho, \quad \Gamma^2_{12} = \Gamma^2_{21} = \frac{1}{\rho}. \tag{5.177}$$

For the introduced orthonormal coordinate system the Ricci rotation coefficients $\Delta_{c,ab}$ are calculated by the formulas

$$\Delta_{2,12} = -\Delta_{2,21} = \frac{1}{\rho}. \tag{5.178}$$

All other components Γ^j_{ik} and $\Delta_{c,ab}$, except specified in equalities (5.177) and (5.178) are equal to zero. The spinor connection coefficients Γ_i for the cylindrical coordinate system are written in the form

$$\Gamma_2 = \frac{1}{2}\gamma^{12}, \quad \Gamma_1 = \Gamma_3 = \Gamma_4 = 0, \tag{5.179}$$

where γ^{12} are calculated in the bases $\boldsymbol{e}_a$.

Equation (5.166) in the cylindrical coordinate system is written as follows

$$\gamma^1\left(\frac{\partial}{\partial\rho} + \frac{1}{2\rho}\right)\psi + \frac{1}{\rho}\gamma^2\frac{\partial}{\partial\varphi}\psi + \gamma^3\frac{\partial}{\partial y^3}\psi + \gamma^4\frac{\partial}{\partial y^4}\psi$$
$$+ \left(\varkappa I + \mathrm{i}\varkappa_j\gamma^j + \frac{\mathrm{i}}{2}\varkappa_{js}\gamma^{js} + \overset{*}{\varkappa}_j\overset{*}{\gamma}{}^j + \overset{*}{\varkappa}\gamma^5\right)\psi = 0. \tag{5.180}$$

The matrices γ^1, γ^2, γ^3, γ^4 in the terms of operator $\gamma^a\nabla_a$ in Eq. (5.180) are calculated in orthonormal bases $\boldsymbol{e}_a$ and do not depend on the variables y^i.

5.11.2 *Spherical Coordinate System in the Pseudo-Euclidean Space*

The variables x^i of the Cartesian coordinate system in the Minkowski space are connected with variables $y^1 = r$, $y^2 = \theta$, $y^3 = \varphi$, y^4 of the spherical coordinate system by the following equalities

$$x^1 = r\sin\theta\cos\varphi, \quad x^2 = r\sin\theta\sin\varphi,$$
$$x^3 = r\cos\theta, \quad x^4 = y^4.$$

The covariant and contravariant components of the metric tensor in the spherical coordinate system are determined by the matrices

$$g_{ij} = \begin{Vmatrix} 1 & 0 & 0 & 0 \\ 0 & r^2 & 0 & 0 \\ 0 & 0 & r^2\sin^2\theta & 0 \\ 0 & 0 & 0 & -1 \end{Vmatrix}, \quad g^{ij} = \begin{Vmatrix} 1 & 0 & 0 & 0 \\ 0 & r^{-2} & 0 & 0 \\ 0 & 0 & (r\sin\theta)^{-2} & 0 \\ 0 & 0 & 0 & -1 \end{Vmatrix}.$$

The scale factors $h^j{}_a$ and $h_i{}^a$ for a nonholonomic orthonormal coordinate system with basis vectors $\boldsymbol{e}_a$, which are directed along the corresponding basis vectors $\boldsymbol{\Im}_i$ of the spherical coordinate system, have the form

$$h_i{}^a = \begin{Vmatrix} 1 & 0 & 0 & 0 \\ 0 & r & 0 & 0 \\ 0 & 0 & r\sin\theta & 0 \\ 0 & 0 & 0 & 1 \end{Vmatrix}, \quad h^j{}_a = \begin{Vmatrix} 1 & 0 & 0 & 0 \\ 0 & r^{-1} & 0 & 0 \\ 0 & 0 & (r\sin\theta)^{-1} & 0 \\ 0 & 0 & 0 & 1 \end{Vmatrix}.$$

For the Christoffel symbols in the spherical coordinate system we have

$$\Gamma^2_{21} = \Gamma^3_{31} = \frac{1}{r}, \quad \Gamma^1_{22} = -r, \quad \Gamma^3_{32} = \operatorname{ctg}\theta,$$
$$\Gamma^1_{33} = -r\sin^2\theta, \quad \Gamma^2_{33} = -\sin\theta\cos\theta. \tag{5.181}$$

The Ricci rotation coefficients $\Delta_{c,ab}$ are calculated by the formulas

$$\Delta_{2,12} = -\Delta_{2,21} = \frac{1}{r}, \quad \Delta_{3,13} = -\Delta_{3,31} = \frac{1}{r},$$
$$\Delta_{3,23} = -\Delta_{3,32} = \frac{\operatorname{ctg}\theta}{r}. \tag{5.182}$$

All components Γ^j_{ik} and $\Delta_{c,ab}$, except noted in Eqs. (5.181) and (5.182), are equal to zero.

Calculation of the spinor connection coefficients Γ_i gives

$$\Gamma_1 = \Gamma_4 = 0, \quad \Gamma_2 = \frac{1}{2}\gamma^{12},$$
$$\Gamma_3 = \frac{1}{2}(-\sin\theta\,\gamma^{31} + \cos\theta\,\gamma^{23}). \tag{5.183}$$

Using the equality (5.183) for Γ_i, we get the writing of the spinor equations (5.166) in the spherical coordinate system

$$\gamma^1\left(\frac{\partial}{\partial r}+\frac{1}{r}\right)\psi+\frac{1}{r}\gamma^2\left(\frac{\partial}{\partial\theta}+\frac{1}{2}\operatorname{ctg}\theta\right)\psi+\frac{1}{r\sin\theta}\gamma^3\frac{\partial}{\partial\varphi}\psi$$
$$+\gamma^4\frac{\partial}{\partial y^4}\psi+\left(\varkappa I+\mathrm{i}\varkappa_j\gamma^j+\frac{\mathrm{i}}{2}\varkappa_{js}\gamma^{js}+\overset{*}{\varkappa}_j\overset{*}{\gamma}{}^j+\overset{*}{\varkappa}\gamma^5\right)\psi=0. \qquad (5.184)$$

The components of spintensors γ^1, γ^2, γ^3, γ^4 and the function ψ in the terms of operator $\gamma^a\nabla_a$ in Eqs. (5.184) are calculated in orthonormal basis $\boldsymbol{e}_a$.

Chapter 6
Exact Solutions of the Nonlinear Spinor Equations

6.1 Einstein–Dirac Equations

The physical space-time in general relativity is the four-dimensional pseudo-Riemannian space V_4 with the metric signature $(+,+,+,-)$. We assume that the space V_4 referred to a coordinate system with the covariant vector basis $\boldsymbol{\vartheta}_i$ and variables x^i, $i = 1, 2, 3, 4$. In the tangent space at each point of the space V_4 we introduce an orthonormal basis (tetrad) $\boldsymbol{e}_a$, $a = 1, 2, 3, 4$, connected with the basis $\boldsymbol{\vartheta}_i$ by the scale factors $h^i{}_a$, $h_i{}^a$:

$$\boldsymbol{e}_a = h^i{}_a \boldsymbol{\vartheta}_i, \qquad \boldsymbol{\vartheta}_i = h_i{}^a \boldsymbol{e}_a.$$

We will denote the indices of tensor components, specified in orthonormal bases $\boldsymbol{e}_a$, by the first letters of the Latin alphabet, a, b, c, d, e, f. The indices of tensor components specified in holonomic bases $\boldsymbol{\vartheta}_i$, will be denoted by the Latin letters i, $j, k, \ldots$.

The Ricci rotation coefficients $\Delta_{a,bc}$ corresponding to the orthonormal tetrads $\boldsymbol{e}_a$, are defined in terms of scale factors by the relation (see Chap. 2)

$$\Delta_{a,bc} = \frac{1}{2}\big[h^j{}_a(\partial_b h_{jc} - \partial_c h_{jb}) + h^j{}_c(\partial_a h_{jb} + \partial_b h_{ja}) - h^j{}_b(\partial_a h_{jc} + \partial_c h_{ja})\big], \tag{6.1}$$

in which $\partial_a = h^i{}_a \partial_i$, $\partial_i = \partial/\partial x^i$.

Let us also give an expression of the components of the Ricci tensor R_{ab}, calculated in the orthonormal bases $\boldsymbol{e}_a$ in terms of the Ricci rotation coefficients

$$R_{ab} = \partial_c \Delta_{b,a}{}^c - \partial_b \Delta_{c,a}{}^c - \Delta_{d,b}{}^c \Delta_{c,a}{}^d + \Delta_{c,d}{}^c \Delta_{b,a}{}^d,$$

V. A. Zhelnorovich, *Theory of Spinors and Its Application in Physics and Mechanics*, https://doi.org/10.1007/978-3-030-27836-6_6

which are obtained by contracting the curvature tensor components (2.44) with components of the metric tensor with respect to the indices b, d. Replacing in the expression of R_{ab} the differentiation operators ∂_a in terms of ∂_i and taking in mind that by virtue of the definition of the Ricci rotation coefficients takes place the identity

$$\Delta_{a,b}{}^{a} = \frac{1}{\sqrt{-g}}\partial_i\left(\sqrt{-g}h^{i}{}_{b}\right), \qquad g = \det||g_{ij}||,$$

expression for R_{ab} can be transformed to a form that will be used in the sequel

$$R_{ab} = \frac{1}{\sqrt{-g}}\partial_j\left[\sqrt{-g}\left(h^{j}{}_{c}\Delta_{b,a}{}^{c} - h^{j}{}_{b}\Delta_{c,a}{}^{c}\right)\right] - \Delta_{f,b}{}^{c}\Delta_{c,a}{}^{f} + \Delta_{c,a}{}^{c}\Delta_{f,b}{}^{f}. \tag{6.2}$$

The Einstein–Dirac equations, describing fields of the spin $1/2$, interacting with the gravitational field, in the framework of the general relativity have the form

$$\begin{aligned} &\gamma^{a}\nabla_a\psi + m\psi = 0,\\ &R_{ab} - \frac{1}{2}Rg_{ab} = \varkappa T_{ab}, \end{aligned} \tag{6.3}$$

where $g_{ab} = \mathrm{diag}(1, 1, 1, -1)$ are the covariant components of the metric tensor in the orthonormal basis $\boldsymbol{e}_a$, R_{ab} are the components of the Ricci tensor, $R = g^{ab}R_{ab}$ is the scalar curvature of the space V_4; $\varkappa = 8\pi G/c^4$, G is the gravitation constant, c is the light velocity; T_{ab} are symmetric components of the energy-momentum tensor of the spinor field

$$T_{ab} = \frac{1}{4}\left(\psi^{+}\gamma_a\nabla_b\psi - \nabla_b\psi^{+}\cdot\gamma_a\psi + \psi^{+}\gamma_b\nabla_a\psi - \nabla_a\psi^{+}\cdot\gamma_b\psi\right). \tag{6.4}$$

The spinor field $\psi(x^i)$ in Eqs. (6.3), (6.4) is specified in some arbitrary generally nonholonomic system of the orthonormal tetrads $\boldsymbol{e}_a$, connected with the holonomic vector basis $\boldsymbol{\partial}_i$ of the Riemannian space by the scale factors $h^{i}{}_{a}$.

System of the Einstein–Dirac equations (6.3), (6.4) is obtained using the variational principle with the Lagrangian

$$\Lambda = -\frac{1}{2\varkappa}R + \frac{1}{2}\left(\psi^{+}\gamma^{i}\nabla_i\psi - \nabla_i\psi^{+}\cdot\gamma^{i}\psi\right) + m\psi^{+}\psi, \tag{6.5}$$

in which the functions $\psi(x^i)$, $\psi^{+}(x^i)$, and $h_i{}^{a}(x^i)$ are varied.

It is easy to see that due to the Dirac equations the invariant $T_a{}^{a}$ of the energy momentum tensor is determined by the relation

$$T_a{}^{a} = \frac{1}{2}\left(\psi^{+}\gamma^{a}\nabla_a\psi - \nabla_a\psi^{+}\cdot\gamma^{a}\psi\right) = -m\psi^{+}\psi \equiv -m\rho\cos\eta. \tag{6.6}$$

Contraction of the Einstein equations in (6.3) with components of the metric tensor g^{ab} with respect to indices a, b gives an equation for the scalar curvature $R = -\varkappa T_a{}^a$ and, by virtue of (6.6)

$$R = \varkappa m \psi^+ \psi = \varkappa m \rho \cos \eta. \tag{6.7}$$

Therefore the Einstein equations in (6.3) can be written also in the form

$$R_{ab} = \varkappa \left(T_{ab} + \frac{1}{2} g_{ab}\, m\rho \cos \eta \right),$$

in which they will be used further.

We face two problems when integrating the Einstein–Dirac equations. The first, purely technical problem stems from the fact that the Einstein–Dirac equations constitute a complex system of nonlinear partial differential equations of the second order for 24 unknown functions.

The second problem is fundamental in nature and stems from the fact that the spinor field functions ψ in the Riemannian space-time can be determined only in certain nonholonomic orthonormal bases (tetrads) that must be specified, or a tetrad gauge is said to be needed. A large number of such gauges are known, and different authors have suggested various gauges. All of these gauges are either noninvariant under transformation of the variables of the observe's coordinate system or are written in the form of differential equations, which complicates the initial system of equations. Physically, all gauges are equivalent, because the Einstein–Dirac equations are invariant under the choice of tetrads. Mathematically, however, using a bad gauge (i.e., additional equations that close the Einstein–Dirac equations) can greatly complicate the equations, while using a good gauge can significantly simplify the equations. In many respects, the problem of choosing a reasonable tetrad gauge stems from the fact that the solutions of the Einstein–Dirac equations have been obtained only for diagonal metrics. Since the basis vectors of a holonomic coordinate system for such metrics are orthogonal, the tetrads associated with the orthogonal holonomic basis of the Riemannian space can be chosen naturally.

In this book, we use a tetrad gauge [91] that is algebraic and, at the same time, is formed in an invariant way. From further it follows that Eqs. (6.3) are simplified if we take as tetrads $\boldsymbol{e}_a$ the proper tetrads $\breve{\boldsymbol{e}}_a$ of the spinor field ψ:

$$\breve{\boldsymbol{e}}_1 = \pi^i\, \boldsymbol{\vartheta}_i, \quad \breve{\boldsymbol{e}}_2 = \xi^i\, \boldsymbol{\vartheta}_i, \quad \breve{\boldsymbol{e}}_3 = \sigma^i\, \boldsymbol{\vartheta}_i, \quad \breve{\boldsymbol{e}}_4 = u^i\, \boldsymbol{\vartheta}_i,$$

determined by the field of the spinor $\boldsymbol{\psi}$ with respect to formulas (3.129), (3.126).

In accordance with this, as the gauge of the tetrads $\boldsymbol{e}_a$ in Eqs. (6.3) and (6.4) we take condition $\boldsymbol{e}_a = \breve{\boldsymbol{e}}_a$. Thus, the scale factors in the Einstein–Dirac equations (6.3) and (6.4) we determine by the matrix

$$h^i{}_a = \breve{h}^i{}_a = \begin{Vmatrix} \pi^1 & \xi^1 & \sigma^1 & u^1 \\ \pi^2 & \xi^2 & \sigma^2 & u^2 \\ \pi^3 & \xi^3 & \sigma^3 & u^3 \\ \pi^4 & \xi^4 & \sigma^4 & u^4 \end{Vmatrix}. \tag{6.8}$$

If the Dirac matrices γ_a and the metric spinor E are determined by equalities (3.24) and (3.25), then the spinor components ψ in the proper basis $\breve{\boldsymbol{e}}_a$ are determined by relations (3.144) as functions of the invariants ρ and η.

In the case under consideration, when the scale factors are determined by matrix (6.8), the contravariant components of metric tensor g^{ij} of the Riemannian space are expressed in terms of π^i, ξ^i, σ^i, and u^i by the relation

$$g^{ij} = \breve{h}^i{}_a \breve{h}^j{}_b g^{ab} = \pi^i \pi^j + \xi^i \xi^j + \sigma^i \sigma^j - u^i u^j, \tag{6.9}$$

and the Einstein–Dirac equations (6.3) are equivalent to the following system of equations (see Chap. 5, Sects. 5.3, 5.8)

$$\begin{aligned} \breve{\partial}_a \ln \rho + \breve{\Delta}_{b,a}{}^b &= 2m\breve{\sigma}_a \sin \eta, \\ \breve{\partial}^a \eta + \frac{1}{2}\varepsilon^{abcd} \breve{\Delta}_{b,cd} &= 2m\breve{\sigma}^a \cos \eta, \\ \breve{R}_{ab} &= \varkappa \left(\breve{T}_{ab} + \frac{1}{2} g_{ab} m\, \rho \cos \eta \right), \end{aligned} \tag{6.10}$$

where $\breve{\partial}_a = \breve{h}^i{}_a \partial_i = \{\pi^i \partial_i, \xi^i \partial_i, \sigma^i \partial_i, u^i \partial_i\}$; $\breve{\Delta}_{b,cd}$ are determined by equalities (3.150). $\breve{R}_{ab}$ are the covariant components of the Ricci tensor, calculated in the proper basis $\breve{\boldsymbol{e}}_a$.

The first two equations in (6.10) are identical to the Dirac equations written in the orthonormal tetrads $\breve{\boldsymbol{e}}_a$. These equations can also be derived from the Dirac equations in system (6.3) by changing the derivatives ∇_a in them using formula (5.105) and by the subsequent simple algebraic transformations. The tetrad components of the energy-momentum tensor $\breve{T}_{ab}$ in the basis $\breve{\boldsymbol{e}}_a$ are determined by the relation

$$\breve{T}_{ab} = \frac{1}{4}\rho \left[-\breve{\sigma}_b \breve{\partial}_a \eta - \breve{\sigma}_a \breve{\partial}_b \eta + \frac{1}{2}\breve{\sigma}_e \left(\breve{\Delta}_{a,cd} \varepsilon_b{}^{cde} + \breve{\Delta}_{b,cd} \varepsilon_a{}^{cde} \right) \right]. \tag{6.11}$$

Expression (6.11) for $\breve{T}_{ab}$ are obtained from relation (5.118) by transition to the basis $\breve{\boldsymbol{e}}_a$ (i.e., by contracting with the coefficients $\breve{h}^i{}_a \breve{h}^j{}_b$ with respect to the indices i, j).

The symbols $\breve{\Delta}_{a,bc}$ in Eqs. (6.10) and (6.11) corresponding to the proper tetrads $\breve{\boldsymbol{e}}_a$ are calculated by formula (6.1) in which the scale factors $\breve{h}^i{}_a$ are determined by the matrix (6.8). The symbols $\breve{\Delta}_{a,bc}$ in an expanded form are defined by formulas (5.114). When integrating the Einstein–Dirac equations it is useful to bear in mind also the matrix of the energy-momentum tensor components which is obtained in conformity with definition (6.11):

$$\breve{T}_{ab} = \frac{1}{4}\rho \begin{Vmatrix} 2\breve{\Delta}_{1,24} & \breve{\Delta}_{2,24} - \breve{\Delta}_{1,14} & -\breve{\partial}_1\eta + \breve{\Delta}_{3,24} & \breve{\Delta}_{4,24} - \breve{\Delta}_{1,12} \\ \breve{\Delta}_{2,24} - \breve{\Delta}_{1,14} & -2\breve{\Delta}_{2,14} & -\breve{\partial}_2\eta - \breve{\Delta}_{3,14} & -\breve{\Delta}_{2,12} - \breve{\Delta}_{4,14} \\ -\breve{\partial}_1\eta + \breve{\Delta}_{3,24} & -\breve{\partial}_2\eta - \breve{\Delta}_{3,14} & -2\breve{\partial}_3\eta & -\breve{\partial}_4\eta - \breve{\Delta}_{3,12} \\ \breve{\Delta}_{4,24} - \breve{\Delta}_{1,12} & -\breve{\Delta}_{2,12} - \breve{\Delta}_{4,14} & -\breve{\partial}_4\eta - \breve{\Delta}_{3,12} & -2\breve{\Delta}_{4,12} \end{Vmatrix}.$$

A system of the differential tensor equations that equivalent to the system of equations (6.10) and (6.11), can be obtained using the variational principle with the Lagrangian

$$\Lambda = -\frac{1}{2\varkappa}R - \frac{1}{2}\rho\breve{\sigma}^a\left(\breve{\partial}_a\eta + \frac{1}{2}\varepsilon_{abcd}\breve{\Delta}^{b,cd}\right) + m\rho\cos\eta,$$

which is identically equal to expression (6.5).

It is easy to see that Eqs. (6.10) are invariant under an arbitrary Lorentz transformation of the vectors of the tetrad $\breve{\boldsymbol{e}}_1$, $\breve{\boldsymbol{e}}_2$, and $\breve{\boldsymbol{e}}_4$ that is independent of the variables x^i.[1] It is obvious that the Riemannian metric g_{ij} does not change under such transformations of the vectors $\breve{\boldsymbol{e}}_1$, $\breve{\boldsymbol{e}}_2$ and $\breve{\boldsymbol{e}}_4$. The spinor components $\breve{\psi}^A$ in the proper basis $\breve{\boldsymbol{e}}_a$ do not change as well since they are defined only by the invariants ρ and η which do not change under the transformations of the proper basis.

Writing the first two equations in (6.10) in the holonomic basis $\boldsymbol{\ni}_i$:

$$\nabla_i \ln\rho + \breve{\Delta}_{j,i}{}^j = 2m\sigma_i \sin\eta,$$

$$\nabla^i\eta + \frac{1}{2}\varepsilon^{ijsm}\breve{\Delta}_{j,sm} = 2m\sigma^i\cos\eta$$

and replacing the Ricci rotation coefficients $\breve{\Delta}_{j,sm}$ by their expression in terms of π_i, ξ_i, σ_i and u_i by formulas (5.106), we obtain a system of the invariant tensor

---

[1] This transformation group of the vectors of the proper basis $\breve{\boldsymbol{e}}_1$, $\breve{\boldsymbol{e}}_2$ and $\breve{\boldsymbol{e}}_4$ is related to a group of the gauge transformations of the spinor components ψ^A in the basis $\boldsymbol{e}_a$:

$$\psi'^A = \alpha\psi^A - \beta\psi^{+A}. \qquad (*)$$

Here α, β are arbitrary complex numbers, satisfying the condition $\dot{\alpha}\alpha - \dot{\beta}\beta = 1$. Transformations (*) were considered in Chap. 3 (see (3.170)).

equations in the components of the proper tetrad vectors of the spinor field ψ [85, 86, 95]

$$\nabla_i \rho\pi^i = 0, \quad \nabla_i \rho\xi^i = 0, \quad \nabla_i \rho\sigma^i = 2m\rho \sin\eta, \quad \nabla_i \rho u^i = 0, \tag{6.12}$$

$$\nabla^i \eta - \frac{1}{2}\varepsilon^{ijms}\left(\pi_j \nabla_m \pi_s + \xi_j \nabla_m \xi_s + \sigma_j \nabla_m \sigma_s - u_j \nabla_m u_s\right) = 2m\sigma^i \cos\eta.$$

The Christoffel symbols in the second equation in (6.12) entering into the covariant derivatives in this equation, disappear because there is fulfilled the alternation with respect to the indices m, s. Taking into account also that for any vector components A^i the identity holds

$$\nabla_i A^i = \frac{1}{\sqrt{-g}}\partial_i(\sqrt{-g}A^i),$$

the system of equations (6.12) can be rewritten in the form

$$\partial_i \sqrt{-g}\rho\pi^i = 0, \qquad \partial_i \sqrt{-g}\rho\sigma^i = 2m\rho\sqrt{-g}\sin\eta,$$

$$\partial_i \sqrt{-g}\rho\xi^i = 0, \qquad \partial_i \sqrt{-g}\rho u^i = 0,$$

$$\sqrt{-g}\, g^{ij}\partial_j \eta - \frac{1}{2}\tilde{\varepsilon}^{\,ijms}\left(\pi_j \partial_m \pi_s + \xi_j \partial_m \xi_s + \sigma_j \partial_m \sigma_s - u_j \partial_m u_s\right)$$

$$= 2m\sigma^i \sqrt{-g}\, \cos\eta, \tag{6.13}$$

where $\tilde{\varepsilon}^{ijks} = \sqrt{-g}\,\varepsilon^{ijks}$ are the Levi-Civita symbols, $\tilde{\varepsilon}^{1234} = -1$.

The quantity $\sqrt{-g}$ in Eqs. (6.13) according to identity (5.99) is expressed in terms of the determinant of the matrix of components π_i, ξ_j, σ_k, and u_s:

$$\sqrt{-g} = \operatorname{mod}\det\begin{Vmatrix} \pi_1 & \xi_1 & \sigma_1 & u_1 \\ \pi_2 & \xi_2 & \sigma_2 & u_2 \\ \pi_3 & \xi_3 & \sigma_3 & u_3 \\ \pi_4 & \xi_4 & \sigma_4 & u_4 \end{Vmatrix} = \operatorname{mod}\left(\tilde{\varepsilon}^{\,ijks}\pi_i \xi_j \sigma_k u_s\right).$$

In a harmonic coordinate system, in which by definition equality $g^{ij}\Gamma^s_{ij} = 0$ is carried out, the first four equations in (6.12) can be written in the form

$$g^{ij}\partial_i \rho\pi_j = 0, \qquad g^{ij}\partial_i \rho\sigma_j = 2m\rho\sin\eta,$$

$$g^{ij}\partial_i \rho\xi_j = 0, \qquad g^{ij}\partial_i \rho u_j = 0.$$

Thus, the Christoffel symbols disappear in all Eqs. (6.12) in harmonic coordinate systems.

We note that in general relativity the components of the vectors π_i, ξ_i, σ_i, u_i are arbitrary functions connected only by the nondegeneracy condition $\det\|h^i{}_a\| \neq 0$.

Equation (6.9) being equivalent to the orthonormality conditions of the tetrad $\breve{\boldsymbol{e}}_a$, in general relativity is the definition of components of the metric tensor and is not a restriction upon π_i, ξ_i, σ_i, and u_i.

Thus, in general relativity the gravitational field and the fermion field of the spin 1/2, interacting with it, are completely described by four arbitrary vector fields $\pi_i(x^j)$, $\xi_i(x^j)$, $\sigma_i(x^j)$, $u_i(x^j)$ and two scalar fields $\rho(x^j)$, $\eta(x^j)$, satisfying Eqs. (6.10).

The initial system of equations (6.3) contains twenty four real unknown functions in $h^i{}_a(x^j)$, $\psi(x^j)$. Equations (6.10) contain eighteen real unknown functions $\rho(x^j)$, $\eta(x^j)$, $\pi_i(x^j)$, $\xi_i(x^j)$, $\sigma_i(x^j)$, and $u_i(x^j)$. Thus, the use of the special tetrad gauge $\boldsymbol{e}_a = \breve{\boldsymbol{e}}_a$ in the Einstein–Dirac equations reduces the number of unknown functions by six units.

In various special problems, along with Eqs. (6.10) one can use the Bianchi and Ricci identities just as is done in the Newman–Penrose formalism. The use of these identities, written in the bases $\breve{\boldsymbol{e}}_a$, in some cases can additionally simplify the integration of Eqs. (6.10).

6.2 General Exact Solution of the Einstein–Dirac Equations in Homogeneous Space

Some particular exact solutions of the Einstein–Dirac equations (6.3) in the Riemannian space whose metric is defined by the relation

$$ds^2 = a^2(t)dx^2 + b^2(t)dy^2 + c^2(t)dz^2 - f^2(t)dt^2$$

with the coefficients $a(t)$, $b(t)$, $c(t)$, $f(t)$, depending only on the single variable $t = x^4$, have been obtained in [3, 29, 51, 52, 59]. Here, we obtain the general exact solution of the Einstein–Dirac equations in the homogeneous Bianchi 1 type Riemannian space [91, 92, 94, 95].

Let us consider a class of the differentiable functions $g_{ij}(x^k)$, depending only on the variable x^4 of the coordinate systems. In this case under the admissible transformations of variables of the coordinate system

$$x^4 = \varphi(y^4), \quad x^\alpha = y^\alpha + \varphi^\alpha(y^4), \tag{6.14}$$

where $\varphi(y^4)$ and $\varphi^\alpha(y^4)$ are arbitrary differentiable functions, covariant components of the metric tensor are transformed according to formulas

$$g'_{\alpha\beta} = g_{\alpha\beta}, \quad g'_{4\alpha} = \frac{\mathrm{d}\varphi}{\mathrm{d}y^4}\, g_{4\alpha} + \frac{\mathrm{d}\varphi^\beta}{\mathrm{d}y^4}\, g_{\alpha\beta}.$$

From these formulas it is clear that by the transformation (6.14) (even at $x^4 = e^4$) the components $g_{4\alpha}$ can be turned to zero if $\det \|g_{\alpha\beta}\| \neq 0$. Therefore, if $\det \|g_{\alpha\beta}\| \neq 0$, then in the Riemannian space one can introduce a synchronous coordinate system in which the following equalities hold for the covariant components of the metric tensor

$$g_{41} = g_{42} = g_{43} = 0, \quad g_{44} = -1.$$

It is obvious that in this case for the contravariant components g^{4i} the equalities $g^{41} = g^{42} = g^{43} = 0, \quad g^{44} = -1$ hold.

Let us assume further that the coordinate system with the variables x^i is synchronous and seek the exact solutions of the system of equations (5.114), (6.10), (6.11), (6.13), (6.2), depending only on the variable $x^4 = t$. The first four equations and the last one in (6.13) for $j = 4$ in this case take the form

$$\begin{aligned} &\partial_4 \sqrt{-g}\rho\pi^4 = 0, \qquad \partial_4 \sqrt{-g}\rho\sigma^4 = 2m\rho\sqrt{-g}\sin\eta, \\ &\partial_4 \sqrt{-g}\rho\xi^4 = 0, \qquad \partial_4 \sqrt{-g}\rho u^4 = 0, \\ &\partial_4 \eta = -2m\sigma^4 \cos\eta. \end{aligned} \tag{6.15}$$

The condition of synchronism of the coordinate system yields

$$g^{44} \equiv \pi^4\pi^4 + \xi^4\xi^4 + \sigma^4\sigma^4 - u^4u^4 = -1. \tag{6.16}$$

Equations (6.15) and (6.16) are a closed system of equations for determining the quantities π^4, ξ^4, σ^4, u^4, η, $\rho\sqrt{-g}$.

From Eqs. (6.15) it follows

$$\sqrt{-g}\,\rho\pi^4 = \text{const}, \quad \sqrt{-g}\,\rho\xi^4 = \text{const}, \quad \sqrt{-g}\,\rho u^4 = \text{const}\,. \tag{6.17}$$

Dividing the second equation in (6.15) by $\sqrt{-g}\rho u^4 = \text{const}$, we find

$$\partial_4 \left(\frac{\sigma^4}{u^4}\right) = 2m\frac{\sin\eta}{u^4}. \tag{6.18}$$

Now from the last equation in (6.15) and from Eq. (6.18) we obtain

$$\frac{d}{d\eta}\left(\frac{\sigma^4}{u^4}\right) = -\frac{1}{\sigma^4 u^4}\,\text{tg}\,\eta. \tag{6.19}$$

As due to integrals (6.17) the following relations are valid

$$\frac{\pi^4}{u^4} = \text{const}, \quad \frac{\xi^4}{u^4} = \text{const},$$

then from the condition of synchronism (6.16) we find

$$\left(\frac{\sigma^4}{u^4}\right)^2 = -\frac{1}{(u^4)^2} + \text{const.}$$

Therefore Eq. (6.19) after multiplication by σ^4/u^4, can be transformed to the form

$$\frac{1}{2}\frac{\mathrm{d}}{\mathrm{d}\eta}\left(\frac{1}{u^4}\right)^2 = \left(\frac{1}{u^4}\right)^2 \operatorname{tg}\eta. \tag{6.20}$$

Equation (6.20) has the integral[2]

$$u^4 = C_u|\cos\eta|, \quad C_u \geqslant 1 - \text{const},$$

by means of which we find the general solution of Eqs. (6.15), (6.16):

$$\frac{C_\rho}{\rho\sqrt{-g}} = \frac{\pi^4}{C_\pi} = \frac{\xi^4}{C_\xi} = \frac{u^4}{C_u} = \frac{1}{\sqrt{1 + C_\sigma^2\cos^2(2mt + \varphi)}}, \tag{6.21}$$

$$\sigma^4 = \frac{\varepsilon C_\sigma \sin(2mt + \varphi)}{\sqrt{1 + C_\sigma^2\cos^2(2mt + \varphi)}}, \qquad \exp \mathrm{i}\eta = \varepsilon \frac{1 + \mathrm{i}C_\sigma \cos(2mt + \varphi)}{\sqrt{1 + C_\sigma^2\cos^2(2mt + \varphi)}}.$$

Here φ, C_π, C_ξ, C_σ, $C_u \geqslant 1$, $C_\rho > 0$ are integration constants; the coefficient ε can take any of the two values $+1$ or -1. Due to the synchronism condition (6.16) the constants C are connected by the relation

$$(C_\pi)^2 + (C_\xi)^2 + (C_\sigma)^2 - (C_u)^2 = -1. \tag{6.22}$$

It is easy to see that due to solution (6.21) are carried out the equalities

$$2m\sigma^4 \sin\eta = \frac{\mathrm{d}}{\mathrm{d}t}\ln u^4, \quad \rho u^4\sqrt{-g} = C_\rho C_u, \tag{6.23}$$

which will be used further.

We now pass to consideration of Eqs. (6.10), (6.11). In the class of the admissible functions depending only on the variable t, for the derivatives $\breve{\partial}_a$ we have $\breve{\partial}_a =$

[2]In an arbitrary coordinate system received from synchronous system by admissible transformations (6.14), this integral has the form

$$C_u^2 g^{44}\cos^2\eta = -(u^4)^2,$$

where g^{44} is the component of the metric tensor.

$\breve{h}^i{}_a \partial_i = \breve{h}^4{}_a \partial_4$. Therefore in the considered class of functions, formula (6.1) for the Ricci rotation coefficients $\breve{\Delta}_{a,bc}$ can be represented in the form

$$\breve{\Delta}_{a,bc} = \frac{1}{2}(h_b s_{ac} - h_c s_{ab} - h_a a_{bc}), \tag{6.24}$$

where by definition

$$\begin{aligned} s_{ab} &= \breve{h}^i{}_a \partial_4 \breve{h}_{ib} + \breve{h}^i{}_b \partial_4 \breve{h}_{ia}, \\ a_{ab} &= \breve{h}^i{}_a \partial_4 \breve{h}_{ib} - \breve{h}^i{}_b \partial_4 \breve{h}_{ia} \end{aligned} \tag{6.25}$$

and to simplify the notation we denote $h_a = \breve{h}^4{}_a$. Taking into account the gauge condition (6.8), we have

$$h_1 = \pi^4, \quad h_2 = \xi^4, \quad h_3 = \sigma^4, \quad h_4 = u^4. \tag{6.26}$$

Thus, the quantities h_a in formula (6.24) are determined by solution (6.21) as functions of the parameter t. Therefore formula (6.24) expresses 24 dependent functions $\breve{\Delta}_{a,bc}$ in terms of the 16 independent functions s_{ab}, a_{ab}.

The quantities s_{ab} are symmetric in the indices a and b, while the quantities a_{ab} are antisymmetric

$$s_{ab} = s_{ba}, \qquad a_{ab} = -a_{ba}.$$

By virtue of definitions (6.25) the quantities a_{ab} and s_{ab} in the synchronous coordinate system satisfy the identities

$$\begin{aligned} h^b s_{ab} = 0, \quad s_a{}^a &= 2\partial_4 \ln \sqrt{-g}, \\ h^b a_{ab} &= -2\partial_4 h_a. \end{aligned} \tag{6.27}$$

From definitions (6.25) it also follows that the symmetric quantities s_{ab} are determined by the derivative of the metric tensor components

$$s_{ab} = \breve{h}^i{}_a \breve{h}^j{}_b \partial_4 g_{ij}.$$

Substituting in the first two equations in (6.10) the Ricci rotation coefficients $\breve{\Delta}_{a,bc}$ according to formula (6.24), we write these equations in the form (when transforming the second equation in (6.10) identity (6.27) should be taking into account)

$$\begin{aligned} h^a \partial_4 \eta - \frac{1}{4} \varepsilon^{abcd} h_d a_{bc} &= 2m\breve{\sigma}^a \cos \eta, \\ \partial_4 h^a + h^a \partial_4 \ln(\rho\sqrt{-g}) &= 2m\breve{\sigma}^a \sin \eta. \end{aligned} \tag{6.28}$$

Contracting the first equation in (6.28) with components of the tensor $\varepsilon_{abcd}h^d$ with respect to the index a, after some transformations in view of the second equation in (6.28), we obtain the following expression for the antisymmetric quantities a_{ab}:

$$a_{ab} = 4m\left[(\sigma_a h_b - \sigma_b h_a)\sin\eta - \varepsilon_{abcd}\sigma^c h^d \cos\eta\right]. \tag{6.29}$$

Thus, relation (6.29) for the antisymmetric coefficients a_{ab} is satisfied due to the first two equations in (6.10).

It is easy to see that derivative $\partial_4 \breve{h}_{ib}$ are expressed in terms of the quantities a_{ab}, s_{ab}. Indeed, taking into account the equality $\breve{h}_i{}^a \breve{h}^j{}_a = \delta_i^j$ we find

$$\begin{aligned}\partial_4 \breve{h}_{ib} &= \breve{h}_i{}^a \breve{h}^j{}_a \partial_4 h_{jb}\\ &= \breve{h}_i{}^a\left[\frac{1}{2}(\breve{h}^j{}_a\partial_4\breve{h}_{jb} + \breve{h}^j{}_b\partial_4\breve{h}_{ja}) + \frac{1}{2}(\breve{h}^j{}_a\partial_4\breve{h}_{jb} - \breve{h}^j{}_b\partial_4\breve{h}_{ja})\right].\end{aligned}$$

From this taking into account definitions (6.25) we obtain

$$\partial_4 \breve{h}_{ib} = \frac{1}{2}\breve{h}_i{}^a(a_{ab} + s_{ab}). \tag{6.30}$$

Let us now consider the Einstein equations in (6.10). For the considered class of unknown functions the definitions (6.11) and (6.2) for the components of the energy-momentum tensor and the Ricci tensor by virtue of Eqs. (6.28) and (6.24) give the following expressions

$$\begin{aligned}\breve{T}_{ab} &= \frac{1}{8}\rho h_c\left(\varepsilon_a{}^{cfe}s_{bf} + \varepsilon_b{}^{cfe}s_{af}\right)\breve{\sigma}_e + \rho m h_a h_b \cos\eta, \qquad (6.31)\\ \breve{R}_{ab} &= \frac{1}{2\sqrt{-g}}\partial_4(\sqrt{-g}\,s_{ab}) - \frac{1}{4}h_a h_b(s_{ef}s^{ef} + 2\partial_4 s_e{}^e) + \frac{1}{4}(s_{ac}a_b{}^c + s_{bc}a_a{}^c).\end{aligned}$$

Using Eqs. (6.27), (6.29) and definitions (6.31) of $\breve{T}_{ab}$ and $\breve{R}_{ab}$, the Einstein equations in (6.10) can be written in the form of an equivalent system of equations

$$\begin{aligned}&\partial_4(\sqrt{-g}\,s_{ab}) - 2\sqrt{-g}\left(m\cos\eta + \frac{1}{8}\varkappa\rho\right)\left(\varepsilon_{cefa}s_b{}^f + \varepsilon_{cefb}s_a{}^f\right)h^c\breve{\sigma}^e\\ &-2m\sqrt{-g}\,(h_a s_{bc} + h_b s_{ac})\breve{\sigma}^c\sin\eta = \varkappa m\rho\sqrt{-g}\,\cos\eta(g_{ab} + h_a h_b),\\ &\qquad (s_a{}^a)^2 - s_{ab}s^{ab} = 8\varkappa\rho m\cos\eta. \qquad (6.32)\end{aligned}$$

The first equation in (6.32) is obtained by contracting the Einstein equations in (6.10) with the tensor components $\delta_c^a + h_c h^a$ with respect to the index a. The second equation in (6.32) is obtained by contracting the Einstein equations in (6.10) with the tensor components $g^{ab} + 2h^a h^b$ with respect to the indices a, b.

The quantity $\rho\sqrt{-g}\cos\eta$ in the right-hand side of the first equation (6.32) due to solution (6.21) is constant

$$\rho\sqrt{-g}\cos\eta = \varepsilon C_\rho. \tag{6.33}$$

Contraction of Eqs. (6.32) with g^{ab} with respect to the indices a, b gives the equation

$$\partial_4\partial_4\sqrt{-g} = \frac{3}{2}\varkappa m\varepsilon C_\rho,$$

whence we find

$$\sqrt{-g} = \frac{3}{4}\varkappa m\varepsilon C_\rho t^2 + ft + n. \tag{6.34}$$

Here f and n are integration constants. If in Eq. (6.34) $\varkappa m \neq 0$, then by mean of a linear transformation of the parameter t the expression for $\sqrt{-g}$ can be transformed to the form

$$\sqrt{-g} = \frac{3}{4}\varkappa m\varepsilon C_\rho(t^2 - a^2). \tag{6.35}$$

where a is a real or imaginary constant $a = |a|$ or $a = |a|\mathrm{i}$, $\mathrm{i} = \sqrt{-1}$.

In this case from Eqs. (6.33) we get an expression for the invariant ρ of the spinor field:

$$\rho = \frac{4\sqrt{1 + C_\sigma^2\cos^2(2mt + \varphi)}}{3\varkappa m\varepsilon(t^2 - a^2)}.$$

The quantity a^2 is defined further (see (6.48)), the constant ε should be chosen in such a way that the condition $\sqrt{-g} > 0$ is fulfilled.

The obtained relations already permits us to write out the solution for the spinor components $\breve{\psi}$ in the proper basis $\breve{\boldsymbol{e}}_a$. If the Dirac metrices γ_a and the metric spinor E are determined by relations (3.24) and (3.25), then, according to definitions (3.144) and the above solution (6.21), (6.35) for ρ, η, $\sqrt{-g}$, we can write for $\breve{\psi}$ in the proper basis $\breve{\boldsymbol{e}}_a$

$$\breve{\psi} = \pm\begin{Vmatrix} 0 \\ \mathrm{i}\sqrt{\dfrac{2}{3\varkappa m}\dfrac{1 + \mathrm{i}C_\sigma\cos(2mt + \varphi)}{t^2 - a^2}} \\ 0 \\ \mathrm{i}\sqrt{\dfrac{2}{3\varkappa m}\dfrac{1 - \mathrm{i}C_\sigma\cos(2mt + \varphi)}{t^2 - a^2}} \end{Vmatrix}. \tag{6.36}$$

Since in the proper basis $\breve{\boldsymbol{e}}_a$ of the spinor ψ the equality $\breve{\sigma}^a = (0, 0, 1, 0)$ is satisfied, and the quantities ρ, η, $\sqrt{-g}$, and h_c in Eqs. (6.32) are already determined as functions of t by relations (6.21) and (6.35), then Eqs. (6.32) are a linear differential equations with variables factors in the unknown functions $\sqrt{-g}\, s_{ab}$.

From solution (6.21) it follows that the quantities h_a, determined by equalities (6.26), can be represented as

$$\{h_1, h_2, h_4\} = \frac{1}{\sqrt{1 + C_\sigma^2 \cos^2(2mt + \varphi)}} \{C_\pi, C_\xi, C_u\}.$$

It is seen from this that the direction of the three-dimensional vector with components h_1, h_2, h_4 does not depend on parameter t. Therefore components h_1, h_2 can be transformed to zero by a t-independent Lorentz transformation of the vectors $\breve{\boldsymbol{e}}_1$, $\breve{\boldsymbol{e}}_2$, $\breve{\boldsymbol{e}}_4$.

In view of the invariancy of the considered equations under such transformations it is sufficient to consider the solutions of Eqs. (6.32) only for $h_1 = h_2 = 0$ (i.e., for $\pi^4 = \xi^4 = 0$ and consequently $(u^4)^2 - (\sigma^4)^2 = 1$). Under this condition from the first equation in (6.32) it follows (here it is necessary to bear in mind also Eq. (6.33))

$$\begin{aligned}
&\partial_4(s_{33}\sqrt{-g}\,) - 4m\sigma^4 \sin\eta\, s_{33}\sqrt{-g} = \varepsilon\varkappa m C_\rho (u^4)^2,\\
&\partial_4(s_{23}\sqrt{-g}\,) - 2m\sigma^4 \sin\eta\, s_{23}\sqrt{-g} - \left(2m\cos\eta + \frac{1}{4}\varkappa\rho\right) u^4 s_{13}\sqrt{-g} = 0,\\
&\partial_4(s_{13}\sqrt{-g}\,) - 2m\sigma^4 \sin\eta\, s_{13}\sqrt{-g} + \left(2m\cos\eta + \frac{1}{4}\varkappa\rho\right) u^4 s_{23}\sqrt{-g} = 0,\\
&\partial_4(s_{11}\sqrt{-g}\,) + 2\left(2m\cos\eta + \frac{1}{4}\varkappa\rho\right) u^4 s_{12}\sqrt{-g} = \varepsilon\varkappa m C_\rho,\\
&\partial_4(s_{22}\sqrt{-g}\,) - 2\left(2m\cos\eta + \frac{1}{4}\varkappa\rho\right) u^4 s_{12}\sqrt{-g} = \varepsilon\varkappa m C_\rho,\\
&\partial_4(s_{12}\sqrt{-g}\,) - \left(2m\cos\eta + \frac{1}{4}\varkappa\rho\right) u^4 (s_{11} - s_{22})\sqrt{-g} = 0,\\
&\partial_4(s_{14}\sqrt{-g}\,) + \left(2m\cos\eta + \frac{1}{4}\varkappa\rho\right) u^4 s_{24}\sqrt{-g} - 2mu^4 \sin\eta\, s_{31}\sqrt{-g} = 0,\\
&\partial_4(s_{24}\sqrt{-g}\,) - \left(2m\cos\eta + \frac{1}{4}\varkappa\rho\right) u^4 s_{14}\sqrt{-g} - 2mu^4 \sin\eta\, s_{23}\sqrt{-g} = 0,\\
&\partial_4(s_{34}\sqrt{-g}\,) - 2m\sin\eta \left(\sigma^4 s_{34} + u^4 s_{33}\right)\sqrt{-g} = \varepsilon\varkappa m C_\rho \sigma^4 u^4,\\
&\partial_4(s_{44}\sqrt{-g}\,) - 4mu^4 \sin\eta\, s_{34}\sqrt{-g} = \varepsilon\varkappa m C_\rho (\sigma^4)^2.
\end{aligned} \tag{6.37}$$

The last four equations in (6.37) due to the identities (6.27) are the consequences of the first six equations. The remaining equations in (6.37) are split into three groups consisting of one, two and three equations which admit consecutive integration. Having determined from these six equations the functions $\sqrt{-g}\, s_{11}$, $\sqrt{-g}\, s_{22}$, $\sqrt{-g}\, s_{33}$, $\sqrt{-g}\, s_{13}$, $\sqrt{-g}\, s_{23}$, $\sqrt{-g}\, s_{12}$, one finds $\sqrt{-g}\, s_{\alpha 4}$ from the first identity in (6.27).

At first we consider the first equation in (6.37). Using the first equality in (6.23), we transform the first equation in (6.37) to the form

$$\partial_4 \left[\frac{s_{33}\sqrt{-g}}{(u^4)^2} \right] = \varepsilon \varkappa m C_\rho .$$

From this, we find by means of the second equality in (6.23)

$$s_{33} = \frac{\rho (u^4)^3}{C_\rho C_u} \int \varepsilon \varkappa m C_\rho dt = \rho (u^4)^3 \left(\frac{\varepsilon \varkappa m}{C_u} t + \frac{2}{3} N \right),$$

where N is an arbitrary constant. An expression for s_{33} conveniently to write in another form. Notice that due to definition (6.35) the equality holds

$$\frac{\varepsilon \varkappa m}{C_u} t = \frac{2}{3 C_\rho C_u} \partial_4 \sqrt{-g}.$$

Therefore the solution for s_{33} can be finally written in the form

$$s_{33} = \frac{2}{3} \rho (u^4)^3 \left(N + \frac{1}{C_\rho C_u} \partial_4 \sqrt{-g} \right). \tag{6.38}$$

In the same way from the last equation in (6.37) we obtained the solution for s_{44}:

$$s_{44} = \frac{2}{3} \rho u^4 \left(\sigma^4\right)^2 \left(N + \frac{1}{C_\rho C_u} \partial_4 \sqrt{-g} \right). \tag{6.39}$$

Solutions (6.38) and (6.39) can also be obtained by considering the difference between the first and last equations in (6.37) and using the identities

$$u^4 s_{34} = \sigma^4 s_{33}, \quad u^4 s_{44} = \sigma^4 s_{34}, \tag{6.40}$$

following in the case under consideration from the first identity in (6.27).

By means of identities (6.40) we obtained also the solution for s_{34}:

$$s_{34} = \frac{2}{3} \rho \sigma^4 (u^4)^2 \left(N + \frac{1}{C_\rho C_u} \partial_4 \sqrt{-g} \right).$$

From the second and third equations in (6.37) we obtain the solution for functions s_{13} and s_{23}:

$$s_{13} = \frac{1}{2}\rho(u^4)^2 A\cos(\zeta + \alpha),$$

$$s_{23} = \frac{1}{2}\rho(u^4)^2 A\sin(\zeta + \alpha),$$

where A and α are arbitrary constants, the quantity ζ is defined by the relation

$$\zeta = \int \left(2m\cos\eta + \frac{1}{4}\varkappa\rho\right) u^4\, dt.$$

Using expressions (2.11) for ρ, u^4 and η, we find for $a \neq 0$

$$\zeta = \varepsilon\,\mathrm{arctg}\left(\frac{\mathrm{tg}(2mt+\varphi)}{\sqrt{1+C_\sigma^2}}\right) + \frac{\varepsilon C_u}{6ma}\ln\left|\frac{t-a}{t+a}\right|. \tag{6.41}$$

If $a = 0$, then we have

$$\zeta = \varepsilon\,\mathrm{arctg}\left(\frac{\mathrm{tg}(2mt+\varphi)}{\sqrt{1+C_\sigma^2}}\right) - \frac{\varepsilon C_u}{3m}\frac{1}{t}. \tag{6.42}$$

By means of identities (6.27) we get also the solutions for s_{14}, s_{24}:

$$s_{14} = \frac{\sigma^4}{u^4}s_{13} = \frac{1}{2}\rho u^4\sigma^4 A\cos(\zeta + \alpha),$$

$$s_{24} = \frac{\sigma^4}{u^4}s_{23} = \frac{1}{2}\rho u^4\sigma^4 A\sin(\zeta + \alpha).$$

It remains to find the functions s_{11}, s_{12}, and s_{22}. For this purpose let us add and subtract the fourth and fifth equations in (6.37). As a result, we obtain the equations

$$\partial_4\left[(s_{11}+s_{22})\sqrt{-g}\right] = 2\varepsilon\varkappa m C_\rho,$$

$$\partial_4\left(\frac{s_{11}-s_{22}}{2}\sqrt{-g}\right) = -2\left(2m\cos\eta + \frac{1}{4}\varkappa\rho\right)u^4 s_{12}\sqrt{-g}. \tag{6.43}$$

A solution of the first equation in (6.43) has the form[3]

$$s_{11}+s_{22} = \frac{2}{3}\rho u^4\left(-N + \frac{2}{C_\rho C_u}\partial_4\sqrt{-g}\right). \tag{6.44}$$

[3]To determine the integration constant in this solution it is necessary to use the second identity in (6.27) and solution (6.38), (6.39).

From the sixth equation in (6.37) and the second equation in (6.43) we obtaine

$$s_{12} = -\frac{1}{2} B\rho u^4 \cos 2(\zeta + \beta),$$
$$s_{11} - s_{22} = B\rho u^4 \sin 2(\zeta + \beta). \tag{6.45}$$

Here B is an arbitrary constant, and function ζ is determined by (6.41) or (6.42). From Eqs. (6.44) and (6.45) we obtain

$$s_{11} = \rho u^4 \left[-\frac{1}{3} N + \frac{2}{3 C_\rho C_u} \partial_4 \sqrt{-g} + \frac{1}{2} B \sin 2(\zeta + \beta) \right],$$
$$s_{22} = \rho u^4 \left[-\frac{1}{3} N + \frac{2}{3 C_\rho C_u} \partial_4 \sqrt{-g} - \frac{1}{2} B \sin 2(\zeta + \beta) \right].$$

Thus, all functions s_{ab} are found. Let us write out completely the obtained solution for s_{ab}:

$$\begin{aligned}
s_{11} &= \rho u^4 \left[-\frac{1}{3} N + \frac{2}{3 C_\rho C_u} \partial_4 \sqrt{-g} + \frac{1}{2} B \sin 2(\zeta + \beta) \right], \\
s_{22} &= \rho u^4 \left[-\frac{1}{3} N + \frac{2}{3 C_\rho C_u} \partial_4 \sqrt{-g} - \frac{1}{2} B \sin 2(\zeta + \beta) \right], \\
s_{33} &= \frac{2}{3} \rho (u^4)^3 \left(N + \frac{1}{C_\rho C_u} \partial_4 \sqrt{-g} \right), \\
s_{44} &= \frac{2}{3} \rho u^4 (\sigma^4)^2 \left(N + \frac{1}{C_\rho C_u} \partial_4 \sqrt{-g} \right), \\
s_{12} &= -\frac{1}{2} \rho u^4 B \cos 2(\zeta + \beta), \quad s_{34} = \frac{2}{3} \rho \sigma^4 (u^4)^2 \left(N + \frac{1}{C_\rho C_u} \partial_4 \sqrt{-g} \right), \\
s_{13} &= \frac{1}{2} \rho (u^4)^2 A \cos(\zeta + \alpha), \quad s_{14} = \frac{1}{2} \rho u^4 \sigma^4 A \cos(\zeta + \alpha), \\
s_{23} &= \frac{1}{2} \rho (u^4)^2 A \sin(\zeta + \alpha), \quad s_{24} = \frac{1}{2} \rho u^4 \sigma^4 A \sin(\zeta + \alpha).
\end{aligned} \tag{6.46}$$

Here A, B, N, α, and β are integration constants; the quantities σ^4, u^4 are determined by relations (6.21); $\varepsilon = \pm 1$; the quantity $\sqrt{-g}$ is determined by equality (6.35); the function ζ are determined by equality (6.41) or (6.42). The general solution of Eqs. (6.37) for $h_1 \neq 0$, $h_2 \neq 0$ is obtained from (6.46) by a constant Lorentz transformation of the vectors $\breve{\boldsymbol{e}}_1$, $\breve{\boldsymbol{e}}_2$, and $\breve{\boldsymbol{e}}_4$.

From (6.46) it follows

$$s_a{}^a = 2\frac{\partial_4\sqrt{-g}}{\sqrt{-g}},$$

$$s_{ab}s^{ab} = \frac{4}{3}\left(\frac{\partial_4\sqrt{-g}}{\sqrt{-g}}\right)^2 + 2\rho^2(u^4)^2\left(\frac{1}{4}A^2 + \frac{1}{4}B^2 + \frac{1}{3}N^2\right). \tag{6.47}$$

Taking into account (6.47), we get that the second equation in (6.32) defines the constant a^2 in expression (6.35) for $\sqrt{-g}$

$$a^2 = \frac{C_u^2}{3\varkappa^2 m^2}\left(\frac{1}{4}B^2 + \frac{1}{4}A^2 + \frac{1}{3}N^2\right) \geqslant 0. \tag{6.48}$$

Relations (6.21), (6.29), and (6.46) determine the Ricci rotation coefficients (6.24) as functions of the variable t.

Relations (6.21), (6.29), (6.35), (6.36), and (6.46) completely determine a first integral of the Einstein–Dirac equations. To obtain the general solution of the Einstein–Dirac equations it is now sufficient to integrate Eqs. (6.30), where a_{ab} and s_{ab} are determined by equalities (6.29) and (6.46). Replacing in Eqs. (6.30) the quantities $\breve{h}_{ib}$, a_{ab}, s_{ab} in terms of π_i, ξ_i, σ_i, u_i according to formulas (6.8), (6.29), and (6.46), after certain transformations we obtain a system of equations for determining the components of the vectors π_i, ξ_i, σ_i, and u_i:

$$\begin{aligned}
\frac{d}{d\tau}\left(u^4\sigma_j - \sigma^4 u_j\right) &= \frac{1}{4}A\left[\pi_j\cos(\zeta+\alpha) + \xi_j\sin(\zeta+\alpha)\right] + \\
&\quad + \left(u^4\sigma_j - \sigma^4 u_j\right)\left(\frac{1}{3\sqrt{-g}}\frac{d}{d\tau}\sqrt{-g} + \frac{1}{3}N\right), \\
\frac{d}{d\tau}(\xi_j + i\pi_j) &= (\xi_j + i\pi_j)\left(\frac{1}{3\sqrt{-g}}\frac{d}{d\tau}\sqrt{-g} - \frac{1}{6}N - i\frac{2m}{\rho}\cos\eta\right) - \\
&\quad - \frac{i}{4}(\xi_j - i\pi_j)B\exp\left[-2i(\zeta+\beta)\right] + \frac{i}{4}A\left(u^4\sigma_j - \sigma^4 u_j\right)\exp[-i(\zeta+\alpha)],
\end{aligned} \tag{6.49}$$

which should be supplemented with the synchronism condition

$$g_{4\alpha} \equiv \sigma_4\sigma_\alpha - u_4 u_\alpha = 0. \tag{6.50}$$

For $d\tau$ in (6.49) we have by definition $d\tau = \rho u^4 dt$. The connection between τ and t can be written in a closed form. Replacing in the equation $d\tau = \rho u^4 dt$ the quantities ρ and u^4 according to solution (6.21), we obtain

$$\tau = \int \rho u^4 dt = \int \frac{C_\rho C_u}{\sqrt{-g}} dt.$$

Replacing in this equation $\sqrt{-g}$ by formula (6.35) and integrating, we obtain (for $a = |a| \neq 0$)

$$\tau = \mu \int \frac{dt}{t^2 - a^2} = \frac{\mu}{2a} \ln \left| \frac{t-a}{t+a} \right| + \tau_0, \tag{6.51}$$

where $\mu = 4\varepsilon C_u / 3\varkappa m$ and τ_0 is an arbitrary real constant.

If $a = 0$, we have

$$\tau = -\frac{\mu}{t} + \tau_0.$$

The integration constant τ_0 is insignificant for the subsequent conclusions; without loss of generality one can put $\tau_0 = 0$. For $j = 4$ Eqs. (6.49) are satisfied identically due to the conditions $h_1 = 0$, $h_2 = 0$ (i.e., $\pi_4 = 0$, $\xi_4 = 0$).

The subsequent solutions of the equations will be formulated with the aid of the variable τ, therefore here we give expression $\sqrt{-g}$ in terms of τ:

$$\sqrt{-g} = \frac{3}{4}\varkappa m \varepsilon C_\rho \left(t^2 - a^2\right) \equiv \begin{cases} \frac{3}{4}\varkappa m C_\rho a^2 \sinh^{-2} \frac{a}{\mu}\tau, & \text{if} \quad \varepsilon = 1, \\ \frac{3}{4}\varkappa m C_\rho a^2 \cosh^{-2} \frac{a}{\mu}\tau, & \text{if} \quad \varepsilon = -1. \end{cases} \tag{6.52}$$

To integrate Eqs. (6.49), let us make in them the change of the unknown functions

$$(\pi_\lambda, \xi_\lambda, \sigma_\lambda, u_\lambda) = \left(\pi_\lambda, \xi_\lambda, \sigma_\lambda, \frac{\sigma_4}{u_4}\sigma_\lambda\right) \to (\pi_\lambda^0, \xi_\lambda^0, \theta_\lambda), \quad \lambda = 1, 2, 3,$$

by formulas

$$\begin{aligned} \xi_\lambda + \mathrm{i}\pi_\lambda &= \left(\xi_\lambda^0 + \mathrm{i}\pi_\lambda^0\right)(\sqrt{-g})^{1/3} \exp\left(-\frac{1}{6}N\tau - \mathrm{i}\zeta\right), \\ \sigma_\lambda &= \theta_\lambda(\sqrt{-g})^{1/3} u^4 \exp\left(-\frac{1}{6}N\tau\right), \\ u_\lambda &= \theta_\lambda(\sqrt{-g})^{1/3} \sigma^4 \exp\left(-\frac{1}{6}N\tau\right). \end{aligned} \tag{6.53}$$

Here ζ is determined by relation (6.41) or (6.42); quantities σ^4, u^4 are determined by the solution (6.21); $\sqrt{-g}$ is determined by equality (6.52). As a result the condition of synchronism (6.50) is satisfied identically, and system of equations (6.49) becomes a system of linear equations with constant coefficients

$$\frac{\mathrm{d}}{\mathrm{d}\tau}(\xi_\lambda^0 + \mathrm{i}\pi_\lambda^0) = \frac{\mathrm{i}}{4}\varkappa(\xi_\lambda^0 + \mathrm{i}\pi_\lambda^0) - \frac{\mathrm{i}}{4}Be^{-2\mathrm{i}\beta}(\xi_\lambda^0 - \mathrm{i}\pi_\lambda^0) + \frac{\mathrm{i}}{4}Ae^{-\mathrm{i}\alpha}\theta_\lambda,$$

$$\frac{\mathrm{d}}{\mathrm{d}\tau}\theta_\lambda = \frac{1}{4}A\left(\pi_\lambda^0 \cos\alpha + \xi_\lambda^0 \sin\alpha\right) + \frac{1}{2}N\theta_\lambda. \tag{6.54}$$

From (6.9) and (6.53) it follows that the spatial components of the metric tensor $g_{\alpha\beta}$ are expressed in terms of π_λ^0, ξ_λ^0, θ_λ as follows

$$g_{\alpha\beta} = \left(\sqrt{-g}\,\right)^{2/3} e^{-\frac{1}{3}N\tau}\left(\pi_\alpha^0\pi_\beta^0 + \xi_\alpha^0\xi_\beta^0 + \theta_\alpha\theta_\beta\right). \tag{6.55}$$

The linear equations (6.54) are solved by a well known method. The characteristic equation corresponding to Eq. (6.54) for each value of the index $\lambda = 1, 2, 3$ is a cubic equation with respect to the eigenvalue q:

$$\det\left\{\frac{1}{4}\begin{Vmatrix} B\sin 2\beta & -B\cos 2\beta + \varkappa & A\cos\alpha \\ -B\cos 2\beta - \varkappa & -B\sin 2\beta & A\sin\alpha \\ A\cos\alpha & A\sin\alpha & 2N \end{Vmatrix} - qI\right\} = 0,$$

which can be written in the form

$$2(N-2q)(16q^2 + \varkappa^2 - A^2 - B^2) + A^2[2N - B\sin 2(\alpha - \beta)] = 0. \tag{6.56}$$

In general, the solution of the cubic equation (6.56) is given by the Cardano formulas. In the general case the solution of Eqs. (6.54) looks rather cumbersome, therefore we here restrict ourselves only to the case $A = 0$, as well as the case when the integration constants are connected by the relation $2N = B\sin 2(\alpha - \beta)$.

Let us consider at first the case when the equations $2N = B\sin 2(\alpha - \beta)$, $A \neq 0$ are carried out. In this case the eigenvalues q have the form $\frac{1}{2}N$, $\frac{1}{4}\sqrt{A^2 + B^2 - \varkappa^2}$, $-\frac{1}{4}\sqrt{A^2 + B^2 - \varkappa^2}$. If $0 < A^2 + B^2 < \varkappa^2$, then the solution of Eqs. (6.54) can be written in the form

$$\begin{aligned}\xi_\lambda^0 + \mathrm{i}\pi_\lambda^0 &= e^{-\mathrm{i}\alpha}\{-AQ_\lambda e^{\frac{1}{2}N\tau} + [-(\varkappa + Be^{2\mathrm{i}(\alpha-\beta)})F_\lambda + 4\mathrm{i}\Lambda G_\lambda]\cos\Lambda\tau \\ &\quad + [-(\varkappa + Be^{2\mathrm{i}(\alpha-\beta)})G_\lambda - 4\mathrm{i}\Lambda F_\lambda]\sin\Lambda\tau\}, \\ \theta_\lambda &= [\varkappa - B\cos 2(\alpha-\beta)]Q_\lambda e^{\frac{1}{2}N\tau} + A(F_\lambda\cos\Lambda\tau + G_\lambda\sin\Lambda\tau).\end{aligned} \tag{6.57}$$

Here F_λ, G_λ, Q_Λ are integration constants, $\Lambda = \frac{1}{4}\sqrt{\varkappa^2 - A^2 - B^2}$.

For the spatial components of the metric tensor, we obtain the following expression using formula (6.55)

$$\begin{aligned}g_{\alpha\beta} &= (\sqrt{-g}\,)^{2/3}e^{-\frac{1}{3}N\tau}\big[Q_\alpha Q_\beta Z^2 e^{N\tau} \\ &\quad + F_\alpha F_\beta(S + M\cos 2\Lambda\tau + 8\Lambda N\sin 2\Lambda\tau) \\ &\quad + G_\alpha G_\beta(S - M\cos 2\Lambda\tau - 8\Lambda N\sin 2\Lambda\tau)\end{aligned}$$

$$+ (F_\alpha G_\beta + F_\beta G_\alpha)(M \sin 2\Lambda\tau - 8\Lambda N \cos 2\Lambda\tau)$$
$$+ (F_\alpha Q_\beta + F_\beta Q_\alpha) 2\varkappa A e^{\frac{1}{2}N\tau} \cos \Lambda\tau$$
$$+ (G_\alpha Q_\beta + G_\beta Q_\alpha) 2\varkappa A e^{\frac{1}{2}N\tau} \sin \Lambda\tau\big], \tag{6.58}$$

where $2N = B \sin 2(\alpha - \beta)$, and we use the following notations for constants

$$M = A^2 + B^2 + \varkappa B \cos 2(\alpha - \beta), \quad S = \varkappa[\varkappa + B \cos 2(\alpha - \beta)],$$
$$Z^2 = A^2 + [\varkappa - B \cos 2(\alpha - \beta)]^2.$$

The quantity $\sqrt{-g}$ in the right-hand side of equality (6.58) is defined by relation (6.52).

Formula (6.58) defines the oscillatory approach to the singular points of the solution.

The metric is particularly simple when the phases α and β satisfy the condition $\alpha - \beta = k\pi$, $k = 0, \pm 1, \pm 2, \ldots$. In this case $N = 0$, constants A and B are arbitrary, and the integration constants F_λ, G_λ, Q_Λ satisfy the equation

$$\varepsilon^{\alpha\beta\lambda} F_\alpha G_\beta Q_\Lambda = \left(\varkappa^2 - A^2 - B^2\right)^{-3/2}, \tag{6.59}$$

where $\varepsilon^{\alpha\beta\lambda}$ is the Levi-Civita symbol. The relation (6.59) expresses the equality between the value of $\sqrt{-g}$ calculated from the solution (6.57) and that according to (6.52).

Obvious, by a constant transformation of the variables x^1, x^2, x^3 it is always possible to reduce components F_α, G_α, Q_α to the form

$$F_\alpha = \left\{(\varkappa^2 - A^2 - B^2)^{-1/2}, 0, 0\right\}, \quad G_\alpha = \left\{0, (\varkappa^2 - A^2 - B^2)^{-1/2}, 0\right\},$$
$$Q_\alpha = \left\{0, 0, (\varkappa^2 - A^2 - B^2)^{-1/2}\right\}.$$

Then the spatial part of the component of the metric tensor is determined by the matrix

$$g_{\alpha\beta} = \frac{(\sqrt{-g}\,)^{2/3}}{16\Lambda^2} \left\| \begin{matrix} S + M \cos 2\Lambda\tau & M \sin 2\Lambda\tau & 2\varkappa A \cos \Lambda\tau \\ M \sin 2\Lambda\tau & S - M \cos 2\Lambda\tau & 2\varkappa A \sin \Lambda\tau \\ 2\varkappa A \cos \Lambda\tau & 2\varkappa A \sin \Lambda\tau & Z^2 \end{matrix} \right\|.$$

If $A^2 + B^2 > \varkappa^2$, then the trigonometric functions in (6.57) are replaced by the hyperbolic ones.

Consider now the case when in Eqs. (6.54) the coefficient A is equal to zero $A = 0$. In this case the equations for θ_λ and for π_λ, ξ_λ are split and for $0 < |B| < \varkappa$ the general solution of Eqs. (6.54) can be written as

$$\theta_\lambda = Q_\lambda(\varkappa - B)\exp\left(\frac{1}{2}N\tau\right), \tag{6.60}$$

$$\xi_\lambda^0 + \mathrm{i}\pi_\lambda^0 = e^{-\mathrm{i}\beta}\{[-(\varkappa + B)F_\lambda + 4\mathrm{i}\Lambda G_\lambda]\cos\Lambda\tau + [-(\varkappa + B)G_\lambda - 4\mathrm{i}\Lambda F_\lambda]\sin\Lambda\tau\},$$

where $\Lambda = \frac{1}{4}\sqrt{\varkappa^2 - B^2}$, while the real integration constants F_α, G_α, Q_α satisfy the equation

$$\varepsilon^{\alpha\beta\lambda}F_\alpha G_\beta Q_\Lambda = (\varkappa^2 - B^2)^{-3/2}.$$

The metric tensor is defined by the components

$$g_{\alpha\beta} = (\sqrt{-g})^{2/3}e^{-N\tau/3}\big[F_\alpha F_\beta(\varkappa + B)(\varkappa + B\cos 2\Lambda\tau) + G_\alpha G_\beta(\varkappa + B)(\varkappa - B\cos 2\Lambda\tau) + (F_\alpha G_\beta + F_\beta G_\alpha)(\varkappa + B)B\sin 2\Lambda\tau + Q_\alpha Q_\beta(\varkappa - B)^2 e^{N\tau}\big]. \tag{6.61}$$

The quantity $\sqrt{-g}$ in the right-hand side of equality (6.61) is defined by Eq. (6.52). If we define the constants F_α, G_α, and Q_α by the relations

$$F_\alpha = \{(\varkappa^2 - B^2)^{-1/2}, 0, 0\}, \quad G_\alpha = \{0, (\varkappa^2 - B^2)^{-1/2}, 0\},$$
$$Q_\alpha = \{0, 0, (\varkappa^2 - B^2)^{-1/2}\},$$

then we find

$$g_{\alpha\beta} = \frac{(\sqrt{-g})^{2/3}e^{-\frac{1}{3}N\tau}}{\varkappa - B}\begin{Vmatrix} \varkappa + B\cos 2\Lambda\tau & B\sin 2\Lambda\tau & 0 \\ B\sin 2\Lambda\tau & \varkappa - B\cos 2\Lambda\tau & 0 \\ 0 & 0 & \dfrac{(\varkappa - B)^2}{\varkappa + B}e^{N\tau} \end{Vmatrix}.$$

The solution of the equations under consideration for $|B| > \varkappa$ is defined from (6.60) by replacement of the trigonometric functions by the hyperbolic ones.

If in Eqs. (6.54) constants A and B are equal to zero $A = B = 0$, then the following solution is obtained

$$\theta_\alpha = \varkappa Q_\alpha \exp\left(\frac{1}{2}N\tau\right),$$

$$\xi_\alpha^0 + \mathrm{i}\pi_\alpha^0 = \varkappa(-F_\alpha + \mathrm{i}G_\alpha)\exp\left(\frac{\mathrm{i}}{4}\varkappa\tau\right).$$

In this case we find the components of the vectors π^i, ξ^i, σ^i, u^i by means of definitions (6.53)

$$\xi_\alpha + \mathrm{i}\pi_\alpha = \varkappa(\sqrt{-g}\,)^{1/3}\left|\frac{t+a\varepsilon}{t-a\varepsilon}\right|^{1/3} \times \frac{C_u\cos(2mt+\varphi) - \mathrm{i}\varepsilon\sin(2mt+\varphi)}{\sqrt{1+C_\sigma^2\cos^2(2mt+\varphi)}}(\mathrm{i}G_\alpha - F_\alpha)\,,$$

$$\sigma_\alpha = \varkappa(\sqrt{-g}\,)^{1/3}\left|\frac{t-a\varepsilon}{t+a\varepsilon}\right|^{2/3}\frac{C_\sigma\sin(2mt+\varphi)}{\sqrt{1+C_\sigma^2\cos^2(2mt+\varphi)}}Q_\alpha,$$

$$u_\alpha = \varkappa(\sqrt{-g}\,)^{1/3}\left|\frac{t-a\varepsilon}{t+a\varepsilon}\right|^{2/3}\frac{C_u}{\sqrt{1+C_\sigma^2\cos^2(2mt+\varphi)}}Q_\alpha,$$

$$\sqrt{-g} = \frac{3}{4}\varkappa m\varepsilon C_\rho\left(t^2 - a^2\right), \tag{6.62}$$

and the metric is defined as follows

$$g_{\alpha\beta} = \varkappa^2(\sqrt{-g})^{2/3}\left[e^{-\frac{1}{3}N\tau}\left(F_\alpha F_\beta + G_\alpha G_\beta\right) + e^{\frac{2}{3}N\tau}Q_\alpha Q_\beta\right]$$
$$\equiv \varkappa^2(\sqrt{-g}\,)^{2/3}\left\{\left|\frac{t+a\varepsilon}{t-a\varepsilon}\right|^{2/3}\left(F_\alpha F_\beta + G_\alpha G_\beta\right) + \left|\frac{t-a\varepsilon}{t+a\varepsilon}\right|^{4/3}Q_\alpha Q_\beta\right\}.$$

If we define the constants F_α, G_α, and Q_α by the relations

$$F_\alpha = \left\{\varkappa^{-1}, 0, 0\right\}, \quad G_\alpha = \left\{0, \varkappa^{-1}, 0\right\}, \quad Q_\alpha = \left\{0, 0, \varkappa^{-1}\right\},$$

then metric (6.63) is defined by the diagonal matrix

$$g_{\alpha\beta} = (\sqrt{-g})^{2/3}\,\mathrm{diag}\left(e^{-\frac{1}{3}N\tau}, e^{-\frac{1}{3}N\tau}, e^{\frac{2}{3}N\tau}\right)$$
$$\equiv (\sqrt{-g})^{2/3}\,\mathrm{diag}\left\{\left|\frac{t+a\varepsilon}{t-a\varepsilon}\right|^{2/3}, \left|\frac{t+a\varepsilon}{t-a\varepsilon}\right|^{2/3}, \left|\frac{t-a\varepsilon}{t+a\varepsilon}\right|^{4/3}\right\}. \tag{6.63}$$

The quantity $\sqrt{-g}$ is determined by formula (6.35).

The solution of the Einstein–Dirac equations (6.3) with the metric tensor (6.63) for $a = 0$ was obtained in [29].

Let us calculate the scalar curvature R of the Riemannian space. It follows from Eqs. (6.7), (6.33) that

$$R = \frac{\varepsilon\varkappa m C_\rho}{\sqrt{-g}}. \tag{6.64}$$

Replacing here $\sqrt{-g}$ by formula (6.35) we find a dependence of curvature on time t:

$$R = \frac{4}{3\left(t^2 - a^2\right)}. \tag{6.65}$$

In the general case, by virtue of the condition of positivity of the invariant $\rho > 0$ (or $\sqrt{-g} > 0$), from Eq. (6.35) we find that function (6.53), (6.57), (6.60), and (6.62) for $\varepsilon = 1$ determine solutions of Eqs. (6.10) only for $|t| > a$, while in the interval $|t| < a$ the solution does not exist. At $\varepsilon = 1$ and $t = \pm a$ the metric is degenerate, and the curvature tensor has a singularity at these points. If $\varepsilon = -1$, then functions (6.53), (6.57), (6.60), and (6.62) determine the solutions of Eqs. (6.10) only in the interval $|t| < a$. Recall that similar solutions of the Einstein equations in the space without matter (such as the Kazner solutions) in the synchronous coordinate system have only single singular point [40]. Thus, the existence of a fermion field in the space-time of results in qualitative changes in solutions of the Einstein equations.

Let us calculate the components of the energy-momentum tensor for the obtained solutions. Replacing quantities s_{ab} in definition (6.31) according to solution (6.46), we obtain for the tetrad components of the energy-momentum tensor $\breve{T}_{ab}$:

$$\breve{T}_{11} = -\breve{T}_{22} = \frac{1}{8}B(\rho u^4)^2 \cos 2(\zeta + \beta), \tag{6.66}$$

$$\begin{aligned}
\breve{T}_{33} &= \rho m(\sigma^4)^2 \cos\eta, & \breve{T}_{44} &= \rho m(u^4)^2 \cos\eta,\\
\breve{T}_{12} &= \frac{1}{8}B(\rho u^4)^2 \sin 2(\zeta + \beta), & \breve{T}_{34} &= \rho m\sigma^4 u^4 \cos\eta,\\
\breve{T}_{23} &= \frac{1}{16}\rho^2(u^4)^3 A\cos(\zeta + \alpha), & \breve{T}_{31} &= -\frac{1}{16}\rho^2(u^4)^3 A\sin(\zeta + \alpha),\\
\breve{T}_{14} &= -\frac{1}{16}(\rho u^4)^2\sigma^4 A\sin(\zeta + \alpha), & \breve{T}_{24} &= \frac{1}{16}(\rho u^4)^2\sigma^4 A\cos(\zeta + \alpha).
\end{aligned}$$

The components T_{ij} of the energy-momentum tensor calculated in the holonomic coordinate system x^i are connected with the tetrad components $\breve{T}_{ab}$ as follows

$$\begin{aligned}
T_{ij} = \breve{h}_i{}^a\breve{h}_j{}^b\breve{T}_{ab} &= \breve{T}_{11}\pi_i\pi_j + \breve{T}_{22}\xi_i\xi_j + \breve{T}_{33}\sigma_i\sigma_j + \breve{T}_{44}u_iu_j\\
&+ \breve{T}_{12}(\pi_i\xi_j + \pi_j\xi_i) + \breve{T}_{23}(\xi_i\sigma_j + \xi_j\sigma_i) + \breve{T}_{13}(\sigma_i\pi_j + \sigma_j\pi_i)\\
&- \breve{T}_{14}(\pi_iu_j + \pi_ju_i) - \breve{T}_{24}(\xi_iu_j + \xi_ju_i) - \breve{T}_{34}(\sigma_iu_j + \sigma_ju_i).
\end{aligned}$$

For component T_{44} in the holonomic coordinate system we have

$$T_{44} = \breve{T}_{33}\sigma_4^2 + \breve{T}_{44}u_4^2 - 2\breve{T}_{34}\sigma_4u_4.$$

Taking into account expression (6.66) for the tetrad components $\breve{T}_{ab}$, the formula for T_{44} can be represented in the form

$$T_{44} = \rho m \cos\eta \left(\sigma_4^2 - u_4^2\right)^2 .$$

From this by virtue of the condition $g_{44} = -1$ for component T_{44}, calculated in the holonomic coordinate system the expression $T_{44} = \rho m \cos\eta$ is obtained. Bearing in mind solution (6.21) and (6.35), we find

$$T_{44} = \frac{\varepsilon m C_\rho}{\sqrt{-g}} = \frac{4}{3\varkappa(t^2 - a^2)}. \tag{6.67}$$

As known, in the classical field theory component T_{44} of the energy-momentum tensor determines the field energy density. From formula (6.67) it follows that for all solutions under consideration the energy density T_{44} is negative in the interval $|\,t\,| < a$ and corresponds then to the value of coefficient $\varepsilon = -1$; solutions in the interval $|\,t\,| < a$ for $\varepsilon = 1$ does not exist. The solutions with the positive density of energy lie in the area $|\,t\,| > a$ and correspond to the value $\varepsilon = 1$, solutions in the interval $|\,t\,| > a$ for $\varepsilon = -1$ does not exist. Thus, for the considered class of solutions at each value of parameter t in the synchronous coordinate system the solution can be only with one sign of the fermion field energy.

If in the solutions obtained here one puts $A = B = N = 0$, $\varkappa = 0$, they turn into solutions of the Dirac equations in flat space-time (in the general case this is a plane wave). It should be noted that the solutions of the Dirac equations in the form of plane waves in pseudo-Euclidean space possesses the positive density T_{44} for $\varepsilon = 1$ and negative density T_{44} for $\varepsilon = -1$ for any values of t. To the solutions of the Dirac equations in a Riemannian space, as follows from the obtained solutions, also corresponds to a positive energy density T_{44} for $\varepsilon = 1$ and a negative one for $\varepsilon = -1$, but in the synchronous coordinate system these solutions lie in the disjoint intervals $|\,t\,| > a$ and $|\,t\,| < a$.

Let us consider transformation of the variables of the observer coordinate system $(x^\alpha, t) \to (x^\alpha, \tau)$ determined by the equation

$$d\tau = \rho u^4 dt.$$

The explicit dependence of the function $\tau(t)$ is given by relations (6.51). A calculation of the quantity $\sqrt{-g}\, g^{4i}$ in the coordinate system with the variables x^α, τ gives

$$(\sqrt{-g}\, g^{4\alpha})' = 0, \quad (\sqrt{-g}\, g^{44})' = \text{const}.$$

From this it follows that the following equation holds in the coordinate system with the variables x^α and τ

$$\left[\partial_j(\sqrt{-g}\,g^{ij})\right]' = \frac{\mathrm{d}}{\mathrm{d}\tau}(\sqrt{-g}\,g^{4i})' = 0.$$

Thus, the coordinate system with the variables x^α, τ is the harmonic system. The transformation from the synchronous coordinate system to the harmonic system in the solutions obtained is carried out by the change of the variable $t \to \tau$ and the usual tensor transformation of components π_i, ξ_i, σ_i, u_i, g_{ij}. In particular, for component T_{44} of the energy-momentum tensor in the harmonic coordinate system is obtained the following expression

$$T_{44} = \frac{3\varkappa m^2 a^2}{4C_u^2}\sinh^{-2}\frac{a}{\mu}\tau \qquad \text{for} \quad \varepsilon = 1,$$

$$T_{44} = -\frac{3\varkappa m^2 a^2}{4C_u^2}\cosh^{-2}\frac{a}{\mu}\tau \qquad \text{for} \quad \varepsilon = -1. \tag{6.68}$$

The solutions with the each value $\varepsilon = 1$ and $\varepsilon = -1$ in the harmonic coordinate system exist in all interval of time $(-\infty < t < +\infty)$. The metric in the harmonic coordinate system has the singular points at $\tau = \pm\infty$ in the case $\varepsilon = -1$, and in the case $\varepsilon = 1$ there are the singular points at $\tau = 0$ and $\tau = \pm\infty$. From (6.68) it follows that the positive density of the energy T_{44} in the harmonic coordinate system corresponds to the value $\varepsilon = 1$.

When transforming $t \to \tau$ the spatial part of the metric does not change.

6.3 Exact Solutions of Some Nonlinear Differential Spinor Equations

As an example of the use of the tensor characteristics of spinors we give below a number of exact solutions of nonlinear spinor equations, used in the theory of elementary particles. At first we note the following identities connecting the first rank spinor $\boldsymbol{\psi}$ and tensors $\boldsymbol{C}$, $\boldsymbol{D}$ determined by the spinor

$$\begin{aligned}
S_i \overset{*}{\gamma}{}^i_{DA}\psi^A &= \left(\Omega e_{DA} + N\gamma^5_{DA}\right)\psi^A,\\
j_i\gamma^i_{DA}\psi^A &= \mathrm{i}\left(\Omega e_{DA} + N\gamma^5_{DA}\right)\psi^A,\\
S_i\gamma^i_{DA}\psi^A &= \mathrm{i}j_i\overset{*}{\gamma}{}^i_{DA}\psi^A,\\
C_i\gamma^i_{DA}\psi^A &= 0,\\
C_i\gamma^i_{DA}\psi^{+A} &= 2\left(\Omega e_{DA} + N\gamma^5_{DA}\right)\psi^A.
\end{aligned} \tag{6.69}$$

The first and second identities in (6.69) are obtained by contracting identities (C.2 d, c) with the spinor components $\psi^{+D}\psi^{E}\psi^{A}$ with respect to the indices D and A. The third identity in (6.69) are obtained by contracting identities (C.2 b) with the spinor components $\psi^{+D}\psi^{E}\psi^{A}$. The fourth identity in (6.69) are obtained by contracting identities (C.2 a) with the spinor components $\psi^{D}\psi^{E}\psi^{A}$. For receiving the last identities in (6.69) it is enough to contract (C.2 c) with components of spinor $\psi^{D}\psi^{E}\psi^{+A}$.

Consider the following nonlinear spinor equations

$$\gamma^i \nabla_i \psi + \lambda(\rho) S^i \overset{*}{\gamma}_i \psi = 0. \tag{6.70}$$

Here $\lambda = \lambda(\rho)$ is the given function of invariant $\rho = \left(S_i S^i\right)^{1/2}$; ∇_i is the symbol of the covariant derivative calculated in generally an arbitrary curvilinear coordinate system with the variables $x^i (i = 1, 2, 3, 4)$ in the Minkowski space.

It follows from identities (6.69) that Eq. (6.70) can be written also in the form

$$\gamma^i \nabla_i \psi - \mathrm{i}\lambda(\rho) j_i \gamma^i \psi = 0 \tag{6.71}$$

or in the form

$$\gamma^i \nabla_i \psi + \lambda(\rho) \left(\Omega I + N\gamma^5\right) \psi = 0. \tag{6.72}$$

The equations for the contravariant components $\bar{\psi} = \|\psi^{+A}\|$ of the conjugate spinor field, corresponding to Eqs. (6.70)–(6.72), are written as follows

$$\gamma^i \nabla_i \bar{\psi} + \lambda(\rho) S^i \overset{*}{\gamma}_i \bar{\psi} = 0,$$

$$\gamma^i \nabla_i \bar{\psi} + \mathrm{i}\lambda(\rho) j^i \overset{*}{\gamma}_i \bar{\psi} = 0,$$

$$\gamma^i \nabla_i \bar{\psi} + \lambda(\rho)\left(\Omega I + N\gamma^5\right)\bar{\psi} = 0.$$

Lagrangian Λ, corresponding to Eqs. (6.70), one can define as

$$\Lambda = \frac{1}{2}\left(\psi^+ \gamma^i \nabla_i \psi - \nabla_i \psi^+ \cdot \gamma^i \psi\right) + \int \rho \lambda(\rho)\, d\rho. \tag{6.73}$$

Due to Eqs. (6.70) we have

$$\Lambda = -\lambda\rho^2 + \int \rho\lambda(\rho)\, d\rho. \tag{6.74}$$

The current and spin of the field described by Eqs. (6.70) are defined, respectively, by the vectors with components $j_s = \mathrm{i}\psi^+ \gamma_s \psi$ and $S_i = \psi^+ \overset{*}{\gamma}_i \psi$. The components of the energy-momentum tensor, corresponding to the Lagrangian (6.73), are defined as follows

$$P_i{}^j = \frac{1}{2}\left(\psi^+\gamma^j\nabla_i\psi - \nabla_i\psi^+\cdot\gamma^j\psi\right) + \delta_i^j\left(\lambda\rho^2 - \int\rho\lambda(\rho)\,d\rho\right).$$

Equations (6.70) for λ = const are the Heisenberg equations [17]. In this case components $P_i{}^j$ are written in the form

$$P_i{}^j = \frac{1}{2}\left(\psi^+\gamma^j\nabla_i\psi - \nabla_i\psi^+\cdot\gamma^j\psi\right) + \frac{1}{2}\lambda S_m S^m\delta_i^j.$$

Here it is used equality $S_m S^m = \rho^2$.

Equation (6.70) and Lagrangian (6.73) are invariant under a group of the transformations

$$\psi' = e^{iM}(aI + b\gamma^5)\psi, \tag{6.75}$$

where M is an arbitrary real number, a and b are real numbers satisfying the equation $a^2+b^2 = 1$. A direct verification shows that the components of the vectors j^i, S^i are invariant under transformation (6.75):

$$j'^i = j^i, \quad S'^i = S^i.$$

The group of the gauge transformations (6.75) has an one-parameter subgroup[4]

$$\psi' = e^{\mathrm{i}\nu\varphi}\left(I\cos\mu\varphi + \gamma^5\sin\mu\varphi\right)\psi \equiv \exp\left[(\mu\gamma^5 + \mathrm{i}\nu I)\varphi\right]\psi, \tag{6.76}$$

where φ is a group parameter, μ and ν are arbitrary real numbers. An infinitesimal transformation, corresponding to the transformation group (6.76), has the form

$$\delta\psi = \left(\mathrm{i}\nu I + \mu\gamma^5\right)\psi\delta\varphi.$$

[4]As known, the exponential function of the matrix argument is defined as

$$\exp\alpha\gamma^5 = I + \sum_{n=1}^{\infty}\frac{1}{n!}(\alpha\gamma^5)^n.$$

Bearing in mind that $(\gamma^5)^2 = -I$, we find

$$\exp\alpha\gamma^5 = I\left(1 - \frac{1}{2!}\alpha^2 + \frac{1}{4!}\alpha^4 - \cdots\right) + \gamma^5\left(\alpha - \frac{1}{3!}\alpha^3 + \frac{1}{5!}\alpha^5 - \cdots\right) = I\cos\alpha + \gamma^5\sin\alpha.$$

According to the Noether theorem due to the invariance of the Lagrangian (6.73) with respect to the transformation group (6.76), the conservation law is fulfilled

$$\nabla_i\left(\nu j^i + \mu S^i\right) = 0,$$

Since the coefficients μ and ν are arbitrary, from this equation it follows:

$$\nabla_i j^i = 0, \quad \nabla_i S^i = 0.$$

Equation (6.70) and Lagrangian (6.73) are invariant also under following group of the gauge transformations[5]

$$\psi' = \alpha\psi - \beta\bar{\psi}. \tag{6.77}$$

Here α and β are arbitrary in general complex numbers satisfying the condition $\dot{\alpha}\alpha - \dot{\beta}\beta = 1$, $\bar{\psi}$ is the column of the contravariant component of the conjugate spinor ψ^{+A}.

Transformation (6.77) is the special case of transformation (3.163) for $\alpha = \dot{\delta}$ and $\beta = \dot{\gamma}$. Using relations (3.163), (3.165), (3.170), (3.171) we find that under transformation (6.77) the invariants Ω, N and the vector components S^i do not change

$$\Omega' = \Omega, \quad N' = N, \quad S'^i = S^i.$$

Group (6.77) has one-parameter subgroup of the form

$$\psi'^A = \psi^A \cosh\varphi + \psi^{+A} e^{\mathrm{i}\chi} \sinh\varphi, \tag{6.78}$$

where φ is a group parameter, χ is an arbitrary real number. The infinitesimal transformation corresponding to the group (6.78) is

$$\delta\psi = e^{\mathrm{i}\chi}\bar{\psi}\delta\varphi, \quad \delta\bar{\psi} = e^{-\mathrm{i}\chi}\psi\delta\varphi.$$

[5]It is easy to show that Eq. (6.70) does not invariant[83] under the Pauli group

$$\psi'^A = \alpha\psi^A - \mathrm{i}\beta\gamma^{5A}{}_B\psi^{+B}, \quad \dot{\alpha}\alpha + \dot{\beta}\beta = 1.$$

However, as is well-known [17], quantized equations

$$\gamma^i\nabla_i\psi + \lambda : S_i\overset{*}{\gamma}{}^i\psi := 0$$

are invariant under the Pauli group. Thus, a symmetry group of nonlinear spinor equations can change at their quantization. We note in this regard that the opinion is sometimes expressed about impossibility changing the symmetry group of the equations at them quantization (see e.g. the article [17] devoted to properties of nonlinear spinor equations).

Using relations (3.170) and (3.171) is easy to find that the components of the vector $e^{-\mathrm{i}\chi}C^j + e^{\mathrm{i}\chi}\dot{C}^j$ do not change under transformation (6.78):

$$\left(e^{-\mathrm{i}\chi}C^j + e^{\mathrm{i}\chi}\dot{C}^j\right)' = e^{-\mathrm{i}\chi}C^j + e^{\mathrm{i}\chi}\dot{C}^j.$$

Due to the invariance of the Lagrangian (6.73) under the transformation group (6.78), the conservation law is fulfilled

$$\nabla_j\left(e^{-\mathrm{i}\chi}C^j + e^{\mathrm{i}\chi}\dot{C}^j\right) = 0.$$

1. Let us consider a transformation $\psi_\circ \to \psi$:

$$\psi = \exp\left[\gamma^5 \int \left(\eta S_i^\circ + \nu j_i^\circ\right) dx^i + \mathrm{i}I \int \left(\nu S_i^\circ + \theta j_i^\circ\right) dx^i\right]\psi_\circ, \tag{6.79}$$

where $\nu = \nu(x^i)$ is an arbitrary differentiable function; $\eta = \eta(x^i)$ and $\theta = \theta(x^i)$ are the differentiable functions connected by the relation $\theta - \eta = \lambda(\rho_\circ)$. Thus, the transformation $\psi_\circ \to \psi$ depends on two arbitrary functions. The components of the vectors j_i°, S_i° and the scalar $\rho_\circ$ are defined by the spinor field $\psi_\circ$:

$$j_i^\circ = \mathrm{i}\overset{+}{\psi_\circ}\gamma_i\psi_\circ, \quad S_i^\circ = \overset{+}{\psi_\circ}\overset{*}{\gamma}_i\psi_\circ, \quad \rho_\circ = \left(S_i^\circ S_\circ^i\right)^{1/2}.$$

Integration in (6.79) is taken over some in general an arbitrary sufficiently smooth path in the Minkowski space. In order to the integrals in (6.79) do not depend on the path of integration, it is necessary and sufficient that the equations

$$\begin{aligned}
\partial_i\left(\eta S_j^\circ + \nu j_j^\circ\right) &= \partial_j\left(\eta S_i^\circ + \nu j_i^\circ\right),\\
\partial_i\left(\nu S_j^\circ + \theta j_j^\circ\right) &= \partial_j\left(\nu S_i^\circ + \theta j_i^\circ\right).
\end{aligned} \tag{6.80}$$

be satisfied. The components of the vectors j_i and S_i are invariant under transformation (6.79)

$$\begin{aligned}
j_i &= \mathrm{i}\overset{+}{\psi}\gamma_i\psi \equiv \mathrm{i}\overset{+}{\psi_\circ}\gamma_i\psi_\circ = j_i^\circ,\\
S_i &= \overset{+}{\psi}\overset{*}{\gamma}_i\psi \equiv \overset{+}{\psi_\circ}\overset{*}{\gamma}_i\psi_\circ = S_i^\circ,
\end{aligned} \tag{6.81}$$

while the invariants $\Omega = \psi^+\psi$ and $N = \psi^+\gamma^5\psi$ of the spinor ψ depend on the integral $\int \left(\eta S_i^\circ + \nu j_i^\circ\right) dx^i$:

$$\begin{aligned} \Omega = &\ \ \Omega_\circ \cos\left[2\int \left(\eta S_i^\circ + \nu j_i^\circ\right) dx^i\right] + N_\circ \sin\left[2\int \left(\eta S_i^\circ + \nu j_i^\circ\right) dx^i\right], \\ N = &-\Omega_\circ \sin\left[2\int \left(\eta S_i^\circ + \nu j_i^\circ\right) dx^i\right] + N_\circ \cos\left[2\int \left(\eta S_i^\circ + \nu j_i^\circ\right) dx^i\right]. \end{aligned} \tag{6.82}$$

Here $\Omega_\circ = \psi_\circ^+\psi_\circ$, $N_\circ = \psi_\circ^+\gamma^5\psi_\circ$.

A calculation of the energy-momentum tensor components $P_i{}^j$ under transformation (6.79) gives

$$\begin{aligned} P_i{}^j = &\frac{1}{2}\left[\psi_\circ^+\gamma^j\nabla_i\psi_\circ - \left(\nabla_i\psi_\circ^+\right)\gamma^j\psi_\circ\right] \\ &+ \left(\eta S_i^\circ + \nu j_i^\circ\right) S_\circ^j + \left(\nu S_i^\circ + \theta j_i^\circ\right) j_\circ^j - \Lambda(\rho_\circ)\delta_i^j, \end{aligned} \tag{6.83}$$

where function $\Lambda = \Lambda(\rho)$ is defined by equality (6.74).

Let us calculate result of action of the operator $\gamma^i\nabla_i$ upon the function ψ, determined by equality (6.79). We have identically

$$\begin{aligned} \gamma^i\nabla_i\psi = &\ \gamma^i\left[\left(\eta S_i^\circ + \nu j_i^\circ\right)\gamma^5 + \mathrm{i}\left(\nu S_i^\circ + \theta j_i^\circ\right) I\right]\psi \\ &+ \exp\left[-\gamma^5\int \left(\eta S_i^\circ + \nu j_i^\circ\right) dx^i + \mathrm{i}I\int \left(\nu S_i^\circ + \theta j_i^\circ\right) dx^i\right]\gamma^j\nabla_j\psi_\circ. \end{aligned} \tag{6.84}$$

By means of identity (6.69), (6.81), we transform the first term on the right-hand side of Eq. (6.84) to the form

$$\gamma^i\left[\left(\eta S_i^\circ + \nu j_i^\circ\right)\gamma^5 + \mathrm{i}\left(\nu S_i^\circ + \theta j_i^\circ\right) I\right]\psi = (\eta - \theta)\, S^i\overset{*}{\gamma}_i\psi = -\lambda S^i\overset{*}{\gamma}_i\psi.$$

Therefore equality (6.84) can be rewritten in the form

$$\begin{aligned} &\gamma^i\nabla_i\psi + \lambda S^i\overset{*}{\gamma}_i\psi \\ &\equiv \exp\left[-\gamma^5\int \left(\eta S_i^\circ + \nu j_i^\circ\right) dx^i + \mathrm{i}I\int \left(\nu S_i^\circ + \theta j_i^\circ\right) dx^i\right]\gamma^j\nabla_j\psi_\circ. \end{aligned} \tag{6.85}$$

If the functions $\psi_\circ$ in (6.79) satisfy the linear equation

$$\gamma^i\nabla_i\psi_\circ = 0, \tag{6.86}$$

then from (6.85) it follows that the functions ψ, determined in terms of $\psi_\circ$ in according to equalities (6.79), are a solution of Eqs. (6.70).

If spintensors γ^i are defined by equalities (3.24), and $\eta =$ const, $\nu =$ const, $\theta =$ const, then the spinor $\boldsymbol{\psi}_\circ$ in solution (6.79) in the Cartesian coordinate system x^i is possible to define, in particular, as follows

$$\psi_\circ = \begin{Vmatrix} \psi_\circ^1(x^3 - x^4) \\ \psi_\circ^2(x^3 + x^4) \\ \psi_\circ^3(x^3 + x^4) \\ \psi_\circ^4(x^3 - x^4) \end{Vmatrix}, \tag{6.87}$$

where $\psi_\circ^A$ are arbitrary differentiable functions of the arguments noted in (6.87) satisfy the relations

$$\begin{aligned} (\eta - \nu)\dot{\psi}_\circ^1 \psi_\circ^2 + (\eta + \nu)\dot{\psi}_\circ^3 \psi_\circ^4 &= \text{const}, \\ (\nu - \theta)\dot{\psi}_\circ^1 \psi_\circ^2 + (\nu + \theta)\dot{\psi}_\circ^3 \psi_\circ^4 &= \text{const}, \end{aligned} \tag{6.88}$$

which, according to definitions (3.65), are equivalent to equalities

$$\eta S_1^\circ + \nu j_1^\circ = \text{const}, \quad \eta S_2^\circ + \nu j_2^\circ = \text{const},$$
$$\nu S_1^\circ + \theta j_1^\circ = \text{const}, \quad \nu S_2^\circ + \theta j_2^\circ = \text{const}.$$

It is not difficult to verify that formulas (6.87), (6.88) and the equalities $\eta =$ const, $\nu =$ const, $\theta =$ const determine in the Cartesian coordinate system some set of partial solutions of Eqs. (6.86), (6.80).

A study of the functional equations (6.88) is quite elementary and we shall not give here a more detailed analysis of these equations. Note only that from these equations it follows or the existence of linear connections between the functions $\psi_\circ^1$ and $\psi_\circ^4$ and between the functions $\psi_\circ^2$ and $\psi_\circ^3$ or their constancy; perhaps also the equality to zero some functions $\psi_\circ^A$ at arbitrariness of the anothers. For example, $\psi_\circ^A$ is possible to determine by the relation

$$\psi_\circ = \begin{Vmatrix} \psi_\circ^1(x^3 - x^4) \\ 0 \\ \psi_\circ^3(x^3 + x^4) \\ 0 \end{Vmatrix} \quad \text{or} \quad \psi_\circ = \begin{Vmatrix} 0 \\ \psi_\circ^2(x^3 + x^4) \\ 0 \\ \psi_\circ^4(x^3 - x^4) \end{Vmatrix},$$

where $\psi_\circ^A$ are arbitrary differentiable functions.

The physical meaning of solutions (6.79) is defined by relations (6.81)–(6.83) also depends on the choice of the functions $\psi_\circ$. In particular, if $\psi_\circ$ and η, ν, θ are constants, then solutions (6.79) determine plane, monochromatic waves. If $\psi_\circ$ to take in the form (6.87), then the vectors j^i and S^i according to formulas (6.81) are represented in the form of an plane progressive wave.

2. Using relations (6.69), similarly to the derivation of solution (6.79), it is possible to show that the spinor equations (6.70) admit exact solutions of the form[6]

$$\psi = \exp\left(\gamma^5 \int \eta S_j^\circ \, dx^j\right)\left(\psi_\circ \cosh \int b_j \, dx^j + \bar{\psi}_\circ e^{i\chi} \sinh \int b_j \, dx^j\right), \quad (6.89)$$

where the components of the vectors S_j°, b_j are defined by the equalities

$$S_j^\circ = \overset{+}{\psi}_\circ \overset{*}{\gamma}_j \psi_\circ, \quad b_j = -\frac{1}{2}\theta\left(e^{-i\chi} C_j^\circ + e^{i\chi} \dot{C}_j^\circ\right). \quad (6.90)$$

In formulas (6.89), (6.90) $\theta = \theta(x^i)$ and $\eta(x^i)$ are arbitrary real differentiable functions, connected by the relation $\theta - \eta = \lambda(\rho_\circ)$; χ is an arbitrary real constant. The vector components C_j° in (6.90) are defined by the field $\psi_\circ$:

$$C_j^\circ = \psi_\circ^T E \gamma_j \psi_\circ.$$

As $\psi_\circ$ in (6.89) and (6.90) we can take any solution of Eqs. (6.86) which satisfies the independence conditions of the integrals in (6.89) of the integration path in the Minkowski space

$$\partial_i b_j = \partial_j b_i, \quad \partial_i\left(\eta S_j^\circ\right) = \partial_j\left(\eta S_i^\circ\right).$$

For solutions (6.89) the vector components S_i are defined by functions $\psi_\circ$

$$S_i = \psi^+ \overset{*}{\gamma}_i \psi \equiv \psi_\circ^+ \overset{*}{\gamma}_i \psi_\circ = S_i^\circ,$$

while the invariants Ω, N of the spinor field ψ are determined by the relations

$$\begin{aligned} \Omega &= \Omega_\circ \cos\left(2\int \eta S_j^\circ \, dx^j\right) + N_\circ \sin\left(2\int \eta S_j^\circ \, dx^j\right), \\ N &= -\Omega_\circ \sin\left(2\int \eta S_j^\circ \, dx^j\right) + N_\circ \cos\left(2\int \eta S_j^\circ \, dx^j\right), \end{aligned} \quad (6.91)$$

in which $\Omega_\circ = \psi_\circ^+ \psi_\circ$, $N_\circ = \psi_\circ^+ \gamma^5 \psi_\circ$.

When transforming it is necessary to use also the relation

$$e^{-i\chi} C_j + e^{i\chi} \dot{C}_j \equiv e^{-i\chi} C_j^\circ + e^{i\chi} \dot{C}_j^\circ,$$

in which the vector components C_j are defined by the functions ψ.

---

[6]The more general solutions of this type see in [79, 82].

If matrices γ^i are determined according to (3.24) and $\lambda = \text{const}$, $\eta = \text{const}$, then the spinor components $\psi_\circ$ in solution (6.89), in particular, is possible to define in the Cartesian coordinate system by formula (6.87), in which arbitrary functions $\psi_\circ^A$ satisfy the relations

$$S_1^\circ = \text{const}, \quad S_2^\circ = \text{const},$$
$$e^{-\mathrm{i}\chi} C_1^\circ + e^{\mathrm{i}\chi} \dot{C}_1^\circ = \text{const}, \quad e^{-\mathrm{i}\chi} C_2^\circ + e^{\mathrm{i}\chi} \dot{C}_2^\circ = \text{const}.$$

3. Let us consider transformation $\psi_\circ \to \psi$:

$$\psi = \exp\left(\gamma^5 \int \eta S_j^\circ \, dx^j\right)$$
$$\times \left[\alpha \psi_\circ \exp\left(\mathrm{i} \int \theta j_s^\circ \, dx^s\right) + \mu \bar{\psi}_\circ \exp\left(-\mathrm{i} \int \theta j_s^\circ \, dx^s\right)\right], \tag{6.92}$$

where α and μ are some generally complex numbers satisfying the condition $\dot{\alpha}\alpha - \dot{\mu}\mu = 1$; $\theta(x^i)$ and $\eta(x^i)$ are arbitrary real differentiable functions, connected by the relation $\theta - \eta = \lambda(\rho_\circ)$; the vector components S_j°, j_s° are expressed in terms of the spinor field $\psi_\circ$ as follows

$$S_j^\circ = \psi_\circ^+ \overset{*}{\gamma}_j \psi_\circ, \quad j_s^\circ = \mathrm{i} \psi_\circ^+ \gamma_s \psi_\circ.$$

The vector components S_j, corresponding to the field ψ, are expressed in terms of the field $\psi_\circ$: $S_j = \psi^+ \overset{*}{\gamma}_j \psi \equiv \psi_\circ^+ \overset{*}{\gamma}_j \psi_\circ$. The invariants Ω and N of the spinor field ψ are expressed in terms of the field $\psi_\circ$ by relations (6.91).

If the functions $\psi_\circ$ in (6.92) satisfy Eq. (6.86) and the independence conditions of integrals in (6.92) from the path of integration in the Minkowsky space

$$\partial_i (\eta S_j^\circ) = \partial_j (\eta S_i^\circ), \quad \partial_i (\theta j_j^\circ) = \partial_j (\theta j_i^\circ),$$

then the functions ψ determined by formulas (6.92), are the exact solution of Eqs. (6.70).

If as the function $\psi_\circ$ take functions (6.87), satisfying relations

$$\eta S_1^\circ = \text{const}, \quad \eta S_2^\circ = \text{const},$$
$$\theta j_1^\circ = \text{const}, \quad \theta j_2^\circ = \text{const},$$

then the vector components j^i and S^i corresponding to such solutions are represented in the form of the plane progressive waves.

Note that if in functions (6.89) and (6.92) we put $\eta = 0$, and in functions (6.79) $\eta = \nu = 0$, then these functions determine solutions of Eqs. (6.70) and in a case when the coefficient λ in Eq. (6.70) depend arbitrarily on the invariants Ω, N of the field ψ.

Transformations (6.79), (6.89) and (6.92) reduce the solution of Eqs. (6.70) to the solution of Eqs. (6.86) and independence conditions of integration in these transformations from the integration path. It is obvious that if in formulas (6.79), (6.89), we put (6.92) $\lambda = 0$, these transformations will determine solutions equations (6.86).

It is clear that functions (6.79), (6.89) and (6.92) for $\lambda = 0$ can be used as functions $\psi_\circ$ in transformations (6.79), (6.89) and (6.92). In this way, we can obtain another exact solutions of Eq. (6.70).

It is easy to show that transformations (6.79), (6.89) and (6.92) applied to the equations

$$\gamma^i \nabla_i \psi + m\psi + \lambda(\rho) S^i \overset{*}{\gamma}_i \psi = 0, \tag{6.93}$$

reduce the solution of Eqs. (6.93) to the solution of the Dirac equation for the functions $\psi_\circ$:

$$\gamma^i \nabla_i \psi_\circ + m\psi_\circ = 0$$

6.4 Integrals of the Differential Equations, Describing Relativistic Models of the Magnetizable Spin Fluids

In this section differential equations, describing some models of the magnetizable spin fluids in electromagnetic field in special relativity theory are considered. The systems of partial differential equations describing these models are usually very complicated, and a solution of these equations is a mathematically very complicated problem. The use of the method of the tensor representation of spinors in some quite general cases allows to simplify integration of the specified equations.

Below, using results of Sects. 6.3 and 6.5 some integrals of the equations defining models of the magnetized spin fluids, characterized by the condition constancy of the module of the vector of specific density of magnetization are formulated. In the following section by means of the found integrals a number of exact solutions of considered equations is established

6.4.1 Relativistic Equations Describing the Magnetizable Spin Fluids

We assume further that the physical space-time is the four-dimensional pseudo-Euclidean Minkowski space referred to a Cartesian inertial coordinate system of an observer. Let us consider models of the spin fluids defined by the Lagrangian[7]

$$\Lambda = \frac{1}{8\pi} F^{ij}\left(\partial_i A_j - \partial_j A_i - \frac{1}{2} F_{ij}\right)$$
$$- M^{ij}\partial_i A_j + \eta u^i M_i - \frac{1}{g} M^i \Omega_i + \Lambda_0(\rho, s, \omega_i, M_i, g^{ij}), \tag{6.94}$$

where g is a phenomenological constant. The functional δW^* in the variational equation (A.49) we determine by the relation

$$\delta W^* = \int\limits_{V_4} \left(-\rho T \delta s + \tau_i{}^j \nabla_j \delta x^i\right) dV_4. \tag{6.95}$$

For the models under consideration the vector of the spin is proportional to the vector of the volume density of the fluid magnetization $gK^i = M^i$. The tensor of the volume density of the internal angular momentum for the considered spin fluids is defined by components $K^{ij} = g^{-1} M^{ij}$.

A complete system of dynamic and kinematic differential equations, corresponding to the considered models of the magnetizable spin fluids, has the form

$$\begin{aligned}
&a. \quad \partial_j H^{ij} = 0, \quad \partial_i F_{jk} + \partial_j F_{ki} + \partial_k F_{ij} = 0,\\
&b. \quad \partial_j P_i{}^j = 0, \quad \partial_i \rho u^i = 0,\\
&c. \quad \rho \frac{d}{d\tau}\frac{1}{\rho} M_i = g \overset{*}{F}_{ij} M^j,\\
&d. \quad \rho T \frac{ds}{d\tau} = -c u^i \nabla_j \tau_i{}^j, \quad \rho T = \frac{\partial \Lambda_0}{\partial s}.
\end{aligned} \tag{6.96}$$

The components of the total energy-momentum tensor $P_i{}^j$ and the components of the effective field tensor $\overset{*}{F}_{ij}$ in Eqs. (6.96) are defined as follows

$$P_i{}^j = P_i{}^j_{(f)} + (p+e) u_i u^j + p\delta_i^j - \tau_i{}^j + \frac{1}{g} M_{is} u^j \frac{d}{d\tau} u^s + \frac{1}{2} S_{is} u^j \frac{d}{d\tau} u^s +$$
$$+ \frac{1}{2} c u^j \nabla_k S_i{}^k - \frac{1}{2} c S^{js} \nabla_i u_s - u^j u_k \left(M_i \frac{\partial \Lambda_0}{\partial M_k} + \omega_i \frac{\partial \Lambda_0}{\partial \omega_k}\right),$$
$$\overset{*}{F}_{ij} = \sigma_i^m \sigma_j^n \left(F_{mn} + \varepsilon_{mnks} u^k \frac{\partial \Lambda_0}{\partial M_s}\right) + \frac{1}{g}\left(u_i \frac{d}{d\tau} u_j - u_j \frac{d}{d\tau} u_i\right). \tag{6.97}$$

---

[7]A variational derivation of the differential equations used here that describe models of the magnetizable spin fluids in an electromagnetic field is contained in Appendix A. About parameters appearing in Lagrangian (6.94) and functional (6.95), see Appendix A.

Here the energy–momentum tensor components of the electromagnetic field $P_{i\,(f)}^{\;j}$ are defined by equality (A.72), for quantities p, e, and S_{ij} in (6.97) according to definitions (A.61) we have

$$p = \rho^2 \frac{\partial \Lambda_0/\rho}{\partial \rho} + M_i \frac{\partial \Lambda_0}{\partial M_i} - \frac{1}{2} B_i M^i,$$
$$e = \Lambda_0 - \frac{1}{2} B_i M^i - \omega_i \frac{\partial \Lambda_0}{\partial \omega_i},$$
$$S_{ij} = \varepsilon_{ijks} u^k \frac{\partial \Lambda_0}{\partial \omega_s}. \tag{6.98}$$

The vector components of the intrinsic rotation Ω^i are eliminated from Eqs. (6.96) to (6.98) by means of the equation

$$B^i = -\frac{1}{g} \Omega^i + \sigma_s^i \frac{\partial \Lambda_0}{\partial M_s}, \tag{6.99}$$

which is obtained from the second equation in (A.59 e).

In Eqs. (6.97) the components M_{ij} of the volume density tensor of the fluid magnetization are connected components M_i of the volume density vector of magnetization by the equality

$$M_{ij} = \varepsilon_{ijks} u^s M^k, \quad M_i = \frac{1}{2} \varepsilon_{ijks} u^s M^{jk}. \tag{6.100}$$

It follows from definition (6.100) that in the proper basis the matrix of components $\breve{M}_{ij}$ is defined only by the components $\breve{M}^1$, $\breve{M}^2$, and $\breve{M}^3$ of three-dimensional vector of the volume density of the fluid magnetization

$$\breve{M}_{ij} = \begin{Vmatrix} 0 & \breve{M}^3 & -\breve{M}^2 & 0 \\ -\breve{M}^3 & 0 & \breve{M}^1 & 0 \\ \breve{M}^2 & -\breve{M}^1 & 0 & 0 \\ 0 & 0 & 0 & 0 \end{Vmatrix}.$$

The set of Eqs. (6.96) contains the first and second pairs of the Maxwell's equations for the electromagnetic field, the energy-momentum equation, the angular momentum equation, the continuity equation for the mass density of the fluid, the temperature equation and the entropy balance equation. In Eqs. (6.96) unknown required functions are the mass density $\rho(x^i)$, the components of the four-dimensional velocity vector of the fluid $u^i(x^j)$, the components of the four-dimensional vector of the volume density of the fluid magnetization $M_i(x^j)$, the temperature $T(x^i)$, the specific entropy density $s(x^i)$ and the components of the tensor electromagnetic field $F_{ij}(x^i)$. The components of the four-dimensional velocity vector u^i and the components of the four-dimensional magnetization vector M_i, appearing in Eq. (6.96), according to their definition are connected by the algebraic equations

$$u_i u^i = -1, \quad u_i M^i = 0. \tag{6.101}$$

To complete the set of Eqs. (6.96)–(6.99) it is necessary to specify the tensor components $\tau_i{}^j$, determining viscosity and heat conductivity of the fluid.

6.4.2 Integrals of the Differential Equations Describing the Magnetizable Spin Fluids

Let us introduce in the Minkowski space the four-component spinor field in the Cartesian coordinate system of the observer by components $\psi(x^i)$ and let $\psi^+(x^i)$ be the components of the conjugate spinor field. For mass density ρ, velocity vector components u^i, vector component magnetization M_i and for entropy s in Eqs. (6.96)–(6.98) we put by definition

$$\rho u^j = \mathrm{i}\psi^+\gamma^j\psi, \quad M_i = m\psi^+\overset{*}{\gamma}_i\psi,$$
$$\rho \exp \mathrm{i}\eta(s) = \psi^+\psi + \mathrm{i}\psi^+\gamma^5\psi. \tag{6.102}$$

Here γ^j are the Dirac matrices, the matrices $\overset{*}{\gamma}_i$, γ^5 are defined by equalities (3.9); m is a constant; $\eta(s)$ is the given differentiable function with the nonzero derivative in all range s.

Definitions (6.102) imply that the fluid density ρ is expressed in terms of the spinor fields ψ, ψ^+ thus

$$\rho = \left[(\psi^+\psi)^2 + (\psi^+\gamma^5\psi)^2\right]^{1/2}.$$

Due to definitions (6.102), Eqs. (6.101) are simply identities, while the components of the magnetization vector M_i satisfy an additional algebraic equation

$$M_i M^i = \rho^2 m^2. \tag{6.103}$$

Parametrization (6.102) imposes no other restrictions on the components ρ, u^j, M_i, and s, except (6.101), (6.103).

Let us consider the following system of equations

$$\gamma^j\partial_j\psi + \left(\varkappa I + \mathrm{i}\varkappa_j\gamma^j + \overset{*}{\varkappa}_j\overset{*}{\gamma}{}^j + \overset{*}{\varkappa}\gamma^5\right)\psi = 0,$$
$$\partial_i F_{jk} + \partial_j F_{ki} + \partial_k F_{ij} = 0, \quad \partial_j\left(F^{ij} - 4\pi M^{ij}\right) = 0. \tag{6.104}$$

where the real coefficients $\varkappa$, $\varkappa_i$, $\overset{*}{\varkappa}_i$, and $\overset{*}{\varkappa}$ entering in Eq. (6.104), are determined by the relations

$$\varkappa = \frac{g}{2mc\rho^2}\Big\{ - \Omega\Big(\rho\frac{\partial \Lambda_0}{\partial \rho} + \frac{1}{2}F_{ij}M^{ij}\Big) + N\Big[\frac{1}{\partial\eta/\partial s}\Big(\frac{\partial \Lambda_0}{\partial s} - \rho T\Big) + \frac{c}{g}\partial_i M^i\Big]\Big\},$$

$$\overset{*}{\varkappa} = \frac{g}{2mc\rho^2}\Big\{ - N\Big(\rho\frac{\partial \Lambda_0}{\partial \rho} + \frac{1}{2}F_{ij}M^{ij}\Big) - \Omega\Big[\frac{1}{\partial\eta/\partial s}\Big(\frac{\partial \Lambda_0}{\partial s} - \rho T\Big) + \frac{c}{g}\partial_i M^i\Big]\Big\},$$

$$\varkappa_i = \frac{g}{2mc\rho}\Big\{\rho G_i + \frac{c}{2}\sigma_{ij}\varepsilon^{jnks}\Big[\frac{\partial \Lambda_0}{\partial \omega^n}\partial_k u_s + \partial_k\Big(\frac{\partial \Lambda_0}{\partial \omega^n}u_s\Big)\Big] + \frac{1}{2}\varepsilon_{jksi}F^{j\kappa}M^s\Big\} + \frac{1}{2mc\rho}\sigma_{ij}\Big(c\partial_s M^{js} + M^{js}\frac{d}{d\tau}u_s\Big) + \partial_i\gamma, \tag{6.105}$$

$$\overset{*}{\varkappa}_i = \frac{g}{2c}\Big(\frac{1}{2}\varepsilon_{ijks}F^{jk}u^s - \frac{\partial \Lambda_0}{\partial M^i}\Big) + \frac{1}{2}\frac{\partial\eta}{\partial s}\partial_i s + \frac{1}{2}\varepsilon_{ijks}u^j\partial^k u^s.$$

Here $\gamma = \gamma(x^i)$ is an arbitrary differentiable function; quantities Ω and N are defined by equalities (3.58) and (3.59); the components ρ, u^i, M_i, s in (6.105) are expressed in terms of the fields ψ and ψ^+ by formulas (6.102); $\sigma_{ij} = g_{ij} + u_i u_j$, g_{ij} are the covariant components of the metric tensor of the space-time. The components M_{ij} are connected with M_i by relation (6.100).

To determine the quantities G_i, entering into coefficients $\varkappa$, we introduce the following equations, linear in G_i:

$$\rho u^j(\partial_j G_i - \partial_i G_j) + \rho T\partial_i s + \partial_j \tau_i{}^j = 0. \tag{6.106}$$

The physical meaning of Eqs. (6.106) and quantities G_i is explained below.

Thus, Eqs. (6.104) and (6.105) with the given functions G_i constitute a system differential equations for determining the spinor field $\psi(x^i)$, temperature $T(x^i)$ and components $F_{ij}(x^i)$ of the electromagnetic field tensor.

Equations (6.104) and (6.105) can be obtained by means of the variational equation (A.49), in which the Lagrangian Λ is determined by the relation

$$\Lambda = \frac{1}{8\pi}F^{ij}\Big(\partial_i A_j - \partial_j A_i - \frac{1}{2}F_{ij}\Big) - M^{ij}\partial_i A_j - \frac{c}{g}M^i\partial_i\eta - \frac{2}{g}M^i\omega_i - \frac{mc}{g}(\psi^+\gamma^i\partial_i\psi - \partial_i\psi^+\cdot\gamma^i\psi) + \Lambda_0(\rho, s, \omega_i, M_i, g^{ij}), \tag{6.107}$$

and the functional δW^* is determined by equality (6.95). The functions ρ, u_i, M_i, and s in (6.107) and in δW^* are expressed in terms of ψ and ψ^+ by formulas (6.102).

If equalities (6.102) are carried out, then by virtue of identity (5.70) the Lagrangian (6.107) identically coincides with Lagrangian (6.94).

Let us show that all equations in (6.96) for arbitrary functions G_i, $\tau_i{}^j$, and T, satisfying Eqs. (6.106) and (6.105) are a corollary of Eqs. (6.104), if the density ρ, the entropy s, the velocity u^i and the fluid magnetization M^i are defined by the fields ψ and ψ^+ by equalities (6.102). For this purpose we consider a system of the real tensor equations (5.78) for $\varkappa_{ij} = 0$ and $\alpha = -mc/g$, which are carried out due to the spinor equations (6.104) and (6.105).

It is easy to see that the fourth equation in (5.78), with the coefficients $\varkappa$, $\varkappa_i$, $\overset{*}{\varkappa}_i$, and $\overset{*}{\varkappa}$ defined by equalities (6.105), can be written as

$$\frac{1}{2}\partial_i S^i - \varkappa N + \overset{*}{\varkappa}\Omega = \frac{1}{2m}\partial_i M^i + \frac{g}{2mc}\left[\frac{1}{\partial\eta/\partial s}\left(-\frac{\partial\Lambda_0}{\partial s} + \rho T\right) - \right.$$
$$\left. - \frac{c}{g}\partial_i M^i\right] = \frac{g}{2mc\partial\eta/\partial s}\left(-\frac{\partial\Lambda_0}{\partial s} + \rho T\right) = 0.$$

From this it follows

$$-\frac{\partial\Lambda_0}{\partial s} + \rho T = 0.$$

This means that the equation for the temperature T in (6.96) is a corollary of Eqs. (6.104) and (6.105).

The first equation in (5.78), according to definitions (5.19) and (6.102), is written in the form

$$\partial_i j^i = \partial_i \rho u^i = 0.$$

This means that the continuity equation in (6.96) also is a corollary of Eqs. (6.104) and (6.105).

We now calculate the components of the antisymmetric tensor $\overset{*}{F}_{ij}$ entering the third equation in (5.78) and determined by formulas (5.75). Substituting $\varkappa_{ij} = 0$ and the quantity $\overset{*}{\varkappa}_i$ determined by (6.105) into definition (5.75) and doing simple transformations we find for components $\overset{*}{F}_{ij}$

$$\overset{*}{F}_{ij} = \frac{c}{g}\left[\varepsilon_{ijks}u^k\left(-2\overset{*}{\varkappa}{}^s + \partial^s\eta\right) - \partial_i u_j + \partial_j u_i\right]$$
$$= \sigma_i^m \sigma_j^n \left(F_{mn} + \varepsilon_{mnks}u^k \frac{\partial\Lambda_0}{\partial M_s}\right) + \frac{1}{g}\left(u_i \frac{d}{d\tau}u_j - u_j \frac{d}{d\tau}u_i\right). \tag{6.108}$$

The components $\overset{*}{F}_{ij}$, determined by formula (6.108), coincide with those determined by formulas (6.97). Thus, the angular momentum equation in (6.96) is a corollary of Eqs. (6.104) and (6.105).

Consider now the second equation (5.78) for $\alpha = -mc/g$, $\varkappa_{ij} = 0$ which is now conveniently written down in the form

$$\partial_j \left[\mathcal{P}_i{}^j - \frac{2mc}{g} \delta_i^j \left(\varkappa \Omega + \varkappa_s j^s + \overset{*}{\varkappa}_s S^s + \overset{*}{\varkappa} N \right) \right] = -\frac{2mc}{g} \left(\varkappa \partial_i \Omega + \varkappa_j \partial_i j^j + \overset{*}{\varkappa}_j \partial_i S^j + \overset{*}{\varkappa} \partial_i N \right). \tag{6.109}$$

Substituting $\varkappa_{ij} = 0$, $\alpha = -mc/g$, and the coefficients $\varkappa$, $\varkappa_i$, $\overset{*}{\varkappa}_i$, $\overset{*}{\varkappa}$ determined by formulas (6.105), into definition (5.67) of components $\mathcal{P}_i{}^j$, for the components of the tensor entering the left-hand side of Eq. (6.109), we find

$$\begin{aligned}
\mathcal{P}_i{}^j &- \frac{2mc}{g} \delta_i^j \left(\varkappa \Omega + \varkappa_s j^s + \overset{*}{\varkappa}_s S^s + \overset{*}{\varkappa} N \right) \\
&= -\frac{c}{g} \left[u_s \partial_i M^{js} + u^j \partial_s M^{is} - M^j \partial_i \eta + u^j \left(M_i u^k - u_i M^k \right) \partial_k \eta \right] \\
&\quad - \frac{2mc}{g} \left[\sigma_i^j \left(\Omega \varkappa + N \overset{*}{\varkappa} \right) + \left(\delta_i^j u^n - \delta_i^n u^j \right) \left(\rho \varkappa_n - \frac{1}{2m} \varepsilon_{npqs} M^{pq} \overset{*}{\varkappa}{}^s \right) \right] \\
&= \frac{c}{g} \left[-u_s \partial_i M^{js} + M^j \partial_i \eta + \frac{1}{c} u^j M_{ik} \frac{d}{d\tau} u^k - \delta_i^j \left(M^k \partial_k \eta + M^{ks} \partial_k u_s \right) \right] \\
&\quad + \frac{2mc}{g} \left(-\delta_i^j \rho u^s \partial_s \gamma + \rho u^j \partial_i \gamma \right) + \rho \frac{\partial \Lambda_0}{\partial \rho} \sigma_i^j + \rho \left(u^j G_i - \delta_i^j G_m u^m \right) \\
&\quad - \frac{1}{4\pi} u^j u_m \left(F_{in} H^{mn} - F^{mn} H_{in} \right) - \frac{1}{2} \sigma_i^j M^{ks} F_{ks} \\
&\quad + \left(M_k \sigma_i^j - M_i u^j u_k \right) \frac{\partial \Lambda_0}{\partial M_k} - u^j \omega_i u_k \frac{\partial \Lambda_0}{\partial \omega_k} + \frac{1}{2} u^j S_{im} \frac{d}{d\tau} u^m + \frac{c}{2} u^j \partial_k S_i{}^k .
\end{aligned} \tag{6.110}$$

Let us transform an expression for components F_i entering the right-hand side of Eq. (6.109)

$$F_i = -\frac{2mc}{g} \left(\varkappa \partial_i \Omega + \varkappa_j \partial_i j^j + \overset{*}{\varkappa}_j \partial_i S^j + \overset{*}{\varkappa} \partial_i N \right),$$

The terms with the electromagnetic field entering into F_i, by means of Maxwell's equations one can transform to the form

$$\frac{1}{2\rho^2}F^{sj}M_{sj}(\Omega\partial_i\Omega + N\partial_i N)$$

$$+\frac{1}{2}\varepsilon_{jksq}\left(-\frac{1}{\rho}F^{jk}M^s\partial_i j^q + mF^{jk}u^s\partial_i S^q\right) \equiv -\frac{1}{2}F^{jk}\partial_i M_{jk}$$

$$= -\partial_j\left[\frac{1}{4\pi}\left(F_{in}H^{jn} - \frac{1}{4}F_{sm}H^{sm}\delta_i^j\right) + \frac{1}{4}\delta_i^j F_{sm}H^{sm}\right]. \qquad (6.111)$$

For the terms in F_i with ρT and G_i taking into account the continuity equation, Eq. (6.106) and the equality

$$\partial_i s = \frac{\Omega\partial_i N - N\partial_i\Omega}{\rho^2\partial\eta/\partial s},$$

which follows from definition (6.102) of $\eta(s)$, we find

$$\frac{T}{\rho\partial\eta/\partial s}(N\partial_i\Omega - \Omega\partial_i N) - G_j\partial_i j^j = \partial_j(\tau_i{}^j + \rho u^j G_i - \delta_i^j\rho u^m G_m). \qquad (6.112)$$

and for the terms in F_i with the function Λ_0:

$$\frac{1}{\rho^2\partial\eta/\partial s}\frac{\partial\Lambda_0}{\partial s}(-N\partial_i\Omega + \Omega\partial_i N) -$$

$$-\frac{c}{2\rho}\sigma_{jq}\varepsilon^{qnks}\left[\frac{\partial\Lambda_0}{\partial\omega^n}\partial_k u_s + \partial_k\left(\frac{\partial\Lambda_0}{\partial\omega^n}u_s\right)\right]\partial_i j^j - \frac{\partial\Lambda_0}{\partial\rho}u_j\partial_i j^j + m\frac{\partial\Lambda_0}{\partial M_j}\partial_i S_j \equiv$$

$$\equiv -\partial_j\left(-\frac{c}{2}S^{jm}\partial_i u_m - \Lambda_0\delta_i^j\right). \qquad (6.113)$$

The terms in F_i with an arbitrary function γ are transformed as follows

$$\partial_q\gamma\partial_i j^q = -\partial_j(\rho u^j\partial_i\gamma - \delta_i^j\rho u^s\partial_s\gamma). \qquad (6.114)$$

The remained terms in F_i can be written in the form

$$-\frac{c}{g\rho^2}(N\partial_i\Omega - \Omega\partial_i N)\partial_j M^j$$

$$-\frac{1}{g\rho}\sigma_{qj}\left(c\partial_s M^{js} + M^{js}\frac{d}{d\tau}u_s\right)\partial_i j^q - \frac{mc}{g}(\partial_q\eta + \varepsilon_{qjks}u^j\partial^k u^s)\partial_i S^q$$

$$= -\partial_j\left\{\frac{c}{g}\left[-M^j\partial_i\eta - M^{js}\partial_i u_s + \delta_i^j\left(M^k\partial_k\eta + M^{ks}\partial_k u_s\right)\right]\right\}. \qquad (6.115)$$

Using relations (6.111)–(6.115), we represent the expression for F_i in the form

$$F_i = -\partial_j N_i{}^j, \tag{6.116}$$

where

$$\begin{aligned}
N_i{}^j = {}& \frac{1}{4\pi}\left(F_{in}H^{jn} - \frac{1}{4}\delta_i^j F_{sm}H^{sm}\right) + \frac{1}{4}\delta_i^j M^{sm}F_{sm} - \tau_i{}^j - \Lambda_0\delta_i^j \\
& + \frac{c}{g}\big[-M^j\partial_i\eta - M^{js}\partial_i u_s + \delta_i^j\big(M^k\partial_k\eta + M^{ks}\partial_k u_s\big)\big] \\
& + \frac{2mc}{g}\big(\rho u^j\partial_i\gamma - \delta_i^j\rho u^s\partial_s\gamma\big) - \frac{c}{2}S^{jm}\partial_i u_m - \rho u^j G_i + \delta_i^j\rho u^m G_m.
\end{aligned} \tag{6.117}$$

Bearing in mind (6.110), (6.116), and (6.117), the second equation in (5.78) can be written down as

$$\partial_j P_i{}^j = 0 \tag{6.118}$$

with components $P_i{}^j = \mathcal{P}_i{}^j - \dfrac{2mc}{g}(\varkappa\Omega + \varkappa_s j^s + \overset{*}{\varkappa}_s S^s + \overset{*}{\varkappa}N)\delta_i^j + N_i{}^j$ defined by the equality

$$\begin{aligned}
P_i{}^j = {}& P_i{}^j{}_{(f)} + (p+e)u_i u^j + p\delta_i^j - \tau_i{}^j + \frac{1}{g}u^j M_{ik}\frac{d}{d\tau}u^k + \frac{1}{2}u^j S_{is}\frac{d}{d\tau}u^s + \\
& + \frac{c}{2}u^j\nabla_k S_i{}^k - \frac{c}{2}S^{js}\nabla_i u_s - u^j u_k\left(M_i\frac{\partial\Lambda_0}{\partial M_k} + \omega_i\frac{\partial\Lambda_0}{\partial\omega_k}\right),
\end{aligned} \tag{6.119}$$

where

$$p = \rho^2\frac{\partial\Lambda_0/\rho}{\partial\rho} + M_i\frac{\partial\Lambda_0}{\partial M_i} - \frac{1}{2}B_i M^i,$$

$$e = \Lambda_0 - \frac{1}{2}B_i M^i - \omega_i\frac{\partial\Lambda_0}{\partial\omega_i}, \qquad S_{ij} = \varepsilon_{ijks}u^k\frac{\partial\Lambda_0}{\partial\omega_s}.$$

A comparison shows that components $P_i{}^j$, determined according to (6.119) satisfying Eq. (6.118), in accuracy coincide with components $P_i{}^j$, determined by formulas (6.97) and entering Eqs. (6.96). This means that the energy-momentum equations in (6.96) are a corollary of Eqs. (6.104) and (6.105).

It is easy to see that the entropy balance equation in (6.96) are obtained by contracting of Eq. (6.106) with components of the velocity vector u^i.

Thus, it is shown that all equations in (6.96) are satisfied by virtue of Eqs. (6.104) and (6.105) provided (6.106), if the mass density, entropy, velocity and magnetization of the fluid are expressed in terms of the fields ψ and ψ^+ by relations (6.102).

It is easy to see that if the coefficients G_i in (6.105) contain derivatives of the determining parameters ρ, u_i, M_i, and s not higher than the second order, then the spinor equations in (6.104) with the coefficients $\varkappa$, $\varkappa_i$, $\overset{*}{\varkappa}_i$, and $\overset{*}{\varkappa}$, defined by equalities (6.105) are the second order differential equations in the components of the spinor ψ and contain the components of the electromagnetic field tensor F_{ij} without derivatives.

It is obvious, if in the system of hydrodynamic equations in (6.96), corresponding to the spinor equations in (6.104), we use formulas (6.102) for ρ, u_i, M_i, and s, then the resulting equations are, in general, third order differential equations in components ψ and first order differential equations in components F_{ij}.

In this sense equation (6.104) can be thought of as the integral of Eqs. (6.96), which ρ, u_i, M_i, and s expressed in terms of ψ and ψ^+ via (6.102). The peculiar feature of these integrals is that for them $M_i M^i = \rho^2 m^2$.

Due to the temperature equation $\rho T = \partial \Lambda_0 / \partial s$ which are a corollary of (6.104), the coefficients $\varkappa$ and $\overset{*}{\varkappa}$ in Eqs. (6.104) can be written as

$$\begin{aligned} \varkappa &= -\frac{g}{2mc\rho^2}\Omega\left(\rho\frac{\partial \Lambda_0}{\partial \rho} + \frac{1}{2}M^{ij}F_{ij}\right) + \frac{N}{2m\rho^2}\partial_i M^i, \\ \overset{*}{\varkappa} &= -\frac{g}{2mc\rho^2}N\left(\rho\frac{\partial \Lambda_0}{\partial \rho} + \frac{1}{2}M^{ij}F_{ij}\right) - \frac{\Omega}{2m\rho^2}\partial_i M^i. \end{aligned} \tag{6.120}$$

Direct verification shows that Eqs. (6.104), in which the coefficients $\varkappa$ and $\overset{*}{\varkappa}$ are determined by equalities (6.120), while coefficients $\varkappa_i$ and $\overset{*}{\varkappa}_i$ by (6.105), are connected by the identity

$$\operatorname{Re}\left\{\psi^+\gamma^5\left[\gamma^j\partial_j\psi + \left(\varkappa I + \mathrm{i}\varkappa_j\gamma^j + \overset{*}{\varkappa}_j\overset{*}{\gamma}{}^j + \overset{*}{\varkappa}\gamma^5\right)\psi\right]\right\} \equiv 0$$

and therefore contain no more than seven independent real equations.

It is clear that if the coefficients $\varkappa$ and $\overset{*}{\varkappa}$ are determined by equalities (6.120), and the coefficients $\varkappa_i$ and $\overset{*}{\varkappa}_i$ are determined by equalities (6.105), then the following system of equations can be taken as a closed system

$$\begin{gathered} \gamma^j\partial_j\psi + \left(\varkappa I + \mathrm{i}\varkappa_j\gamma^j + \overset{*}{\varkappa}_j\overset{*}{\gamma}{}^j + \overset{*}{\varkappa}\gamma^5\right)\psi = 0, \\ \partial_i F_{jk} + \partial_j F_{ki} + \partial_k F_{ij} = 0, \quad \partial_j\left(F^{ij} - 4\pi M^{ij}\right) = 0, \\ \rho T = \frac{\partial \Lambda_0}{\partial s}. \end{gathered} \tag{6.121}$$

The formulas for coefficients $\varkappa$, $\varkappa_i$, $\overset{*}{\varkappa}_i$, and $\overset{*}{\varkappa}$ in Eqs. (6.121) can be written in the form

$$
\begin{aligned}
\varkappa &= \frac{g}{2mc\rho}\left(-\Omega\frac{\partial\Lambda_0}{\partial\rho} + \frac{c}{g\rho}N\partial_i M^i\right),\\
\overset{*}{\varkappa} &= \frac{g}{2mc\rho}\left(-N\frac{\partial\Lambda_0}{\partial\rho} - \frac{c}{g\rho}\Omega\partial_i M^i\right),\\
\varkappa_i &= \frac{g}{2mc\rho}\left\{\rho G_i - M_{ij}E^j + \frac{c}{2}\sigma_{ij}\varepsilon^{jnks}\left[\frac{\partial\Lambda_0}{\partial\omega^n}\partial_k u_s + \partial_k\left(\frac{\partial\Lambda_0}{\partial\omega^n}u_s\right)\right]\right\}\\
&+ \frac{1}{2mc\rho}\sigma_{ij}\left(c\partial_s M^{js} + M^{js}\frac{d}{d\tau}u_s\right) + \partial_i\gamma,\\
\overset{*}{\varkappa}_i &= \frac{g}{2c}\left(B^i - \frac{\partial\Lambda_0}{\partial M^i}\right) + \frac{1}{2}\frac{\partial\eta}{\partial s}\partial_i s + \frac{1}{2}\varepsilon_{ijks}u^j\partial^k u^s.
\end{aligned}
\tag{6.122}
$$

Here the components of the four-dimensional vector of electric strength E^j and the components of the four-dimensional vector of magnetic induction B_i are defined by equalities

$$E^j = u_m F^{jm}, \quad B_i = \frac{1}{2}\varepsilon_{ijks}F^{jk}u^s.$$

Using the second identity in (6.69) and the identity

$$\frac{1}{2}\varepsilon_{jksi}F^{jk}M^s \equiv -M_{ij}E^j - \frac{1}{2}u_i F_{jk}M^{jk},$$

it is easy to show that Eq. (6.104) with coefficients (6.105) and Eq. (6.121) with coefficients (6.122) are equivalent.

Let us give now some interpretation of Eqs. (6.106). Consider functional δW^* determined by relation (6.95) corresponding to Eqs. (6.96). Let us replace in expression (6.95) of the functional δW^* the variations δs by the local variations ∂s according to the formula $\delta s = \partial s + \delta x^i\partial_i s$ (see (A.30)). As a result we get

$$\delta W^* = \int\limits_{V_4}\left[-\rho T\partial s + \delta x^i\left(-\rho T\partial_i s - \partial_j\tau_i{}^j\right) + \partial_j\left(\tau_i{}^j\delta x^i\right)\right]dV_4. \tag{6.123}$$

Taking into account equation (6.106) we transformed formula (6.123) for δW^* to the form

$$\delta W^* = \int\limits_{V_4}\left[-\rho T\partial s + \rho u^j\left(\partial_j G_i - \partial_i G_j\right)\delta x^i\right]dV_4 + \int\limits_{V_4}\partial_j\left(\tau_i{}^j\delta x^i\right)dV_4. \tag{6.124}$$

Using formulas (A.38) and (A.43) for variations of quantities ρ and u^i, expression (6.124) for the functional δW^* we write in the form

$$\delta W^* = \int\limits_{V_4} (-\rho T \partial s - G_i \partial \rho u^i)\, dV_4 + \int\limits_{\Sigma_3} (\tau_i{}^j + \rho u^j G_i - \delta_i^j \rho u^m G_m) \delta x^i n_j d\sigma, \tag{6.125}$$

where n_j are the components of a unit vector of the outward normal to the surface Σ_3, bounding the region V_4. $d\sigma$ is an invariant element of the three-dimensional surface Σ_3.

The purpose of the carried out transformation of the functional δW^* consists of replacement in the volume part of the functional δW^* the variations δx^i by the local variations of the fluid density $\partial \rho$ and local variations Of the velocity vector components ∂u^i which can be expressed in terms of the variations of the functions $\partial \psi$ and $\partial \psi^+$. Such replacement is possible when conditions (6.106) are fulfilled.

Thus, performance of Eqs. (6.106) is a condition of the identity of the functionals (6.95) and (6.125). The physical meaning of quantities G_i is defined by expression (6.125) of the functional δW^*.

If quantities G_i that entering coefficients $\varkappa_i$ and satisfy Eqs. (6.106), are given as functions of the determining parameters of the fluid and the field, then the system of equations (6.104), (6.105) is complete. A definition of quantities G_i is related to the concrete thermodynamic and mechanical formulation of the problem.

We now consider some simple cases connected with the concrete assignment of the functions G_i, T, $\tau_i{}^j$ satisfying Eq. (6.106).

6.4.3 Adiabatic Processes

If the quantities G_i, T, $\tau_i{}^j$ are defined by the equalities

$$T = \frac{\mathrm{d}}{\mathrm{d}\tau}\lambda, \quad G_i = -c\lambda \partial_i s, \quad \tau_i{}^j = 0, \tag{6.126}$$

where $\lambda = \lambda(x^i)$ is an arbitrary differentiable function, c is the light velocity in vacuum, that Eq. (6.106) is carried out identically if

$$\mathrm{d}s/\mathrm{d}\tau = 0. \tag{6.127}$$

This case corresponds to adiabatic processes in an ideal magnetizable fluid. Note that Eqs. (6.96) in the presence of Eqs. (6.126) and (6.127) are obtained by means of the holonomic variational equation

$$\delta \int\limits_{V_4} \left(\Lambda + \lambda \rho \frac{\mathrm{d}s}{\mathrm{d}\tau} \right) dV_4 = 0,$$

where function Λ is considered as a Lagrange multiplier corresponding to Eq. (6.127).

6.4.4 Isentropic Processes

Equations (6.106) are carried out identically, if to put

$$s = \text{const}, \quad \tau_i{}^j = 0, \quad G_i = 0. \tag{6.128}$$

This case corresponds to isentropic processes in an ideal magnetizable fluid. Equations (6.96) in the presence conditions (6.128) are obtained by means of the holonomic variational equation

$$\delta \int\limits_{V_4} \left[\Lambda - \rho T(s - s_0\right] dV_4 = 0,$$

in which $s_0 =$ const and ρT is considered as an arbitrary varied function.

6.5 Non-steady Exact One-Dimensional Solutions for Relativistic Models of Spin Fluids

Consider Eq. (6.96) for isentropic processes in the magnetizable spin fluid when the equalities are carried out

$$s = s_0 = \text{const}, \quad \tau_i{}^j = G_i = 0.$$

For function Λ_0 in the Lagrangian (6.94) we accept

$$\Lambda_0 = \frac{2}{g} M_i \omega^i + \Lambda_m(\rho, M),$$

where $\Lambda_m(\rho, M)$ is the given function, M is the module of the volume density vector of the fluid magnetization $M = \left(M_i M^i\right)^{1/2} \equiv \rho m$. The components of tensors $P_i{}^j$ and $\overset{*}{F}_{ij}$ in Eqs. (6.96) are defined in the considered case by the equalities

$$P_i{}^j = \frac{1}{4\pi}\left[F_{in}H^{jn} - \frac{1}{4}\delta_i^j F_{sm}H^{sm} - u^j u_m \left(H_{in}F^{mn} - H^{mn}F_{in}\right)\right] +$$
$$+ (p+e)u_i u^j + p\delta_i^j - \frac{c}{g}\left(u_k \partial_i M^{jk} + u^j \partial_k M_i{}^k\right),$$
$$\overset{*}{F}_{ij} = \sigma_i^s \sigma_j^n \left(F_{sn} - \frac{1}{M}\frac{\partial \Lambda_m}{\partial M} M_{sn}\right) - \frac{c}{g}\left(\partial_i u_j - \partial_j u_i\right), \tag{6.129}$$

in which

$$p = \rho^2 \frac{\partial \Lambda_m/\rho}{\partial \rho} + M \frac{\partial \Lambda_m}{\partial M} - \frac{1}{2} B_i M^i,$$

$$e = \Lambda_m - \frac{1}{2} B_i M^i. \tag{6.130}$$

Let us write out the system of equations (6.104) corresponding to Eq. (6.96) and the isentropic condition

$$\gamma^j \left(\partial_j + \frac{ig}{2mc} \frac{\partial \Lambda_m}{\partial \rho} u_j \right) \psi + \frac{1}{2m\rho^2} \partial_j M^j (NI - \Omega \gamma^5) \psi +$$
$$+ \frac{g}{2c} \left[-\frac{i}{M} M_{sj} E^j \gamma^s + \left(B_j - \frac{1}{M} \frac{\partial \Lambda_m}{\partial M} M_j \right) \overset{*}{\gamma}{}^j \right] \psi = 0,$$
$$\partial_j \left(F^{ij} - 4\pi M^{ij} \right) = 0, \quad \partial_i F_{jk} + \partial_j F_{ki} + \partial_k F_{ij} = 0,$$
$$s = \text{const}. \tag{6.131}$$

Note that the system of equations (6.131) is a quasilinear system of the first order in ψ, whereas Eqs. (6.96), (6.129) corresponding to it are a nonlinear system of the second order (relatively ψ).

By means of the first and second identity in (6.69) the spinor equations in (6.131) can be written in a simpler form[8]

$$\gamma^j \partial_j \psi - \frac{g}{2mc\rho} \left(\frac{\partial \Lambda_m}{\partial \rho} + m \frac{\partial \Lambda_m}{\partial M} \right) (\Omega I + N \gamma^5) \psi$$

[8]If Λ_m depends only on the fluid density ρ, there is no electromagnetic field $E_i = B_i = 0$ and an additional equation $\partial_i M^i = 0$ is fulfilled, then the spinor equations (6.132) for $\lambda(\rho) = -\frac{g}{2mc\rho} \frac{d\Lambda_m}{d\rho}$ coincide with Eqs. (6.70). Therefore, Eqs. (6.70) describe a spin fluid, defined by the equations

$$\partial_j \widetilde{P}_i{}^j = 0, \quad \partial_i \rho u^i = 0, \quad \partial_i M^i = 0,$$
$$\rho \frac{d}{d\tau} \left(\frac{1}{\rho} M_i \right) = g \widetilde{F}_{ij} M^j, \quad \widetilde{F}_{ij} = \overset{*}{F}_{ij} + \frac{c}{g} \varepsilon_{ijks} u^k \partial^s \eta,$$
$$\widetilde{P}_i{}^j = P_i{}^j + \frac{c}{g} (M^s u_i u^j - M_i u^s u^j + M^j \delta_i^s) \partial_s \eta,$$

in which $P_i{}^j$ and $\overset{*}{F}_{ij}$ are defined according to (6.129) (without terms with an electromagnetic field). The quantity η in this case is considered as a Lagrange multiplier corresponding to the equation $\partial_i M^i = 0$; the additional terms in $\breve{P}_i{}^j$ and $\breve{F}_{ij}$ are related with the introduction into the Lagrangian of the term $cg^{-1} M^i \partial_i \eta$ with the Lagrange multiplier η. These equations determine hydrodynamic analogy of the theory of the elementary particles described by Eq. (6.70). In particular, for the Heisenberg equation ($\lambda =$ const) we have $\Lambda_m = -\lambda mcg^{-1} \rho^2$ and for the pressure p we obtain $p = -\lambda mcg^{-1} \rho^2$. A condition of positivity of the pressure (or sound velocity) for $g < 0$ gives $\lambda > 0$.

$$+\frac{1}{2m\rho^2}\partial_j M^j(NI-\Omega\gamma^5)\psi+\frac{g}{2c}\left(-\frac{\mathrm{i}}{M}M_{sj}E^j\gamma^s+B_j\overset{*}{\gamma}{}^j\right)\psi=0. \tag{6.132}$$

We establish further some exact solutions of the system of equations (6.131).

1. It is not difficult to verify that Eqs. (6.131) admit an exact solution

$$E^i=\text{const},\quad B^i=\frac{B}{M}\overset{\circ}{M}{}^i, \tag{6.133}$$

$$\psi=\psi_\circ\exp\left\{\frac{\mathrm{i}g}{2Mc}\left[\left(B_jM^j-\rho\frac{\partial\Lambda_m}{\partial\rho}-M\frac{\partial\Lambda_m}{\partial M}\right)_\circ\overset{\circ}{u}_i+\overset{\circ}{M}_{ij}E^j\right]x^i\right\},$$

where $\psi_\circ$ is an arbitrary constant spinor field; B is an arbitrary constant; brackets $(\ )_\circ$ mean that parameters ρ, M^i, being in brackets, are calculated for $\psi=\psi_\circ$; $\overset{\circ}{u}_i$, $\overset{\circ}{M}_i$ are the components of the velocity vector and the magnetization vector determined by the field $\psi_\circ$. Using formulas (6.102) we find $\rho=\rho_0$, $u^i=u^i_0$, $M^i=M^i_0$, and $s=s_0$. Thus, solution (6.133) determines translatory motion with the constant velocity of the uniformly magnetized spin fluid in a constant electromagnetic field.

Determining the matrices γ_i by formulas (3.24) we take the spinor components $\psi_\circ$ in solution (6.133) in the form

$$\psi^1_\circ=0,\quad \psi^2_\circ=\sqrt{\frac{1}{2}\rho_\circ}\,\exp\left[\frac{\mathrm{i}}{2}\eta(s_\circ)\right],$$

$$\psi^3_\circ=0,\quad \psi^4_\circ=\sqrt{\frac{1}{2}\rho_\circ}\,\exp\left[-\frac{\mathrm{i}}{2}\eta(s_\circ)\right], \tag{6.134}$$

where $\rho_\circ$ and $s_\circ$ are arbitrary real constants. Then, according to formulas (6.102), for the fields ρ, u^i, M^i, s we get

$$\rho=\rho_\circ,\quad u^1=u^2=u^3=0,\quad u^4=1,$$

$$s=s_\circ,\quad M^1=M^2=M^4=0,\quad M^3=\rho_\circ m.$$

Thus, if the matrices γ_i and the functions $\psi_\circ$ in solutions (6.133) are defined by equalities (3.24) and (6.134), then solution (6.133) of Eqs. (6.131) determines a solution of Eqs. (6.96) and (6.129), for which in the coordinate system x^i the fluid is at rest and uniformly magnetized along the axis x^3.

2. Let us seek a solution of Eqs. (6.131), for which the vector of the electric strength is equal to zero and the magnetic displacement vector is proportional to magnetization vector of the fluid

$$E^i=0,\quad B^i=\frac{B}{M}M^i. \tag{6.135}$$

The quantity B in (6.135) is considered as a required function of the variables x^i. In the presence of equalities (6.135) the spinor equations in (6.131) can be written in the form

$$\gamma^j \left[\partial_j + \frac{\mathrm{i}g}{2mc} \left(\frac{\partial \Lambda_m}{\partial \rho} + m \frac{\partial \Lambda_m}{\partial M} - mB \right) u_j \right] \psi + \frac{1}{2m\rho^2} \partial_j M^j (NI - \Omega \gamma^5) \psi = 0. \qquad (6.136)$$

To transform the spinor equations in (6.131) to the form (6.136) we should use identities (6.69).

We represent the function ψ in Eqs. (6.136) in the form

$$\psi = \psi_\circ \exp \left[-\frac{\mathrm{i}g}{2mc} \int \left(\frac{\partial \Lambda_m}{\partial \rho} + m \frac{\partial \Lambda_m}{\partial M} - mB \right) u_j d x^j \right]. \qquad (6.137)$$

The necessary and sufficient condition for the independence of the integral in (6.137) from the path of integration in the space-time is written in the form

$$\partial_i \left[\left(\frac{\partial \Lambda_m}{\partial \rho} + m \frac{\partial \Lambda_m}{\partial M} - mB \right) u_j \right] = \partial_j \left[\left(\frac{\partial \Lambda_m}{\partial \rho} + m \frac{\partial \Lambda_m}{\partial M} - mB \right) u_i \right]. \qquad (6.138)$$

Replacing the functions ψ in terms of $\psi_\circ$ in Eqs. (6.136) by formula (6.137), we get an equation for $\psi_\circ$

$$\gamma^i \partial_i \psi_\circ + \frac{1}{2m\rho_\circ^2} \partial_j M_\circ^j (N_\circ I - \Omega_\circ \gamma^5) \psi_\circ = 0, \qquad (6.139)$$

Here the components $\rho_\circ$, $\Omega_\circ$, $N_\circ$, and $M_\circ^i$ are expressed in terms of $\psi_\circ$ by the same formulas, as ρ, Ω, N, M^i are expressed in terms of ψ.

Determining the matrices γ_i by formulas (3.24) we will seek one-dimensional non-steady solution of (6.138) and (6.139) under the condition $s = s_\circ =$ const in the form

$$\psi_\circ^1 = 0, \quad \psi_\circ^2 = \sqrt{f_2} \exp \mathrm{i}\varphi_2,$$
$$\psi_\circ^3 = 0, \quad \psi_\circ^4 = \sqrt{f_4} \exp \mathrm{i}\varphi_4, \qquad (6.140)$$

where f_2, f_4, φ_2, and φ_4 depend only on the variables x^3, x^4. The tensor fields $\rho_\circ$, $u^i_\circ$, $M^i_\circ$, and $s_\circ$, corresponding to the spinor field $\psi_\circ$, determined by equalities (6.140), have a special form

$$\rho^2 = \rho_\circ^2 = 4 f_2 f_4, \quad u^1_\circ = u^2_\circ = 0, \quad M^1_\circ = M^2_\circ = 0,$$
$$M^3_\circ = \rho m u^4_\circ = m(f_2 + f_4), \quad M^4_\circ = \rho m u^3_\circ = m(f_4 - f_2),$$
$$\eta(s_\circ) = \varphi_2 - \varphi_4. \tag{6.141}$$

If $u^i_\circ$ and $M^i_\circ$ are determined according to (6.141) and equalities (6.135) are carried out, then for components F_{ij} of the electromagnetic field tensor we have

$$F_{ij} = B\left(\delta^1_i \delta^2_j - \delta^1_j \delta^2_i\right).$$

Bearing in mind equalities (6.140) and (6.141), the spinor equation (6.139) and conditions (6.138) can be written in form of equations for the functions φ_2, φ_4, f_2, and f_4:

$$(\partial_3 + \partial_4)\varphi_4 = 0, \quad (\partial_3 - \partial_4)\varphi_2 = 0,$$
$$\partial_3(f_2 - f_4) = \partial_4(f_2 + f_4),$$
$$\partial_3\left[\frac{1}{\rho}\left(\frac{\partial \Lambda_m}{\partial \rho} + m\frac{\partial \Lambda_m}{\partial M} - mB\right)(f_2 + f_4)\right]$$
$$= \partial_4\left[\frac{1}{\rho}\left(\frac{\partial \Lambda_m}{\partial \rho} + m\frac{\partial \Lambda_m}{\partial M} - mB\right)(f_2 - f_4)\right]. \tag{6.142}$$

The condition $s =$ const in (6.131) passes in the equation

$$\varphi_2 - \varphi_4 = \eta(s_\circ) = \text{const}, \tag{6.143}$$

while from the Maxwell equations in (6.131) it follows $B =$ const.

Thus, for definition of the functions φ_2, φ_4, f_2, and f_4 we have a system of equations (6.142) and (6.143).

From Eqs. (6.142) and (6.143) it follows a solution for φ_2 and φ_4:

$$\varphi_4 = -\frac{1}{2}\eta(s_\circ) + C_1, \quad \varphi_2 = \frac{1}{2}\eta(s_\circ) + C_1, \tag{6.144}$$

where C_1 is an arbitrary real constant.

Further we will seek solutions for which the functions f_2 and f_4 depend only on the density ρ:

$$f_2 = f_2(\rho), \quad f_4 = f_4(\rho).$$

In this case the equations for the functions f_2, f_4 in (6.142) can be rewritten in the form

$$\frac{\mathrm{d}}{\mathrm{d}\rho}\left[\left(\frac{\partial \Lambda_m}{\partial \rho}+m\frac{\partial \Lambda_m}{\partial M}-mB\right)\frac{f_2+f_4}{\rho}\right]\partial_3\rho$$
$$=\frac{\mathrm{d}}{\mathrm{d}\rho}\left[\left(\frac{\partial \Lambda_m}{\partial \rho}+m\frac{\partial \Lambda_m}{\partial M}-mB\right)\frac{f_2-f_4}{\rho}\right]\partial_4\rho,$$
$$\frac{\mathrm{d}(f_2-f_4)}{\mathrm{d}\rho}\partial_3\rho-\frac{\mathrm{d}(f_2+f_4)}{\mathrm{d}\rho}\partial_4\rho=0. \tag{6.145}$$

It is obvious that necessary and sufficient condition for the compatibility of Eqs. (6.145) is the fulfillment of the equation

$$\frac{\mathrm{d}(f_2-f_4)}{\mathrm{d}\rho}\frac{\mathrm{d}}{\mathrm{d}\rho}\left[\left(\frac{\partial \Lambda_m}{\partial \rho}+m\frac{\partial \Lambda_m}{\partial M}-mB\right)\frac{f_2-f_4}{\rho}\right]$$
$$=\frac{\mathrm{d}(f_2+f_4)}{\mathrm{d}\rho}\frac{\mathrm{d}}{\mathrm{d}\rho}\left[\left(\frac{\partial \Lambda_m}{\partial \rho}+m\frac{\partial \Lambda_m}{\partial M}-mB\right)\frac{f_2+f_4}{\rho}\right], \tag{6.146}$$

which can be transformed to the form

$$\frac{\mathrm{d}f_2}{\mathrm{d}\rho}\frac{\mathrm{d}f_4}{\mathrm{d}\rho}+\frac{1}{4}\left(\frac{a^2}{c^2}-1\right)=0. \tag{6.147}$$

Here the quantity a with the dimensions of velocity is defined by the equality

$$a=\left(\rho c^2\frac{\dfrac{\partial^2 \Lambda_m}{\partial \rho^2}+2m\dfrac{\partial^2 \Lambda_m}{\partial \rho\partial M}+m^2\dfrac{\partial^2 \Lambda_m}{\partial M^2}}{\dfrac{\partial \Lambda_m}{\partial \rho}+m\dfrac{\partial \Lambda_m}{\partial M}-mB}\right)^{\frac{1}{2}}. \tag{6.148}$$

Eliminating the function f_4 from Eq. (6.147) with the aid of the equation $\rho^2 = 4f_2f_4$ (see (6.141)), for the function f_2 we get an ordinary differential equation

$$\rho\left(\frac{\mathrm{d}f_2}{\mathrm{d}\rho}\right)^2-2f_2\frac{\mathrm{d}f_2}{\mathrm{d}\rho}-\frac{1}{\rho}\left(\frac{a^2}{c^2}-1\right)f_2^2=0,$$

from which it follows

$$\frac{\mathrm{d}f_2}{\mathrm{d}\rho}=\frac{f_2}{\rho}\left(1\pm\frac{a}{c}\right). \tag{6.149}$$

The general solution of Eq. (6.149) has the form

$$f_2 = \frac{1}{2}\rho \exp\left(\pm \int \frac{a}{c}\frac{d\rho}{\rho}\right). \tag{6.150}$$

Here an arbitrary constant is included in the sign of the indefinite integral. For f_4 we have

$$f_4 = \frac{\rho^2}{4f_2} = \frac{1}{2}\rho \exp\left(\mp \int \frac{a}{c}\frac{d\rho}{\rho}\right). \tag{6.151}$$

Thus, by virtue of equalities (6.137), (6.140), (6.144), (6.150), and (6.151) for the considered exact solutions of Eqs. (6.131) the functions ψ^A have the form

$$\begin{aligned}
\psi^1 &= 0, \quad \psi^3 = 0,\\
\psi^4 &= \sqrt{\frac{1}{2}\rho}\ \exp\frac{1}{2}\left[\mp \int \frac{a}{c}\frac{d\rho}{\rho}\right.\\
&\quad \left. - \frac{\mathrm{ig}}{mc}\int\left(\frac{\partial \Lambda_m}{\partial \rho} + m\frac{\partial \Lambda_m}{\partial M} - mB\right)u_j dx^j - \mathrm{i}\eta(s_\circ)\right],\\
\psi^2 &= \sqrt{\frac{1}{2}\rho}\ \exp\frac{1}{2}\left[\pm \int \frac{a}{c}\frac{d\rho}{\rho}\right.\\
&\quad \left. - \frac{\mathrm{ig}}{mc}\int\left(\frac{\partial \Lambda_m}{\partial \rho} + m\frac{\partial \Lambda_m}{\partial M} - mB\right)u_j dx^j + \mathrm{i}\eta(s_\circ)\right],
\end{aligned} \tag{6.152}$$

Due to Eqs. (6.142), the integrals in formulas (6.152) do not depend on the path of integration in the space-time.

To determine the density $\rho(x^3, x^4)$ one can use Eqs. (6.145) in which due to (6.146) there is only one independent equation

$$\frac{\mathrm{d}(f_2 - f_4)}{\mathrm{d}\rho}\partial_3\rho - \frac{\mathrm{d}(f_2 + f_4)}{\mathrm{d}\rho}\partial_4\rho = 0. \tag{6.153}$$

Taking into account the found expressions (6.150) and (6.151) for the functions f_2, f_4 and connection (6.141) of the components of the velocity vector $u^i = u^i_\circ$ with the quantities f_2, f_4, Eq. (6.153) can be transformed to the following form

$$\left(u^4 \mp \frac{a}{c}u^3\right)\frac{\partial \rho}{\partial x^4} + \left(u^3 \mp \frac{a}{c}u^4\right)\frac{\partial \rho}{\partial x^3} = 0. \tag{6.154}$$

From (6.154) it follows

$$x^3 = x^4 \frac{u^3 \mp \dfrac{a}{c} u^4}{u^4 \mp \dfrac{a}{c} u^3} + F(\rho), \tag{6.155}$$

where $F(\rho)$ is an arbitrary differentiable function of the fluid density ρ. Formula (6.155) determines an implicit dependence of the mass density ρ on the variables x^3 and x^4.

Relations (6.152) and (6.155) and the equality

$$F_{ij} = B\left(\delta_i^1 \delta_j^2 - \delta_j^1 \delta_i^2\right), \quad B = \text{const}$$

completely determine a series of the exact solutions of Eqs. (6.131).

Solutions (6.152) and (6.155) of Eqs. (6.131) determine solutions of Eqs. (6.96), (6.129) by formulas (6.102) in the form of Riemannian waves, for which the components of the four-dimensional vectors of the velocity and magnetization are defined as follows

$$u^1 = u^2 = 0, \quad u^3 = \frac{1}{\rho m} M^4 = \sinh\left(\mp \int \frac{a}{c} \frac{d\rho}{\rho}\right),$$

$$M^1 = M^2 = 0, \quad u^4 = \frac{1}{\rho m} M^3 = \cosh\left(\mp \int \frac{a}{c} \frac{d\rho}{\rho}\right).$$

The quantity a defined by formula (6.148) is the front velocity of the wave. Expression (6.148) for a is possible to write also in the form

$$a = \left[\frac{c^2}{p+e}\left(\rho \frac{\partial p}{\partial \rho} + M_i \frac{\partial p}{\partial M_i} + \frac{1}{2} M_i B^i\right)\right]^{1/2}, \tag{6.156}$$

where the pressure p is the same, as in formula (6.130).

If $B = 0$ and Λ_m does not depend on the magnetizations, then expression (6.156) passes into the well known formula[40] for the wave velocity in the relativistic theory of the perfect compressible fluid.

Similarly (6.134)–(6.152) one can find that Eqs. (6.131) admit exact solutions of the form

$$\psi^2 = 0, \quad \psi^4 = 0, \quad F_{ij} = B\left(\delta_i^1 \delta_j^2 - \delta_j^1 \delta_i^2\right),$$

$$\psi^3 = \sqrt{\frac{1}{2}\rho} \exp \frac{1}{2}\left[\pm \int \frac{a}{c} \frac{d\rho}{\rho} - \frac{\mathrm{i}g}{mc} \int \left(\frac{\partial \Lambda_m}{\partial \rho} + m \frac{\partial \Lambda_m}{\partial M} - mB\right) u_j dx^j - \mathrm{i}\eta(s_\circ)\right],$$

$$\psi^1 = \sqrt{\frac{1}{2}\rho}\, \exp\frac{1}{2}\Bigg[\mp\int\frac{a}{c}\frac{d\rho}{\rho} \\ -\frac{ig}{mc}\int\left(\frac{\partial\Lambda_m}{\partial\rho}+m\frac{\partial\Lambda_m}{\partial M}-mB\right)u_j dx^j + i\eta(s_\circ)\Bigg], \tag{6.157}$$

Here B is an arbitrary real constant. For solutions (6.157) we have

$$u^1 = u^2 = 0, \quad u^3 = -\frac{1}{\rho m}M^4 = \sinh\left(\mp\int\frac{a}{c}\frac{d\rho}{\rho}\right), \\ M^1 = M^2 = 0, \quad u^4 = -\frac{1}{\rho m}M^3 = \cosh\left(\mp\int\frac{a}{c}\frac{d\rho}{\rho}\right). \tag{6.158}$$

Relations (6.158) also determine solutions of Eqs. (6.96) and (6.129) in the form of Riemannian waves, for which dependence of density ρ on the variables x^3 and x^4 is defined by Eq. (6.155).

Appendix A
Relativistic Models of Spin Fluids in Electromagnetic Field

The description of various real phenomena and processes occurring in nature is related to the need to introduce certain mathematical characteristics and mathematical models of these phenomena and processes. To solve a wide range of theoretical and applied problems in field theory and in the mechanics of a continuous medium, it is sufficient to use well-known and classical models—model of the electromagnetic field, models of ideal compressible and incompressible liquids, model of the Navier–Stokes viscous fluid, models of ideal elastic and plastic media. However, there are experiences and conditions (strong electric and magnetic fields, high temperatures and pressures, large gradients of deformations, etc.), in which even usual physical systems show properties that cannot be described by specified (and generally known) models. On the other hand, many new materials have appeared in recent times, whose behavior can not be described on the basis of known classical models and representations. In many processes, new effects and new properties of materials are essential and determinative. This makes actual the problem of creating mathematical models of continuous media with complicated properties and characteristics.

Below we obtain equations describing the relativistic models of spin fluids interacting with an electromagnetic field. One of the applications of the theory presented in the book is obtaining of integrals and exact solutions of such equations (see Chap. 6, Sect. 6.3). Note also that the relativistic models of the spin fluids discussed below are of interest also because the spin fluids can be regarded as a torsion source in the theory of gravitation.

The method used here to obtain dynamic equations for the spin fluids by means of introduction into a Lagrangian the special arguments (related to the Ricci rotation coefficients of the Cosserat continuum) makes it possible to describe the presence in the fluid of an intrinsic angular momentum in the framework of a holonomic variational equation.

V. A. Zhelnorovich, *Theory of Spinors and Its Application in Physics and Mechanics*, https://doi.org/10.1007/978-3-030-27836-6

A.1 The Determining Parameters of the Spin Fluids and Electromagnetic Field

A.1.1 Kinematic Characteristics of the Fluids

Let us introduce in the Minkowski space generally speaking an arbitrary curvilinear coordinate system of an observer with the variables x^i $(i = 1, 2, 3, 4)$ and the covariant vector basis $\boldsymbol{\vartheta}_i$; we also introduce a Lagrange coordinate system with the variables ξ^i and the covariant vector basis $\widehat{\boldsymbol{\vartheta}}_i$. By definition of the Lagrangian coordinate system, the coordinates ξ^1, ξ^2, ξ^3 are constant for individual points of the fluid, the coordinate ξ^4 varies along the world line of the point. Any two Lagrangian coordinate systems with the variables ξ^i and ξ'^i as coordinate systems for which the coordinate line ξ^4 coincide with the world line of the fluid point are connected by the transformation

$$\xi'^4 = f(\xi^\alpha, \xi^4), \quad \xi'^\alpha = f^\alpha(\xi^\beta), \quad \alpha, \beta = 1, 2, 3.$$

Let us assume that the motion law of individual points of the fluid is determined by smooth functions

$$x^i = x^i(\xi^1, \xi^2, \xi^3, \xi^4). \tag{A.1}$$

Function (A.1) for the constant variables ξ^1, ξ^2, ξ^3 determine in the Minkowski space the world line of an individual fluid point with the Lagrangian coordinates ξ^1, ξ^2, ξ^3.

We will denote the tensor components in the Lagrangian coordinate system by the symbol $\widehat{\ }$. Recalculation of the tensor components given in the observer's coordinate system to the Lagrangian coordinate system is carried out by the usual tensor formulas with the coefficients of transformation $x^i{}_p$ and $\xi^p{}_i$ determined by the equalities

$$x^i{}_p = \frac{\partial x^i}{\partial \xi^p}, \quad \xi^p{}_j = \frac{\partial \xi^p}{\partial x^j}.$$

It is obvious that the components $x^i{}_p$ and $\xi^p{}_i$ satisfy the equations

$$x^i{}_p \xi^p{}_j = \delta^i_j, \quad x^i{}_p \xi^q{}_i = \delta^q_p. \tag{A.2}$$

The metric tensor $\boldsymbol{g}$ we define in the observer's coordinate system by the contravariant components g^{ij} and in a Lagrangian coordinate system by the contravariant components $\widehat{g}^{pq}$:

$$\boldsymbol{g} = g^{ij}\boldsymbol{\vartheta}_i\boldsymbol{\vartheta}_j = \widehat{g}^{pq}\widehat{\boldsymbol{\vartheta}}_p\widehat{\boldsymbol{\vartheta}}_q.$$

We have

$$g^{ij} = x^i{}_p x^j{}_q \widehat{g}^{pq}, \quad \widehat{g}^{pq} = \xi^p{}_i \xi^q{}_j g^{ij}.$$

The covariant components of the metric tensor g_{ij} and $\widehat{g}_{ij}$ are defined by the matrices inverse to g^{ij} and $\widehat{g}^{ij}$

$$\|g_{ij}\| = \|g^{ij}\|^{-1}, \quad \|\widehat{g}_{ij}\| = \|\widehat{g}^{ij}\|^{-1}$$

Along with the actual physical Minkowski space we will regard also the space of initial states as the metric space defined on the manifold ξ^i with the metric $\overset{\circ}{g}_{ij}$. The tensor components in the space of initial states we will denote further by the symbol "$\overset{\circ}{}$". In many cases[1] as the metric $\overset{\circ}{g}_{ij}$ we can consider the metric of the actual space at some value $\xi^4 = \xi^4_\circ$:

$$\overset{\circ}{g}_{ij} = \widehat{g}_{ij}(\xi^\alpha, \xi^4_\circ), \quad \xi^4_\circ = \text{const}.$$

From the definition it follows $\partial \overset{\circ}{g}_{ij} / \xi^4 = 0$.

Let us define some main kinematic and dynamic characteristics of the spin fluid.

The contravariant components of the four-dimensional dimensionless velocity vector of individual points of the fluid $\boldsymbol{u} = u^i\, \boldsymbol{\mathcal{Э}}_i$ in the observer's coordinate system are defined by the formula

$$u^i = \frac{dx^i}{ds} = \frac{1}{\sqrt{-\widehat{g}_{44}}} x^i{}_4. \tag{A.3}$$

Here $ds = \sqrt{-\widehat{g}_{44}}\, d\xi^4$ is an arch element of the world line of the individual point of the fluid. Due to definition (A.3) we have

$$u_i u^i = g_{ij} u^i u^j = -\left(\widehat{g}_{44}\right)^{-1} g_{ij} x^i{}_4 x^j{}_4 = -\left(\widehat{g}_{44}\right)^{-1} \widehat{g}_{44} = -1.$$

Thus,

$$u_i u^i = -1. \tag{A.4}$$

From definition (A.3) and Eqs. (A.2) it follows that the contravariant components $\widehat{u}^k$ and covariant components $\widehat{u}_s$ of the four-dimensional velocity vector in the Lagrange coordinate system have the form

$$\widehat{u}^k = \xi^k{}_i u^i = (-\widehat{g}_{44})^{-1/2}\, \delta^k_4 = \left\{0,\ 0,\ 0,\ \frac{1}{\sqrt{-\widehat{g}_{44}}}\right\},$$

$$\widehat{u}_s = \widehat{g}_{sk} \widehat{u}^k = \widehat{g}_{4s} \left(-\widehat{g}_{44}\right)^{-1/2} = \left\{\frac{\widehat{g}_{14}}{\sqrt{-\widehat{g}_{44}}},\ \frac{\widehat{g}_{24}}{\sqrt{-\widehat{g}_{44}}},\ \frac{\widehat{g}_{34}}{\sqrt{-\widehat{g}_{44}}},\ -\sqrt{-\widehat{g}_{44}}\right\}.$$

[1] In particular, for classical models of perfect fluids and ideal elastic bodies.

The contravariant components of the four-dimensional vorticity vector $\boldsymbol{\omega} = \omega^i \boldsymbol{\partial}_i$ in the observer's coordinate system are defined by the relation

$$\omega^i = \frac{c}{2}\varepsilon^{ijks} u_j \nabla_k u_s, \tag{A.5}$$

in which ε^{ijks} are the components of the Levi-Civita pseudotensor, c is the velocity light in vacuum; ∇_k is the symbol of the covariant derivative calculated in the observer's coordinate system. From definition (A.5) it follows that the four-dimensional vorticity vector and four-dimensional velocity vector are orthogonal $u_i \omega^i = 0$.

A.1.2 The Proper Basis of the Individual Points of Continuous Medium

With the field of the four-dimensional velocity vector $\boldsymbol{u}(x^i)$ we can relate some generally non-holonomic system of orthonormal bases $\breve{\boldsymbol{e}}_a$, whose fourth vector $\breve{\boldsymbol{e}}_4$ coincides with the four-dimensional velocity vector

$$\breve{\boldsymbol{e}}_4 = \boldsymbol{u} = u^i \boldsymbol{\partial}_i, \tag{A.6}$$

and the vectors $\breve{\boldsymbol{e}}_1$, $\breve{\boldsymbol{e}}_2$, and $\breve{\boldsymbol{e}}_3$, generally speaking, are arbitrary. Orthonormal bases $\breve{\boldsymbol{e}}_a$ are called the proper bases for individual points of the fluid.

The components $\breve{u}^a$ of the four-dimensional velocity vector $\boldsymbol{u}$, calculated in the proper basis $\breve{\boldsymbol{e}}_a$, have the form $\breve{u}^a = \{0, 0, 0, 1\}$. Therefore three-dimensional velocity of an individual point of the fluid, calculated in the proper basis, is equal to zero. Thus, the individual point of the fluid is at rest relative to the proper basis.

The proper basis of the individual point of the fluid $\breve{\boldsymbol{e}}_a$ is determined up to an arbitrary orthogonal transformation of the vectors $\breve{\boldsymbol{e}}_1$, $\breve{\boldsymbol{e}}_2$, and $\breve{\boldsymbol{e}}_3$.

We introduce the four-dimensional tensor $\widetilde{\boldsymbol{\varepsilon}} = \varepsilon^{ijk} \boldsymbol{\partial}_i \boldsymbol{\partial}_j \boldsymbol{\partial}_k$ by the contravariant components ε^{ijk} that are antisymmetric with respect to all indices i, j, and k:

$$\varepsilon^{ijk} = \varepsilon^{ijks} u_s, \tag{A.7}$$

where ε^{ijks} are the components of the four-dimensional Levi-Civita pseudotensor.

It is easy to see that in the proper basis $\breve{\boldsymbol{e}}_i$ the spatial components $\varepsilon^{\alpha\beta\lambda}$ (α, β, λ = 1, 2, 3) define the three-dimensional Levi-Civita pseudotensor $\varepsilon^{123} = 1$, and components ε^{ijk}, when at least one of the indices i, j, k is equal 4, are equal to zero.

Direct calculation taking into account definition (A.7) and relations in the footnote on p. 130 shows that the following relations are valid

$$\varepsilon_{ijk}\varepsilon^{smn} = \begin{Vmatrix} \sigma_i^s & \sigma_i^m & \sigma_i^n \\ \sigma_j^s & \sigma_j^m & \sigma_j^n \\ \sigma_k^s & \sigma_k^m & \sigma_k^n \end{Vmatrix}, \qquad \begin{array}{l} \varepsilon_{ijk}\varepsilon^{imn} = \sigma_j^m\sigma_k^n - \sigma_j^n\sigma_k^m, \\ \varepsilon_{ijk}\varepsilon^{ijm} = 2\sigma_k^m, \quad \varepsilon_{ijk}\varepsilon^{ijk} = 6, \end{array}$$

where $\sigma_i^j = \delta_i^j + u_i u^j$. Due to (A.4) the tensor components σ_i^j satisfy the equations

$$u_j\sigma_i^j = 0, \quad u^i\sigma_i^j = 0, \quad \sigma_i^j\sigma_j^s = \sigma_i^s.$$

The use of the components of the four-dimensional tensors σ_i^j and ε^{ijk} simplifies and makes more obvious the tensor transformations related to the spacelike vectors and tensors. By means of the tensor $\widetilde{\boldsymbol{\varepsilon}}$ definition (A.5) for the components of the four-dimensional vorticity vector is written in the form

$$\omega^i = \frac{c}{2}\varepsilon^{ijk}\nabla_j u_k.$$

A.1.3 The Mass Density of the Fluid

We introduce the mass density of the fluid ρ by the formula

$$\rho = \rho_\circ\sqrt{\overset{\circ}{\sigma}/\widehat{\sigma}}, \quad \overset{\circ}{\sigma} = \operatorname{mod}\ \det\|\overset{\circ}{\sigma}_{\alpha\beta}\|,$$
$$\widehat{\sigma} = \operatorname{mod}\ \det\|\widehat{\sigma}_{\alpha\beta}\|, \tag{A.8}$$

in which the invariant $\rho_\circ$ is given and depends only on the variables ξ^1, ξ^2, ξ^3 of the Lagrangian coordinate system. The components of tensors $\widehat{\sigma}_{\alpha\beta}$, $\overset{\circ}{\sigma}_{\alpha\beta}$ are defined by the relations

$$\widehat{\sigma}_{\alpha\beta} = \widehat{g}_{\alpha\beta} + \widehat{u}_\alpha\widehat{u}_\beta, \quad \overset{\circ}{\sigma}_{\alpha\beta} = \overset{\circ}{g}_{\alpha\beta} + \overset{\circ}{u}_\alpha\overset{\circ}{u}_\beta.$$

It is easy to show that the mass density of the fluid determined by formula (A.8), identically satisfies the continuity equation

$$\nabla_i\rho u^i = 0.$$

A direct check shows that the components of the velocity vector u^i and the mass density ρ for the fixed individual fluid points with coordinates ξ^1, ξ^2, ξ^3 do not depend on the choice of the Lagrangian coordinate ξ^4 or, more precisely, are invariant under transformations

$$\xi'^4 = f(\xi^\alpha, \xi^4), \quad \xi'^\alpha = \xi^\alpha,$$

where f is an arbitrary differentiable function. From definition (A.8) it follows also that the fluid density ρ is invariant under the group of the general transformations of the variables x^i (i.e., ρ is the four-dimensional scalar).

A.1.4 The Microstructure of a Fluid

It is said that a microstructure of a fluid is given if at each point of a fluid there is defined a four-dimensional orthonormal basis (tetrad) $\boldsymbol{e}_a$ connected with the basis vectors $\boldsymbol{\vartheta}_i$ of the observer's coordinate system by the scale factors $h_i{}^a$, $h^i{}_a$:

$$\boldsymbol{e}_a = h^i{}_a e_i, \quad \boldsymbol{\vartheta}_i = h_i{}^a \boldsymbol{e}_a.$$

The four-dimensional rotation of the tetrad $\boldsymbol{e}_a$ in the transition from a point with coordinates x^k to a point with coordinates $x^k + dx^k$ are defined by the Ricci rotation coefficients

$$d\boldsymbol{e}_a = dx^k \Delta_{k,ab} \boldsymbol{e}^b,$$

which are expressed in terms of the scale factors $h^j{}_a$ by the relation

$$\Delta_{k,ab} = \frac{1}{2} g_{ij} \big(h^i{}_b \nabla'_k h^j{}_a - h^i{}_a \nabla'_k h^j{}_b \big).$$

Here the symbol of the covariant derivative ∇'_k acts only upon the indices relating to the coordinate system x^i. Along with the Ricci rotation coefficients $\Delta_{k,ab}$ we will use also coefficients

$$\Delta_{k,ij} = h_i{}^a h_j{}^b \Delta_{k,ab} = \frac{1}{2} g_{ab} \big(h_i{}^a \nabla'_k h_j{}^b - h_j{}^b \nabla'_k h_i{}^a \big). \tag{A.9}$$

It is obvious that when the individual point of the fluid moves along the world line and the increment of the coordinates of the point is determined by the relation $dx^i = u^i ds = cu^i d\tau$, the relative rotation of the vectors $\boldsymbol{e}_a$ can be determined by the formula

$$d\boldsymbol{e}_a = \Omega_{ab} \boldsymbol{e}^b d\tau, \quad \Omega_{ab} = cu^k \Delta_{k,ab}.$$

The components

$$\Omega_{ij} = h_i{}^a h_j{}^b \Omega_{ab} = cu^k \Delta_{k,ij} = \frac{1}{2} g_{ab} \left(h_i{}^a \frac{d}{d\tau} h_j{}^b - h_j{}^b \frac{d}{d\tau} h_i{}^a \right), \tag{A.10}$$

where $d/d\tau = cu^i \nabla'_i$ is the symbol of the derivative with respect to the proper time, in the general case can be represented in the form

$$\Omega_{ij} = -u_i a_j + u_j a_i + \varepsilon_{ijk} \Omega^k. \tag{A.11}$$

The components of the four-dimensional vectors a_i and Ω^s by definition satisfy the equations

$$u^i a_i = 0, \quad u_s \Omega^s = 0 \tag{A.12}$$

and are expressed in terms of Ω_{ij} by the equalities[2]

$$\Omega^s = \frac{1}{2} \varepsilon^{sij} \Omega_{ij}, \quad a_i = -u^j \Omega_{ij}. \tag{A.13}$$

In connection with physical applications, the vector $\boldsymbol{\Omega} = \Omega^i \boldsymbol{\vartheta}_i$ defined in the observer's coordinate system by the components Ω^i, calculated by equality (A.13), is called the four-dimensional vector of the internal rotation of the fluid.

Consider the case when the system of tetrads $\boldsymbol{e}_a$ is such that on the world line of the individual point of the fluid the tetrads $\boldsymbol{e}_a$ are connected by the Fermi-Walker transport. In this case by the definition of the Fermi–Walker transport the scale factors $h_i{}^a$ satisfy the equation (see (2.82))

$$\frac{d}{d\tau} h_i{}^a = h_j{}^a \left(u_i \frac{d}{d\tau} u^j - u^j \frac{d}{d\tau} u_i \right)$$

and direct calculation by formula (A.10) shows that the tensor components Ω_{ij} for such system of tetrads are represented in the form

$$\Omega_{ij} = -u_i \frac{d}{d\tau} u_j + u_j \frac{d}{d\tau} u_i.$$

Thus, if the tetrads $\boldsymbol{e}_a$, defining the fluid microstructure, are transported along the world line of the point according to Fermi–Walker, then the components of the vector of internal rotation $\boldsymbol{\Omega}$ are equal to zero $\Omega^s = 0$, and the components a_i in formula (A.11) define the four-dimensional acceleration vector

$$a_i = \frac{d}{d\tau} u_i.$$

[2]Relations (A.11)–(A.13) are valid for the components of any antisymmetric tensor of the second rank Ω_{ij}.

Let us calculate the Ricci rotation coefficients $\breve{\Delta}_{k,ij}$ for the system of tetrads $\breve{\boldsymbol{e}}_a$, which are the proper bases for the individual points of the fluid. We denote the scale factors connecting bases $\boldsymbol{\ni}_i$ and $\breve{\boldsymbol{e}}_a$ by the symbol $\breve{h}^i{}_a$, i.e., $\breve{\boldsymbol{e}}_a = \breve{h}^i{}_a \boldsymbol{\ni}_i$.

By definition of the proper basis (A.6) we have

$$\breve{h}^i{}_4 = u^i. \tag{A.14}$$

Contracting equality (A.9) written for the proper bases $\breve{\boldsymbol{e}}_a$, with the velocity vector components $u^i = \breve{h}^i{}_4$, we find

$$u^j \breve{\Delta}_{k,ij} = \breve{h}^j{}_4 \breve{\Delta}_{k,ij} = -\nabla'_k \breve{h}_{i4} = -\nabla_k u_i.$$

Thus, in the presence of equality (A.14) the Ricci rotation coefficients $\breve{\Delta}_{k,ij}$ satisfy the equality $u^j \breve{\Delta}_{k,ij} = -\nabla_k u_i$. Therefore the Ricci rotation coefficients $\breve{\Delta}_{k,ij}$, corresponding to the proper bases, can be represented in the form

$$\breve{\Delta}_{k,ij} = -u_i \nabla_k u_j + u_j \nabla_k u_i + \varepsilon_{ijs} \breve{\Delta}_k{}^s, \tag{A.15}$$

where the quantities $\breve{\Delta}_k{}^s$ by definition satisfy the equation

$$u_s \breve{\Delta}_k{}^s = 0. \tag{A.16}$$

For the quantities $\breve{\Omega}_{ij} = cu^k \breve{\Delta}_{k,ij}$ by virtue of definition (A.15) we have

$$\breve{\Omega}_{ij} = -u_i \frac{d}{d\tau} u_j + u_j \frac{d}{d\tau} u_i + \varepsilon_{ijs} \breve{\Omega}^s.$$

Here $\breve{\Omega}^s = cu^k \breve{\Delta}_k{}^s$. From Eq. (A.16) it follows $u_s \breve{\Omega}^s = 0$.

A.1.5 The Intrinsic Angular Momentum of the Fluid

The intrinsic angular momentum of the fluid in special relativity can be described by the second rank tensor $\boldsymbol{K}$, defined in an observer's coordinate system with the vector basis $\boldsymbol{\ni}_i$ by antisymmetric components $K^{ij} = -K^{ji}$, and in the proper basis $\breve{\boldsymbol{e}}_a$ by components $\breve{K}^{ab} = -\breve{K}^{ba}$. Thus,

$$\boldsymbol{K} = K^{ij} \boldsymbol{\ni}_i \boldsymbol{\ni}_j = \breve{K}^{ab} \breve{\boldsymbol{e}}_a \breve{\boldsymbol{e}}_b.$$

According to the definition, the spatial part of the components $\breve{K}^{ab}$ of tensor $\boldsymbol{K}$, calculated in the proper basis, defines the usual three-dimensional axial vector of the volume density of the intrinsic angular momentum of the fluid, while the components $\breve{K}^{4a}$ are equal to zero

$$\breve{K}^{ab} = \begin{Vmatrix} 0 & \breve{K}_3 & -\breve{K}_2 & 0 \\ -\breve{K}_3 & 0 & \breve{K}_1 & 0 \\ \breve{K}_2 & -\breve{K}_1 & 0 & 0 \\ 0 & 0 & 0 & 0 \end{Vmatrix}. \tag{A.17}$$

In the observer's coordinate system all six components K^{ij} in the general case are nonzero, but due to the fact that in the proper basis the components $\breve{K}^{ab}$ have the special form (A.17), the components K^{ij} are connected by the equation $u_j K^{ij} = 0$, which in the proper basis can be written in the form $\breve{K}^{\alpha 4} = 0$, $\alpha = 1, 2,3$.

Instead of the volume density tensor of the intrinsic angular momentum $\boldsymbol{K} = K^{ij}\boldsymbol{\partial}_i\boldsymbol{\partial}_j$ can use also the four-dimensional volume density vector of the intrinsic angular momentum $\mathcal{K} = K^i\boldsymbol{\partial}_i$, defined in the observer's coordinate system by the components

$$K^i = \frac{1}{2}\varepsilon^{ijk}K_{jk}. \tag{A.18}$$

Due to definition (A.18) the components K^i of the vector of the intrinsic angular momentum satisfy the invariant equation $u_i K^i = 0$.

Due to definition (A.18) the four-dimensional vector $\mathcal{K} = K^i\boldsymbol{\partial}_i = \breve{K}^a\breve{\boldsymbol{e}}_a$ is defined by three spatial components $\breve{K}^\alpha$ in the proper basis, while the fourth component $\breve{K}^4$ in the proper basis is equal to zero $\breve{K}^a = (\breve{K}^1,\ \breve{K}^2,\ \breve{K}^3,\ 0)$.

Contracting equality (A.18) with components ε_{ijk} with respect to the index i, we find expression for the components of the volume density tensor of the intrinsic angular momentum in terms of the vector components K^i in observer's coordinate system $K^{ij} = \varepsilon^{ijs}K_s$.

A.1.6 Electromagnetic Parameters

The main characteristic of the magnetization and dielectric polarization of a fluid in an electromagnetic field is the volume density antisymmetric tensor of the magnetization and the dielectric polarization

$$M^{ij}\boldsymbol{\partial}_i\boldsymbol{\partial}_j = \breve{M}^{ab}\breve{\boldsymbol{e}}_a\breve{\boldsymbol{e}}_b.$$

By definition, in the proper basis the components of the volume density tensor of the magnetization and the dielectric polarization of the fluid are determined by the antisymmetric matrix

$$\breve{M}^{ab} = \begin{Vmatrix} 0 & \breve{M}_3 & -\breve{M}_2 & \breve{P}^1 \\ -\breve{M}_3 & 0 & \breve{M}_1 & \breve{P}^2 \\ \breve{M}_2 & -\breve{M}_1 & 0 & \breve{P}^3 \\ -\breve{P}^1 & -\breve{P}^2 & -\breve{P}^3 & 0 \end{Vmatrix}, \tag{A.19}$$

where $\breve{M}_\alpha$ are the components of the three-dimensional volume density vector of the fluid magnetization in the proper basis, $\breve{P}^\alpha$ are the components of the three-dimensional volume density vector of the fluid polarization in the proper basis.

For the description of the magnetization and dielectric polarization of the fluid instead of the antisymmetric tensor with components M^{ij} it is possible and in some cases is convenient to use the four-dimensional vector of the volume density of the magnetization $\boldsymbol{M} = M^i \boldsymbol{\partial}_i = \breve{M}^a \breve{\boldsymbol{e}}_a$ and the four-dimensional vector of the volume density of the dielectric polarization of the fluid $\boldsymbol{P} = P^i \boldsymbol{\partial}_i = \breve{P}^a \breve{\boldsymbol{e}}_a$, whose components M^i and P^i in the observer's coordinate system are defined in terms of the tensor components M^{ij}

$$M^i = \frac{1}{2}\varepsilon^{ijk} M_{jk}, \quad P^i = -u_j M^{ij}, \tag{A.20}$$

where u_j are the covariant components of the velocity vector of the individual fluid points. Unlike the tensor components M^{ij}, the components of four-dimensional vectors M^i and P^i are not arbitrary, and satisfy the invariant algebraic equations

$$u_i M^i = 0, \quad u_i P^i = 0, \tag{A.21}$$

From Eqs. (A.21) it follows that the components $\breve{M}^4$ and $\breve{P}^4$ calculated in the proper basis are equal to zero $\breve{M}^4 = \breve{P}^4 = 0$, and the components $\breve{M}^\alpha$ and $\breve{P}^\alpha$ ($\alpha = 1, 2, 3$) coincide with components of the three-dimensional magnetization and dielectric polarization vectors

$$\breve{M}^i = \left(\breve{M}^1, \breve{M}^2, \breve{M}^3, 0\right), \quad \breve{P}^i = \left(\breve{P}^1, \breve{P}^2, \breve{P}^3, 0\right),$$

which determine matrix (A.19).

The tensor components M^{ij} in the observer's coordinate system are expressed in terms of the components of the four-dimensional vectors of the magnetization and dielectric polarization

$$M^{ij} = -u^i P^j + u^j P^i + \varepsilon^{ijk} M_k. \tag{A.22}$$

The electromagnetic field in the fluid is described by the electromagnetic field tensor $\boldsymbol{F} = F^{ij} \boldsymbol{\partial}_i \boldsymbol{\partial}_j$, defined in the observer's coordinate system by the components F^{ij}, antisymmetric with respect to the indices i, j. The covariant components F_{ij} of the electromagnetic field tensor can be determined by means of the vector potential A_i:

$$F_{ij} = \nabla_i A_j - \nabla_j A_i \equiv \partial_i A_j - \partial_j A_i,$$

The component A_4 is related to the potential φ of the electric field, $A_4 = -\varphi$.

In each inertial Cartesian coordinate system the matrix of the components F_{ij} can be represented in the form

$$F_{ij} = \begin{Vmatrix} 0 & B^3 & -B^2 & E_1 \\ -B^3 & 0 & B^1 & E_2 \\ B^2 & -B^1 & 0 & E_3 \\ -E_1 & -E_2 & -E_3 & 0 \end{Vmatrix},$$

where E_α are the components of the three-dimensional vector of the electric strength, B^α are the components of the three-dimensional vector of the magnetic induction. If the electromagnetic field is considered in a medium, then one can define also the four-dimensional vector $\boldsymbol{B} = B^i \boldsymbol{\mathcal{E}}_i$ of the magnetic induction and the four-dimensional vector $\boldsymbol{E} = E^i \boldsymbol{\mathcal{E}}_i$ of the electric strength by the components B^i and E^i:

$$B^i = \frac{1}{2}\varepsilon^{ijk} F_{jk}, \quad E^i = u_j F^{ij}. \tag{A.23}$$

It is obvious that the components of the vectors E^i and B^i, determined by relations (A.23), satisfy the equations

$$u_i B^i = 0, \quad u_i E^i = 0,$$

and, in the observer's coordinate system

$$F_{ij} = u_i E_j - u_j E_i + \varepsilon_{ijk} B^k. \tag{A.24}$$

To describe the electromagnetic field in the fluid, we also introduce the electromagnetic induction tensor $\boldsymbol{\mathcal{H}} = H^{ij} \boldsymbol{\mathcal{E}}_i \boldsymbol{\mathcal{E}}_j$, whose components are defined by

$$H^{ij} = F^{ij} - 4\pi M^{ij}. \tag{A.25}$$

In an inertial Cartesian coordinate system the matrix of the tensor components H^{ij} are represented in the form

$$H^{ij} = \begin{Vmatrix} 0 & H^3 & -H^2 & -D_1 \\ -H^3 & 0 & H^1 & -D_2 \\ H^2 & -H^1 & 0 & -D_3 \\ D_1 & D_2 & D_3 & 0 \end{Vmatrix},$$

where D_α are the components of the three-dimensional electric displacement vector, H^α are the components of the three-dimensional magnetic strength vector.

The four-dimensional electric displacement vector $\boldsymbol{D} = D^i \boldsymbol{\partial}_i$ and the four-dimensional strength vector of the magnetic field $\boldsymbol{H} = H^i \boldsymbol{\partial}_i$ in the observer's coordinate system are defined by the components D^i and H^i:

$$D^i = u_j H^{ij}, \quad H^i = \frac{1}{2} \varepsilon^{ijk} H_{jk}. \tag{A.26}$$

From Eqs. (A.20), (A.23), (A.25), and (A.26) it follows

$$D^i = E^i + 4\pi P^i, \quad H^i = B^i - 4\pi M^i.$$

It is easy to see that due to definitions (A.26) the components D^i and H^i satisfy the invariant equations

$$u_i D^i = 0, \quad u_i H^i = 0.$$

It is obvious that in the proper basis the components $\breve{D}^4$ and $\breve{H}^4$ of the vectors $\boldsymbol{D} = D^i \boldsymbol{\partial}_i = \breve{D}^a \breve{\boldsymbol{e}}_a$ and $\boldsymbol{H} = H^i \boldsymbol{\partial}_i = \breve{H}^a \breve{\boldsymbol{e}}_a$ are equal to zero

$$\breve{D}^a = \left(\breve{D}^1, \breve{D}^2, \breve{D}^3, 0\right), \quad \breve{H}^a = \left(\breve{H}^1, \breve{H}^2, \breve{H}^3, 0\right).$$

Let us give also expression of the components H^{ij} in terms of the components D^i, H^i in the observer's coordinate system

$$H^{ij} = u^i D^j - u^j D^i + \varepsilon^{ijk} H_k. \tag{A.27}$$

A.2 Variations of the Determining Parameters

We denote by the symbol $\mu'(x^i)$ the value of an arbitrary function $\mu(x^i)$ in the varied state. Assuming that μ' differs little from μ, the variation of the function $\mu(x^i)$ we determine by the equality, considered up to the small quantities of the first order

$$\partial \mu = \mu'(x^i) - \mu(x^i). \tag{A.28}$$

From definition (A.28) it follows

$$\partial \frac{\partial}{\partial x^i} \mu = \frac{\partial}{\partial x^i} \mu' - \frac{\partial}{\partial x^i} \mu = \frac{\partial}{\partial x^i} \partial \mu.$$

Therefore and for the covariant derivatives of the components of tensors or spinors is carried out the equality[3]

$$\partial \nabla_i \mu = \nabla_i \partial \mu. \tag{A.29}$$

For the components μ of tensors and spinors along with the variation $\partial \mu$ we define also the variation $\delta \mu$:

$$\delta \mu = \partial \mu + \delta x^i \nabla_i \mu. \tag{A.30}$$

Here δx^i are the variations of the law of the fluid motion which are the components of a four-dimensional vector

$$\delta x^i = x'^i(\xi^j) - x^i(\xi^j).$$

It is obvious that if μ are the components of tensors (or spinors), then variations $\partial \mu$ and $\delta \mu$ determine the tensors (or spinors) the same rank, as μ.

Using definition (A.30) of the variation δ, for the variation of the derivatives $x^i{}_j = \partial x^i / \partial \xi^j$, $\nabla_i \mu$ one can find

$$\delta x^i{}_j = \partial \delta x^i / \partial \xi^j = x^s{}_j \nabla_s \delta x^i,$$
$$\delta \nabla_j \mu = \nabla_j \delta \mu - \nabla_j \delta x^i \nabla_i \mu. \tag{A.31}$$

Let us now calculate the variation of the vector components $\widehat{A}_i$, given in the Lagrange coordinate system

$$\delta \widehat{A}_i = \delta\big(x^s{}_i A_s\big) = x^s{}_i \delta A_s + A_s \delta x^s{}_i = x^s{}_i \big(\delta A_s + A_j \nabla_s \delta x^j\big).$$

The variation

$$\delta_L A_s = \delta A_s + A_j \nabla_s \delta x^j \tag{A.32}$$

is called the absolute variation (or the Lie variation) of the covariant components of the vector A_s. Thus, we have

$$\delta \widehat{A}_i = x^s{}_i \delta_L A_s. \tag{A.33}$$

---

[3]This equality is carried out in the special relativity under the assumption that the metric of the space-time does not vary.

Similarly, for the variation of the contravariant components $\widehat{A}^i$ one can find

$$\delta \widehat{A}^i = \xi^i{}_s \delta_L A^s, \tag{A.34}$$

where $\delta_L A^s$ is the absolute variation of the contravariant components A^s:

$$\delta_L A^s = \delta A^s - A^j \nabla_j \delta x^s. \tag{A.35}$$

In the general case the absolute variation of the tensor components $\mu^{\mathcal{A}}$ ($\mathcal{A}$ is a generalized index) is defined by the relation

$$\delta_L \mu^{\mathcal{A}} = \delta \mu^{\mathcal{A}} - F^{\mathcal{A}j}_{\mathcal{B}i} \mu^{\mathcal{B}} \nabla_j \delta x^i, \tag{A.36}$$

in which components $F^{\mathcal{A}j}_{\mathcal{B}i} \mu^{\mathcal{B}}$ the same, as in the definition of the covariant derivative

$$\nabla_k \mu^{\mathcal{A}} = \partial_k \mu^{\mathcal{A}} + F^{\mathcal{A}j}_{\mathcal{B}i} \mu^{\mathcal{B}} \Gamma^i_{jk}.$$

By definition, for the components $\overset{\circ}{g}_{ij}$ of the metric tensor and the components $\overset{\circ}{u}_i$ of the velocity vector in the space of the initial states, we have

$$\delta \overset{\circ}{g}_{ij} = 0, \quad \delta \overset{\circ}{u}_i = 0. \tag{A.37}$$

Let us establish now formulas for the variation of the velocity vector and the fluid density. Using definition (A.3), we find an expression for the variation of the contravariant components of the four-dimensional velocity vector

$$\delta u^j = \delta \frac{x^j{}_4}{\sqrt{-\widehat{g}_{44}}} = \left(-\widehat{g}_{44}\right)^{-1/2} \delta x^j{}_4 + \left(-\widehat{g}_{44}\right)^{-3/2} g_{pq} x^j{}_4 x^p{}_4 \delta x^q{}_4.$$

Replacing here the variations $\delta x^q{}_4$ by formula (A.31), we obtain

$$\delta u^j = \sigma^j_i u^s \nabla_s \delta x^i. \tag{A.38}$$

In the same way it is possible to find formulas for the variation of the velocity vector components given in the Lagrangian coordinate system

$$\begin{aligned} \delta \widehat{u}^i &= \widehat{u}^i u_j u^s \nabla_s \delta x^j, \\ \delta \widehat{u}_i &= x^p{}_i (\sigma_{pq} u^s + \delta^s_p u_q) \nabla_s \delta x^q \end{aligned} \tag{A.39}$$

and for the variation of the metric tensor components calculated in the Lagrangian coordinate system

$$\delta\widehat{g}_{ij} = x^p{}_i x^q{}_j \big(\delta^s_p g_{mq} + \delta^s_q g_{mp}\big)\nabla_s \delta x^m. \tag{A.40}$$

Denote $\widehat{g} = \det \|\widehat{g}_{ij}\|$. Then

$$\delta\sqrt{-\widehat{g}} = -\frac{1}{2\sqrt{-\widehat{g}}}\,\delta\widehat{g} = \frac{1}{2}\sqrt{-\widehat{g}}\,\widehat{g}^{ij}\,\delta\widehat{g}_{ij}.$$

Taking into account formula (A.40) we obtain

$$\delta\sqrt{-\widehat{g}} = \sqrt{-\widehat{g}}\,\nabla_i \delta x^i. \tag{A.41}$$

Using the obtained formulas for the variations $\delta\widehat{u}_i$ and $\delta\widehat{g}_{ij}$, we find

$$\delta\widehat{\sigma}_{ij} = x^s{}_i x^m{}_j \big(\sigma_{ns}\sigma^k_m + \sigma_{nm}\sigma^k_s\big)\nabla_k \delta x^n. \tag{A.42}$$

For mass density ρ, according to (A.8) and (A.37), we have

$$\delta\rho = -\frac{1}{2}\rho\,\widehat{g}^{ij}\,\delta\widehat{\sigma}_{ij}.$$

Replacing here the variations $\delta\widehat{\sigma}_{ij}$ by formula (A.42), we finally obtain

$$\delta\rho = -\rho\sigma_i^j\,\nabla_j \delta x^i. \tag{A.43}$$

The variation of the Ricci rotation coefficients $\Delta_{k,ij}$, which determine the microstructure of the fluid, is obtained by varying equality (A.9)

$$\delta\Delta_{k,ij} = \nabla_k \delta\varphi_{ij} - \Delta_{k,i}{}^n \delta\varphi_{jn} + \Delta_{k,j}{}^n \delta\varphi_{in} - \Delta_{s,ij}\nabla_k \delta x^s, \tag{A.44}$$

where the antisymmetric quantities $\delta\varphi_{ij}$ are defined by the formula

$$\delta\varphi_{ij} = \frac{1}{2}\big(h_{ia}\delta h_j{}^a - h_{ja}\delta h_i{}^a\big). \tag{A.45}$$

In the sequel the absolute variation will be also used

$$\delta_L\varphi_{ij} = \delta\varphi_{ij} + \frac{1}{2}\big(g_{si}\nabla_j \delta x^s - g_{js}\nabla_i \delta x^s\big), \tag{A.46}$$

This formula is obtained if we replace $\delta h_j{}^a$ in (A.45) by the absolute variation

$$\delta_L h_j{}^a = \delta h_j{}^a + h_s{}^a \nabla_j \delta x^s.$$

Let us also give an expression for the variation of quantities Ω_{ij} defined by relations (A.10):

$$\delta\Omega_{ij} = \frac{d}{d\tau}\delta\varphi_{ij} - \Omega_i{}^n\delta\varphi_{jn} + \Omega_j{}^n\delta\varphi_{in} + \Omega_{ij}u^m u_s\nabla_m\delta x^s, \tag{A.47}$$

which is obtained by contracting relation (A.44) with components cu^k.

The invariant element dV_4 of the four-dimensional region of the Minkowski space is determined by

$$dV_4 = \sqrt{-g}\,dx^1dx^2dx^3dx^4 = \sqrt{-\hat{g}}\,d\xi^1d\xi^2d\xi^3d\xi^4$$

Bearing in mind (A.41), we get

$$\delta dV_4 = d\xi^1d\xi^2d\xi^3d\xi^4\delta\sqrt{-\hat{g}} = \nabla_i\delta x^i dV_4. \tag{A.48}$$

A.3 The Variational Equation

To obtain the dynamic equations and the state equations, describing spin fluids and the electromagnetic field, we will use the following variational equation [63]

$$\delta\int\limits_{V_4}\Lambda dV_4 + \delta W^* = 0. \tag{A.49}$$

Here V_4 is an arbitrary four-dimensional region of the Minkowski space; dV_4 is the invariant element of the volume V_4; Λ is the Lagrange function for the fluid and field which is the four-dimensional scalar; δW^* is a given functional introduced to account for external influences and irreversible processes in the fluid. In Eq. (A.49), the determining parameters of the fluid and the field whose function is the Lagrangian Λ, and the region V_4 are varied. The variations of the determining parameters in Eq. (A.49) are assumed to be equal to zero on the three-dimensional surface Σ_3 bounding the region V_4.

We will consider models of the magnetizable and polarizable fluids with the intrinsic angular momentum, described by the Lagrangian[4]

$$\Lambda = \frac{1}{8\pi}F^{ij}\left(\nabla_i A_j - \nabla_j A_i - \frac{1}{2}F_{ij}\right) - M^{ij}\nabla_i A_j + \eta u^i M_i + \lambda u^i P_i + \Lambda_m\big(\rho, s, \omega_i, \Omega_i, M_i, P_i, g^{ij}\big). \tag{A.50}$$

[4]The more general models of the magnetizable and polarizable media have been considered in [96].

Here F_{ij} are the components of the electromagnetic field tensor; A_i are the components of the vector potential of the electromagnetic field; M_i are the components of the four-dimensional volume density vector of the fluid magnetization, P_i are the components of the four-dimensional volume density vector of the dielectric polarization of the fluid. M^{ij} are the components of the volume density antisymmetric tensor of the magnetization and dielectric polarization of the fluid related to the components of the four-dimensional vectors M_i and P_i by Eq. (A.22). u^i are the components of the four-dimensional dimensionless velocity vector of individual points of the fluid; ω_i are the components of the four-dimensional vorticity vector; Ω_i are the components of the four-dimensional vector of the internal rotation; ρ is the mass density of the fluid; s is the specific density of entropy. Coefficients λ and η in the Lagrangian (A.50) are the Lagrange multipliers.

In the Lagrangian (A.50) the term $-M^{ij}\nabla_i A_j$ determines an interaction of the magnetized and polarized fluid with the electromagnetic field. The quantity

$$\Lambda_f = \frac{1}{8\pi}F^{ij}\left(\nabla_i A_j - \nabla_j A_i - \frac{1}{2}F_{ij}\right)$$

is the Lagrangian of the electromagnetic field. The terms with Lagrange multipliers λ and η ensure the fulfillment of Eqs. (A.21). Λ_m is the Lagrangian of the matter, defined as a scalar function of the arguments noted in (A.50).

The functional δW^*, entering the variational equation (A.49), in general case is defined by the equality

$$\delta W^* = \int\limits_{V_4}\Big(- \rho T\delta s + \tau_i{}^j\nabla_j\delta x^i - Q_i\delta x^i - \frac{1}{c}i^j\delta_L A_j$$

$$+ \Pi^i\delta_L P_i + L^i\delta_L M_i - \frac{1}{2}R^{ij}\delta_L\varphi_{ij}\Big)dV_4. \qquad \text{(A.51)}$$

Here T is the quantity playing the temperature role in equilibrium processes. i^j are the components of the conduction current vector, c is the light velocity in vacuum. The absolute variation δ_L is determined by equalities (A.32) and (A.46). The use of the quantities Π^i and L^i in the functional δW^* is related to consideration of irreversible processes of magnetization and polarization in the fluid. Q_i are the components of the external volume forces vector. By definition, the vectors determined by the components Π^i, L^i, Q^i, and i^j, are spacelike:

$$u_i Q^i = 0, \quad u_i\Pi^i = 0, \quad u_i L^i = 0, \quad u_i i^i = 0.$$

The tensor components R^{ij} in functional (A.51) defining the relaxation spin processes, we determine by the equality

$$R^{ij} = \varepsilon^{ijk}R_k, \quad u^s R_s = 0. \qquad \text{(A.52)}$$

The tensor components $\tau_i{}^j$ in functional (A.51) determine a viscosity and thermal conduction in the fluid. The components of an arbitrary tensor of the second rank $\tau_i{}^j$ can always be represented in the form

$$\tau_i{}^j = s_i{}^j - \frac{1}{c}u_i q^j - cG_i u^j + \tau u_i u^j, \tag{A.53}$$

where c is the light velocity, and the components $s_i{}^j$, q^j, and G_i by definition satisfy the equations

$$\begin{aligned} u^i s_i{}^j &= 0, \quad u_j s_i{}^j = 0, \\ u_j q^j &= 0, \quad u^i G_i = 0, \end{aligned} \tag{A.54}$$

Using (A.54) it is easy to find an expression τ, q^j, G_i, and $s_i{}^j$ in terms of $\tau_i{}^j$:

$$\begin{aligned} s_i{}^j &= \sigma_i^m \sigma_n^j \tau_m{}^n, \quad q^j = c\sigma_n^j u^i \tau_i{}^n, \\ G_i &= \frac{1}{c}\sigma_i^m u_j \tau_m{}^j, \quad \tau = u^i u_j \tau_i{}^j. \end{aligned}$$

Let us assume further that the contravariant components of the tensor $\tau^{mj} = g^{mi}\tau_i{}^j$ in δW^* are symmetric $\tau^{mj} = \tau^{jm}$ and satisfy the invariant equation $u_i u_j \tau^{ij} = 0$ which means that the component τ^{44}, calculated in the proper basis, is equal to zero. In this case formula (A.53) for $\tau_i{}^j$ can be written in the form

$$\tau_i{}^j = s_i{}^j - \frac{1}{c}\left(u_i q^j + u^j q_i\right),$$

where the contravariant components of the viscous stress tensor s^{ij} are symmetric $s^{ij} = s^{ji}$.

The quantities $\tau_i{}^j$, Π^i, L^i, R^{ij}, and i^j entering in functional δW^*, must be given as functions of the determining parameters of the fluid and field or to be defined from the specified equations.

A.4 Dynamic Equations for Spin Fluids

Using formulas for variations of the determining parameters (A.29)–(A.48), it is possible to calculate the variation of the action integral. We have

$$\delta \int\limits_{V_4} \Lambda dV_4 = \int\limits_{V_4} \delta\Lambda dV_4 + \int\limits_{V_4} \Lambda \delta dV_4 = \int\limits_{V_4} \left(\delta\Lambda + \Lambda \nabla_i \delta x^i\right) dV_4. \tag{A.55}$$

Let us give expressions for the variations of the separate terms entering into the action integral. For the variation of the action integral for the electromagnetic field we have

$$\frac{1}{8\pi}\delta\int\limits_{V_4} F^{ij}\left(\nabla_i A_j - \nabla_j A_i - \frac{1}{2}F_{ij}\right)dV_4 = \int\limits_{V_4}\left\{\frac{1}{8\pi}\delta F^{ij}(\nabla_i A_j - \nabla_j A_i\right.$$

$$\left.-F_{ij}) + \frac{1}{4\pi}\nabla_j F^{ij}\delta_L A_i + \delta x^i\left[\frac{1}{4\pi}\nabla_j\left(F^{js}F_{is} - \frac{1}{4}F_{sm}F^{sm}\delta_i^j\right)\right]\right\}dV_4$$

$$-\int\limits_{\Sigma_3}\left[\frac{1}{4\pi}F^{ij}\delta_L A_i + \frac{1}{4\pi}\left(F^{js}F_{is} - \frac{1}{4}F_{sm}F^{sm}\delta_i^j\right)\delta x^i\right]n_j\, d\sigma. \qquad (A.56)$$

Here n_j are the components of a unit vector of the outward normal to the three-dimensional surface Σ_3, bounding the four-dimensional region V_4. $d\sigma$ is the invariant element of the surface Σ_3.

The variation of the part of the action integral with the terms $-M^{ij}\nabla_i A_j$ hase the form

$$\delta\int\limits_{V_4}(-M^{ij}\nabla_i A_j)dV_4 = \int\limits_{V_4}\left\{-\nabla_j M^{ij}\delta_L A_i\right. \qquad (A.57)$$

$$+\delta x^i\nabla_j\left[-M^{js}F_{is} + \frac{1}{2}M_{sm}F^{sm}\delta_i^j + u^j u_m\left(F^{mn}M_{in} - F_{in}M^{mn}\right)\right]$$

$$\left.+u_i F^{ij}\delta P_j + \frac{1}{2}\varepsilon^{ksij}F_{ks}u_i\delta M_j\right\}dV_4 - \int\limits_{\Sigma_3}\left\{-M^{ij}\delta_L A_i\right.$$

$$\left.+\left[-M^{js}F_{is} + \frac{1}{2}M_{sm}F^{sm}\delta_i^j + u^j u_m\left(F^{mn}M_{in} - F_{in}M^{mn}\right)\right]\delta x^i\right\}n_j\, d\sigma.$$

To simplify the writing, the equation $F_{ij} = \nabla_i A_j - \nabla_j A_i$ is used under the transformation of terms in the surface part of expressions (A.56) and (A.57), which does not affect the form of the Euler equations obtained below.

The variation of the action integral with the Lagrange multipliers are written as follows

$$\delta\int\limits_{V_4}\left(\eta u^i P_i + \lambda u^i M_i\right)dV_4 = \int\limits_{V_4}\left\{u^i P_i\delta\eta + u^i M_i\delta\lambda + \eta u^i\delta P_i + \lambda u^i\delta M_i\right.$$

$$\left.-\delta x^i\nabla_j\left[\sigma_i^j\left(\eta u^s P_s + \lambda u^s M_s\right) + u^j\left(\eta P_i + \lambda M_i\right)\right]\right\}dV_4$$

$$+\int\limits_{\Sigma_3}\left[\sigma_i^j\left(\eta u^s P_s + \lambda u^s M_s\right) + u^j\left(\eta P_i + \lambda M_i\right)\right]\delta x^i n_j\, d\sigma. \qquad (A.58)$$

From the variational equation (A.49), taking into account formulas for variations (A.29)–(A.48), expressions (A.55)–(A.58) for the variation of separate parts of the action integral and definition (A.51) for the functional δW^*, we find the following system of differential Euler's equations[5]

$$
\begin{aligned}
&a. \quad \nabla_j H^{ij} = \frac{4\pi}{c} i^i, \\
&b. \quad F_{ij} = \nabla_i A_j - \nabla_j A_i, \\
&c. \quad \nabla_j P_i{}^j = Q_i, \\
&d. \quad \rho \frac{d}{d\tau}\left(\frac{1}{\rho} K^{ij}\right) = K^i{}_n \Omega^{nj} - K^j{}_n \Omega^{ni} + R^{ij}, \\
&e. \quad E^i = \sigma^i_s \frac{\partial \Lambda_m}{\partial P_s} + \Pi^i, \qquad B^i = \sigma^i_s \frac{\partial \Lambda_m}{\partial M_s} + L^i, \\
&f. \quad \rho T = \frac{\partial \Lambda_m}{\partial s}, \\
&g. \quad u^i M_i = 0, \qquad u^i P_i = 0, \\
&h. \quad \lambda = u_i \frac{\partial \Lambda_m}{\partial M_i}, \qquad \eta = u_i \frac{\partial \Lambda_m}{\partial P_i}.
\end{aligned}
\tag{A.59}
$$

Here $\sigma_i^j = \delta_i^j + u_i u^j$; the components of the tensors K_{ij} and $P_i{}^j$ in Eqs. (A.59) are defined by the relations

$$
K_{ij} = \varepsilon_{ijks} u^k \frac{\partial \Lambda_m}{\partial \Omega_s}, \tag{A.60}
$$

$$
\begin{aligned}
P_i{}^j = {} & \frac{1}{4\pi}\left[F_{in} H^{jn} - \frac{1}{4}\delta_i^j F_{sm} H^{sm} - u^j u_m \left(H_{in} F^{mn} - H^{mn} F_{in}\right)\right] \\
& + (p+e) u_i u^j + p\delta_i^j + \frac{1}{2} u^j S_{is} \frac{d}{d\tau} u^s + \frac{1}{2} c u^j \nabla_k S_i{}^k - \frac{1}{2} c S^{js} \nabla_i u_s \\
& + u^j K_{is} u_k \Omega^{ks} + \frac{1}{2}\left(R_i{}^j + L_i M^j - L^j M_i + \Pi_i P^j - \Pi^j P_i\right) - \overset{*}{s}_i{}^j \\
& + \frac{1}{c}\left(u_i q^j + u^j q_i\right) - u^j u_k \left(M_i \frac{\partial \Lambda_m}{\partial M_k} + P_i \frac{\partial \Lambda_m}{\partial P_k} + \omega_i \frac{\partial \Lambda_m}{\partial \omega_k} + \Omega_i \frac{\partial \Lambda_m}{\partial \Omega_k}\right),
\end{aligned}
$$

[5]The equations for fluids with an intrinsic angular momentum considered here have been obtained in [77]. Analogous equations were obtained in [43] in the case when the four-dimensional vector of the electric field is equal to zero, for a special type of internal energy and using non-holonomic terms in the variational equation to describe the intrinsic angular momentum. The simplest relativistic model of a fluid with the intrinsic angular momentum (dust) have been obtained in [71]. See also [19, 33].

where the quantities p, e, $\overset{*}{s}_i{}^j$, and S_{ij} are defined as follows

$$
\begin{aligned}
p &= \rho^2 \frac{\partial \Lambda_m/\rho}{\partial \rho} + \frac{1}{4} F_{ij} M^{ij}, \\
e &= \Lambda_m - \frac{1}{4} F_{ij} M^{ij} - \omega_i \frac{\partial \Lambda_m}{\partial \omega_i} - \Omega_i \frac{\partial \Lambda_m}{\partial \Omega_i}, \\
S_{ij} &= \varepsilon_{ijks} u^k \frac{\partial \Lambda_m}{\partial \omega_s}, \\
\overset{*}{s}_i{}^j &= s_i{}^j + \frac{1}{2}\left(L_i M^j + L^j M_i + \Pi_i P^j + \Pi^j P_i\right).
\end{aligned} \tag{A.61}
$$

To the system of dynamic equations (A.59) it is necessary to add the continuity equation

$$\nabla_i \rho u^i = 0.$$

which is satisfied by virtue of the definition of the fluid density.

The equations "a" in (A.59) are the second pair of Maxwell's equations for electromagnetic fields. The vector potential of the electromagnetic field A_i enters into the equations system (A.59) only in the equation $F_{ij} = \nabla_i A_j - \nabla_j A_i$. Therefore the components A_i can be excluded from Eqs. (A.59) by omitting the equations $F_{ij} = \nabla_i A_j - \nabla_j A_i$, but adding the first pair of Maxwell's equations to the system of equations (A.59)

$$\nabla_i F_{jk} + \nabla_j F_{ki} + \nabla_k F_{ij} = 0,$$

which are carried out identically due to the equations $F_{ij} = \nabla_i A_j - \nabla_j A_i$.

The equation "c" in (A.59) is the equation for the energy-momentum of the field and fluid.

The equation "d" in (A.59) obtained due to the variations $\delta\varphi_{ij}$, is the equation for the tensor of the intrinsic angular momentum. Contracting the equation "d" in (A.59) with components $\varepsilon_{skij} u^k$ with respect to the indices i and j, we get the equation

$$\rho \frac{d}{d\tau} \frac{1}{\rho} K_i = K^j \left(-\varepsilon_{ijk} \Omega^k + u_i \frac{d}{d\tau} u_j - u_j \frac{d}{d\tau} u_i \right) + R_i$$

for the vector components of the spin K^i, determined by the relation

$$K^i = \frac{1}{2} \varepsilon^{ijs} K_{js} = -\sigma^i_j \frac{\partial \Lambda_m}{\partial \Omega_j}.$$

The contraction of the equation "d" in (A.59) with the velocity vector components u_j gives

$$K^{ij}\frac{d}{d\tau}u_j = K^{is}\Omega_{ks}u^k. \tag{A.62}$$

By means of Eq. (A.62) it is possible to exclude the term with Ω_{ks} in expression (A.60) for the components of the energy-momentum tensor $P_i{}^j$. Taking into account also that the scalar Lagrangian Λ satisfies the identity[6]

$$u_k\left(M_i\frac{\partial \Lambda_m}{\partial M_k} + P_i\frac{\partial \Lambda_m}{\partial P_k} + \omega_i\frac{\partial \Lambda_m}{\partial \omega_k} + \Omega_i\frac{\partial \Lambda_m}{\partial \Omega_k}\right) \equiv 0, \tag{A.63}$$

the expression (A.60) for $P_i{}^j$ can be written as follows

$$\begin{aligned} P_i{}^j &= \frac{1}{4\pi}\left[F_{in}H^{jn} - \frac{1}{4}\delta_i^j F_{sm}H^{sm} - u^j u_m\left(H_{in}F^{mn} - H^{mn}F_{in}\right)\right] + p\delta_i^j \\ &+(p+e)u_i u^j + \frac{1}{2}u^j S_{is}\frac{d}{d\tau}u^s + \frac{1}{2}cu^j\nabla_k S_i{}^k - \frac{1}{2}cS^{js}\nabla_i u_s + u^j K_{is}\frac{d}{d\tau}u^s \\ &+\frac{1}{2}\left(R_i{}^j + L_i M^j - L^j M_i + \Pi_i P^j - \Pi^j P_i\right) - \overset{*}{s}_i{}^j + \frac{1}{c}\left(u_i q^j + u^j q_i\right). \end{aligned}$$

A.5 The Energy-Momentum Tensor of the Electromagnetic Field in Continuous Media

Let P^{ij} be the contravariant components of the total energy-momentum tensor of the fluid and the electromagnetic field, satisfying the equations

$$\begin{aligned} \nabla_j P^{ij} &= Q^i, \\ \nabla_k\left(r^i P^{jk} - r^j P^{ik} + S^{ijk}\right) &= r^i Q^j - r^j Q^i, \end{aligned} \tag{A.64}$$

in which r^i are the components of the radius vector of a point in the Minkowski space, S^{ijk} are the components of the intrinsic angular momentum tensor.

In general case, the components of the energy-momentum tensor P^{ij} is possible to represent in the form

$$P^{ij} = -p^{ij} + \frac{1}{c}u^i\varepsilon^j + cu^j g^i + (\rho c^2 + \rho U)u^i u^j, \tag{A.65}$$

[6]Identity (A.63) is obtained by equating to zero variation of the action integral under an arbitrary transformation of the variables x^i of the observer's coordinate systems.

where the four-dimensional scalar ρU, the components of the four-dimensional vectors ε^j, g^i and the components of the four-dimensional second rank tensor p^{ij} are defined in terms of P^{ij} by the equalities

$$\rho U = -\rho c^2 + u_i u_j P^{ij}, \quad p^{ij} = -\sigma_k^i \sigma_s^j P^{ks},$$
$$\varepsilon^j = -c u_i \sigma_s^j P^{is}, \quad g^i = -\frac{1}{c}\sigma_s^i u_j P^{sj}. \tag{A.66}$$

Due to definitions (A.66) the components ε^j, g^i, and p^{ij} satisfy the equations

$$u_j p^{ij} = 0, \quad u_i p^{ij} = 0, \quad u_i g^i = 0, \quad u_j \varepsilon^j = 0. \tag{A.67}$$

It is easy to see that in the proper basis the components $\breve{\varepsilon}^4$ and $\breve{g}^4$ are equal to zero; the components $\breve{p}^{ij}$ in the proper basis are defined by the three-dimensional spatial matrix

$$\breve{\varepsilon}^j = (\breve{\varepsilon}^1, \breve{\varepsilon}^2, \breve{\varepsilon}^3, 0), \quad \breve{g}^i = (\breve{g}^1, \breve{g}^2, \breve{g}^3, 0),$$
$$\breve{p}^{ij} = \begin{Vmatrix} \breve{p}^{\alpha\beta} & 0 \\ 0 & 0 \end{Vmatrix}, \quad \alpha, \beta = 1, 2, 3.$$

From equalities (A.65) and (A.67) it follows that the matrix of the contravariant components of the energy-momentum tensor in the proper basis has the form

$$\breve{P}^{ij} = \begin{Vmatrix} -\breve{p}^{11} & -\breve{p}^{12} & -\breve{p}^{13} & c\breve{g}^1 \\ -\breve{p}^{21} & -\breve{p}^{22} & -\breve{p}^{23} & c\breve{g}^2 \\ -\breve{p}^{31} & -\breve{p}^{32} & -\breve{p}^{33} & c\breve{g}^3 \\ c^{-1}\breve{\varepsilon}^1 & c^{-1}\breve{\varepsilon}^2 & c^{-1}\breve{\varepsilon}^3 & \rho c^2 + \rho U \end{Vmatrix}.$$

According to definition of the energy-momentum tensor, the scalar quantity ρU entering in (A.65), is the volume density of the energy of the fluid and electromagnetic field in the proper basis; the four-dimensional vector with components ε^j is the volume density vector of the energy flux of the fluid and field; the four-dimensional vector with components g^i is the volume density vector of the momentum of the fluid and field; p^{ij} are the components of the four-dimensional stress tensor.

In the general case, the components P^{ij} of the total energy-momentum tensor of the fluid and the field can be written as the sum of the energy-momentum tensor components $P^{ij}_{(m)}$ of the fluid and the energy-momentum tensor components $P^{ij}_{(f)}$ of the electromagnetic field; the components S^{ijk} of the total tensor of the intrinsic angular momentum for the fluid and field can be represented as the sum of the tensors components of the intrinsic angular momentum of the fluid $S^{ijk}_{(m)}$ and field $S^{ijk}_{(f)}$:

$$P^{ij} = P^{ij}_{(m)} + P^{ij}_{(f)}, \quad S^{ijk} = S^{ijk}_{(m)} + S^{ijk}_{(f)}. \tag{A.68}$$

Taking into account formula (A.68), Eqs. (A.64) for the energy-momentum and intrinsic angular momentum can be written in the form

$$\nabla_j P^{ij}_{(m)} = F^i + Q^i,$$
$$\nabla_k \big(r^i P^{jk}_{(m)} - r^j P^{ik}_{(m)} + S^{ijk}_{(m)}\big) = h^{ij} + r^i (F^j + Q^j) - r^j (F^i + Q^i),$$

where the components F^i of the four-dimensional force vector and the components h^{ij} of the four-dimensional internal torque tensor acting on the fluid, are defined by the relations

$$F^i = -\nabla_j P^{ij}_{(f)}, \quad h^{ij} = P^{ij}_{(f)} - P^{ji}_{(f)} - \nabla_k S^{ijk}_{(f)}.$$

Similarly (A.65), the energy-momentum tensor components $P^{ij}_{(m)}$ and $P^{ij}_{(f)}$ we represent in the form

$$P^{ij}_{(m)} = -p^{ij}_{(m)} + \frac{1}{c} u^i \varepsilon^j_{(m)} + c u^j g^i_{(m)} + (\rho c^2 + \rho U_{(m)}) u^i u^j,$$
$$P^{ij}_{(f)} = -p^{ij}_{(f)} + \frac{1}{c} u^i \varepsilon^j_{(f)} + c u^j g^i_{(f)} + U_{(f)} u^i u^j. \tag{A.69}$$

The quantities $\rho U_{(m)}$, $U_{(f)}$, $\varepsilon^j_{(m)}$, $\varepsilon^j_{(f)}$, $g^i_{(m)}$, $g^i_{(f)}$, $p^{ij}_{(m)}$, and $p^{ij}_{(f)}$ in formulas (A.69) are determined by the equalities

$$\begin{aligned}
\rho U_{(m)} &= -\rho c^2 + u_i u_j P^{ij}_{(m)}, & U_{(f)} &= u_i u_j P^{ij}_{(f)},\\
p^{ij}_{(m)} &= -\sigma^i_k \sigma^j_s P^{ks}_{(m)}, & p^{ij}_{(f)} &= -\sigma^i_k \sigma^j_s P^{ks}_{(f)},\\
\varepsilon^j_{(m)} &= -c u_i \sigma^j_s P^{is}_{(m)}, & \varepsilon^j_{(f)} &= -c u_i \sigma^j_s P^{is}_{(f)},\\
g^i_{(m)} &= -\frac{1}{c} \sigma^i_s u_j P^{sj}_{(m)}, & g^i_{(f)} &= -\frac{1}{c} \sigma^i_s u_j P^{sj}_{(f)}.
\end{aligned} \tag{A.70}$$

Comparing Eqs. (A.65) and (A.69) and taking into account equalities (A.68), we find

$$\rho U = \rho U_{(m)} + U_{(f)}, \quad p^{ij} = p^{ij}_{(m)} + p^{ij}_{(f)},$$
$$\varepsilon^j = \varepsilon^j_{(m)} + \varepsilon^j_{(f)}, \quad g^i = g^i_{(m)} + g^i_{(f)}.$$

In the macroscopic phenomenological theory of the electromagnetic field it is assumed that the electromagnetic field has no an intrinsic angular momentum

$$S^{ijk}_{(f)} = 0. \tag{A.71}$$

As for the energy-momentum tensor of the electromagnetic field in the medium, several different definitions have been proposed for $P^{ij}_{(f)}$. The Minkowski definition and the Abraham definition are most known and spread (see e.g., [26–28, 64]). Expression (A.60) received above for the total energy-momentum tensor of the fluid and electromagnetic field shows that under the assumption $S^{ijk}_{(f)} = 0$ as the energy-momentum tensor of the electromagnetic field in the medium one can take the tensor defined in an observer's coordinate system by the components [78][7]

$$P^{ij}_{(f)} = \frac{1}{4\pi}\left[F^{i}{}_{n}H^{jn} - \frac{1}{4}g^{ij}F_{sm}H^{sm} - u^{j}u_{m}\left(H^{i}{}_{n}F^{mn} - H^{mn}F^{i}{}_{n}\right)\right]. \tag{A.72}$$

Substituting the components of the tensors F^{ij} and H^{ij} by formulas (A.24) and (A.27) in (A.72), we obtain the expression for the components $P^{ij}_{(f)}$ in terms of the components of the four-dimensional vectors E^i, D^i, H^i, and B^i of the electric and magnetic strength and induction

$$P^{ij}_{(f)} = \frac{1}{4\pi}\Big[\Big(\frac{1}{2}g^{ij} + u^{i}u^{j}\Big)\big(E_m D^m + B_m H^m\big) - E^i D^j - H^i B^j + \big(u^i\varepsilon^{jms} + u^j\varepsilon^{ims}\big)E_m H_s\Big]. \tag{A.73}$$

The components ε^{ims} in (A.73) are defined by equality (A.7).

Let us consider some simple properties of the tensor determined by components (A.72). It is easy to see that due to definition (A.72) the components $P^{ij}_{(f)}$ identically satisfy the invariant equations

$$P_{i}{}^{i}{}_{(f)} = 0, \quad u_j\big(P^{ij}_{(f)} - P^{ji}_{(f)}\big) = 0.$$

The volume density of the electromagnetic energy $U_{(f)}$, in accordance with definitions (A.70) and (A.72), is defined by the equality

$$U_{(f)} = \frac{1}{8\pi}\big(E_i D^i + H_i B^i\big) = \frac{1}{16\pi}\big(F_{ij}H^{ij} + 4u_j u^m F^{ij}H_{im}\big). \tag{A.74}$$

---

[7] The components of this energy-momentum tensor in the pseudoeuclidean space with the metric signature $(-,-,-,+)$ are defined as follows

$$P^{ij}_{(f)} = -\frac{1}{4\pi}\left[F^{i}{}_{n}H^{jn} - \frac{1}{4}g^{ij}F_{sm}H^{sm} + u^{j}u_{m}\left(H^{i}{}_{n}F^{mn} - H^{mn}F^{i}{}_{n}\right)\right].$$

It is obvious that the energy $U_{(f)}$, determined by formula (A.74), is the four-dimensional scalar. In the proper basis of the individual fluid point the quantity $U_{(f)}$ is written in the form

$$U_{(f)} = \frac{1}{8\pi}\left(\breve{\boldsymbol{E}} \cdot \breve{\boldsymbol{D}} + \breve{\boldsymbol{B}} \cdot \breve{\boldsymbol{H}}\right),$$

where $\breve{\boldsymbol{E}}$, $\breve{\boldsymbol{H}}$ are the three-dimensional vectors of the electric and magnetic strength; $\breve{\boldsymbol{D}}$ and $\breve{\boldsymbol{B}}$ are the three-dimensional vectors of the dielectric and magnetic induction in the proper basis.

The four-dimensional vector of the volume density of the energy flux of the electromagnetic field, corresponding to the energy-momentum tensor with components (A.72), is defined in the observer's coordinate system by the components

$$\varepsilon^{j}_{(f)} = -\frac{c}{4\pi}\sigma^{js} u_i F^{im} H_{sm} = \frac{c}{4\pi}\varepsilon^{jms} E_m H_s.$$

In the proper basis $\breve{\varepsilon}^{4}_{(f)} = 0$, and the components $\breve{\varepsilon}^{1}_{(f)}$, $\breve{\varepsilon}^{2}_{(f)}$, and $\breve{\varepsilon}^{3}_{(f)}$ define the three-dimensional Poynting vector of the energy flux density of the electromagnetic field

$$\breve{\varepsilon}^{j}_{(f)} = \left(\breve{\varepsilon}^{1}_{(f)}, \breve{\varepsilon}^{2}_{(f)}, \breve{\varepsilon}^{3}_{(f)}, 0\right) = \left\{\frac{c}{4\pi}\breve{\boldsymbol{E}} \times \breve{\boldsymbol{H}}, 0\right\}.$$

The four-dimensional vector of volume density of the electromagnetic field momentum, corresponding to the tensor with components (A.72), is related to the vector of the energy flux volume density by the multiplier c^{-2}:

$$g^{j}_{(f)} = \frac{1}{c^2}\varepsilon^{j}_{(f)}.$$

Thus, in the proper basis $\breve{g}^{4}_{(f)} = 0$, while the components $\breve{g}^{1}_{(f)}$, $\breve{g}^{2}_{(f)}$, $\breve{g}^{3}_{(f)}$ determine the three-dimensional vector of the volume density of the electromagnetic field momentum, proportional to the three-dimensional vector of the energy flux volume density of the electromagnetic field

$$\breve{g}^{j}_{(f)} = \left(\breve{g}^{1}_{(f)}, \breve{g}^{2}_{(f)}, \breve{g}^{3}_{(f)}, 0\right) = \left\{\frac{1}{4\pi}\breve{\boldsymbol{E}} \times \breve{\boldsymbol{H}}, 0\right\}.$$

The four-dimensional stress tensor of the electromagnetic field is determined by the components $p^{ij}_{(f)}$:

$$\begin{aligned} p^{ij}_{(f)} &= -\frac{1}{4\pi}\sigma^{i}_{k}\sigma^{j}_{s}\left(F_{n}{}^{s} H^{nk} - \frac{1}{4}g^{sk} F_{qm} H^{qm}\right) \\ &= \frac{1}{4\pi}\left[E^i D^j + H^i B^j - \frac{1}{2}\sigma^{ij}\left(E_m D^m + H_m B^m\right)\right]. \end{aligned}$$

In the proper basis $\breve{p}^{\alpha 4}_{(f)} = \breve{p}^{4\alpha}_{(f)} = 0$, while the spatial components $\breve{p}^{\alpha\beta}_{(f)}$ determine the Maxwell stress tensor of the electromagnetic field

$$\breve{p}^{\alpha\beta}_{(f)} = \frac{1}{4\pi}\left[\breve{E}^{\alpha}\breve{D}^{\beta} + \breve{H}^{\alpha}\breve{B}^{\beta} - \frac{1}{2}\delta^{\alpha\beta}\big(\breve{E}_{\lambda}\breve{D}^{\lambda} + \breve{H}_{\lambda}\breve{B}^{\lambda}\big)\right].$$

Using Maxwell's equations, the expression for the components F^i of the vector of the ponderomotive force acting on the fluid from the electromagnetic field, one can transform to the form

$$F^i = -\nabla_j P^{ij}_{(f)} = \frac{1}{c} i_j F^{ij} + \frac{1}{16\pi}\big(F_{js}\nabla^i H^{js} - H^{js}\nabla^i F_{js}\big)$$
$$+ \frac{\rho}{4\pi c}\frac{d}{d\tau}\left[\frac{1}{\rho}u^m\big(H^{in}F_{mn} - H_{mn}F^{in}\big)\right].$$

In the non-relativistic approximation the expression for the spatial components F_α of the force vector has the form

$$F_\alpha = \frac{1}{c}\varepsilon_{\alpha\beta\lambda} i^{\beta} B^{\lambda} + \frac{1}{8\pi}\big(- E_{\lambda}\partial_{\alpha} D^{\lambda} + D^{\lambda}\partial_{\alpha} E_{\lambda} - H_{\lambda}\partial_{\alpha} B^{\lambda} + B^{\lambda}\partial_{\alpha} H_{\lambda}\big)$$
$$+ \frac{\rho}{4\pi c}\varepsilon_{\alpha\beta\lambda}\frac{d}{dt}\left[\frac{1}{\rho}\big(D^{\beta}B^{\lambda} - E^{\beta}H^{\lambda}\big)\right],$$

where the components of the three-dimensional vectors B^λ, H_λ, E_λ, and D^λ are determined in an inertial Cartesian observer's coordinate system; $\varepsilon_{\alpha\beta\lambda}$ are the components of the three-dimensional pseudotensor Levi-Civita.

The tensor of the ponderomotive torque corresponding to the energy-momentum tensor with components (A.72), in the observer's coordinate system is defined by the components $h^{ij}_{(f)}$:

$$h^{ij}_{(f)} = P^{ij}_{(f)} - P^{ji}_{(f)} = \frac{1}{4\pi}\sigma^{i}_{s}\sigma^{j}_{m}\big(F_n{}^{s}H^{nm} - F_n{}^{m}H^{ns}\big)$$
$$= \frac{1}{4\pi}\big(- E^iD^j + E^jD^i - H^iB^j + H^jB^i\big). \qquad \text{(A.75)}$$

The components $\breve{h}^{ij}_{(f)}$ in the proper basis are defined by the three-dimensional spatial matrix

$$\breve{h}^{ij}_{(f)} = \begin{Vmatrix} 0 & \mathcal{M}^3 & -\mathcal{M}^2 & 0 \\ -\mathcal{M}^3 & 0 & \mathcal{M}^1 & 0 \\ \mathcal{M}^2 & -\mathcal{M}^1 & 0 & 0 \\ 0 & 0 & 0 & 0 \end{Vmatrix},$$

where $\mathcal{M}^\alpha$ are the components of the three-dimensional ponderomotive torque which is well corresponding to the known experimental data

$$\boldsymbol{\mathcal{M}} = \{\mathcal{M}^\alpha\} = \frac{1}{4\pi}(\breve{\boldsymbol{B}} \times \breve{\boldsymbol{H}} + \breve{\boldsymbol{D}} \times \breve{\boldsymbol{E}}).$$

From definition (A.75) it follows

$$u_j h^{ij}_{(f)} = 0.$$

In conclusion we note that from the second equation in (A.64), which taking into account (A.71) can be written in the form

$$P^{ji} - P^{ij} + \nabla_k S^{ijk} = P^{ji}_{(m)} - P^{ij}_{(m)} + \nabla_k S^{ijk}_{(m)} - h^{ij} = 0,$$

it follows that under condition (A.71) if a medium is described by a model without intrinsic moments, $S^{ijk}_{(m)} = 0$, and if the energy-momentum tensor components of the medium are symmetric, then by virtue of the equations describing the considered model, the ponderomotive torque tensor is equal to zero

$$h^{ij} = P^{ji}_{(f)} - P^{ij}_{(f)} = 0.$$

In the application to the models of the fluids considered above, at $S^{ijk}_{(m)} = 0$, when the fluid has no intrinsic angular momenta, the components $P^{ij}_{(f)}$, defined by equality (A.72), become symmetric by virtue of the dynamic equations (A.59) and in this case coincide with the components of the Abraham tensor, if the function Λ_m in the Lagrangian is defined in such a way that the corresponding energy-momentum tensor of the fluid is symmetric. Thus, for the indicated classes of models, the use of the Abraham tensor corresponds to neglecting the intrinsic angular momentum of the medium.

It is known that in real media the magnetization is related to the intrinsic angular momentum of the medium. Therefore, in such media the physically determined energy-momentum tensor of the field, generally speaking, must be asymmetric.

A.5.1 The Minkowski Tensor

The energy-momentum tensor of the electromagnetic field by Minkowski in an observer's coordinate system is defined by the components

$$S^{ij} = \frac{1}{4\pi}\left(F^{i}{}_{n}H^{jn} - \frac{1}{4}g^{ij}F_{sm}H^{sm}\right).$$

The components S^{ij} are not symmetric at all $S^{ij} \neq S^{ji}$.

The energy, four-dimensional vector of the energy flux, and four-dimensional stress tensor of the electromagnetic field, determined by the Minkowski tensor, are the same as those determined by the tensor with components $P^{ij}_{(f)}$ (see (A.72)):

$$U_{(M)} = U_{(f)}, \quad \varepsilon^{j}_{(M)} = \varepsilon^{j}_{(f)}, \quad p^{ij}_{(M)} = p^{ij}_{(f)}.$$

The four-dimensional vector of the volume density of the electromagnetic field momentum in a medium, corresponding to the Minkowski energy-momentum tensor, is defined by the components

$$g^{i}_{(M)} = -\frac{1}{c}\sigma^{i}_{q}u_{j}S^{qj} = \frac{1}{4\pi c}\varepsilon^{iks}D_{k}B_{s}.$$

In the proper basis, this vector has the components

$$\breve{g}^{i}_{(M)} = \left(\breve{g}^{1}_{(M)}, \breve{g}^{2}_{(M)}, \breve{g}^{3}_{(M)}, 0\right) = \left\{\frac{1}{4\pi}\breve{\boldsymbol{D}} \times \breve{\boldsymbol{B}}, 0\right\},$$

where $\breve{\boldsymbol{D}}$ and $\breve{\boldsymbol{B}}$ are the three-dimensional vectors of electric and magnetic induction of the field in the proper basis. For the components of the ponderomotive force vector, calculated by the Minkowski tensor, in virtue of Maxwell's equations, we have

$$F^{i} = -\nabla_{j}S^{ij} = \frac{1}{c}i_{j}F^{ij} + \frac{1}{16\pi}\left(F_{js}\nabla^{i}H^{js} - H^{js}\nabla^{i}F_{js}\right).$$

The tensor of the ponderomotive torque corresponding to the Minkowski energy-momentum tensor, is defined in the observer's coordinate system by the components $h^{ij}_{(M)} = S^{ij} - S^{ji}$. In the proper basis the components $\breve{h}^{ij}_{(M)}$ are determined by the matrix

$$\breve{h}^{ij}_{(M)} = \begin{Vmatrix} 0 & \mathcal{M}^{3} & -\mathcal{M}^{2} & -\mathcal{L}^{1} \\ -\mathcal{M}^{3} & 0 & \mathcal{M}^{1} & -\mathcal{L}^{2} \\ \mathcal{M}^{2} & -\mathcal{M}^{1} & 0 & -\mathcal{L}^{3} \\ \mathcal{L}^{1} & \mathcal{L}^{2} & \mathcal{L}^{3} & 0 \end{Vmatrix}.$$

Here

$$\mathcal{M} = \{\mathcal{M}^\alpha\} = \frac{1}{4\pi}(\breve{\boldsymbol{B}} \times \breve{\boldsymbol{H}} + \breve{\boldsymbol{D}} \times \breve{\boldsymbol{E}}),$$

$$\mathcal{L} = \{\mathcal{L}^\alpha\} = \frac{1}{4\pi}(\breve{\boldsymbol{D}} \times \breve{\boldsymbol{B}} - \breve{\boldsymbol{E}} \times \breve{\boldsymbol{H}}),$$

$\breve{\boldsymbol{E}}$, $\breve{\boldsymbol{H}}$ are the three-dimensional vectors of the electric and magnetic strength in the proper basis.

A.5.2 The Abraham Tensor

The Abraham energy-momentum tensor of the electromagnetic field in a medium is the symmetric part of the tensor with components (A.72). Thus, in an observer's coordinate system the components A^{ij} of the Abraham tensor are defined by the equality

$$A^{ij} = \frac{1}{2}\left(P^{ij}_{(f)} + P^{ji}_{(f)}\right).$$

The energy, four-dimensional vector of the momentum and four-dimensional vector of the energy flux of the electromagnetic field, determined by the Abraham tensor, are the same as those determined by the tensor with components $P^{ij}_{(f)}$:

$$U_{(A)} = U_{(f)}, \quad \varepsilon^j_{(A)} = \varepsilon^j_{(f)}, \quad g^i_{(A)} = g^i_{(f)}.$$

The stress tensor components of the electromagnetic field corresponding to the Abraham tensor, are the symmetric part of the stress tensor components corresponding to the energy-momentum tensor with components (A.72):

$$p^{ij}_{(A)} = \frac{1}{2}\left(p^{ij}_{(f)} + p^{ij}_{(f)}\right) = \frac{1}{8\pi}\big[E^i D^j + E^j D^i + H^i B^j + H^j B^i \\ - \sigma^{ij}\left(E_m D^m + H_m B^m\right)\big].$$

From the symmetry properties of the Abraham tensor it follows that the tensor of the ponderomotive torque, determined by the Abraham tensor, is equal to zero

$$h^{ij}_{(A)} = A^{ij} - A^{ji} = 0.$$

It is easy to see that if the medium does not have a magnetization and a dielectric polarization $M^i = P^i = 0$, then the components of the Minkowski tensor, the components of the Abraham tensor, and the components (A.72) coincide

$$P^{ij}_{(f)} = S^{ij} = A^{ij} = \frac{1}{4\pi}\left(F_n{}^i F^{nj} - \frac{1}{4}g^{ij} F_{sm} F^{sm}\right).$$

The components of the tensors $P^{ij}_{(f)}$ and A^{ij} identically coincide also in the case when the equations connecting E^i, D^i and H^i, B^i have the form $D^i = \varepsilon E^i$, $B^i = \mu H^i$.

A.6 Equations of the Heat Influx and Entropy Balance

The equation of heat influx in the observer's coordinate system is obtained by contracting of the energy-momentum equation with the velocity vector components

$$u_i \nabla_j (P^{ij}_{(m)} + P^{ij}_{(f)}) = 0. \tag{A.76}$$

Using definitions (A.69) and (A.72) for the components of the energy-momentum tensors $P^{ij}_{(m)}$ and $P^{ij}_{(f)}$, the equation of the heat influx (A.76) by identical transformations (using the continuity equation for the fluid density ρ) can be represented in the form

$$\begin{aligned} &\rho\frac{d}{d\tau}\left[U_{(m)} + \frac{1}{8\pi\rho}(E_j D^j + B_j H^j)\right] \\ &\quad = -\nabla_j\left(\varepsilon^j_{(m)} + \frac{c}{4\pi}\varepsilon^{jks} E_k H_s\right) - c\left(g^i_{(m)} + \frac{1}{4\pi c}\varepsilon^{iks} E_k H_s\right)\frac{d}{d\tau}u_i \\ &\quad + c\left[p^{ij}_{(m)} - \frac{1}{8\pi}\sigma^{ij}(E_k D^k + B_k H^k) + \frac{1}{4\pi}(E^i D^j + H^i B^j)\right] d_j u_i. \end{aligned} \tag{A.77}$$

The quantities $U_{(m)}$, $\varepsilon^j_{(m)}$, $g^i_{(m)}$, and $p^{ij}_{(m)}$ in the equation of the heat influx (A.77) are defined by equalities (A.70), and $d_i = \sigma_i^j \partial_j$. Using the Maxwell and continuity equations, Eq. (A.77) can be written also in the form

$$\rho\frac{d}{d\tau}U_{(m)} = -\nabla_j \varepsilon^j_{(m)} - c g^i_{(m)}\frac{d}{d\tau}u_i + c p^{ij}_{(m)} d_j u_i + q, \tag{A.78}$$

where the energy influx q from the electromagnetic field to the fluid is defined by the equality

$$q = cu_i \nabla_j P^{ij}_{(f)} = i^i u^j F_{ij} + \frac{1}{16\pi}\left(H^{ij}\frac{d}{d\tau}F_{ij} - F_{ij}\frac{d}{d\tau}H^{ij}\right) + \frac{1}{4\pi}u^m\left(H^{in}F_{mn} - H_{mn}F^{in}\right)\frac{d}{d\tau}u_i. \quad \text{(A.79)}$$

Expression (A.79) for the energy influx q by means of the components of the four-dimensional vectors of electric and magnetic strength and induction one can rewrite in a more symmetrical form

$$q = i_j E^j + \frac{1}{8\pi}\left(E_i\frac{d}{d\tau}D^i - D^i\frac{d}{d\tau}E_i + H_i\frac{d}{d\tau}B^i - B^i\frac{d}{d\tau}H_i\right).$$

In particular, if E_i, D_i, H_i, and B_i are connected by the classical relations $D_i = \varepsilon E_i$, $B_i = \mu H_i$, then for q we get

$$q = i_j E^j + \frac{1}{8\pi}\left(E_i E^i\frac{d\varepsilon}{d\tau} + H_i H^i\frac{d\mu}{d\tau}\right). \quad \text{(A.80)}$$

From Eq. (A.80) it follows that if the magnetic permeability μ and dielectric permittivity ε do not depend on the proper time τ, then the heat influx q is defined only by the Joule heat $q = i^j E_j$.

A calculation of the energy influx of the electromagnetic field to the medium by the Minkowski tensor gives

$$q_{(M)} = cu_i \nabla_j S^{ij} = i^i u^j F_{ij} + \frac{1}{16\pi}\left(H^{ij}\frac{d}{d\tau}F_{ij} - F_{ij}\frac{d}{d\tau}H^{ij}\right)$$
$$\equiv i_j E^j + \frac{1}{8\pi}\left(E_i\frac{d}{d\tau}D^i - D^i\frac{d}{d\tau}E_i + H_i\frac{d}{d\tau}B^i - B^i\frac{d}{d\tau}H_i\right) + \frac{1}{8\pi}\varepsilon_{iks}\left(E^k H^s - D^k B^s\right)\frac{d}{d\tau}u^i.$$

If $D^i = \varepsilon E^i$, $B^i = \mu H^i$, then for $q_{(M)}$ the following expression is obtained

$$q_{(M)} = i_j E^j + \frac{1}{8\pi}\left(E_i E^i\frac{d\varepsilon}{d\tau} + H_i H^i\frac{d\mu}{d\tau}\right) + \frac{1-\varepsilon\mu}{8\pi}\varepsilon_{iks}E^k H^s\frac{d}{d\tau}u^i.$$

Thus, the energy influx of the electromagnetic field to the medium $q_{(M)}$ corresponding to the Minkowski energy-momentum tensor, depends on acceleration of the medium. The physical sense of the last member with acceleration $du^i/d\tau$

in expression for the heat influx $q_{(M)}$, calculated by the Minkowski tensor, is represented problematic.

A calculation of the electromagnetic energy influx by the Abraham tensor gives

$$q_{(A)} = i^j E_j + \frac{1}{8\pi}\left(E_i \frac{d}{d\tau} D^i - D^i \frac{d}{d\tau} E_i + H_i \frac{d}{d\tau} B^i - B^i \frac{d}{d\tau} H_i\right) + \frac{1}{8\pi}\left(E^i D^j - E^j D^i + H^i B^j - H^j B^i\right)\omega_{ij},$$

where ω_{ij} are the components of the four-dimensional vorticity tensor. If E_i, D_i, H_i, B_i are connected by the relations $D_i = \varepsilon E_i$, $B_i = \mu H_i$, then the expression for $q_{(A)}$ coincides with expression (A.80) for heat influx corresponding to the tensor (A.72).

Differentiating the function $U_{(m)}$ in the equation of the heat influx (A.78) and performing transformations taking into account definitions (A.70) of the quantities $\varepsilon^j_{(m)}$, $g^i_{(m)}$, $p^{ij}_{(m)}$ and the Euler equations (A.59), Eq. (A.78) can be transformed to the form

$$\rho T \frac{ds}{d\tau} = -cu_i \nabla_j \left[\overset{*}{s}{}^{ij} - \frac{1}{2} R^{ij} - \frac{1}{c}\left(u^i q^j + u^j q^i\right)\right] - \frac{1}{2} R^{ij} \Omega_{ij} + i^j E_j + L^i \left(\frac{d}{d\tau} M_i + \varepsilon_{ijk} M^j \omega^k\right) + \Pi^i \left(\frac{d}{d\tau} P_i + \varepsilon_{ijk} P^j \omega^k\right), \tag{A.81}$$

where $\Omega_{ij} = cu^k \Delta_{k,ij}$, the components of the viscous stress tensor $\overset{*}{s}{}^{ij}$ are determined by equality (A.61).

Equation (A.81) in some cases can be considered as the entropy balance equation.

A transformation of Eq. (A.78) to the form (A.81) is quite cumbersome, therefore we give simpler derivation of Eq. (A.81) directly from the variational equation. Let us consider the variational equation (A.49), assuming that the variations of the independent parameters are defined as follows

$$\delta x^i = u^i \delta\eta, \quad \partial\varphi_{ij} = 0, \quad \partial s = 0,$$
$$\partial A_i = 0, \quad \partial P_i = 0, \quad \partial M_i = 0.$$

Here $\delta\eta$ is an arbitrary function (variation). In this case from the formulas for variations (A.38) and (A.43) it follows

$$\partial\rho = 0, \quad \partial u^i = 0.$$

For the functional δW^*, determined by equality (A.51), for the considered special class of the variations we find

$$\delta W^* = \int \delta\eta \Big\{ -\frac{1}{c}\rho T \frac{ds}{d\tau} + \frac{1}{c} i^j u^s F_{js} + \frac{1}{c} L^i \left(\frac{d}{d\tau} M_i + \varepsilon_{ijk} M^j \omega^k \right)$$

$$+ \frac{1}{c} \Pi^i \left(\frac{d}{d\tau} P_i + \varepsilon_{ijk} P^j \omega^k \right) - \frac{1}{2c} R^{ij} \Omega_{ij} - u_i \nabla_j \Big[\tau^{ij} + \frac{1}{2} (L^i M^j + L^j M^i$$

$$+ \Pi^i P^j + \Pi^j P^i - R^{ij}) \Big] \Big\} dV_4 + \int u_i \left(\tau^{ij} - \frac{1}{2} R^{ij} - i^j A^i \right) \delta\eta n_j \, d\sigma. \quad \text{(A.82)}$$

Also for the variation of the action integral we have

$$\delta \int \Lambda dV_4 = \int [\partial \Lambda + \nabla_i (\Lambda \delta x^i)] dV_4 = \int \nabla_i (\Lambda u^i \delta\eta) dV_4. \quad \text{(A.83)}$$

Taking into account the expressions (A.82) and (A.83) for the functional δW^* and the variation of the action integral, it is easy to verify that from the variational equation (A.49) with the variation $\delta\eta$, equal to zero on the surface Σ_3 of the region V_4, it follows the entropy balance equation (A.81).

Using Eqs. (A.52) and (A.54), the equation of entropy balance (A.81) one can write also in the form

$$\rho T \frac{ds}{d\tau} = i^j E_j - \nabla_i q^i - \frac{1}{c} q^i \frac{d}{d\tau} u_i + c \overset{*}{s}{}^{ij} \nabla_j u_i - R^i (\Omega_i - \omega_i)$$

$$+ L^i \left(\frac{d}{d\tau} M_i + \varepsilon_{ijk} M^j \omega^k \right) + \Pi^i \left(\frac{d}{d\tau} P_i + \varepsilon_{ijk} P^j \omega^k \right).$$

Appendix B
Proper Bases and Invariant Internal Energy in the Theory of Electromagnetic Field

B.1 Definition of the Proper Basis of the Electromagnetic Field

Let us consider in the Minkowski pseudoeuclidean space, referred to a Cartesian coordinate system of the observer with the vector basis $\mathfrak{I}_i$, a free electromagnetic field, described by antisymmetric components of the tensor $F^{ij} = -F^{ji}$, satisfying the Maxwell equations

$$\partial_j F^{ij} = 0, \quad \partial_j \overset{*}{F}{}^{ij} = 0,$$

where $\overset{*}{F}{}^{ij} = \frac{1}{2}\varepsilon^{ijms} F_{ms}$ are the components of the dual tensor of the electromagnetic field.

The energy-momentum tensor of the free electromagnetic field is defined by the components

$$P_i{}^j = \frac{1}{4\pi}\left(F_{in}F^{jn} - \frac{1}{4}\delta_i^j F_{sm}F^{sm}\right) \equiv \frac{1}{8\pi}\left(F_{in}F^{jn} + \overset{*}{F}_{in}\overset{*}{F}{}^{jn}\right), \tag{B.1}$$

which by virtue of the Maxwell equations satisfy the conservation law

$$\partial_j P_i{}^j = 0.$$

In the general case the tensor of the electromagnetic field has two independent invariants J_1 and J_2:

$$J_1 = \frac{1}{2}F_{ij}F^{ij}, \quad J_2 = \frac{1}{2}F_{ij}\overset{*}{F}{}^{ij}. \tag{B.2}$$

V. A. Zhelnorovich, *Theory of Spinors and Its Application in Physics and Mechanics*, https://doi.org/10.1007/978-3-030-27836-6

From definition (B.1) it follows that the trace of the energy-momentum tensor of the electromagnetic field is equal to zero, and the square of the matrix $\| P_i{}^j \|$ is proportional to the unit matrix [53, 72]

$$P_i{}^i = 0, \quad P_i{}^j P_j{}^m = U^2 \delta_i^m .$$

Here the quantity U is expressed in terms of the invariants J_1 and J_2 of the electromagnetic field tensor

$$U = \frac{1}{8\pi}\sqrt{J_1^2 + J_2^2} = \frac{1}{16\pi}\sqrt{\left(F_{ij}F^{ij}\right)^2 + \left(F_{ij}\overset{*}{F}{}^{ij}\right)^2}. \tag{B.3}$$

Let us consider the eigenvalue problem of the matrix of the components of the energy-momentum tensor $P_i{}^j a^i = \lambda a^j$. If at least one invariant J_1 or J_2 of the electromagnetic field tensor is not equal to zero $J_1^2 + J_2^2 \neq 0$, then four orthonormal eigenvectors exist with componentsπ^i, ξ^i, σ^i, u^i, satisfying the equations

$$\begin{aligned} P_i{}^j \pi^i &= U\pi^j, \quad P_i{}^j \sigma^i = -U\sigma^j, \\ P_i{}^j \xi^i &= U\xi^j, \quad P_i{}^j u^i = -U u^j, \end{aligned} \tag{B.4}$$

in which the eigenvalue U is defined by the relation (B.3).

The solution of Eqs. (B.4) for the eigenvectors can be written down in the parametrical form. In order to write down such solution, we introduce a four-component spinor field ψ (playing the role of the parametrization), so that the components of the electromagnetic field tensor F_{js} are expressed in terms of ψ by the relationship of the form

$$F_{js} = \frac{\mathrm{i}}{2}\psi^+ \left(\gamma_j\gamma_s - \gamma_s\gamma_j\right)\psi. \tag{B.5}$$

Then Eqs. (B.4) have the following solution for π^i, ξ^i, σ^i, u^i:

$$\begin{aligned} \rho\pi^i &= \mathrm{Im}(\psi^T E\gamma^i\psi), \\ \rho\xi^i &= \mathrm{Re}(\psi^T E\gamma^i\psi), \\ \rho\sigma^i &= \psi^+\gamma^i\gamma^5\psi, \\ \rho u^i &= \mathrm{i}\psi^+\gamma^i\psi, \\ \rho\exp \mathrm{i}\eta &= \psi^+\psi + \mathrm{i}\psi^+\gamma^5\psi. \end{aligned} \tag{B.6}$$

while the eigenvalue U is related to the invariant ρ of the spinor field ψ:

$$U = \frac{1}{8\pi}\rho^2.$$

The vectors with components π^i, ξ^i, σ^i, u^i represent the proper basis of the spinor ψ.

Indeed, if relations (B.5), (B.6) are carried out, then it follows from identity (m) in (3.60) that the components of the energy-momentum tensor (B.1) can be written in the form[8]

$$P_i{}^j = U\left[\delta_i^j + 2\left(u_i u^j - \sigma_i \sigma^j\right)\right]. \tag{B.7}$$

In the proper basis we have

$$\breve{P}_a{}^b = \begin{Vmatrix} U & 0 & 0 & 0 \\ 0 & U & 0 & 0 \\ 0 & 0 & -U & 0 \\ 0 & 0 & 0 & -U \end{Vmatrix}.$$

The contraction of equality (B.7) with components of the vectors $\pi^i, \xi^i, \sigma^i, u^i$ gives Eq. (B.4).

Consider a two-parametrical spinor field

$$\psi' = \exp\frac{\mathrm{i}}{2}\left(\alpha\gamma^5 + \varphi I\right)\psi \equiv e^{\frac{\mathrm{i}}{2}\varphi}\left(I\cosh\frac{\alpha}{2} + \mathrm{i}\gamma^5\sinh\frac{\alpha}{2}\right)\psi, \tag{B.8}$$

where I is the unit four-dimensional matrix; α, φ are arbitrary real parameters. It is easy to show that the components of the electromagnetic field tensor F_{js}, determined by the equality (B.5), do not change under the transformation (B.5) $\psi \rightarrow \psi'$, while the components of the proper vectors (B.6) are transformed as follows[9]

$$\begin{aligned} \pi_i' &= \pi_i\cos\varphi + \xi_i\sin\varphi, & \sigma_i' &= \sigma_i\cosh\alpha - u_i\sinh\alpha, \\ \xi_i' &= -\pi_i\sin\varphi + \xi_i\cos\varphi, & u_i' &= -\sigma_i\sinh\alpha + u_i\cosh\alpha. \end{aligned} \tag{B.9}$$

Equations (B.9) determine the orthogonal transformation in the plane of the vectors $\pi^i\boldsymbol{\vartheta}_i$, $\xi^i\boldsymbol{\vartheta}_i$ through an angle φ and hyperbolic rotation in the plane of the vectors $\sigma^i\boldsymbol{\vartheta}_i$, $u^i\boldsymbol{\vartheta}_i$ through an angle α. The eigenvalue U that is a the four-dimensional invariant evidently does not change under transformations (B.8), (B.9).

The orthonormal basis $\breve{\mathbf{e}}_a = \{\pi^i\boldsymbol{\vartheta}_i, \xi^i\boldsymbol{\vartheta}_i, \sigma^i\boldsymbol{\vartheta}_i, u^i\boldsymbol{\vartheta}_i\}$, defined by Eqs. (B.4), we shall call the proper basis of the electromagnetic field. The proper basis $\breve{\mathbf{e}}_a$ introduced here is defined by the electromagnetic field up to transformation (B.9).

---

[8]In Chap. 3 the components of tensor (B.5) are denoted by the symbol M_{js}. The left part of the equation (m) in (3.60) represents the quantity $4\pi P^{is}$.

[9]Transformation (B.8) is the particular case of the gauge transformation (3.163).

The tensor components F_{js} are expressed in an arbitrary coordinate system of the observer in terms of the vector components of the proper basis (see Eq. (3.139))

$$F_{js} = \Omega\left(\pi_j \xi_s - \pi_s \xi_j\right) + N\left(\sigma_j u_s - \sigma_s u_j\right), \tag{B.10}$$

where $\Omega = \psi^+ \psi$, $N = \psi^+ \gamma^5 \psi$. The invariants Ω and N are connected with the invariants J_1, J_2 by the relations

$$\Omega^2 = \frac{1}{2}\left(J_1 + \sqrt{J_1^2 + J_2^2}\,\right), \quad N^2 = \frac{1}{2}\left(-J_1 + \sqrt{J_1^2 + J_2^2}\,\right), \tag{B.11}$$

or

$$J_1 = \Omega^2 - N^2, \quad J_2 = 2\Omega N. \tag{B.12}$$

It is seen from (3.140) that the components of the electromagnetic field tensor $\breve{F}^{ab}$ in the proper basis (B.6) are defined by the matrix

$$\breve{F}^{ab} = \begin{Vmatrix} 0 & \Omega & 0 & 0 \\ -\Omega & 0 & 0 & 0 \\ 0 & 0 & 0 & N \\ 0 & 0 & -N & 0 \end{Vmatrix}. \tag{B.13}$$

Directly from definition (B.10) and due to the orthonormality conditions of the components π^i, ξ^i, σ^i, u^i it follows that the vectors with components $u^i + \sigma^i$, $u^i - \sigma^i$, $\xi^i + \mathrm{i}\pi^i$, $\xi^i - \mathrm{i}\pi^i$ are the eigenvectors of the electromagnetic field tensor with eigenvalues, respectively, $-N$, N, $-\mathrm{i}\Omega$, $\mathrm{i}\Omega$:

$$\begin{aligned} F_{js}\left(u^s + \sigma^s\right) &= -N\left(u_j + \sigma_j\right), \\ F_{js}\left(u^s - \sigma^s\right) &= N\left(u_j - \sigma_j\right), \\ F_{js}\left(\xi^s + \mathrm{i}\pi^s\right) &= -\mathrm{i}\Omega\left(\xi_j + \mathrm{i}\pi_j\right), \\ F_{js}\left(\xi^s - \mathrm{i}\pi^s\right) &= \mathrm{i}\Omega\left(\xi_j - \mathrm{i}\pi_j\right). \end{aligned} \tag{B.14}$$

The eigenvalues Ω and N are expressed in terms of the electromagnetic field tensor by equalities (B.11), (B.2).

B.2 The Invariant Definition of the Electromagnetic Energy

If $P_i^{\ j}$ are the energy-momentum tensor components of certain physical system, then by definition the quantity $\varepsilon = -P_4^{\ 4}$ is the volume density of energy of the physical system under consideration. It is obvious that the quantity $P_4^{\ 4}$ depends essentially on the choice of the coordinate system in which the components $P_i{}^j$ are calculated. In accordance with definition (B.1) we have for energy ε of the electromagnetic field

$$\varepsilon = -P_4{}^4 = \frac{1}{8\pi}\left(E_\alpha E^\alpha + B_\alpha B^\alpha\right), \tag{B.15}$$

where E_α are the components of the three-dimensional electric strength vector, B_α are the components of the three-dimensional magnetic induction vector, calculated in the observer's coordinate system. Components E_α and B^α define the matrix of the components of the electromagnetic field tensor

$$F_{ij} = \begin{Vmatrix} 0 & B^3 & -B^2 & E_1 \\ -B^3 & 0 & B^1 & E_2 \\ B^2 & -B^1 & 0 & E_3 \\ -E_1 & -E_2 & -E_3 & 0 \end{Vmatrix}.$$

Together with the energy ε the concept of internal energy U that is a four-dimensional scalar and is independent of the coordinate system choice is also introduced in thermodynamics. Introduction of the quantity U in known theories is related with the use of special proper bases. In order to introduce the internal energy of a physical system it turns out to be sufficient to determine the field of a certain unit timelike vector $\boldsymbol{u} = u^i \boldsymbol{\vartheta}_i$ which is an eigenvector of the energy-momentum tensor. Then the internal energy is determined in terms of the components of the energy-momentum tensor by the relation of the form

$$U = u_j u^i P_i{}^j. \tag{B.16}$$

Usually in mechanics of continuous media the velocity vector of the individual points of a medium is taken as the vector $\boldsymbol{u}$.

From definitions (B.16), (B.7) it follows that the quantity U, calculated according to (B.3), determines the volume density of the electromagnetic energy, which can be considered as an analog of the internal energy in mechanics of continuous medium. Bearing in mind definitions (B.12), (B.13) we find that in the proper basis expression (B.3) for U hase the form [87]

$$U = \frac{1}{16\pi}\sqrt{\left(F_{ij}F^{ij}\right)^2 + \left(F_{ij}\overset{*}{F}{}^{ij}\right)^2} = \frac{1}{8\pi}\left(\breve{E}^2 + \breve{B}^2\right),$$

where $\breve{E} = (0, 0, N)$, $\breve{B} = (0, 0, \Omega)$ are the three-dimensional magnetic field induction and electric field vectors determined in the proper basis of the electromagnetic field. Thus, the internal energy U of the electromagnetic field is defined here as the field energy (B.15) calculated in the proper basis.

If both invariants J_1 and J_2 of the electromagnetic field tensor are equal to zero $J_1 = J_2 = 0$, then there exists the isotropic vector field $\boldsymbol{u}$ such that $P_i{}^j = u_i u^j$,

$u_i u^i = 0$. In this case $U = 0$ and one can assume that the internal energy of the electromagnetic field is equal to zero.

B.3 The Maxwell Equations in the Proper Basis

The Maxwell equations in the proper basis of the electromagnetic field $\breve{\boldsymbol{e}}_a$ are written as follows

$$\breve{\partial}_b \breve{F}^{ab} + \breve{\Delta}_{b,e}{}^a \breve{F}^{eb} + \breve{\Delta}_{b,e}{}^b \breve{F}^{ae} = 0,$$
$$\breve{\partial}_b \breve{F}^{*ab} + \breve{\Delta}_{b,e}{}^a \breve{F}^{*eb} + \breve{\Delta}_{b,e}{}^b \breve{F}^{*ae} = 0, \tag{B.17}$$

where the components of the electromagnetic field tensor $\breve{F}^{ab}$ are calculated in the basis $\breve{\boldsymbol{e}}_a$. $\breve{F}^{*ab} = \frac{1}{2}\varepsilon^{abcd}\breve{F}_{ab}$. $\breve{\Delta}_{a,bc}$ are the Ricci rotation coefficients corresponding to the bases $\breve{\boldsymbol{e}}_a$; $\breve{\partial}_b$ is the symbol of the partial derivative in the direction of the vectors of the basis $\breve{\boldsymbol{e}}_a$.

Taking into account definition (B.13) for the component F^{ab} in the basis $\breve{\boldsymbol{e}}_a$, the Maxwell equations (B.17) can be transformed to the form

$$\begin{aligned}
\breve{\partial}_1 \eta &= \breve{\Delta}_{3,42} - \breve{\Delta}_{4,32}, & \breve{\partial}_1 \ln U &= \breve{\Delta}_{4,14} + \breve{\Delta}_{3,31},\\
\breve{\partial}_2 \eta &= \breve{\Delta}_{4,31} - \breve{\Delta}_{3,41}, & \breve{\partial}_2 \ln U &= \breve{\Delta}_{4,24} + \breve{\Delta}_{3,32},\\
\breve{\partial}_3 \eta &= \breve{\Delta}_{1,24} - \breve{\Delta}_{2,14}, & \breve{\partial}_3 \ln U &= \breve{\Delta}_{1,13} + \breve{\Delta}_{2,23},\\
\breve{\partial}_4 \eta &= \breve{\Delta}_{1,23} - \breve{\Delta}_{2,13}, & \breve{\partial}_4 \ln U &= \breve{\Delta}_{1,14} + \breve{\Delta}_{2,24}.
\end{aligned} \tag{B.18}$$

The invariant η is connected with the invariants of the electromagnetic field Ω, N by the relation

$$\Omega + \mathrm{i}N = \sqrt{\Omega^2 + N^2}\, \exp \mathrm{i}\eta.$$

Equation (B.18) can be written also directly in the components of the vectors of the proper tetrad π^i, ξ^i, σ^i, u^i. For this it is sufficient to substitute the Ricci rotation coefficients $\breve{\Delta}_{a,bc}$ in Eqs. (B.18) by formulas (3.150). As a result of identical transformations we obtain the system of equations

$$\begin{aligned}
\breve{\partial}_1 \eta &= -u^i \breve{\partial}_3 \xi_i + \sigma^i \breve{\partial}_4 \xi_i, & \partial_j\left(U\pi^j\right) + U\pi^j \breve{\partial}_2 \xi_j &= 0,\\
\breve{\partial}_2 \eta &= u^i \breve{\partial}_3 \pi_i - \sigma^i \breve{\partial}_4 \pi_i, & \partial_j\left(U\xi^j\right) + U\xi^j \breve{\partial}_1 \pi_j &= 0,\\
\breve{\partial}_3 \eta &= \pi^i \breve{\partial}_2 u_i - \xi^i \breve{\partial}_1 u_i, & \partial_j\left(U\sigma^j\right) - U\sigma^j \breve{\partial}_4 u_j &= 0,\\
\breve{\partial}_4 \eta &= \pi^i \breve{\partial}_2 \sigma_i - \xi^i \breve{\partial}_1 \sigma_i, & \partial_j\left(U u^j\right) + U u^j \breve{\partial}_3 \sigma_j &= 0.
\end{aligned} \tag{B.19}$$

The last equation in system (B.19) is identical with the energy equation $u^i \partial_j P_i{}^j = 0$.

B.4 The Maxwell Equations in the Proper Null Basis

Let us introduce the null complex basis

$$\breve{\boldsymbol{e}}_1^\circ = l^i \boldsymbol{\ni}_i, \quad \breve{\boldsymbol{e}}_2^\circ = n^i \boldsymbol{\ni}_i, \quad \breve{\boldsymbol{e}}_3^\circ = m^i \boldsymbol{\ni}_i, \quad \breve{\boldsymbol{e}}_4^\circ = \dot{m}^i \boldsymbol{\ni}_i,$$

with components l^i, n^i, m^i, $\dot{m}^i$ expressed in terms of the vector components of the proper basis of the electromagnetic field

$$\sqrt{2}\, l^i = u^i + \sigma^i, \quad \sqrt{2}\, m^i = \pi^i - \mathrm{i}\xi^i,$$
$$\sqrt{2}\, n^i = u^i - \sigma^i, \quad \sqrt{2}\, \dot{m}^i = \pi^i + \mathrm{i}\xi^i. \tag{B.20}$$

From Eqs. (B.14) it follows that the vectors $\breve{\boldsymbol{e}}_a^\circ$ with components (B.20) are the eigenvalue vectors of the electromagnetic field tensor with components F_{js}. Using equalities (B.9) it is not difficult to show that the proper null basis $\breve{\boldsymbol{e}}_a^\circ$ of the electromagnetic field is defined up to the two-parameter transformation

$$l_i' = l_i \exp(-\alpha), \quad n_i = n_i \exp \alpha, \quad m_i' = m_i \exp \mathrm{i}\varphi.$$

The Maxwell equations (B.18) in notations (3.152) of the spin-coefficients take the form

$$D \ln U = -\varrho - \dot{\varrho}, \quad D\eta = \mathrm{i}(-\varrho + \dot{\varrho}),$$
$$\Delta \ln U = \mu + \dot{\mu}, \quad \Delta\eta = \mathrm{i}(\mu - \dot{\mu}),$$
$$\delta \ln U = -\tau + \dot{\pi}, \quad \delta\eta = \mathrm{i}(-\tau - \dot{\pi}). \tag{B.21}$$

Equations (B.21) can also be written as follows

$$\varrho = -DG, \quad \mu = \Delta G, \quad \tau = -\delta G, \quad \pi = \dot{\delta} G,$$

where for the complex scalar G we have $G = \frac{1}{2}(\ln U - \mathrm{i}\eta)$.

Appendix C
The Bilinear Identities Connecting the Dirac Matrices

Let us give here the complete set of the invariant algebraic identities, connecting the bilinear products of the matrices $\boldsymbol{\gamma}$, which are often used in various calculations. These identities are obtained by contracting the Pauli identity (3.21) with the components of spintensors γ^i, γ^{ij}, $\overset{*}{\gamma}{}^i$, γ^5 and transformations with the aid of identity (3.11).

$$
\begin{aligned}
4e_{DE}e_{BA} &= e_{DA}e_{BE} + \gamma_{iDA}\gamma^i_{BE} - \frac{1}{2}\gamma_{ijDA}\gamma^{ij}_{BE} + \overset{*}{\gamma}_{iDA}\overset{*}{\gamma}{}^i_{BE} - \gamma^5_{DA}\gamma^5_{BE},\\
4e_{DE}\gamma^i_{BA} &= \gamma^i_{DA}e_{BE} + e_{DA}\gamma^i_{BE} + \gamma^{si}_{DA}\gamma_{sBE} - \gamma_{sDA}\gamma^{si}_{BE} \qquad \text{(C.1)}\\
&\quad + \frac{1}{2}\varepsilon^{ijks}\left(\gamma_{ksDA}\overset{*}{\gamma}_{jBE} + \overset{*}{\gamma}_{jDA}\gamma_{ksBE}\right) - \gamma^5_{DA}\overset{*}{\gamma}{}^i_{BE} + \overset{*}{\gamma}{}^i_{DA}\gamma^5_{BE},\\
4e_{DE}\gamma^{ij}_{BA} &= \delta^{ij}_{ks}\Big(\frac{1}{2}\gamma^{ks}_{DA}e_{BE} + \frac{1}{2}e_{DA}\gamma^{ks}_{BE} - \gamma^k_{DA}\gamma^s_{BE} - \overset{*}{\gamma}{}^k_{DA}\overset{*}{\gamma}{}^s_{BE}\\
&\quad + \gamma^{kn}_{DA}\gamma^s{}_{nBE}\Big) - \varepsilon^{ijks}\Big(\gamma_{kDA}\overset{*}{\gamma}_{sBE} + \overset{*}{\gamma}_{sDA}\gamma_{kBE} + \frac{1}{2}\gamma_{ksDA}\gamma^5_{BE} +\\
&\quad + \frac{1}{2}\gamma^5_{DA}\gamma_{ksBE}\Big),\\
4e_{DE}\overset{*}{\gamma}{}^i_{BA} &= \overset{*}{\gamma}{}^i_{DA}e_{BE} + e_{DA}\overset{*}{\gamma}{}^i_{BE} + \gamma^5_{DA}\gamma^i_{BE} - \gamma^i_{DA}\gamma^5_{BE} + \overset{*}{\gamma}_{jDA}\gamma^{ij}_{BE}\\
&\quad - \gamma^{ij}_{DA}\overset{*}{\gamma}_{jBE} - \frac{1}{2}\varepsilon^{ijks}\left(\gamma_{sDA}\gamma_{jkBE} + \gamma_{jkDA}\gamma_{sBE}\right),\\
4e_{DE}\gamma^5_{BA} &= \gamma^5_{DA}e_{BE} + e_{DA}\gamma^5_{BE} + \overset{*}{\gamma}_{iDA}\gamma^i_{BE} - \gamma^i_{DA}\overset{*}{\gamma}_{iBE}\\
&\quad - \frac{1}{4}\varepsilon_{ijks}\gamma^{ij}_{DA}\gamma^{ks}_{BE},\\
4\gamma^i_{DE}\gamma^5_{BA} &= \overset{*}{\gamma}{}^i_{DA}e_{BE} - e_{DA}\overset{*}{\gamma}{}^i_{BE} + \gamma^5_{DA}\gamma^i_{BE} + \gamma^i_{DA}\gamma^5_{BE} - \overset{*}{\gamma}_{jDA}\gamma^{ij}_{BE}
\end{aligned}
$$

V. A. Zhelnorovich, *Theory of Spinors and Its Application in Physics and Mechanics*, https://doi.org/10.1007/978-3-030-27836-6

$$- \gamma^{ij}_{DA}\overset{*}{\gamma}_{jBE} + \frac{1}{2}\varepsilon^{ijks}(-\gamma_{sDA}\gamma_{jkBE} + \gamma_{jkDA}\gamma_{sBE}),$$

$$4\gamma^{ij}_{DE}\gamma^{5}_{BA} = \delta^{ij}_{ks}\left(\frac{1}{2}\gamma^{ks}_{DA}\gamma^{5}_{BE} + \frac{1}{2}\gamma^{5}_{DA}\gamma^{ks}_{BE} - \gamma^{k}_{DA}\overset{*}{\gamma}{}^{s}_{BE} - \overset{*}{\gamma}{}^{s}_{DA}\gamma^{k}_{BE}\right)$$

$$+ \varepsilon^{ijks}\left(-\gamma^{n}{}_{kDA}\gamma_{nsBE} + \frac{1}{2}\gamma_{ksDA}e_{BE} + \frac{1}{2}e_{DA}\gamma_{ksBE}\right.$$

$$\left. - \gamma_{kDA}\gamma_{sBE} - \overset{*}{\gamma}_{kDA}\overset{*}{\gamma}_{sBE}\right),$$

$$4\overset{*}{\gamma}{}^{i}_{DE}\gamma^{5}_{BA} = -\gamma^{i}_{DA}e_{BE} + e_{DA}\gamma^{i}_{BE} + \gamma^{5}_{DA}\overset{*}{\gamma}{}^{i}_{BE} + \overset{*}{\gamma}{}^{i}_{DA}\gamma^{5}_{BE} + \gamma_{jDA}\gamma^{ij}_{BE}$$

$$+ \gamma^{ij}_{DA}\gamma_{jBE} + \frac{1}{2}\varepsilon^{ijks}(\gamma_{ksDA}\overset{*}{\gamma}_{jBE} - \overset{*}{\gamma}_{jDA}\gamma_{ksBE}),$$

$$4\gamma^{5}_{DE}\gamma^{5}_{BA} = -e_{DA}e_{BE} + \gamma_{iDA}\gamma^{i}_{BE} + \frac{1}{2}\gamma_{ijDA}\gamma^{ij}_{BE} + \overset{*}{\gamma}_{iDA}\overset{*}{\gamma}{}^{i}_{BE} + \gamma^{5}_{DA}\gamma^{5}_{BE},$$

$$4\gamma^{i}_{DE}\gamma^{j}_{BA} = \gamma^{ij}_{DA}e_{BE} - e_{DA}\gamma^{ij}_{BE} + \gamma^{i}_{DA}\gamma^{j}_{BE} + \gamma^{j}_{DA}\gamma^{i}_{BE} - \overset{*}{\gamma}{}^{i}_{DA}\overset{*}{\gamma}{}^{j}_{BE}$$

$$- \overset{*}{\gamma}{}^{j}_{DA}\overset{*}{\gamma}{}^{i}_{BE} + \gamma^{ik}_{DA}\gamma^{j}{}_{kBE} + \gamma^{jk}_{DA}\gamma^{i}{}_{kBE} + g^{ij}\left(e_{DA}e_{BE} - \gamma_{sDA}\gamma^{s}_{BE}\right.$$

$$\left. - \frac{1}{2}\gamma_{ksDA}\gamma^{ks}_{BE} + \overset{*}{\gamma}_{sDA}\overset{*}{\gamma}{}^{s}_{BE} + \gamma^{5}_{DA}\gamma^{5}_{BE}\right) + \varepsilon^{ijks}\left(-\gamma_{kDA}\overset{*}{\gamma}_{sBE}\right.$$

$$\left. + \overset{*}{\gamma}_{sDA}\gamma_{kBE} + \frac{1}{2}\gamma_{ksDA}\gamma^{5}_{BE} - \frac{1}{2}\gamma^{5}_{DA}\gamma_{ksBE}\right),$$

$$4\gamma^{m}_{DE}\gamma^{ij}_{BA} = \delta^{ij}_{ks}\left[\frac{1}{2}\gamma^{ks}_{DA}\gamma^{m}_{BE} + \frac{1}{2}\gamma^{m}_{DA}\gamma^{ks}_{BE} - \gamma^{sm}_{DA}\gamma^{k}_{BE} - \gamma^{k}_{DA}\gamma^{sm}_{BE}\right.$$

$$+ g^{sm}(-\gamma^{k}_{DA}e_{BE} + e_{DA}\gamma^{k}_{BE} + \overset{*}{\gamma}{}^{k}_{DA}\gamma^{5}_{BE} + \gamma^{5}_{DA}\overset{*}{\gamma}{}^{k}_{BE} + \gamma^{nk}_{DA}\gamma_{nBE}$$

$$\left. + \gamma_{nDA}\gamma^{nk}_{BE}) + \frac{1}{2}\varepsilon^{smpq}(\overset{*}{\gamma}{}^{k}_{DA}\gamma_{pqBE} - \gamma_{pqDA}\overset{*}{\gamma}{}^{k}_{BE})\right]$$

$$+ \varepsilon^{ijks}\left[\delta^{m}_{k}(-\overset{*}{\gamma}_{sDA}e_{BE} - e_{DA}\overset{*}{\gamma}_{sBE} - \gamma_{sDA}\gamma^{5}_{BE} + \gamma^{5}_{DA}\gamma_{sBE})\right.$$

$$\left. + \overset{*}{\gamma}_{sDA}\gamma^{m}{}_{kBE} - \gamma^{m}{}_{kDA}\overset{*}{\gamma}_{sBE}\right],$$

$$4\gamma^{i}_{DE}\overset{*}{\gamma}{}^{j}_{BA} = \gamma^{i}_{DA}\overset{*}{\gamma}{}^{j}_{BE} + \gamma^{j}_{DA}\overset{*}{\gamma}{}^{i}_{BE} + \overset{*}{\gamma}{}^{j}_{DA}\gamma^{i}_{BE} + \overset{*}{\gamma}{}^{i}_{DA}\gamma^{j}_{BE} - \gamma^{ij}_{DA}\gamma^{5}_{BE}$$

$$- \gamma^{5}_{DA}\gamma^{ij}_{BE} + g^{ij}\left(\gamma^{5}_{DA}e_{BE} - e_{DA}\gamma^{5}_{BE} - \overset{*}{\gamma}_{sDA}\gamma^{s}_{BE} - \gamma^{s}_{DA}\overset{*}{\gamma}_{sBE}\right.$$

$$\left. + \frac{1}{4}\varepsilon_{ijks}\gamma^{ij}_{DA}\gamma^{ks}_{BE}\right) + \varepsilon^{ijks}\left(\frac{1}{2}\gamma_{ksDA}e_{BE} + e_{DA}\gamma_{ksBE} + \gamma_{kDA}\gamma_{sBE}\right.$$

$$- \overset{*}{\gamma}_{kDA}\overset{*}{\gamma}_{sBE}\Big) - \frac{1}{2}\big(\varepsilon^{imks}\gamma^{j}{}_{mDA} + \varepsilon^{jmks}\gamma^{i}{}_{mDA}\big)\gamma_{ksBE},$$

$$4\overset{*}{\gamma}{}^{i}_{DE}\overset{*}{\gamma}{}^{j}_{BA} = \gamma^{ij}_{DA}e_{BE} - e_{DA}\gamma^{ij}_{BE} - \gamma^{i}_{DA}\gamma^{j}_{BE} - \gamma^{j}_{DA}\gamma^{i}_{BE} + \overset{*}{\gamma}{}^{i}_{DA}\overset{*}{\gamma}{}^{j}_{BE}$$

$$+ \overset{*}{\gamma}{}^{j}_{DA}\overset{*}{\gamma}{}^{i}_{BE} + \gamma^{ik}_{DA}\gamma^{j}{}_{kBE} + \gamma^{jk}_{DA}\gamma^{i}{}_{kBE} + g^{ij}\Big(e_{DA}e_{BE} + \gamma_{sDA}\gamma^{s}_{BE}$$

$$- \frac{1}{2}\gamma_{ksDA}\gamma^{ks}_{BE} - \overset{*}{\gamma}_{sDA}\overset{*}{\gamma}{}^{s}_{BE} + \overset{5}{\gamma}_{DA}\overset{5}{\gamma}_{BE}\Big) + \varepsilon^{ijks}\Big(\gamma_{kDA}\overset{*}{\gamma}_{sBE}$$

$$- \overset{*}{\gamma}_{sDA}\gamma_{kBE} + \frac{1}{2}\gamma_{ksDA}\overset{5}{\gamma}_{BE} - \frac{1}{2}\overset{5}{\gamma}_{DA}\gamma_{ksBE}\Big),$$

$$4\overset{*}{\gamma}{}^{m}_{DE}\gamma^{ij}_{BA} = \delta^{ij}_{ks}\Big[\frac{1}{2}\gamma^{ks}_{DA}\overset{*}{\gamma}{}^{m}_{BE} + \frac{1}{2}\overset{*}{\gamma}{}^{m}_{DA}\gamma^{ks}_{BE} - \gamma^{sm}_{DA}\overset{*}{\gamma}{}^{k}_{BE} - \overset{*}{\gamma}{}^{k}_{DA}\gamma^{sm}_{BE}$$

$$+ g^{sm}\big(- \overset{*}{\gamma}{}^{k}_{DA}e_{BE} + e_{DA}\overset{*}{\gamma}{}^{k}_{BE} - \gamma^{k}_{DA}\overset{5}{\gamma}_{BE} - \overset{5}{\gamma}_{DA}\gamma^{k}_{BE} + \gamma^{nk}_{DA}\overset{*}{\gamma}_{nBE}$$

$$+ \overset{*}{\gamma}_{nDA}\gamma^{nk}_{BE}\big) + \frac{1}{2}\varepsilon^{smpq}\big(\gamma_{pqDA}\gamma^{k}_{BE} - \gamma^{k}_{DA}\gamma_{pqBE}\big)\Big]$$

$$+ \varepsilon^{ijks}\big[\delta^{m}_{k}\big(\gamma_{sDA}e_{BE} + e_{DA}\gamma_{sBE} - \overset{*}{\gamma}_{sDA}\overset{5}{\gamma}_{BE} + \overset{5}{\gamma}_{DA}\overset{*}{\gamma}_{sBE}\big)$$

$$- \gamma_{sDA}\gamma^{m}{}_{kBE} + \gamma^{m}{}_{kDA}\gamma_{sBE}\big],$$

$$4\gamma^{ij}_{DE}\gamma^{ks}_{BA} = (g^{ik}g^{js} - g^{is}g^{jk})\big(- e_{DA}e_{BE} + \overset{5}{\gamma}_{DA}\overset{5}{\gamma}_{BE}\big)$$

$$+ \varepsilon^{ijks}\big(e_{DA}\overset{5}{\gamma}_{BE} + \overset{5}{\gamma}_{DA}e_{BE}\big) + \delta^{ij}_{mn}\delta^{ks}_{pq}\Big[\frac{1}{4}\gamma^{pq}_{DA}\gamma^{mn}_{BE} - \gamma^{mp}_{DA}\gamma^{qn}_{BE}$$

$$+ g^{mp}\big(\gamma^{qn}_{DA}e_{BE} - e_{DA}\gamma^{qn}_{BE} + \gamma^{q}_{DA}\gamma^{n}_{BE} + \overset{*}{\gamma}{}^{q}_{DA}\overset{*}{\gamma}{}^{n}_{BE} + \gamma^{q}{}_{sDA}\gamma^{sn}_{BE}\big)\Big]$$

$$+ \delta^{ij}_{pn}\varepsilon^{kspq}\big(- \overset{*}{\gamma}_{qDA}\gamma^{n}_{BE} + \gamma_{qDA}\overset{*}{\gamma}{}^{n}_{BE} - \overset{5}{\gamma}_{DA}\eta^{n}{}_{qBE}\big)$$

$$+ \delta^{ks}_{pq}\varepsilon^{ijqn}\big(\gamma^{p}_{DA}\overset{*}{\gamma}_{nBE} - \overset{*}{\gamma}{}^{p}_{DA}\gamma_{nBE} + \gamma^{p}{}_{nDA}\overset{5}{\gamma}_{BE}\big)$$

$$+ \varepsilon^{kspq}\varepsilon^{ijmn}\Big[- \frac{1}{4}\gamma_{pqDA}\gamma_{mnBE} + g_{mp}\big(\overset{*}{\gamma}_{qDA}\overset{*}{\gamma}_{nBE} + \gamma_{qDA}\gamma_{nBE}\big)\Big].$$

Note also following useful identities which are the consequence of identities (C.1).

The Scalar and Pseudo-Scalar Identities

$$a. \qquad \gamma_{iDE}\gamma^{i}_{BA} - \overset{*}{\gamma}_{iDE}\overset{*}{\gamma}{}^{i}_{BA} = -\big(\gamma_{iDA}\gamma^{i}_{BE} - \overset{*}{\gamma}_{iDA}\overset{*}{\gamma}{}^{i}_{BE}\big),$$

$$b. \qquad \gamma_{iDE}\overset{*}{\gamma}{}^{i}_{BA} + \overset{*}{\gamma}_{iDE}\gamma^{i}_{BA} = -\big(\gamma_{iDA}\overset{*}{\gamma}{}^{i}_{BE} + \overset{*}{\gamma}_{iDA}\gamma^{i}_{BE}\big),$$

$$c.\qquad \gamma_{iDE}\gamma^{i}_{BA} - e_{DE}e_{BA} - \gamma^{5}_{DE}\gamma^{5}_{BA} = = -\left(\gamma_{iDA}\gamma^{i}_{BE} - e_{DA}e_{BE} - \gamma^{5}_{DA}\gamma^{5}_{BE}\right),$$

$$d.\qquad \overset{*}{\gamma}_{iDE}\overset{*i}{\gamma}_{BA} - e_{DE}e_{BA} - \gamma^{5}_{DE}\gamma^{5}_{BA} = = -\left(\overset{*}{\gamma}_{iDA}\overset{*i}{\gamma}_{BE} - e_{DA}e_{BE} - \gamma^{5}_{DA}\gamma^{5}_{BE}\right),$$

$$e.\qquad \gamma_{iDE}\overset{*i}{\gamma}_{BA} + e_{DE}\gamma^{5}_{BA} - \gamma^{5}_{DE}e_{BA}) = -\left(\gamma_{iDA}\overset{*i}{\gamma}_{BE} + e_{DA}\gamma^{5}_{BE} - \gamma^{5}_{DA}e_{BE}\right),$$

$$f.\qquad \overset{*}{\gamma}_{iDE}\gamma^{i}_{BA} - e_{DE}\gamma^{5}_{BA} + \gamma^{5}_{DE}e_{BA}) = -\left(\overset{*}{\gamma}_{iDA}\gamma^{i}_{BE} - e_{DA}\gamma^{5}_{BE} + \gamma^{5}_{DA}e_{BE}\right),$$

$$g.\qquad \frac{1}{2}\gamma_{ijDE}\gamma^{ij}_{BA} + e_{DE}e_{BA} - \gamma^{5}_{DE}\gamma^{5}_{BA} = -\left(\frac{1}{2}\gamma_{ijDA}\gamma^{ij}_{BE} + e_{DA}e_{BE} - \gamma^{5}_{DA}\gamma^{5}_{BE}\right),$$

$$h.\qquad \frac{1}{4}\varepsilon_{ijks}\gamma^{ij}_{DE}\gamma^{ks}_{BA} + e_{DE}\gamma^{5}_{BA} + \gamma^{5}_{DE}e_{BA} = -\left(\frac{1}{4}\varepsilon_{ijks}\gamma^{ij}_{DA}\gamma^{ks}_{BE} + e_{DA}\gamma^{5}_{BE} + \gamma^{5}_{DA}e_{BE}\right). \tag{C.2}$$

The Vector and Pseudo-Vector Identities

$$\overset{*}{\gamma}_{jDE}\gamma^{ij}_{BA} + \gamma^{ij}_{DE}\overset{*}{\gamma}_{jBA} + \gamma^{5}_{DE}\gamma^{i}_{BA} + \gamma^{i}_{DE}\gamma^{5}_{BA} = -\left(\overset{*}{\gamma}_{jDA}\gamma^{ij}_{BE} + \gamma^{ij}_{DA}\overset{*}{\gamma}_{jBE} + \gamma^{5}_{DA}\gamma^{i}_{BE} + \gamma^{i}_{DA}\gamma^{5}_{BE}\right), \tag{C.3}$$

$$e_{DE}\overset{*}{\gamma}_{iBA} + \overset{*}{\gamma}_{iDE}e_{BA} + \frac{1}{2}\varepsilon_{ijks}\left(\gamma^{s}_{DE}\gamma^{jk}_{BA} + \gamma^{jk}_{DE}\gamma^{s}_{BA}\right) = -\left[e_{DA}\overset{*}{\gamma}_{iBE} + \overset{*}{\gamma}_{iDA}e_{BE} + \frac{1}{2}\varepsilon_{ijks}\left(\gamma^{s}_{DA}\gamma^{jk}_{BE} + \gamma^{jk}_{DA}\gamma^{s}_{BE}\right)\right],$$

$$\gamma_{jDE}\gamma^{ij}_{BA} + \gamma^{ij}_{DE}\gamma_{jBA} - \gamma^{5}_{DE}\overset{*i}{\gamma}_{BA} - \overset{*i}{\gamma}_{DE}\gamma^{5}_{BA} = -\left(\gamma_{jDA}\gamma^{ij}_{BE} + \gamma^{ij}_{DA}\gamma_{jBE} - \gamma^{5}_{DA}\overset{*i}{\gamma}_{BE} - \overset{*i}{\gamma}_{DA}\gamma^{5}_{BE}\right),$$

$$\frac{1}{2}\varepsilon_{ijks}\left(\overset{*s}{\gamma}_{DE}\gamma^{jk}_{BA} + \gamma^{jk}_{DE}\overset{*s}{\gamma}_{BA}\right) - e_{DE}\gamma_{iBA} - \gamma_{iDE}e_{BA} = -\left[\frac{1}{2}\varepsilon_{ijks}\left(\overset{*s}{\gamma}_{DA}\gamma^{jk}_{BE} + \gamma^{jk}_{DA}\overset{*s}{\gamma}_{BE}\right) - e_{DA}\gamma_{iBE} - \gamma_{iDA}e_{BE}\right].$$

Bibliography

1. Abraham, P.: On spinors in n-dimensions. J. Math. Phys. **3**, 1135–1139 (1962)
2. Bade, W., Jehle, H.: An introduction to spinors. Rev. Mod. Phys. **25**, 714–728 (1953)
3. Bagrov, V.G., Istomin, A.D., Obukhov, V.V.: Cosmological solutions of Einstein–Dirac equations. Gravit. Cosmol. **2**, 117–121 (1996)
4. Belinfante, F.: On the covariant derivative of tensor-undors. Phys. **7**, 305–324 (1940)
5. Berdichevsky, V.L.: Variational Principles of Continuum Mechanics. Springer, Berlin/Heidelberg (2009)
6. Bergman, P.G.: Two-component spinors in general relativity. Phys. Rev. **107**, 624–629 (1957)
7. Bogoliubov, N.N., Shirkov. D.V.: Introduction to the Theory of Quantized Fields. Nauka, Moscow (1976)
8. Brauer, R., Weyl, H.: Spinors in n-dimensions. Amer. J. Math. **7**, 425–449 (1935)
9. Brill, D., Wheeler, J.: Interaction of neutrinos and gravitational fields. Rev. Mod. Phys. **29**, 465–479 (1957)
10. de Broglie, L.: L'electron Magnetique. Hermann, Paris (1934)
11. Cartan, E.: Les groupes projectifs qui ne laissent invariante aucune multiplicité plane. Bull. Soc. Math. de France. **41**, 53–96 (1913)
12. Cartan, E.: The Theory of Spinors. Dover Publications, New York (1981)
13. Case, K.M.: Biquadratic spinor identities. Phys. Rev. **97**, 810–823 (1955)
14. Chandrasekhar, S.: The solution of Dirac's equations in Kerr geometry. Proc. Roy. Soc. Lond. **A 349**, 571–575 (1976)
15. Crawford, J.P.: On the algebra of Dirac bispinor densities: factorization and inversion. J. Math. Phys. **26**, 1439–1441 (1985)
16. Dubrovin, B., Fomenko, A., Novikov S.: Modern geometry- methods and applications, vol. 1–3. Springer, New York (1985)–(1995)
17. Dürr, H., Heisenberg, W., Mitter, H., Schlieder, S., Yamazaki K.: Zur Theorie der Elementarteilchen. Zeitschrift für Naturforschung. **A14**, 441–485 (1959)
18. Ebner, D.: Classification of relativistic particles according to the representation theory of the eight nonisomorphic simply connected covering groups of the full Lorenz group. Gen. Relativ. Gravit. **8**, 15–28 (1977)
19. Ferrarese, G.: Sulla formulazione intrinseca della dinamica dei continui alla Cosserat. Annali di Mat. Pure ed Applicata. **108**, 109–124 (1976)
20. Fock, V.: Dirac wave equation and Riemann geometry. J. Phys. Radium. **10**, 392–405 (1929)
21. Fock, V.: Geometrisierung der Diracschen Theorie des Elektrons. Zs. für Physik. **57**, 261–277 (1929)
22. Fock, V.: The Theory of Space Time and Gravitation. Pergamon Press Ltd., Pergamon (1964)

V. A. Zhelnorovich, *Theory of Spinors and Its Application in Physics and Mechanics*, https://doi.org/10.1007/978-3-030-27836-6

23. Fock, V., Ivanenko, D.: Zur Quantengeometrie. Phys. Zeit. **30**, 648–651 (1929)
24. Fock V., Iwanenko D.: Géometrie quantique linéaire et déplacement paralléle. Compt. Rend. Acad Sci. Paris. **188**, 1470–1472 (1929)
25. Gel'fand, I.M., Minlos, R.A., Shapiro, Z.J.: Representation of the rotation group and of the Lorentz group, and their applicatins. Fizmatgiz, Moscow (1958); English transl.: Macmillan (1963)
26. Ginsburg, L.: The relativistic theory of excited spin states of the proton and the neutron. Phys. Rev. **63**, 1–12 (1943)
27. Ginzburg, L.: The laws of conservation of energy and momentum in emission of electromagnetic waves (photons) in a medium and the energy-momentum tensor in macroscopic electrodynamics. Sov. Phys. Usp. **16**, 434–439 (1973)
28. Ginzburg, L., Ugarov, A.: Remarks on forces and the energy-momentum tensor in macroscopic electrodynamics Sov.Phys. Usp. **19**, 94–101 (1976)
29. Golubiatnikov, A.N.: Tensor and spinor invariants of subgroups of the Lorentz group and construction of exact soltions of the Einstein–Dirac equations. Ph.D. thesis, Moscow state University (1970) (in Russian)
30. Golubyatnikov, A.N.: Model of the neutrino in the general theory of relativity. Sov. Phys. Dokl. **15**, 451–453 (1970)
31. Griffiths, J.B., Newing, R.A.: Tetrad equations for the two-component neutrino field in general relativity. J. Phys. A: Gen. Phys. **3**, 269–273 (1970)
32. Gürsey, F.: Relativistic kinematics of a classical point particle in spinor form. Nuovo Cimento. **5**, 784–809 (1957)
33. Halbwachs, F.: Theorie Relativiste des Fluides à Spin. Paris, Gauthier-Villars (1960)
34. Hara, O.: Space-time approach to the grand unified theory of the quarks and the leptons. Nuovo Chimento. **75A**, 17–38 (1983)
35. Kaempffer, F.A.: Spinor electrodynamics as a dynamics of currents. Phys. Rev. D. **23**, 918–921 (1981)
36. Kent, R.D., Szamosi, G.: Spinor equations of motion in curved space-time. Nuovo Cimento. **64 B**, 67–80 (1981)
37. Klauder, J.: Linear representation of spinor fields by antisimmetric tensors. J. Math. Phys. **5**, 1204–1214 (1964)
38. Kofink, W.: Zur Mathematik der Diracmatriesen. Math. Z. **51**, 702–711 (1949)
39. Kurdgelaidze, D.F.: On the nonlinear theory of elementary particles. J. Exp. Theor. Phys. **11**, 339–346 (1960)
40. Landau, L.D., Lifshitz, E.M.: Fluid Mechanics. Pergamon Press, London; Addison-Wesley Pub. Co., Reading, MA (1987)
41. Lichnerowicz, A.: Spineurs harmonicues. Compt. Rend. Paris. **257**, 7–28 (1963)
42. Luer, C.P., Rosenbaum, M.: Spinor connections in general relativity. J. Math. Phys. **15**, 1120–1137 (1974)
43. Maugin, G.A.: Sur les fluides relativistes à spin. Ann. Inst. H. Poincaré. **5**, 41–68 (1974)
44. Mickelsson, J.: On a relation between Maxwell and Dirac theories. Lett. Math. Phys. **6**, 221–230 (1982)
45. Mickelsson, J.: The vector form of the neutrino equation and the photon neutrino duality. J. Math. Phys. **26**, 2346–2349 (1985)
46. Newman, E., Penrose, R.: An approach to gravitational radiation by a method of spin coefficients. J. Math. Phys. **3**, 566–578 (1962)
47. Ogievetskii, V.I., Polubarinov, I.V.: Spinors in gravitation theory. J. Exp. Theor. Phys. **21**, 1093–1100 (1965)
48. Pauli, W.: Die Allgemeinen Prinzipien der Wellenmechanik. Handbuch der Phys. **24**, t. 1 (1933)
49. Pauli, W.: Contributions mathématicue à la theorie des matrices de Dirac. Ann. Inst. H. Poincaré. **6**, 109–136 (1936)
50. Penrose, R., Rindler, W.: Spinors and Space-Time, vol. 1. Cambridge University Press, Cambridge (1986)

51. Platania, G.: Rosania, R.: Exact Bianci type-1 solutions of the Einstein equations with Dirac field. Europhys. Lett. **37**, 585–590 (1997)
52. Radford, C.J., Klotz, A.N.: An exact solution of the Dirac–Einstein equations. J. Phys. A: Math. Gen. **16**, 317–320 (1983)
53. Rainich, G.I.: Electrodynamics in the general relativity theory. Trans. Am. Math. Soc. **27**, 106–136 (1925)
54. Rashevskii, P.K.: The theory of spinors. Usp. Mat. Nauk. **10**, 3–110 (1955) (in Russian)
55. Rashevskii, P.K.: Riemannian Geometry and Tensor Analysis. Nauka, Moscow (1967) (in Russian)
56. Reifler, F.: A vector wave equation for neutrinos. J. Math. Phys. **25**, 1088–1092 (1984)
57. Reifler, F.: A vector model for electroweak interactions. J. Math. Phys. **26**, 542–550 (1985)
58. Reifler, F., Morris, R. Hestenes' tetrad and spin connections. Int. J. Theor. Phys. **44**, 1307–1324 (2005)
59. Rybakov, Y.P., Saha, B., Shikin, G.N.: Nonlinear spinor fields in Bianchi-I-type space: exact self-consistent solutions. Izvestia VUZov Phys. **37**, 40–45 (1994) (in Russian)
60. Schouten, J.A., Struik, D.J.: Einfuhrung in die Neueren Methoden der Differentialgeometrie. Bd. 1, 2. Noordhoff, Groningen (1938)
61. Schweber, S.S.: An Introduction to Relativistic Quantum Field Theory. Row, Peterson and Co., New York (1961)
62. Sedov, L.I.: Introduction to the Mechanics of a Continuous Medium. Addison-Wesley Pub. Co., Reading, MA (1965)
63. Sedov, L.I.: Mathematical methods for constructing new models of continuous media. Russ. Math. Surv. **20**, 123–182 (1965)
64. Skobel'tsyn, D.V.: The momentum-energy tensor of the electromagnetic field. Sov. Phys. Usp. **16**, 381–401 (1973)
65. Takabayasi, T.: Relativistic hidrodynamics equalent to the Dirac equation. Prog. Theor. Phys. **13**, 222–224 (1955)
66. Takabayasi, T.: Symmetrical structure and conservation relations of four-component neutrino field. Nucl. Phys. **6**, 477–488 (1958)
67. Takahashi, Y.: Reconstructions of a spinor via Fierz identities. Phys. Rev. D, **26**, 2169–2171 (1982)
68. Takahashi, Y.: A Spinorization of the Frenet-Serret equation. Prog. Theor. Phys. **70**, 1466–1467 (1983)
69. Takahashi, Y.: The Fierz identities—a passage between spinors and tensors. J. Math. Phys. **24**, 1783–1790 (1983)
70. Takahashi, Y., Okuda, K.: A spinorization of a constrained vector system and a spinor reconstruction theorem. Fortschr. Phys. **31**, 511–534 (1983)
71. Weyssenhoff, I., Raabe, A.: Acta Phys. Polon. **9**, 7–18 (1947)
72. Wheeler, J.A.: Neutrinos, Gravitation, and Geometry. In: Radicati, L.A. (ed.) Rendiconti della Scuola internazionale di fisica "Enrico Fermi" Corso XI, pp. 67–196. Bologna, Zanichelli (1960)
73. Whittaker, E.T.: On the Relations of the tensor-calculus to the Spinor-calculus. Proc. R. Soc. **A 158**, 38–46 (1937)
74. Zhelnorovich, V.A.: Spinor as an invariant. J. Appl. Math. Mech. **30**, 1289–1300 (1966)
75. Zhelnorovich, V.A.: Representation of spinors by real and complex tensor aggregates. Theor. Math. Phys. **2**, 66–77 (1970)
76. Zhelnorovich, V.A.: Spinor field as the fusion of tensor fields. Mosc. Univ. Phys. Bull. **13**, 705–714 (1972)
77. Zhelnorovich, V.A.: Equations for fluids with internal magnetic and mechanical moments. Fluid Dyn. **9**, 847–850 (1974). Maik Nauka/Interperiodica Publishing (Russian Federation)
78. Zhelnorovich, V.A.: Energy-momentum tensor for the electromagnetic field of a magnetized and polarized liquid having an internal angular momentum. Sov. Phys. Dokl. **19**, 499–501 (1975)

79. Zhelnorovich, V.A.: Exact one-dimensional nonstationary solutions of the Heysenberg equations. Dokl. Math. **17**, 124–127 (1976). Maik Nauka/Interperiodica Publishing (Russian Federation)
80. Zhelnorovich, V.A.: A tensor description of fields with half-integer spin. Sov. Phys. Dokl. **24**, 899–901 (1979)
81. Zhelnorovich, V.A.: On spinor representations of pseudo-orthogonal groups. Dokl. Math. **22**, 337–340 (1980). Maik Nauka/Interperiodica Publishing (Russian Federation)
82. Zhelnorovich, V.A.: Theory of Spinors and Its Application in Physics and Mechanics. Moscow, Nauka (1982) (in Russian)
83. Zhelnorovich, V.A.: Hydrodynamic interpretation of nonlinear relativistic theories of elementary particles. Sov. Phys. Dokl. **28**, 338–341 (1983)
84. Zhelnorovich, V.A.: Tensor representation of spinor fields. In: Sedov, L.I. (ed.) Macroscopic Theories of Matter and Fields: A Thermodynamic Approach, pp. 224–261. Mir Publishers, Moscow (1983)
85. Zhelnorovich, V.A.: Invariant description of fermionic fields by scalar and vector fields. Sov. Phys. Dokl. **32**, 272–274 (1987)
86. Zhelnorovich, V.A.: Dirac equations in the general theory of relativity. Sov. Phys. Dokl. **32**, 726–728 (1987)
87. Zhelnorovich, V.A.: Proper bases and internal energy in electromagnetic field theory. Sov. Phys. J. **32**, 760–764 (1989)
88. Zhelnorovich V.A.: Complex vector triads in the theory of spinors in Minkowski space. Sov. Phys. Dokl. **35**, 245–247 (1990)
89. Zhelnorovich, V.A.: Invariant tensor description of massless spinor fields. Sov. Phys. Dokl. **37**, 34–36 (1992)
90. Zhelnorovich, V.A.: Derivatives for spinor fields in three-dimensional space and some applications. Sov. Phys. Dokl. **38**, 490–492 (1993)
91. Zhelnorovich, V.A.: An exact general solution of the Einstein-Dirac equations in homogeneous spaces. Dokl. Math. **53**, 87–91 (1996)
92. Zhelnorovich, V.A.: Cosmological solutions of Einstein–Dirac equations. Gravit. Cosmol. **2**, 109–116 (1996)
93. Zhelnorovich, V.A.: On Dirac equations in the formalism of spin coefficients. Gravit. Cosmol. **3**, 97–99 (1997)
94. Zhelnorovich, V.A.: Some exact solutions of the Einstein-Dirac equations with the cosmological constant. Gravit. Cosmol. **6**, 1–10 (2000)
95. Zhelnorovich, V.A.: A general exact solution of the Einstein-Dirac equations with the cosmological term in homogeneous space. J. Exp. Theor. Phys. **98**, 619–628 (2004). Maik Nauka/Interperiodica Publishing (Russian Federation)
96. Zhelnorovich, V.A.: Mechanics of Magnetizable and Polarizable Media with Microstructure. Time to Live, Moscow (2015) (in Russian)
97. Zund, J.D.: A memoir on the projective geometry of spinors. Ann. Mat. Pura Appl. Ser. 4, **110**, 29–136 (1976)